APPLIED STRATIGRAPHY

## TOPICS IN GEOBIOLOGY

http://www.springeronline.com/sgw/cda/frontpage/0,11855,4-40109-69-33109783-0,00.html

**Series Editors: Neil H. Landman,** American Museum of Natural History, New York, New York
**Douglas S. Jones,** University of Florida, Gainesville, Florida

*Current volumes in this series*

Volume 23: **Applied Stratigraphy**
Eduardo A. M. Koutsoukos
Hardbound, ISBN, 1-4020-2632-3, 2005

Volume 22: **The Geobiology and Ecology of *Metasequoia***
Ben A. LePage, Christopher J. Williams and Hong Yang
Hardbound, ISBN, 1-4020-2631-5, 2004

Volume 21: **High-Resolution Approaches in Stratigraphic Paleontology**
Peter J. Harries
Hardbound, ISBN 1-4020-1443-0, September 2003

Volume 20: **Predator-Prey Interactions in the Fossil Record**
Patricia H. Kelley, Michal Kowalewski, Thor A. Hansen
Hardbound, ISBN 0-306-47489-1, January 2003

Volume 19: **Fossils, Phylogeny, and Form**
Jonathan M. Adrain, Gregory D. Edgecombe, Bruce S. Lieberman
Hardbound, ISBN 0-306-46721-6, January 2002

Volume 18: **Eocene Biodiversity**
Gregg F. Gunnell
Hardbound, ISBN 0-306-46528-0, September 2001

Volume 17: **The History and Sedimentology of Ancient Reef Systems**
George D. Stanley Jr.
Hardbound, ISBN 0-306-46467-5, November 2001

Volume 16: **Paleobiogeography**
Bruce S. Lieberman
Hardbound, ISBN 0-306-46277-X, May 2000

Volume 15: **Environmental Micropaleontology**
Ronald E. Martin
Hardbound, ISBN 0-306-46232-X, July 2000

Volume 14: **Neogene, Paleontology of the Manonga Valley, Tanzania**
Terry Harrison
Hardbound, ISBN 0-306-45471-8, May 1997

Volume 13: **Ammonoid Paleobiology**
Neil H. Landman, Kazushige Tanabe, Richard Arnold Davis
Hardbound, ISBN 0-306-45222-7, May 1996

A Continuation Order Plan is available for this series. A continuation order will bring delivery of each new volume immediately upon publication. Volumes are billed only upon actual shipment. For further information please contact the publisher.

# Applied Stratigraphy

Edited by

EDUARDO A.M. KOUTSOUKOS
*PETROBRAS-CENBES, Cidade Universitária, Ilha do Fundao, Rio de Janeiro, RJ, Brazil*

Springer

A C.I.P. Catalogue record for this book is available from the Library of Congress.

ISBN 1-4020-2632-3 (HB)
ISBN 1-4020-2763-X (e-book)

---

Published by Springer,
P.O. Box 17, 3300 AA Dordrecht, The Netherlands.

Sold and distributed in North, Central and South America
by Springer,
101 Philip Drive, Norwell, MA 02061, U.S.A.

In all other countries, sold and distributed
by Springer,
P.O. Box 322, 3300 AH Dordrecht, The Netherlands.



Cover illustrations:

Inserts:
Thin section of fossil wood showing false and true growth rings, Irati Formation (Upper Permian), *Fig. 9.1A of L. S. R. Alves and M. Guerra-Sommer, photo by L. S. R. Alves.*

Trace fossil association characteristic of the Teredolites ichnofacies, *Fig. 13.4 by S. George Pemberton and James A. MacEachern.*

Center:
GEOLOGY AND PALAEONTOLOGY, Vincent Brooks Day & Son, Lith., by Duncan, Peter Martin (editor), 1896, Cassell's Natural History, 6 vols. in 3, London, ca. 380 p., 26 cm, wood-engraved plate, *photo by E. A. M. Koutsoukos.*

*Printed on acid-free paper*



Printed in the Netherlands.

"Indeed, what is there that does not appear marvelous when it comes to our knowledge for the first time? How many things, too, are looked upon as quite impossible until they have been actually effected?"

**Pliny the Elder** (23 AD - 79 AD), Natural History.

•

"Let us suppose that an ichthyologist is exploring the life of the ocean. He casts a net into the water and brings up a fishy assortment. Surveying his catch, he proceeds in the usual manner of a scientist to systematise what it reveals. He arrives at two generalisations:
(1) No sea-creature is less than two inches long.
(2) All sea-creatures have gills.
These are both true of his catch, and he assumes tentatively that they will remain true however often he repeats it.

In applying this analogy, the catch stands for the body of knowledge which constitutes physical science, and the net for the sensory and intellectual equipment which we use in obtaining it. The casting of the net corresponds to observation; for knowledge which has not been or could not be obtained by observation is not admitted into physical science.

An onlooker may object that the first generalisation is wrong. "There are plenty of sea-creatures under two inches long, only your net is not adapted to catch them." The icthyologist dismisses this objection contemptuously. "Anything uncatchable by my net is *ipso facto* outside the scope of icthyological knowledge. In short, "what my net can't catch isn't fish." Or--to translate the analogy--"If you are not simply guessing, you are claiming a knowledge of the physical universe discovered in some other way than by the methods of physical science, and admittedly unverifiable by such methods. You are a metaphysician. Bah!"

**Sir Arthur Eddington** (1882-1944), *The Philosophy of Physical Science,* Ann Arbor Paperbacks, The University of Michigan Press, 1958, p. 16.

# Contents

## Aims & Scope *Topics in Geobiology* Book Series

**Topics in Geobiology** series treats geobiology – the broad discipline that covers the history of life on Earth. The series aims for high quality, scholarly volumes of original research as well as broad reviews. Recent volumes have showcased a variety of organisms including cephalopods, corals, and rodents. They discuss the biology of these organisms – their ecology, phylogeny, and mode of life – and in addition, their fossil record – their distribution in time and space.

Other volumes are more theme based such as predator-prey relationships, skeletal mineralization, paleobiogeography, and approaches to high resolution stratigraphy, that cover a broad range of organisms. One theme that is at the heart of the series is the interplay between the history of life and the changing environment. This is treated in skeletal mineralization and how such skeletons record environmental signals and animal-sediment relationships in the marine environment.

The series editors also welcome any comments or suggestions for future volumes.

**Series Editors:**
Douglas S. Jones dsjones@flmnh.ufl.edu
Neil H. Landman landman@amnh.org

# Contributors

**Laureen Sally da Rosa Alves**
Universidade do Estado do Rio de Janeiro (UERJ)
Faculdade de Geologia
Departamento de Estratigrafia e Paleontologia
Rua São Francisco Xavier, 524/4001-A
Maracanã
20559-900 Rio de Janeiro, RJ
Brazil

**Marie-Pierre Aubry**
Department of Geology
Rutgers State University of New Jersey
Piscataway NJ 00000
U.S.A.

**Nilo Chagas de Azambuja Filho**
PETROBRAS-CENPES
Cidade Universitária, Quadra 7
Ilha do Fundão
21949-900 Rio de Janeiro, RJ
Brazil

**David J. Batten**
Institute of Geography and Earth Sciences
University of Wales
Aberystwyth SY23 3DB
U.K.

**Arthur J. Boucot**
Department of Zoology
Oregon State University
Corvallis, Oregon 97331-2914
U.S.A.

**Rodolfo Dino**
PETROBRAS-CENPES
Cidade Universitária, Quadra 7
Ilha do Fundão
21949-900 Rio de Janeiro, RJ
Brazil

**Tania Lindner Dutra**
Programa de Pos-Graduação
Geologia – Unisinos
Av. Unisinos, 950
93022-000 , São Leopoldo, RS
Brazil

**David Ewen**
British Petroleum (BP)
Farburn Industrial Estate
Dyce, Aberdeen, AB21 7PB
U.K.

**Felix M. Gradstein**
Geology Museum
University of Oslo
N-0316 Oslo
Norway

**Margot Guerra-Sommer**
Universidade Federal do Rio Grande do Sul (UFRGS)
Instituto de Geociências
Caixa Postal 15001
91501-570 Porto Alegre, RS
Brazil

**Peter Heavey**
British Petroleum (BP)
Godesetdalen 8
4065 Stavanger
Norway

**Michael Holz**
Universidade Federal do Rio Grande do Sul (UFRGS)
Instituto de Geociências
Caixa Postal 15001
91501-570 Porto Alegre, RS
Brazil

**Robert Wynn Jones**
British Petroleum (BP)
Building H
Chertsey Road
Sunbury-on-Thames
Middlesex, TW16 7LN
U.K.

**Eduardo A. M. Koutsoukos**
PETROBRAS-CENPES
Cidade Universitária, Quadra 7
Ilha do Fundão
21949-900 Rio de Janeiro, RJ
Brazil

**Stanislas Loboziak**
*(Deceased)*
*Ex* Université des Sciences et Technologies de Lille (U.S.T.L.)
Sciences de la Terre
Villeneuve d'Ascq
France

**Stephen Lowe**
British Petroleum (BP)
Sunbury-on-Thames
Middlesex, TW16 7LN
U.K.

**James A. MacEachern**
Earth Sciences
Simon Fraser University,
Burnaby, British Columbia, V5A 1S6
Canada

**José Henrique Gonçalves de Melo**
PETROBRAS-CENPES
Cidade Universitária, Quadra 7
Ilha do Fundão
21949-900 Rio de Janeiro, RJ
Brazil

**Paul Milner**
British Petroleum (BP)
Godesetdalen 8
4065 Stavanger
Norway

**Hilary Clement Olson**
Institute for Geophysics
The University of Texas at Austin
4412 Spicewood Springs,
Bldg. 600
Austin, Texas, 78759
U.S.A.

**Simon Payne**
British Petroleum (BP)
Farburn Industrial Estate
Dyce, Aberdeen, AB21 7PB
U.K.

**S. George Pemberton**
Ichnology Research Group
Department of Earth & Atmospheric Sciences
University of Alberta
Edmonton, Alberta T6G 2E3
Canada

**Martin A. Perlmutter**
ChevronTexaco Inc.
Energy Technology Company
4800 Fournace Place
Bellaire, Texas 77401
U.S.A.

**Geoffrey Playford**
Department of Earth Sciences/School of Physical Sciences
University of Queensland
Brisbane 4072
Australia

**René Rodrigues**
Universidade do Estado do Rio de Janeiro (UERJ)
Faculdade de Geologia
Rua São Francisco Xavier 524, Bloco A, Sala 2016
Maracanã
20559–900 Rio de Janeiro, RJ
Brazil

**Cesar Leandro Schultz**
Universidade Federal do Rio Grande do Sul (UFRGS)
Instituto de Geociências
Caixa Postal 15001
91501-570 Porto Alegre, RS
Brazil

**Marcello Guimarães Simões**
Instituto de Biociências
Universidade Estadual Paulista (UNESP) – Botucatu
Caixa Postal 510
18618-000 Botucatu, SP
Brazil

**Maurice Streel**
Université de Liège
Services associés de Paléontologie, Paléobotanique et Paléopalynologie,
Sart-Tilman, Bât. B 18,
B-4000 Liège
Belgique

**Darrin T. Stead**
Wellstrat Services Ltd
1 Castle Grange, Wrexham Road
Caergwrle, Wrexham LL12 9HL
U.K.

**Peter R. Thompson**
Computational Biochronology
P.O. Box 261039
Plano, Texas 75026
U.S.A.

**John A. Van Couvering**
Micropaleontology Press,
256 Fifth Ave.
New York, NY 10001
U.S.A.

# Foreword: The Age of Applied Stratigraphy

Few, if any, fundamental disciplines in the Earth Sciences have seen so many dramatic changes and developments as stratigraphy. Its beginnings can be linked to the very earliest human observations of Earth processes, and to philosophical enquires and speculations about the nature of natural phenomena.

Nearly 200 years ago, William Smith studied fossils collected from successive levels of sedimentary strata, and enunciated the first major principle of stratigraphy by stating "the deeper, or lower, layers of rock must be older than the layers of rocks which lie above them." Subsequently, the discipline has come to be applied progressively, and indispensably, to nearly all branches of the Earth Sciences: ranging among such diverse studies as the meticulous investigation of archeological sites, the geological evolution of sedimentary basins, the study of ancient ecosystems, and the origin and evolution of life.

Applied stratigraphic researches have increased spectacularly during the last decades of the 20th century, especially in response to intensive exploration and exploitation of mineral- and hydrocarbon-bearing sedimentary sequences conducted globally in both continental and marine settings. In particular, the past two decades have witnessed a major renaissance in stratigraphy, through the integration of biostratigraphy, magnetostratigraphy, isotope stratigraphy, and seismic-reflection data within two entirely new disciplines, cyclostratigraphy and sequence stratigraphy.

Currently, at the beginning of the new millennium, international efforts are concentrating on the development of interactive, integrated stratigraphic databases that are to be made readily accessible to the international geoscience community via the internet. Such initiatives are vital steps in promoting global scientific cooperation, coupled with the dissemination of well-defined stratigraphic standards. Moreover, exciting opportunities and challenges for earth scientists will undoubtedly arise in ensuing decades from as yet largely unforeseen or unrealized innovations and new applications in cognate scientific fields.

This book aims to incorporate many of the major aspects and essential elements underpinning the modern applications and perspectives of stratigraphy. It focuses on traditional and innovative techniques and how these can be utilized in the reconstruction of the geological history of sedimentary basins and in solving manifold geological problems and phenomena. Each chapter reviews the historical background; includes a synopsis of study principles and methodology; and discusses recent developments and significant applications. These sections are followed by selected case histories that demonstrate the applications and efficacy of stratigraphic and related techniques.

Conceptually, the book consists of four parts. The introductory chapter *(Evolution of a Concept)* provides a historical background to the breadth and diversity of stratigraphic studies, whose roots lie at the very origin of all Earth Science. The second part *(The Search for Patterns: Ordering the Framework)* commences with an overview of chronostratigraphy as applied to the study of regionally extensive stratigraphic sections.

The subsequent chapters review and elucidate current paleontological applications in biochronostratigraphy, event stratigraphy, paleoenvironmental syntheses, and paleobiogeographic reconstructions of Phanerozoic marine and continental sedimentary basins, thus providing insights into ancient ecosystems and their evolution through geological time. The third part *(The Search for Clues: Analyzing and Sequencing the Record)* presents comprehensive and authoritative surveys of diverse geoscience disciplines applied to the analysis of the stratigraphic record, including correlation, paleoclimatic and paleoenvironmental reconstructions, sequence stratigraphy, cyclostratigraphy, and "biosteering." The fourth part *(Modelling the Record)* discusses the development of quantitative stratigraphy and graphic correlation techniques, both of key importance to the refinement of chronostratigraphic frameworks as these pertain to interactive stratigraphic databases and basin modelling.

Stratigraphy has evolved and matured to constitute the most multifaceted and complex of disciplines within the Earth Sciences, with data deriving from and contributing to an impressive array of geological and paleobiological researches, both applied and "pure." These include such endeavours as charting the course and complexities of the evolution of life through time, understanding how ancient ecosystems developed and operated, and furnishing data pivotal to strategic mineral exploration. It is hoped that this book will provide the reader with key insights into all these aspects and applications. Supplemental reading can be found in the extensive reference lists.

Rio de Janeiro
May 2004

Eduardo A. M. Koutsoukos

# Acknowledgements

For their careful reviews which helped significantly improve and shape the final chapters, we are deeply grateful to: Frits P. Agterberg, Marion K. Bamford, Michael J. Benton, Robert B. Blodgett, José Fernando Bonaparte, Carlton Brett, Gerhard C. Cadée, Kelly Dempster, Thomas W. Dignes, Murray K. Gingras, Felix M. Gradstein, Yngve Grahn, Garry D. Jones, Michal Kowalewski, Ernesto L.C. Lavina, Alain Le Hérissé, Jennifer C. McElwain, José Henrique G. Melo, Geoffrey Playford, Paul Potter, Jamie Powell, Eugenio V. dos Santos Neto, Peter Szatmari, Peter Thompson, John Utting, Ron F. Waszczak, Julia S. Wellner, Reed Wicander, John-Paul Zonneveld, and one reviewer who chooses to remain anonymous.

I am grateful to Kenneth Howell, publishing editor at Kluwer Academic/Plenum Publishers, New York, for his invitation and incentive to write the book and all the help at the planning stage of the book. In addition, I thank, Anna Besse-Lototskaya, our present publishing editor, and Judith Terpos, Kluwer Academic Publishers, Dordrecht, for their assistance in editorial tasks and in the final stages of typesetting.

I am also deeply grateful to Laureen Sally da Rosa Alves for her dedicated assistance in the careful editing of each chapter and respective reference lists.

Part I

# Evolution of a Concept

Chapter 1

# Stratigraphy: Evolution of a Concept

EDUARDO A. M. KOUTSOUKOS
PETROBRAS-CENPES, Cidade Universitária, Quadra 7, Ilha do Fundão,
21949-900 Rio de Janeiro, RJ, Brazil.



## 1.1 Introduction

Stratigraphy, from the Latin *stratum* and the Greek *graphia*, is the study of rock successions and the correlation of geological events and processes in time and space. It is a fundamental science of all geological studies, allowing to reconstruct the sequence of events of Earth history, and the evolution of life on Earth.

Stratigraphy is a science as old as philosophy itself, originated with the early observations by the ancients of Earth's natural phenomena and their philosophical speculations on the nature of Earth's structure and processes. From Central Asia to Greece, to Egypt, different views were adopted among the various ancient civilizations according to their individual perception of the natural world, and tied to the prevailing religious



*E.A.M. Koutsoukos (ed.), Applied Stratigraphy,* 3-19.

and philosophical doctrines and myths. The Chinese regarded the Earth as eternal and immutable; to the Indians it changed according to infinite cycles of creation of the universe. To the Western cultures, particularly the Greeks, the Earth was continuously changing over time, and its processes were controlled by natural laws. In that scenario powerful enquiring minds established the foundations of the scientific method of investigation by careful observation of the physical world and related natural phenonema, and tended to treat scientific theories as extensions of philosophy.

## 1.2 The Ancient Greeks and Romans: The Naturalism

The major contribution of the early observations made by the ancient Ionian philosophers on the Earth's structure and processes is that they were attributed to natural phenomena and not to supernatural causes. However, many of the ideas expressed by the ancients on the natural history of the Earth were not to resurface until the Renaissance.

The term "fossil" comes from the ancient Latin *fossilis*, meaning virtually anything dug from the Earth. Thus, originally it was used to indicate not just "fossils" that resembled living organisms, as we would define them today, but also minerals, crystals, rocks, and even geological strata. At those early times two hypotheses had been proposed for the origin of "fossils" that resembled living organisms. According to the Greek philosopher Aristotle (384–322 B.C.), precursor of the inorganic theory for the origin of fossils, forms resembling extant organisms present in rocks were produced by a "formative force" (*vis formativa*), i.e., produced by natural causes imitating living shapes which naturally grew in the rocks. Aristotle's thoughts on Earth sciences can be found in his treatise *Meteorology*, where he discusses the nature of the Earth and the oceans.

However, the organic view of the origin of fossils, that is the idea that they were indeed the remains of once living beings, appeared even before Aristotle. In the 6$^{th}$ and 5$^{th}$ Century B.C. both Pythagoras (ca. 582–ca. 507 B.C.) and Herodotus (ca. 484?–425? B.C.) mentioned marine shells which occurred in mountains and on land places far from the sea, arguing these to be the remains of organisms which once thrived in former seas that later withdrew and became land. Herodotus thought the disc-shaped nummulitid foraminifera (genus *Nummulites*) found in the Eocene limestones of which the Sphinx and the pyramids of Egypt were built, were lentils fed to the slaves who built the pyramids which had accidentally spilled and turned to stone. The Greek Xenophanes of Colophon, Ionia (ca. 570–ca. 480 B.C.), a pre-Socratic Greek philosopher and Pythagorean, reported the occurrence of marine fossils on mountain tops and quarries. For Xenophanes this was proof that the physical arrangement of the Earth changes with time, and that the remains of extant forms were buried in the dried sea mud owing to the blending of land and sea in ancient times.

The Romans assimilated the more practical scientific accomplishments of the Greeks, but added little. Pliny the Elder (Caius Plinius Secundus; ca. 23–79 A.D.), Roman naturalist and encyclopedist killed by the eruption of the Vesuvius on 23–25 August 79 A.D., wrote a major encyclopedia of natural sciences (*Historia naturalis*). Pliny's compilation consisted of 37 volumes and contained a summary of ancient knowledge on the nature of the physical universe, a work mostly summing up what the ancient Greek authors had written. In the early centuries after Christ the idea of the organic origin of fossils took root and spread, although in different form with respect to the early Greeks. According to the Roman Carthaginian, theologian and Christian apologist Tertullian (Quintus Septimius Florens Tertullianus; ca. 155–222) fossil shells found in the mountains were proof of the Universal Flood described in the Old Testament. Both the controversy between the organic and inorganic views and the *Universal Flood* theory would survive until the beginning of the 18th Century.

With the collapse of the Roman Empire in the 5th Century and the coming of the Dark Ages, many of the ancient scientific works passed into the hands of the Muslims, who by the 7th and 8th Century had extended their influence through much of the world surrounding the Mediterranean. All of the Greek works were translated into Arabic, and commentaries were added. The Arabs thus preserved the scientific works of the ancients and added to them, introducing also other contributions from Asia. This body of learning first began to be discovered by Europeans in the 11th Century.

## 1.3 The Middle Ages and Renaissance: The Principles

During the Middle Ages and the Renaissance, the Aristotelian School, through a Muslim, the Persian philosopher and physician Ibn-Sina (Avicenna, 980–1037), influenced many European scholars who attributed the occurrence of fossils to failed abortive attempts of a natural creative "plastic force" *(vis plastica*, or *virtus formativa)* to shape living beings in a process of spontaneous generation of life.

The scientist, philosopher, and theologian Albertus Magnus (ca. 1206?–1280), born at Lauingen, Swabia, and later Leonardo da Vinci (1452–1519), were the first to correctly speculate on the nature of fossils as the remains of once-living ancient organisms. The *Leicester Code* is a collection of numerous manuscripts with handwritten notes of Leonardo's scientific and technical observations. Among them there are drawings of rock formations and of various fossil shells (mostly Cenozoic mollusks), which are probably recollections of his experiences and observations on the hills of Tuscany, Romagna or the Po River plain, during his service as an engineer and artist at the court of Lodovico Sforza, Duke of Milan, from 1482 to 1499. From his notes Leonardo appears to have noticed the mechanisms of sedimentary deposition on mountains and rivers, the role that rivers play in the erosion of land, and the principles of the law of superposition, which would later be demonstrated fully by Nicholaus Steno in 1669. He also appear to have grasped that distinct layers of rocks and fossils could be traced over long distances, and that these layers were formed at different times. After Leonardo's death his notes were scattered to libraries and collections all over Europe. While portions of Leonardo's technical treatises on painting were published as early as 1651, the scope and caliber of much of his scientific work remained unknown until the 19th Century. Yet his geological and paleontological observations and theories foreshadow many later breakthroughs. Nearly three hundred years later, the rediscovery and elaboration of these principles would make possible modern stratigraphy and geological mapping.

Georgius Agricola, latinized form of Georg Bauer (1494–1555), a physician and geologist born in Glauchau, province of Saxony, made fundamental contributions to stratigraphic geology, mineralogy, structural geology, and paleontology. His greatest work, *De Re Metallica* ("On the Nature of Metals"), posthumously published in 1556, is a systematic study of ore deposits and of strata, and was to remain the standard text on mining geology for two centuries. In his profusely illustrated book Agricola noted that rocks were laid down in definite layers, or strata, and that these layers occurred in a consistent order and could be traced over a wide area. Agricola's observations would become important in understanding the arrangement and origins of the rocks of the Earth. Agricola is considered the founder of geology as a discipline.

In the mid-16$^{th}$ Century the first engravings of fossils were published by the Swiss physician Conrad Gessner (1516–1565).

Nicholaus Steno, latinized form of Niels Stensen (1638–1686), a Danish anatomist and geologist, who in Italy was converted to Roman Catholic faith and became a Roman Catholic prelate, pointed out the true origin of geological strata and of fossils. He wrote the first real geological treatise in 1667, while living in Tuscany, Italy. In 1669 he was the first recorded person to apply to the study of a sedimentary rock outcrop what is now referred to as **Steno's law of superposition** (which states that layers of rock are arranged in a time sequence, with the oldest on the bottom and the youngest on the top, unless later processes disturb this arrangement), his most famous contribution to geology. In addition, Steno postulated other general principles of Stratigraphy: the **principle of original horizontality**, which states that rock layers form in the horizontal position, and any deviations from this position are due to the rocks being disturbed later; and the principles of **strata continuity** (material forming any stratum was continuous over the surface of the Earth unless some other solid bodies stood in the way) and **cross-cutting relationships** (if a body or discontinuity cuts across a stratum, it must have formed after that stratum). The data and conclusions of Steno's work on the formation of rock layers and fossils were crucial to the development of modern geology, and were enough to have earned him the title of "Father of Stratigraphy".

Steno's contemporaries, the British natural scientists John Ray (1628–1705), Robert Hooke (1635–1703) and John Woodward (1668–1728), also argued that fossils were the remains of once-living animals and plants. However, the opinion was still universal that fossils represented life destroyed by the Universal Flood, a theory championed especially by the Swiss naturalist Johann Jakob Scheuchzer (1672–1733). Robert Hooke was perhaps the greatest experimental scientist of the seventeenth century. He was the first person to examine fossils with a microscope, to note close similarities between the structures of fossil and living wood and mollusc shells, and to observe, two and a half centuries before Darwin, that the fossil record documents the appearance and extinction of species in the history of life on Earth. Hooke believed that the Biblical Flood had been too short in time to account for all fossils, and suggested that earthquakes had likely destroyed ancient life forms. John

Ray always supported the theory that fossils were once living organisms, buried in liquid rock that then cooled, but was reluctant to accept the idea of extinction, explaining that fossils which did not resemble any living organism were due to our ignorance of the full range of extant forms. Ray expressed fully his belief in the "natural theology" view of studying the natural world as God's creation, a doctrine which remained influential for well over a century after his death. John Woodward related fossils to specific rock formations and atempted to classify them. In 1695 he published *Essay Toward a Natural History of the Earth*, which advanced a theory to explain stratification and the fossils embedded in them by the deposit of debris out of the Flood.

## 1.4 The Eighteenth and Nineteenth Centuries: The Dilemma of Catastrophism *versus* Uniformitarism and Gradualism

At the beginning of the 18th Century the organic nature of fossils as remains of ancient beings was decisively established, and the inorganic theory abandoned. However, the Biblical Flood theory persisted up to the threshold of the 19th Century.

The Swiss naturalist Charles Bonnet (1720–1793) in his work *Principles of Catastrophism*, suggested that at periodic intervals throughout Earth's history all living things have been destroyed by catastrophes or cataclysms (e.g. floods or earthquakes) and they accounted for fossils. He thought that after each catastrophe life forms would be replaced by an entirely different changed population, and that all creatures would rise one level, so after a future catastrophe man would be angels and apes, man. Bonnet was the first to use the term *evolution*. Catastrophism is becoming more actual now that the various Ediacaran, latest Ordovician (Hirnantian), Permian–Triassic, and Cretaceous-Paleogene boundary mass-extinction events are better understood.

The theoretical foundation for much of modern geology was postulated by the Scottish geologist James Hutton (1726–1797) in his *Theory of the Earth* (1795), where he stated that "the Earth must be millions of years old". He first advanced the basic concept of what became known as the Uniformitarian Principle, which holds that the geologic forces and processes that shaped the Earth in the geologic past were referrable to the same ones still in operation on the Earth's surface and could be observed directly. Hutton believed that igneous processes were the chief agent in rock formation, thus representing the Plutonist (or Vulcanist) view, i.e., that some rocks had formed from molten magma either deep in the Earth or from volcanoes. In contrast the so-called Neptunist view maintained that the origin of all rocks was aqueous, related to a primeval ocean and its subsidence, and thus began the Neptunist–Plutonist controversy. In 1751 the French geologist, botanist, and natural historian Jean-Étienne Guettard (1715–1786) was the first to recognize the volcanic nature of the Puy de la Nugère and the lava flow descended from it near Volvic, as well as some seventeen other neighboring volcanic craters and domes in the Auvergne region, central France, and thus became the first known Vulcanist and founder of the school, though years later he proposed that basalt originated as a precipitate out of an aqueous fluid. This controversial Neptunist theory was supported by the German geologist Abraham Gottlob Werner (1750–1817), and also by the German lyric poet, novelist, dramatist, and scientist Johann Wolfgang von Goethe (1749–1832). Werner, who first demonstrated the chronological succession of rocks and the concept of the geological time scale, believed that rock strata were either sediments originally deposited at the bottom of the sea or were crystallized deposits precipitated from sea water, but could not explain the origin of insoluble igneous rocks. Goethe and the German writer Friedrich von Schiller (1759–1805) satirized attempts of attributing basalts in Germany to volcanoes, seeing in this a patriotic attempt to equal Italy; Goethe also satirized the Vulcanist theory in his drama Faust (1808), attributing mountain buildings to the parts of the devil. The Uniformitarian doctrine was further simplified and popularized by the British geologist John Playfair (1748–1819) in his *Illustrations of the Huttonian Theory of the Earth* (1802), which elucidated the methods and principles of uniformitarianism, establishing it as the foundation of the new science of geology.

William Smith (1769–1839), English engineer and canal builder, studied fossils collected from sedimentary rocks. He was the first to recognize the importance of fossils for the historical investigation of Earth's strata, and introduced the **principle of faunal succession**, that different sedimentary rock units contain distinct fossil assemblages. Smith noted that the sequence of fossils in any given stratigraphic record follows a specific order, as a result of evolution; and that the same sequence can be found in isolated strata elsewhere, and thus correlated between them. The principle is still applied today in biostratigraphic correlations, although within the limits of biogeographic distribution of index fossil species. Smith published the first large-scale geological map in 1814–1815, of southern England and Wales, using for the first time the principle of fossil succession as a tool for mapping rocks by their stratigraphic order, and not necessarily by their composition.

However, for most of the late 18$^{th}$ and mid-19$^{th}$ Century, as a consequence of apparently contradicting current religious beliefs (e.g. the accepted biblical chronology and the Flood), the uniformitarism doctrine was largely overshadowed by the opposite one of catastrophism, which also stated that during these catastrophic events the Earth's surface, such as mountains and valleys, would be shaped. Catastrophism was more easily correlated with religious doctrines, and as a consequence remained for some time the interpretation of the Earth's history adopted by the great majority of geologists. The French naturalist Georges L. Cuvier (1769–1832) was one of its major supporter. Cuvier suggested that four main worldwide catastrophes had occurred, the last one being the Biblical Flood. The taxonomic classification scheme introduced by the Swedish botanist Carolus Linnaeus (1707–1778) in his *Systema Naturae* (1735), was extended by Cuvier to fossils, which he recognized as organic remains of extinct animals. He is therefore known as the founder of paleontology as a science separate from geology. However, Cuvier rejected the theory of evolution and Jean-Baptiste Lamarck's (1744–1829) theory of inheritance of acquired characteristics, proposed in *Zoological Philosophy* (1809). He believed that new life forms would be created after periodic sea-level changes; in his view some animals died and some survived, but none evolved. In 1811 Cuvier wrote with the French geologist, mineralogist, and chemist, Alexandre Brongniart (1770–1847), the work *Essai sur la géographie mineralogique des environs de Paris*, in which a system of stratigraphy was developed that relied on the use of fossils for the precise dating of strata, introducing into France William Smith's principle of faunal sucession and method of field work.

During the 1830's fossils were first recognized for use in age correlation of rocks by the German geologist and paleontologist Friedrich August Quenstedt (1809–1889) through his work on Jurassic "time-rock units" defined by ammonites. Quenstedt recognized the base unit, the biostratigraphic "zone", characterized by a particular assemblage of fossils. By the late 1830s, most of the presently known geologic periods had been established based on their fossil content and their observed relative stratigraphic position (see Fig. 1.1, and Moore, 1955, for a review). The twofold modern subdivision of the Cenozoic Era (from the Greek word kainos: recent) in Paleogene and Neogene (from palaeos: ancient, neos: new, and genos: birth/born) goes back to combination of subdivisions proposed by C. Lyell (1833), R. Hoernes

| Era | System | Author & Year | Place first recognized |
|---|---|---|---|
| Cenozoic | Neogene | Hoernes, 1853 | Austria |
| | Paleogene | Hoernes, 1866 | Austria? |
| Mesozoic | Cretaceous | J.J. D'Omalius d'Halloy, 1822 | France |
| | Jurassic | Alexander von Humboldt, 1795<br>Leopold von Buch, 1839 | France, Switzerland, Germany, England |
| | Triassic | Friedrich August von Alberti, 1834 | Germany |
| Paleozoic | Permian | Murchison, 1841 | Russia |
| | Carboniferous | W.D. Conybeare and W. Phillips, 1822 | Northern England |
| | Devonian | Murchison and Sedgwick, 1839 | Southern England |
| | Silurian | Roderick Impey Murchison, 1835 | Welsh Borderland |
| | Ordovician | Charles Lapworth, 1879 | Wales |
| | Cambrian | Adam Sedgwick, 1835 | North Wales |

*Figure 1.1 Phanerozoic systems and their original definitions.*

(1853), and C. F. Naumann (1866) – the terms Tertiary and Quaternary are no longer recommended, as being antiquated like Primary and Secondary, all described by Giovanni Arduino, in northern Italy, in the 1760's.

Cuvier's successors, as d'Orbigny, Agassiz, and Barrande, still maintained the catastrophic theory well into the 19th Century. The naturalist and paleontologist Alcide Dessalines d'Orbigny (1802–1857) published in 1850 *"Prodrome de Paléontologie stratigraphique universelle"*, a major treatise comprising a catalogue list of 18,000 fossil species, and proposed a subdivision of the geological record in 27 stages. His most important work was the founding of the science of stratigraphic paleontology based on observations of exposed fossil-bearing strata in the Paraná Basin, southern Brazil, reported in his work *Voyages dans l'Amerique méridionale*, published in several parts between 1835 and 1847. D'Orbigny's study of foraminifera, pollen grains and spores found in sedimentary rocks for the purpose of dating stages began the science of micropaleontology. The Swiss-American naturalist Jean Louis Rodolphe Agassiz (1807–1873) promoted and defended Cuvier's geological catastrophism and classification of living and fossil animals. His study of glaciers revolutionized geology. The French paleontologist Joachim Barrande (1799–1883) studied fossil remains and their distribution in the various strata in Bohemia. The results of his extensive studies on the Silurian system of Bohemia are contained in his great work *"Système silurien de centre de la Bohême"*, published in 22 volumes from 1852 to his death.

Uniformitarianism finally became widely accepted as a result of the work of the Scottish geologist Sir Charles Lyell (1797–1875), author of the three-volume *Principles of Geology* (1830–1833), published through 11 editions between 1830 and 1872, which presented and popularized James Hutton's work and uniformitarianism. The uniformitarian (uniformity of natural laws and geological processes) and gradualist (uniformity of rates) views expressed in Lyell's work probably influenced the formulation of Charles Darwin's (1809–1882) theory of evolution and facilitated its acceptance. Darwin's theory of evolution through gradual variation and natural selection was published in his revolutionary work *On the Origin of Species by Means of Natural Selection, or the Preservation of Favoured Races in the Struggle for Life* (1859), often abbreviated to *The Origin of Species*, which was a turning-point for the evolution theory and also greatly influenced Geology in the late XIXth Century. Lyell's Uniformitarism and Gradualism and Darwin's theory would dominate the Earth Sciences for nearly 150 years.

Darwin's theory explained Smith's principle of faunal succession which, combined with Steno's law of superposition, allowed the application of fossils to stratigraphic dating and correlation, and the modern conceptualization of the biozone in 1856 by one of Friedrich Quenstedt's student, the German stratigrapher Albert Oppel (1831–1865). Oppel devised a scheme to divide geologic formations into zones based on the overlapping stratigraphic range of two fossil species (defining what are presently known as Oppel Zones). The progressive establishing of locally defined biostratigraphic zonal schemes led to the rapid development of an improved relative time scale and the emerging of the standard subdivisions of the modern Chronostratigraphic Scale, which has been continuously refined since (e.g. Moore, 1955; Heirtzler *et al.*, 1968; Harland *et al.*, 1982, 1990; Berggren and Van Couvering, 1978; Berggren *et al.*, 1995b; Gradstein *et al.*, 1999; website of the International Commission on Stratigraphy – ICS/IUGS – www.stratigraphy.org).

## 1.5 The Twentieth Century & Beyond: The Age of Applied Stratigraphy

The history of stratigraphy during the 20th Century is largely the history of the individual specialized branches as they developed into the traditional and new techniques by which they are recognized today.

### 1.5.1 Plate Tectonics

The earliest hint of plate tectonics was made around 1800 by one of Abraham G. Werner's most famous student, the German naturalist and explorer Alexander von Humboldt (1769–1859). Humboldt first suggested that the South American and African continents had once been joined, as apparent in their complimentary coastlines, but this proposal was largely ignored by the scientific community of that time. It would not be until the early 20th Century, in 1912, when the German astronomer, meteorologist and geophysicist Alfred Wegener (1880–1930) published his first works (*Die Entstehung der Kontiente*, Petermanns Mitteilungen, 1912, pp. 185–195, 253–256, and 305–309; and a somewhat different version with the same title in Geologische Rundschau, Vol. 3, No. 4, 1912, pp. 276–292) outlining his theory of "continental drift". After 1912, Wegener's work was interrupted first by an expedition to Greenland and then by the First World War. In 1915 Wegener published the first edition of *Die Entstehund der Kontinente und Ozeane (The Origin of Continents and Oceans).* In this book Wegener claimed that the continents had once been connected and formed a single supercontinent mass called Pangaea (from the Greek for "all the Earth"), about 300 million years ago, which had since split into pieces that have drifted to their present positions. As supporting evidence for the proposed theory Wegener noted the often matched large-scale geological features on separated continents, such as the close similarity of strata and fossils between Africa and South America and the close fit between their coastlines, and that fossils found in certain places often indicated past climates utterly different from today's. Wegener's revolutionary theory of continental drift took decades to win general acceptance among scientists, remaining controversial until the 1960's. For most of that lapse of time stratigraphy was to stand still, with no significant progress. In 1959 and 1962, Harry Hammond Hess proposed the sea-floor spreading or plate tectonics theory, subsequentely confirmed by Vine and Matthews (1963), which complemented Wegener's continental drift theory, and gave a much needed renewed impetus into the science of stratigraphy. Nevertheless, Wegener's basic insights remain sound nowadays and the same lines of supporting evidence are being continuously complemented and expanded by ongoing research.

### 1.5.2 Geochronology – Radiometric Stratigraphy

Geochronology, the science of absolute dating of rocks and determining the time sequence of geological events in Earth's history, particularly by radiometric dating, developed largely at the turn of the 20th Century and during its first three decades with the advent of atomic and nuclear physics and quantum theory (e.g. Holmes, 1911; see also Hole, 1998, for a review). It provided the framework of absolute time within which the relative chronostratigraphic scale could be calibrated.

### 1.5.3 Magnetostratigraphy

The work of Rutten (1959) presented a chronological scale of polarity reversals of the Earth's magnetic field based on K–Ar radiometric dating in a sequence of volcanic rocks, and gave birth to the new science of magnetostratigraphy. Harrison and Funnel (1964) discovered that magnetic polarity reversals (chrons) are also recorded in marine deposits, which further improved the applicability of the technique. Subsequent works aimed to match and calibrate the reversals with conventional stratigraphic tools (ISSC, 1979; Tarling, 1983; Galbrun, 1984), and use the unique non-periodic pattern of reversals to date and correlate diferent rock sequences. In combining the marine magnetic anomalies measured over the sea-floor record in the South Atlantic spreading profile with their dates of chrons on land, Jim Heirtzler and colleagues in 1968 laid the foundation for the modern timescale based on Cretaceous through Paleogene marine magnetic anomalies, also known as the Geomagnetic Polarity Time Scale (GPTS). The Cande and Kent (1995) GPTS is the currently accepted timescale that is in most widespread use.

### 1.5.4 Stratigraphic Classification, Terminology and Procedure

In 1976 the International Subcommission on Stratigraphic Classification (ISSC) of the International Commission on Stratigraphy (ICS) published the first edition of the *International Stratigraphic Guide* (edited by Hollis D. Hedberg), as a means to promote international agreement on the principles of stratigraphic classification, terminology, and rules of procedure. In 1983 the North American Commission on Stratigraphic Nomenclature proposed a version of the stratigraphic code, which expanded considerably its original scope. The standard international stratigraphic classification was finally approved in 1987 by the ISSC, and updated in 1994 in a second edition of the Guide (Salvador, 1994). An abridged version was published by Murphy and Salvador (2000), and made available on the ICS website. General comments on stratigraphic principles and procedures have also been presented by various authors, such as Reading (1978), Ager (1984), Blatt *et al.* (1991), and Whittaker *et al.* (1991), among others.

### 1.5.5 Facies Stratigraphy

The term and concept of stratigraphic facies (from Latin: appearance, aspect, face, form), meaning the combined lithological and paleontological characteristics of a stratigraphic section, were introduced in 1838 by the Swiss geologist and paleontologist Amanz Gressly (1814–1865) from his studies in the Jura Mountains. Gressly's pioneer contributions on the genesis and applications of sedimentary facies, stratigraphic correlations, and paleogeographic reconstructions are fundamental to modern stratigraphy (Cross and Homewood, 1997). Later in the course of the 19$^{th}$ Century the term was assigned to a variety of descriptive meanings by geologists, paleontologis and ecologists, which somehow confused the original definition. Facies analysis in the modern sense restored the concept to its original meaning, aiming at the description, interpretation and reconstruction of the depositional and paleogeographic setting of sedimentary units, combining lithological and paleontological data (Reading, 1978, 1996; Walker, 1979, 1992; Walker and James, 1992). It provides the basic framework to reconstruct the environmental evolution of the stratigraphic record through time (see Pirrie, 1998, for a review).

### 1.5.6 Quantitative Stratigraphy

Various graphical, numerical and experimental methods applied to refining stratigraphic resolution and basin modelling studies, have been continually developed since the 1960's (e.g. Shaw, 1964; Ager, 1973; Miller, 1977; Van Hinte, 1978, 1982; Gradstein *et al.*, 1985; Mann and Lane, 1995; Harbaugh *et al.*, 1999; Paola *et al.*, 2001). These techniques have jointly the greatest potential to achieve the finest biostratigraphic resolution possible in correlating different rock sequences, in studies of regional *versus* global correlation of geological events, in helping to reconstruct the geological history of sedimentary successions, and in petroleum reservoir correlation and modelling. The methods are greatly assisted by the universal adaptation of microcomputers to digital programming with colour graphics output.

### 1.5.7 Sequence Stratigraphy

Modern stratigraphy had a major impetus by the mid-20$^{th}$ Century, with the increase of petroleum exploration activities, the development of new technologies (e.g. of seismic reflection data in the 1970s) and the application of stratigraphic models to petroleum research (e.g. Sloss, 1962). In 1949 L. L. Sloss and coworkers coined the term sequence to represent a set of sedimentary cycles limited by unconformities. The notion of unconformity-bounded stratigraphic units received further support in the late 1950s with the works of H. E. Wheeler (1958, 1959a and b), who also introduced the concept of the chrono-lithostratigraphic chart. In 1963 Sloss consolidated the term stratigraphic sequence and its usage in regional chronostratigraphic correlations. These studies provided the basic framework to the later formulation of the sequence models, which was to incorporate the use of high-quality seismic-reflection data in modelling subsurface stratal patterns and general geometry, and the expected seismic reflection features of different lithofacies asssociations.

In a series of publications starting in the late 1970s, Peter Vail, coworkers and colleagues presented a revolutionary stratigraphic method of basin analysis for what became known as "Sequence Stratigraphy" (e.g. AAPG Memoir 26, edited by Payton, 1977; Vail, 1987; Van Wagoner *et al.*, 1987, 1988, 1990, 1991; Posamentier *et al.*, 1988, 1992; Posamentier and Vail, 1988; Vail *et al.*, 1991; Schlager, 1992; Walker and James, 1992; and Posamentier and Allen, 1994, among others). Sequence models constitute a powerful tool for unraveling basin-fill history, and as such have been applied to most stratigraphic studies of basin modelling. The method is based on the study of the relationships between global relative sea-level changes and large-scale sedimentary cycles within time-equivalent depositional successions bounded above and below by a significant gap in the stratigraphic record, i.e., by surfaces of erosion (unconformity-bounded units) or nondeposition. Suess (1906) was the first to propose that sea-level changes could be global. The global eustatic sea-level variation curve proposed by Vail *et al.* (1977a, b), and later refined by Haq *et al.* (1987, 1988) and Ross and Ross (1988), for the Phanerozoic sequences, was based on the approximate correlation of seismic sequences from a number of passive continental margins. In 1989 Galloway proposed the model of genetic sequences bounded by maximum-flooding surfaces, which implied a certain discrepancy with the unconformity-bounded depositional sequences of Vail *et al.* (1977a) and Van Wagoner *et al.* (1987), based essentialy on seismic stratigraphy. Galloway's approach, based mostly on sedimentological interpretation of depositional systems, facies relationships and geometries, is particularly significant in stratigraphic successions with little or no available seismic data, due to difficulties in marking and tracing regional unconformities.

Despite some controversies behind the main theoretical basis for the sequence stratigraphy paradigm (e.g. Miall, 1991, 1994, 1997), the method brought about a major revolution in the science of stratigraphy, leading to new research to be carried out on complex clastic and carbonate successions around the world. By gathering within a single stratigraphic framework information derived from diverse disciplines of sedimentary geology, such as seismic stratigraphy, biostratigraphy, paleoecology, paleogeography, and sedimentology, among others, the sequence models permitted a much broader, integrated and sharper research approach in basin analysis.

### 1.5.8 Episodic and Cyclic Sedimentation: Event Stratigraphy and Cyclostratigraphy

In the past decades of the 20th Century new theories developed in the geoscience community which represent a synthesis of Lyell's Uniformitarism and Gradualism combined with a revival of Cuvier's Catastrophism, recognizing that both play a significant role in geological processes and the evolution of life. Theories such as the *actualistic catastrophism* (Hsü, 1983), the *punctualism* (Gould and Elderedge, 1977; Gould, 1984; Goodwin and Anderson, 1985), and the *episodic sedimentation* (Dott, 1983), are fundamented on the assumption that most of the stratigraphic record was produced during episodic events, and that abrupt environmental changes have modulated speciation and mass extinctions (e.g. Signor and Lipps, 1982; Flessa, 1986; Hallam, 1989a, b). Major catastrophic events, such as extraterrestrial impacts (e.g. Alvarez *et al.*, 1980; McLaren and Goodfellow, 1990; Becker *et al.*, 2001) and cataclysmic volcanic activity (e.g. McLean, 1985; Courtillot, 2000; Wignall, 2001) are also thought to have greatly affected the evolution of life on Earth.

In 1982, G. Einsele and A. Seilacher discussed extensively the processes of cyclic and event sedimentation, introducing the principles of what would be later known as *Event Stratigraphy* (e.g. Kauffman, 1987, 1988; Walliser, 1996; Einsele, 1998). The method deals with the integrated study of episodic and short-term sedimentary and biotic processes in the stratigraphic record, and has the potential to improve substantially the resolution of geological correlations.

Rhythmic stratigraphic cycles observed in pelagic siliciclastic and carbonate sequences have been related to the so-called "Milankovitch cycles", after the Serbian astrophysicist Milutin Milankovitch (1879–1958) who in 1941 presented a firm mathematical basis that related periodic variations in Earth's rotational and orbital motions (eccentricity, obliquity, precession) to long-term climate changes. However, the hypothesis of astronomically forced climate cycles was advanced already in the 19$^{th}$ Century to the Pleistocene ice ages by the French mathematician Alphonse Joseph Adhémar (1797–1862), in his work *Les Revolutions de la mer* (1842), and by the Scottish geologist James Croll (1821–1890), who in the 1860's and '70s proposed an Astronomical Theory of the Ice Ages, subsequently published in his *Climate and Time* (1875) and *Climate and Cosmology* (1885). These orbital-forced cycles control the intensity of seasonal and latitudinal distribution of solar radiation (insolation) reaching the planet's surface, and directly influence global climate, depositional processes and biotic productivity (e.g. Fischer and Arthur, 1977; Bottjer *et al.*, 1986; Fischer, 1986, 1991; Schwarzacher, 1987; Fischer and Bottjer, 1991; Weedon, 1993; Satterley, 1996; Perlmutter *et al.*, 1998). The detailed investigation of regular cyclic patterns in the stratigraphic record produced by the interaction of tectonic and Milankovitch-type climatic processes is the study of a new branch of stratigraphy named Cyclostratigraphy (Schwarzacher, 1993; Fischer, 1993, 1995; Gale, 1998). The method allows a way for estimating the time span of biozones and the magnitude of unconformities, improvement of the stratigraphic framework, and for a better understanding of sedimentary and climatic processes (e.g. Perlmutter and Matthews, 1989, 1992). For instance, the modern Neogene timescale now depends on precise orbital tuning of marine and continental cyclic sequences, and evolved into an astronomically tuned (polarity) timescale (APTS), which proved to be far more precise and accurate (e.g. Hilgen, 1991a, b; Wilson, 1993; Shackleton *et al.*, 1995; Hilgen *et al.*, 1995; Zachariasse, 1999).

### 1.5.9 Chemostratigraphy

Chemostratigraphy is a relatively new technique, developed mainly during the last decade (e.g. Humphreys *et al.*, 1991; Ehrenberg and Siring, 1992; Pearce and Jarvis, 1992, 1995; Racey *et al.*, 1995; Pearce *et al.*, 1999). It uses the primary geochemical variation in the whole-rock elemental composition of siliciclastic sediments and sedimentary rocks to correlate stratigraphic sequences, as well as to gather inferences on basin paleotectonic history, source rock lithologies, depositional pathways, and paleoclimates.

### 1.5.10 Isotope Stratigraphy (Sr, C, O)

Over the last decade there have been a substantial number of works concerned with defining the Sr isotopic evolution of the oceans during the Jurassic, Cretaceous and Cenozoic (see Hess *et al.*, 1986, and McArthur, 1998, for a review). Such a high-resolution Sr-isotope curve can be used as a global correlation tool and, over some intervals, have a stratigraphic resolution superior to that of biostratigraphy.

Recent research on isotope stratigraphy has also been forefront in defining and refining carbon- and oxygen-isotope curves for the Mesozoic and Cenozoic, based on the analysis of carbonate rocks and fossils and of terrestrial organic matter (e.g. Holser, 1984; Faure, 1986; Shackleton, 1985; Holser and Margaritz, 1989). The oxygen-isotope curve has been primarily used for estimating the Cenozoic record of water-mass temperatures (e.g. Frakes *et al.*, 1992; McCauley and DePaolo, 1997). When of characteristic shape and form, the carbon-isotope curve can be used for intercontinental correlation (e.g. for sections across the Cenomanian–Turonian boundary; Kuhnt *et al.*, 1990; Gale *et al.*, 1993; Pratt *et al.*, 1994), as well as to allow inferences on patterns of the long-term organic carbon cycle (e.g. Scholle and Arthur, 1980; Arthur *et al.*, 1985), and hence to indicate important periods of petroleum source-rock deposition.

## 1.6 Future Perspectives

The growth of applied stratigraphy in the 20$^{th}$ Century has been unprecedented. Whereas a particular problem might have been studied by a single investigator a century ago, or by a small group of scientists just a few decades ago, today such a problem is dealt in by a multidisciplinary legion of highly experient researchers.

In the foreseeable future stratigraphy holds out many promises, as well as a number of scientific problems. Research on isotope stratigraphy will include modelling strontium-, carbon- and oxygen-isotope variations determined from analysis of fossils in an attempt to understand significant oceanographic and other variables that have controlled the chemistry of ancient oceans. High-resolution chemostratigraphy will be used to improve stratigraphic correlation in barren siliciclastic beds and between continental and marine sequences, and to add important information to the reconstruction of a basin depositional history and paleoclimatic evolution. Cyclostratigraphy will involve the comparison of orbital time-series derived from different sequences in an attempt to improve the absolute resolution of the geological time scales, to identify and quantify the magnitude of gaps in the stratigraphic record, and to delineate paleoclimatic belts. Sequence stratigraphy will aim to document paterns of sequence development in conjunction with modelling studies at regional scale, as well as used in high-resolution studies applied to oil and gas reservoir correlation and modelling (e.g. Cross *et al.*, 1993). Advanced theorectical and experimental studies on fractal geometry and complexity theory will be applied to modelling of complex natural systems, such as ecosystems and anisotropies in oil and water reservoirs. Altogether, these studies will be complemented by the development of highly refined frameworks of biostratigraphy, paleoecology, and event stratigraphy, and further integrated into interactive chronostratigraphic and stratigraphic databases to allow global comparisons of sedimentary cycles and events on continental and marine sections, to analyse the relationship between environmental changes, biotic evolution and extinction, and to identify relationships between sea-level changes and major perturbations in Earth's climatic, oceanographic and sedimentary systems.

Among the many challenges faced by geoscientists are practical and critical ones, some of profound sociological implications, such as how to best explore and exploit Earth's natural resources, by attaining a sustainable development of human communities and simultaneousy preserve Earth's ecosystems' biodiversity. In this broad sense the science of stratigraphy has recently turned towards a more profound holistic non-traditional approach, by gathering information from nearly all disciplines in the geosciences to collectively aiming to provide an unified picture of Earth history. In a way similar to how its early foundations were laid upon in ancient times, the modern since aims to understand and reconstruct Earth's history on a planetary context, the origin and evolution of life on Earth and, ultimately, predict and help to preserve its future (Fig. 1.2).

**The Ancient Greeks and Romans:**
**The Naturalism**

| | |
|---|---|
| Pythagoras (ca. 582–ca. 507 B.C.) | Aristotle (384–322 B.C.) |
| Xenophanes of Colophon (ca. 570–ca. 480 B.C.) | Pliny the Elder (ca. 23–79 AD) |
| | Tertullian (ca. 155–222) |

**The Middle Ages and Renaissance:**
**The Principles**

| | |
|---|---|
| Avicenna (980–1037) | John Ray (1628–1705) |
| Albertus Magnus (ca. 1206?–1280) | Robert Hooke (1635–1703) |
| Leonardo da Vinci (1452–1519) | Nicholaus Steno (1638–1686) |
| Georgius Agricola (1494–1555) | John Woodward (1668–1721) |
| Conrad Gessner (1516–1565) | Johann Jakob Scheuchzer (1672–1733) |

*Figure 1.2 Timeline of stratigraphic thought.*

## The Eighteenth and Nineteenth Centuries: The dillema of Catastrophism *versus* Uniformitarism and Gradualism

| | |
|---|---|
| Carolus Linnaeus (1707-1778) | William D. Conybeare (1787-1857) |
| Charles Bonnet (1720-1793) | Friedrich A. von Alberti (1795-1878) |
| James Hutton (1726-1797) | Sir Charles Lyell (1797-1875) |
| John Playfair (1748 1819) | Friedrich August Quenstedt (1809-1889) |
| Abraham Gottlob Werner (1750-1817) | Roderick I. Murchison (1792-1871) |
| Georges L. Cuvier (1769-1832) | Joachim Barrande (1799-1883) |
| Alexander von Humboldt (1769-1859) | Alcide Dessalines d Orbigny (1802-1857) |
| Alexandre Brongniart (1770 1847) | Jean Louis Rodolphe Agassiz (1807-1873) |
| William Phillips (1775-1828) | Charles Darwin (1809-1882) |
| Jean-Baptiste Lamarck (1744-1829) | Amanz Gressly (1814-1865) |
| William Smith (1769-1839) | Albert Oppel (1831-1865) |
| Leopold von Buch (1774-1853) | Charles Lapworth (1842-1920) |
| J.J. d'Omalius d'Halloy (1783-1875) | Johannes Walther (1860-1937), |
| William Buckland (1784-1856) | Walther's law of correlation of facies (1894) |
| Adam Sedgwick (1785-1873) | |

## The Twentieth Century & Beyond: The Age of Applied Stratigraphy

***Plate tectonics***

- Alfred Wegener (1880-1930; 1915)
- Hess (1959, 1962)
- Vine and Matthews (1963)
- Hallam (1973)
- Miller (1983)

***Chronostratigraphy and Biochronology***

- Shaw (1964)
- Kauffman and Hazel (1977)
- Berggren and Van Couvering (1978)
- Haq *et al.* (1987, 1988)
- Harland *et al.* (1982, 1990)
- Berggren *et al.* (1995b)
- Gradstein *et al.* (1995)

***Magnetostratigraphy***

- Rutten (1959)
- Harrison and Funnel (1964)
- Galbrun (1984)

***Facies Stratigraphy***

- Amanz Gressly (1814-1865)
- Reading (1978, 1996)
- Walker (1979, 1992)
- Walker and James (1992)
- Pirrie (1998)

***Stratigraphic Terminology, Classification and Concepts***

- Ager (1973, 1984)
- NACSN (1974, 1983)
- Hedberg (1976)
- ISSC (1987)
- Blatt *et al.* (1991)
- Whittaker *et al.* (1991)
- Walker and James (1992)
- Salvador (1994)

*Figure 1.2 Continued*

***Quantitative Stratigraphic Analysis and Modelling***

Shaw (1964)
Ager (1973)
Miller (1977)
Van Hinte (1978, 1982)
Gradstein *et al.* (1985)
Mann and Lane (1995)
Harbaugh *et al.* (1999)

***Sequence Stratigraphy***

Sloss (1949, 1962, 1963)
Wheeler (1958, 1959a,b)
Payton (1977)
Vail *et al.* (1977a, b)
Haq *et al.* (1987)
Van Wagoner *et al.* (1987, 1988, 1990, 1991)
Posamentier *et al.* (1988, 1992)
Ross and Ross (1988)
Galloway (1989)
Schlager (1992)
Walker and James (1992)
Miall (1991, 1994, 1997)

***Event Stratigraphy and Cyclostratigraphy***

Milutin Milankovitch (1879-1958)
Fischer and Arthur (1977)
Gould and Elderedge (1977)
Einsele and Seilacher (1982)
Signor and Lipps (1982)
Dott Jr. (1983)
Hsü (1983)
Gould (1984)
Fischer (1986, 1991, 1993, 1995)
Goodwin and Anderson (1985)
Kauffman (1987, 1988)
Bottjer *et al.* (1986)
Flessa (1986)
Hallam (1989 a, b)
Perlmutter and Matthews (1989, 1992)
Schwarzacher (1987, 1993)
Fischer and Bottjer (1991)
Weedon (1993)
Walliser (1996)
Einsele (1998)
Gale (1998)
Perlmutter *et al.* (1998)

***Chemostratigraphy***

Humphreys *et al.* (1991)
Ehrenberg and Siring (1992)
Pearce and Jarvis (1992, 1995)
Racey *et al.* (1995)
Pearce *et al.* (1999)

***Isotope Stratigraphy***

Scholle and Arthur (1980)
Holser (1984)
Arthur *et al.* (1985)
Shackleton (1985)
Frakes *et al.* (1992 )
McCauley and DePaolo (1997)
McArthur (1998)

***Interactive Stratigraphic Databases***

ICS/SIS (http://www.stratigraphy.org)
CHRONOS (2002) (http://www.chronos.org)
NEPTUNE (http://www.neptune.ethz.ch/)

*Figure 1.2 Continued*


## 1.7 Acknowledgements

The manuscript benefited greatly from reviews by Felix M. Gradstein (Geology Museum, University of Oslo) and Peter Szatmari (Petrobras-Cenpes, Rio de Janeiro).

Part II

# The Search for Patterns: Ordering the Framework

Chapter 2

# Buried Time: Chronostratigraphy as a Research Tool

MARIE-PIERRE AUBRY[1] and JOHN A. VAN COUVERING[2]
[1] Department of Geology, Rutgers State University of New Jersey, Piscataway NJ 08854, USA.
[2] Micropaleontology Press, 256 Fifth Avenue, New York, NY 10001, USA.



## 2.1 Introduction

Improvements in stratigraphic dating methods have transformed chronostratigraphy into a practical tool that reveals unexpected realities in the place of earlier conventions. As strata become more precisely ordered in time, applied chronostratigraphy allows us to unravel cause and effect across facies boundaries and gaps, to resolve the context of past changes, and to find relationships between the proxy records of vanished forces found in different lines of evidence. The advances in stratigraphic time-analysis have in turn resulted in the reformulation of some of the basic concepts of stratigraphy and geochronology, as for instance the operational assumption that conformable sections are also depositionally continuous until proven otherwise. In this chapter we emphasize the need for integrated stratigraphy as the essential foundation for greater precision in interpretations of regionally extensive stratigraphic sections. Paradoxically, as temporal interpretations become more easily visualized on first inspection, it is more difficult to hold them intellectually separate from the objective evidence, and a new terminology is required to clarify this basic distinction in discussions. As an example of problems still to be overcome, we review the discrepancies between lower upper Miocene magnetobiostratigraphic correlations in different sections, as reported by Berggren *et al.* (1995c) that result in an unstable early late Miocene biochronology.

Without conscious effort, every Earth scientist understands Steno's First Law – that superposed strata represent the passage of time. Beyond this, even the first geological maps irresistibly imply a lengthy history, simply by showing that countless successive



*E.A.M. Koutsoukos (ed.), Applied Stratigraphy,* 23-53.

strata have accumulated in formations that are themselves seen to be superimposed. The extrapolated concept of chronostratigraphy – that intervals of geological time could be defined in terms of accumulated strata – was soon grasped by Alcide d'Orbigny, who in his monumental study of stratigraphy only 15 years after the publication of William Smith's great map of England, proposed the concept of the "stage" to embody the concept of a worldwide synoptic perspective that coincided with deposition of a given body of strata (our translation):

> "In summary, rigorous application of the general and specialized principles of geology to the sedimentary layers which constitute the Earth's crust, lead to the understanding that these layers form distinct superposed stages, characterized by a specific fauna; that each fauna has clear and definite limits; and that the occurrence of a significant number of species that are limited to and characteristic of these stages always permits them to be distinguished, whatever the different mineralogic [i.e. lithologic] compositions that the strata presently show. Indeed, whereas the study of superposition and concordance of stratification of the geological stages alone often gives excellent results when the stages are superposed without stratigraphic gaps, such [physical] studies cease to provide positive evidence when intermediate stages are missing, as we see on a multitude of points on our planet." (D'Orbigny, 1849, pp. 7, 8.)

That this was not merely a prefiguration of faunal zones, as defined in a group of related species and thereby bounded by diachronous paleoecological limits, is clear from the further explanation of stages in volume 2 (our translation):

> "A stage, for us, is an epoch completely identical to that of the present. It is a steady state of ancient nature during which there were, as nowadays, continents and seas, with terrestrial plants and animals, marine plants and animals, and in the seas coastal and pelagic animals at all depths. For a stage to be complete it must show a collection of terrestrial and marine beings that represent a whole epoch, with development analogous to the one which we see today on Earth." (D'Orbigny, 1851, pp. 256–257.)

This crucial insight, which was in all respects the modern ideal of isochronous synchronization of the geological record, was an attractive philosophical concept and quickly became part of geological theory. Unfortunately, it was far ahead of the means to accomplish it in practise. Thus, for the next century and more, global chronostratigraphy was essentially a stratigraphic belief system, undisturbed by testable statements that went beyond the very broadest generalizations.

It is interesting to note, in this regard, the almost completely unappreciated contribution of 19$^{th}$-century power politics to the creation of the Standard Time Scale. European colonial and commercial expansion provided support for vigorous geological surveys of the overseas territories that focused on exploitable resources as much as on fossil beds and historical reconstructions. These studies demanded regional and intercontinental correlations. Given the imperialistic ethic, it seemed only logical to require that the foreign strata should all be correlated to European units, no matter how overstretched the comparisons of paleofaunas and lithofacies, or how forced the recognition of trans-regional events. The rapid increase of non-European "equivalent strata" in the successive editions of Lyell's textbooks is a clear witness to this process.

By the late XIX$^{th}$ century, European units that had been originally lithologic or paleontologic subdivisions in a hodgepodge of different local conceptions – the period, series, system, epoch, and so on – had spread outward under the wing of imperialism to cover the globe with what can only be called presumptive time-equivalent correlations. With a little trimming and fitting, these extravagant chronostratigraphic hypotheses were combined in a unified hierarchical system that became the Standard Time Scale. In the absence of direct tests of synchrony, correlation rested almost universally on comparative paleontology, in which often the merest resemblances were held to be simultaneous. Even where inter-regional correlation was relatively reliable, as for instance in upper Paleozoic and Mesozoic ammonite-bearing sequences, there was nothing to distinguish chronostratigraphy and biostratigraphy in practise. To this day, the stratigraphic surfaces of first and last appearances of guide fossils are the standard of isochrony in Paleozoic and Mesozoic marine sequences.

Coming full circle, it is a curious twist of fate that the stage - the first and only element in the time scale to be originally defined as a global chronostratigraphic unit – should now be the only unit in the modern Standard Time Scale hierarchy that is under attack for lacking worldwide validity. A strongly expressed view that the stage is "too brief" for global application is, oddly, the outgrowth of improvements in geochronology. When modern planktonic microfossil biostratigraphy and radiometric dating revealed gross errors in long-standing regional correlations of mid-Cenozoic stages (Van Couvering and Miller, 1971; Berggren and Van Couvering, 1978; Hardenbol and Berggren, 1978), it was felt by some stratigraphers that if correlation was that imprecise, then only the broader divisions in the Standard Time Scale could have global validity. Of course, this overlooks the rather obvious fact that the same methods employed in debunking the old correlations are available for new correlations. Nevertheless, the anti-stage pessimism has resulted in the strange situation, as we write, that Cenozoic stages, whose boundaries can now be dated by multiple criteria, are dismissed from their role as the basic element in the chronostratigraphic hierarchy [e.g. compare Premoli Silva and Jenkins, 1998, with Brinkhuis and Visscher, 1995], while Mesozoic and Paleozoic stages continue to be accepted as fundamental divisions of the global time scale even though their boundaries are still traced in poorly controlled biostratigraphy.

In actual fact, the rapid expansion of alternative methods of assessing stratigraphic time, ranging from models of astronomical cycles to standardized magnetostratigraphic and stable isotope profiles, has already been shown to enhance, rather than supplant biostratigraphic definitions (e.g. Hilgen and Langereis, 1994; Lourens *et al.*, 1996). These developments in geochronology (in the sense of Berggren and Van Couvering, 1978, if not in the sense of geochronometry as defined by Aubry *et al.*, 1988) are being extended to ever older levels, so that stage-level precision in global correlation has become commonplace and practical in the Cenozoic, and is beginning to challenge simplistic biostratigraphy in earlier levels.

The sharpening of geochronological tools during the past 50 years, aside from causing a temporary disenchantment with stage correlations, has had wholly beneficial results. The overall lesson to be learned, when we review these developments, is that regional stratigraphic analyses may be far more complex than had been previously thought. For example, the idea that prevailed until the late seventies, that mirror-image facies on opposite sides of epicontinental basins were synchronous pairs, is no longer credible. The simplistic picture of undisturbed "layer-cake" accumulation in the deep-sea stratigraphic record is vanishing as well. Overall, Derek Ager's agnostic view (1981) that the stratigraphic record is fundamentally incomplete, and that preserved strata are at best an uneven and often biased record of geological history, is being vindicated.

It is fair to credit the growth and acceptance of new chronostratigraphic concepts – temporal interpretation of sections, for example, and temporal versus stratigraphic resolution (see Aubry, 1995, 1998) – almost entirely to the flowering of government-funded research on deep ocean stratigraphy that began in the mid-sixties. Access to the condensed but essentially continuous record preserved in the Earth's most extensive facies, that of deep-sea laminites, allowed a new emphasis on precise, globe-spanning correlations with integrative studies of extraordinary microfossil, magnetic and isotopic records. Of course, not all of the applications in modern chronostratigraphy emerged from the blue volumes of the DSDP and ODP. For instance, while modern marine geology depends heavily on worldwide biozonation of calcareous planktonic microfossils and the standard paleomagnetic model for geochronological control, the basis of the planktonic foraminiferal and coccolith biostratigraphies (e.g. Bolli, 1966a; Blow, 1979; Martini, 1971), as well as the fundamental principles of magnetostratigraphy (see review in Opdyke and Channell, 1997) were developed in land sections. Similarly, techniques of tephrostratigraphic correlations of aeolian and volcanic dust-falls, which play a vital role in marine stratigraphy (deMenocal and Brown, 1999), were developed as an essentially continental discipline focused on East Africa (Brown *et al.*, 1992; deMenocal and Brown, 1999) and western North America (Sarna-Wojcicki and Davis, 1991). On the other hand, Milankovitch's prediction that climatic cycles would prove to be linked to long-term astronomical cycles, while it has been spectacularly documented in cyclostratigraphic analyses of marine sections exposed on land (Hilgen *et al.*, 1997; Krijgsman *et al.*, 1999; Hilgen and Krijgsman, 1999), was first proved to affect the geological record on the basis of stable isotope variations in deep-sea cores (Emiliani, 1972; Hayes, Imbrie and Shackleton, 1976). In almost every instance, however, no matter whether dating technologies originally were demonstrated on the basis of evidence in land sections or in oceanic cores, they have been developed to the highest level in the service of marine geology. The notable exception is Neogene (Miocene–Pleistocene) radiometric, fission-track and trapped-charge dating, which continues to be spearheaded by work on hominid-bearing sequences in Africa (e.g. Pickford *et al.*, 1991).

With regard to developments in the deep-sea context, the astronomically-calibrated stable isotopes series obtained from DSDP cores have been used to generate a tightly controlled model curve that is applicable to Cenozoic strata around the world (Shackleton *et al.*, 2000). To give another example, advances in time–distance analysis of the paleomagnetic reversals in the basaltic floor of the oceans has been the key to two-decimal precision in the calibration of magnetostratigraphy (Cande and Kent, 1992, 1995). The dated model of paleomagnetic reversals, while not a "clock" per se, offers a record of frequent, readily observed events, in both marine and continental facies, that are more reliable indicators of age than almost any other stratigraphic evidence. Finally, we note that there are other techniques, such as time-analytical modeling of conformable sedimentary sections in drill cores (Sadler, 1981) that originated wholly within deep-ocean studies.

## 2.2 The Effect of Deep-Sea Studies on Modern Chronostratigraphy

Since the advent of deep-sea studies, Upper Jurassic to Recent chronostratigraphy has been revolutionized by the quality of data available from the unaltered, essentially continuous record of highly fossiliferous strata on the ocean floor. Even distorted by cryptic disconformities and selectively preserved information, the overall completeness and information content in these sequences far exceeds that available on land. Unfortunately, no older seafloor has survived the Wilson Cycle, and pre-Oxfordian chronostratigraphy has not been advanced in the same way. The most dramatic consequence of the explosion of data from deep-sea cores was a sudden shift of the center of chronostratigraphic reference from northwest Europe and the Mediterranean, where the primary units of the Standard Time Scale had been defined almost entirely with reference to disconformities between epicontinental synthems, into the worldwide arena of the deep-sea.

As a result of this shift in focus, the Mesozoic and Cenozoic stages in European classical sequences, as well as stages in local chronostratigraphies of the Paratethys, Japan, New Zealand, and North America, have ceased to be vehicles of direct correlation, although they remain the basic elements of formal chronostratigraphy (Hedberg, 1976). The main reason, of course, is that the stratigraphic and paleontological criteria that had traditionally defined most stage boundaries over the decades were not applicable in the deep-sea context. In addition, the classical criteria were demonstrably less well suited for long distance correlation when compared to the new standards (as noted above). Thus, the weight of global correlation now rests almost entirely on criteria developed from study of the long cores obtained by the DSDP–ODP program. The two most important components are the network of global biozones in planktonic marine microfossils (e.g. Srinivasan and Kennett, 1981; Berggren *et al.*, 1995c; Okada and Bukry, 1980; SanFilippo *et al.*, 1985) and the model of paleomagnetic reversals that make up the Global Polarity Time Scale or GPTS (Berggren *et al.*, 1985a–c; Aubry *et al.*, 1988; Cande and Kent, 1992, 1995) and its outgrowth, the Integrated Magnetobiochronologic Scale or IMBS (Berggren *et al.*, 1985a–c, 1995c). A third, and increasingly important geochronological tool from the same cores is the climatically influenced record of carbon and oxygen isotope ratios, calibrated according to Milankovitch orbital cycles (Shackleton *et al.*, 1985; Shackleton, 1987, 2000).

Almost as soon as high-precision deep-sea chronostratigraphy began to develop, it became a new basis for correlation, and in some instances a stimulus for redefinition, in important open-marine sequences exposed on land, primarily in the Mediterranean, Caribbean, Gulf Coast, and southeastern Pacific regions. Diagnostic deep-sea criteria were more difficult to identify, however, in the updip facies where virtually all classical chronostratigraphic units are typified. For this reason, workers were forced to locate most classical chronostratigraphic boundaries in the deep-sea sections by guesswork, pending further information. The placeholder boundaries, however, soon became deeply entrenched in the rapidly-expanding oceanographic literature. Predictably, as "further information" from the stratotypes has developed, it has generally been found that the deep-sea concepts were miscorrelated to a greater or lesser degree. Because of the ascendancy of deep-sea studies, this has led to calls to abandon the long-established stratotypes on land. The most notable example, so far, is the situation with regard to the Paleocene–Eocene boundary (Aubry, 2000; Aubry and Berggren, 2000a, b).

In terrestrial sequences, the application of recent advances in chronostratigraphy has met with mixed success. While magnetostratigraphic chronology is readily applicable in many nonmarine contexts, opportunities for cyclostratigraphic analysis are limited by the fact that thick laminite sequences with relatively clear and complete records of climatic cycles are rare on land (but see Krijgsman *et al.*, 1996). With regard to the fine-grained global biochronology that has been developed in marine micropaleontology, there is almost no relationship at all with terrestrial paleontological data. It is a plain fact that while marine and nonmarine successions are physically juxtaposed in interfingering relationships at the edge of epicontinental basins, this does not translate to consistent or reliable relationships between the continental fossil record and that of the deep-ocean basins. For this reason, the sound marine–terrestrial correlations that are a basic requirement in our quest to comprehend the modalities of global change depend almost entirely on paleomagnetic, cyclostratigraphic–stable isotopic, and radiometric evidence.

Below, we briefly review difficulties in coordinating the deep-sea and continental time scales that arise from the application of certain non-biostratigraphic techniques in the interpretation of the deep-sea record. Radioisotopic dating is not discussed here, mainly because of its relatively limited role in calibrating deep-sea stratigraphic sequences, and also because it is the principal subject of other parts of this work. We then discuss briefly the strength and weakness of the current IMBS (Berggren *et al.*, 1995c).

## 2.3 Chronostratigraphic Method: Age Assignments and Correlation

Stratigraphy has two main strategies for placing strata in a time context. One is to assign an age to a stratum or body of rock. This is normally a relative age in a superpositional context. While isolated exposures of volcanogenic strata can be dated radioisotopically without a superpositional context, in most instances radioisotopic (and other direct chronometric) analyses can normally supplement, and only occasionally modify, assignments of relative age in a timescale. This is to say, that a relative age is positioned in a time-ordered framework or model of sequential events – e.g. a biozonal scheme, an isotopic curve, a paleoclimatic or eustatic series, a geomagnetic polarity stratigraphy, or to the boundaries in a time scale – but it has no intrinsic value in absolute time. The absolute or numerical ages that we ascribe to a relatively dated level are derived secondarily according to geochronometric calibration of the model in which the relative age is fixed, or (rarely and accidentally) by direct dating of the stratum itself.

The other method of dating is simply to correlate the rock body to another, the age of which has been established by the first strategy. As more and more dates are assigned to the stratigraphic record, however, correlation may be confused with mere *synchronization:* i.e., two rock bodies independently dated to the same age. In the absence of any corroborative data, mere synchrony should be treated as an untested assumption, and not as conclusive proof of stratigraphical equivalence.

The calibrated model of geomagnetic polarity stratigraphy, generally termed the paleomagnetic time scale (PMTS) or global polarity time scale (GPTS), has become a major means of dating and correlation of marine and terrestrial deposits in recent decades (e.g. Tauxe *et al.*, 1994; Krijgsman *et al.*, 1996; Cande and Kent, 1992). The characteristic logs of black and white stripes would appear to be without uncertainty, but Flynn and Tauxe (1998) warn of the uneven quality of published magnetostratigraphic data and of the shortcomings of magnetostratigraphic correlations when the data are poorly documented or assessed. Two tests are advised to determine the reliability of magnetic data, one being histograms of the parameters used to determine polarity, and the other being the "Jacknife parameter", or *J*, defined by Tauxe and Gallet (1991). Yet, even when both tests are satisfactory, there is no guarantee that the pattern observed has a global significance. As an example, Flynn and Tauxe (1998) consider that the magnetic record of the upper Paleocene–lower Eocene deposits recovered from OPD Hole 690B on Maud Rise is of superior quality, according to the very high test values of the measurements. Nevertheless, the identification of the chrons in the section (Spiess, 1990) conflicts with the biostratigraphy, as discussed by Aubry *et al.* (1996). In particular, the identification of Chron C24n.1n-3n in the interval with predominantly normal polarity between 138 and 155 mbsf (Spiess, 1990) in this corehole was inconsistent with the simultaneous determination that this interval should be assigned calcareous nannofossil Zone NP10 (Pospichal and Wise, 1990; see also Aubry *et al.*, 1996). As Flynn and Tauxe (1998) concluded, it is possible that, despite its apparent high quality, the magnetic record at Site 690 includes overprinted intervals, and Ali and Hailwood (1998) further suggested that this magnetic record might be affected by disturbed remanences. Detailed analysis subsequently led Ali *et al.* (2000) to conclude that core splitting induced overprinting of the magnetic record at Site 690. The clear message of this and other examples is that even good-quality magnetostratigraphy benefits from testing against other dating methods, wherever evidence is available (see discussion in Aubry, 1995).

Despite significant shortcomings, even in the deep-sea record, biostratigraphy has been the backbone of stratigraphy since William Smith, as well as the backbone of geochronology since Charles Darwin. Its one great advantage over most features commonly found in strata is that, when undertsood as the physical record of organic evolution, the fossil record is inherently self-ordering. In theory, one can tell by inspection which of any two related fossils lying on a table is the older, simply because one will be more evolved than the other. Put in other words, biostratigraphy contains evidence of progressive, non-repeated changes in species and communities. In this regard it is like the progressive decay of unstable isotopes or organic molecules, or progressive accumulation of trapped charges: it records a non-iterative series, in which each variation in its properties can be identified with a single point in time. Unlike the time-dependent isotopic or cumulative systems, however, the *precision* of biostratigraphy – in this instance, the recognition of evolutionary and population changes – is not a percentage of age, and is thus unaffected by distance from the present. In addition, its *accuracy* – the amount by which errors in morphologic, taxonomic or stratigraphic determinations distort the reported range limit of a taxon – is comparable to the 3 to 5% analytical error in radiochronology, as a percentage of the total age-range of the fossil taxon in question.

By comparison, radioisotopic dating is actually applicable only locally and relatively rarely to sedimentary strata, and overwhelmingly to continental strata in volcanically active settings. Biostratigraphy, however, applies to all fossiliferous rocks from all settings. Again, certain other methods such as stable isotope stratigraphy (particularly $\delta^{18}O$ stratigraphy), lithologic cyclostratigraphy, and magnetic susceptibility stratigraphy may allow more accurate and more finely resolved correlation. Nevertheless, biostratigraphy – notably marine planktonic micro-biostratigraphy – remains the primary tool of broad long distance correlations, simply because of its abundant data and simplicity of application.

Unfortunately even the most fully developed and tested biostratigraphy is often perceived by non-paleontologists as an unreliable tool in chronostratigraphy. As McClelland *et al.* (1996, p. 213) expressed it, "magnetostratigraphic correlations should be constructed independently of biostratigraphy" – meaning marine microfossil biostratigraphy. This is both regrettable and understandable. Regrettable, because it expresses a lack of understanding of the essence of biostratigraphy, and its fundamental role in stratigraphic interpretation and correlation. Understandable, because biostratigraphic studies are not always conducted with the same rigor as that which supports the conclusions in physical and chemical geochronology. Here, we do not refer to simple matters of interpretation, such as the differing taxonomic concepts among authors that result in slight differences in the positioning of biozonal boundaries. We refer instead to a common tendency to uncritically accept the apparent temporal significance of the reported lowest and highest occurrences of marker species in different deep-sea cores, which gives rise to a discouraging picture of unpredictable and extensive diachrony in the boundaries of oceanic (planktonic) biozones. The basis of this credulity among paleontologists is the widespread belief that deep-sea stratigraphic sections are essentially continuous, without missing or condensed intervals of any significance that would affect the observed distribution. With the fossil record seen as an essentially complete and truthful testimony, it has also been the general rule to explain any inconsistent relationships between observed range limits and other stratigraphic markers, such as magnetic reversals or isotopic excursions, as examples of biostratigraphic diachrony. As discussed elsewhere (Aubry, 1995), such inconsistencies should always be suspected, before anything else, as being the effect of unconformities. Even minor hiatuses not only abbreviate the observed stratigraphic range of taxa, but by removing slice after small slice of the record can also significantly distort the stratigraphy of magnetochrons or isotopic cycles without leaving a detectable mark.

## 2.4 Biostratigraphy and Chronostratigraphy

A variety of precisely-worded terms, commonly reduced to acronyms, has been developed by different authors to meet the need for distinguishing clearly between the objectively recorded features of biostratigraphy in the strata and their well-considered interpretation in chronostratigraphy. In this work, we use FAD and LAD, standing for "first appearance datum" and "last appearance datum", to mean the actual moments of evolution and final extinction in the history of a given taxon (Berggren and Van Couvering, 1978). The acronyms LO and HO, for "lowest occurrence" and "highest occurrence", are used here for the two-dimensional stratigraphic surfaces defined by the observed lowest occurrences and observed highest occurrences, respectively, of a given taxon in the rock record. The stratigraphic surface that we call an LO is in fact a composite of stratigraphic points, each of which is a reliably observed first occurrence of the taxon in a given section or group of sections. Each specific point of observation is here called a LOD for "lowest occurrence datum". Similarly, the HO surface is a composite of individual reliable observations of last occurrences, for which we use the term HOD for "highest occurrence datum." The use of "datum", in these terms, refers to a single point, whether in time or space.

Our prime example of the need to distinguish between the observed and the inferred, as discussed below, is the way that *biozone* and the *biochronozone* have become functionally identical in the context of globally correlatable fossil groups, and with planktonic microfossils in particular. The obvious fact cannot be stressed too often: All biostratigraphy is based on inherently inaccurate local observations. Even marine microfossils, despite their ubiquitous presence in every stratum of the most common types of marine facies, are subject not only to the general accidents of deposition and preservation but also to the particular accidents involved in making and studying an individual sample, that bias the records of all fossils (Signor and Lipps, 1982; Ager, 1981). Given this basic limitation, we still can evaluate local occurrence data according to how closely it approaches the potentially observable (if not actual) beginning and end of a species' appearance in the stratigraphic record. A large array of methodologies, such as graphic correlation techniques (Shaw, 1964; Edwards, 1984), abundance patterns of taxa (Keller, 1983), integrated biostratigraphy based on multiple planktonic microfossils (Moore *et al.*, 1978), and above all the depth–age plots now becoming common in DSDP/ODP reports (Barron *et al.*, 1985; Aubry, 1991, 1995), permit us to judge how much of the possible record is missing, and in particular to determine whether an observed local range is artificially shortened, either in a disconformity or in some other process that removes fossil evidence from the view of the scientist.

Accepting that, with sufficient care, it is possible to correct biostratigraphy to a large extent for the fundamentally discontinuous nature of even the most fine-layered and apparently undisturbed accumulations in the stratigraphic record, we turn to the chronostratigraphic potential of a unique group of fossils: the marine planktonic groups that document what seem to be exceedingly rapid, or "geologically instantaneous", worldwide population changes in the only globe-spanning biofacies, that of the upper waters of the world ocean. The LOs and HOs in such fossil groups represent one of two kinds of temporal significance (Figs. 2.1, 2.2). We can consider that the worldwide LO of a species has *global temporal significance* if it corresponds closely to the time horizon of the species' First Appearance Datum (FAD). Similarly, the HO of a species has *global temporal significance* if it can be shown to be close to the Last Appearance Datum (LAD), the time horizon defined by the extinction of the species. Even though global planktonic microfossil zones are the closest approach in biostratigraphy to true chronostratigraphic units, it must be strongly stressed that a biozone – even a global biozone defined by "geologically instantaneous" LO and HO horizons as boundaries – is not even theoretically isochronous. It is always a three-dimensional body of physical strata characterized by its paleontological content (Hedberg, Ed, 1976). On the other hand, a biochronozone is a chronostratigraphic unit, with absolutely isochronous boundaries, that is defined by ideal, unique FADs and LADs. It happens, however, that the only way to correlate a biochronozone is with the same biostratigraphic information – LOs and HOs – that define the biozone based on the same species or assemblages. In the special case of marine planktonic biostratigraphy, where biozones are defined with LOs and HOs of global temporal significance, a biozone and a biochronozone based on the same range criteria can become functionally indistinguishable. We would argue that the only exception is where a significant stratigraphic gap occurs within the biozone. In Figure 2.2, compare Zone BB in section (a) with Zone BB in sections (b) and (c). Biozone BB in section (b), as in section (c), is not a biochronozone, even though its lower and upper boundaries are the same as in section (a).

In pre-Cenozoic chronostratigraphy, we find biozones that have somehow become part of the idealized, isochronous time-scale hierarchy, but without losing their paleontological definition. These might best be called were-zones (i.e., biozones transformed by moonlight into something unnatural). Like planktonic microfossil biozones, these units are also based on widely-distributed and rapidly-dispersing open-ocean groups – ammonites in particular, but also conodonts and graptolites – and in the absence of any realistic competition have been the unchallenged standard criteria for inter-regional correlations for many decades. Over the years their real nature has been distorted by simplistic usage, with the LOs and HOs of key ammonites treated as FADs and LADs – that is to say, instantaneous time horizons – and the biozone boundaries entered as divisions of the standard time scale (Ogg *et al.*, 1984, 1991; Gradstein *et al.*, 1995; Hardenbol *et al.*, 1998). Here again we encounter what we would call "functional isochrony, " where the available geochronological tools are incapable of resolving the diachrony in correlated stratigraphic data. The conflation of biozonation and biogeochronology in these groups is therefore in no way different

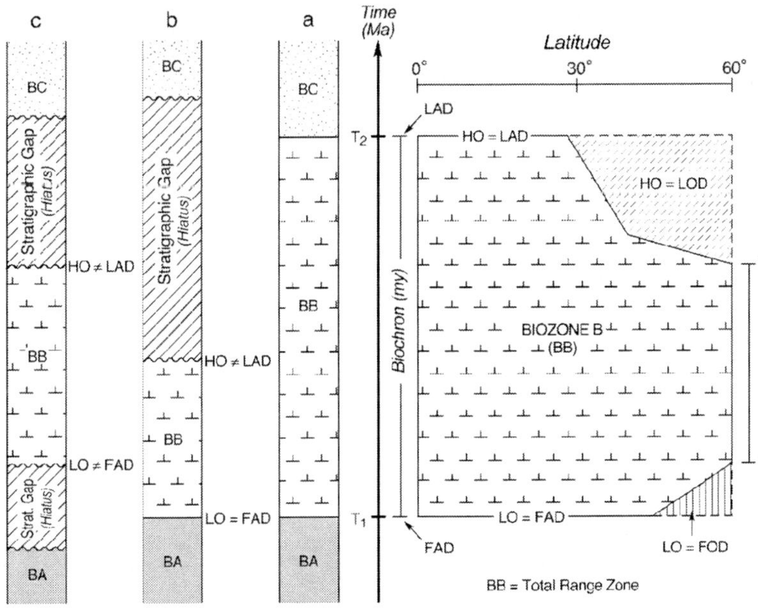


*Figure 2.1 Lowest (LO) and highest (HO) occurrences of species may or may not have a temporal significance (slightly modified from Aubry, 1995, fig. 2).*

*The LO of a taxon has a global temporal significance in sections where it can be shown to correspond to the First Appearance Datum (FAD) of the species. The age of the horizon where the LO is recorded is then T1, the time (in Ma) of the evolutionary appearance of the species. Similarly, the HO of a taxon has global temporal significance in sections where it can be shown to correspond to the Last Appearance Datum (LAD) of the species. The age of the horizon where the HO is recorded is then T2, the age of the extinction of the species.*

*In stratigraphic sections located beyong 30° latitude (right of time axis), the LO and HO do not correspond to the FAD and LAD of the species, and thus do not have global temporal significance. However, they may have regional temporal significance. These LO and HO thus may constitute regional datums, referred to as LOD and HOD. The LO in a given section corresponds to the time of the arrival (immigration) of the species at a given latitude, which is expressed as its LOD. At 60° latitude the LOD has an age T'1. Between 30 and 60° latitude, the LOD has different ages comprised between T1 and T'1 (T'1 < T1). The HO in a given section corresponds to the departure (emigration) of the species from a given latitude, a datum expressed as the LOD of the species. At 60° latitude, this LOD has an age T'2. Between 60 and 30° latitude, the LOD has ages comprised between T'2 and T2 (T'2 > T2).*

*In sections (a), (b) and (c), some LO and HO of a given species have temporal significance whereas others do not (left of time axis). In section (a), both the LO and HO have global temporal significance. In section (b), only the LO has global temporal significance, the HO having no age value, the upper range of the species being truncated by an unconformity. In section (c), neither the LO nor the HO have temporal significance because the stratigraphic range of the species is truncated both base and top. Note that the LO and HO at latitude 60° and the LO and HO in section (c) have the same relationships with respect to the time axis. Diachrony confers regional temporal siginificance whereas unconformities deprive LOs and HOs of temporal meaning.*

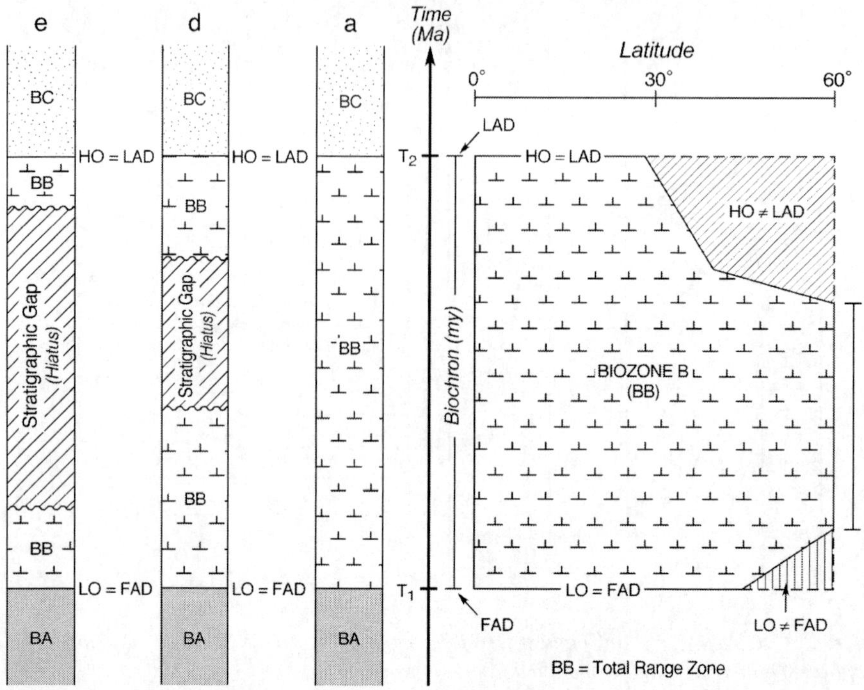


*Figure 2.2 Relationships between biozones and biochronozones. Biozone BB (a total range zone) is present at all latitudes between 0 and 60°. It is also present in sections (a), (d) and (e). However, it corresponds to the biochronozone only between latitude 0 and 30°, and in section (a). In sections (d) and (e), Biozone BB does not correspond to the biochronozone despite the fact that the LO and HO of the marker species correspond to its FAD and LAD.*

from that which we describe in the case of Cenozoic planktonic microfossils, except that the diachrony of apparently isochronous LOs and HOs in deeper time is significantly more difficult to resolve and may thus be significantly greater, than in younger strata. It is to be expected, however, that as the standards for pre-Cenozoic biostratigraphic correlation are raised, both by comparison with rapidly-improving magnetostratigraphy and cyclostratigraphy, and by more rigorous internal tests for range integrity, the "functionally isochronous" stratigraphic evidence that determines the boundaries of the were-zones will be replaced by more refined chronostratigraphic criteria developed in studies of the GSSP. A similar re-evaluation, to be sure, will inevitably transform the similarly misunderstood Cenozoic planktonic microfossil global biozones. It is not overly optmistic, we believe, to imagine a future in which the usefulness of global biozones of all ages will be enhanced and strengthened, rather than diminished, by a more realistic and nuanced appreciation of their real nature.

A recognizably diachronous LO (Fig. 2.1, right of time axis) includes LODs of demonstrably different ages, just as a diachronous HO includes HODs of more than one age. This is, of course, the normal manifestation of a biozone. It is only the unusual expansiveness of tropical–temperate marine planktonic ecofacies that consistently generate fossil records in which diachroneity in correlations is not normally resolvable (Johnson and Nigrini, 1985; Dowsett, 1989; Spencer-Cervato *et al.*, 1994; Berggren *et al.*, 1995a, c). For example, diachrony has widely (and rightly) been accepted as a universal characteristic in continental biostratigraphic correlation, following breakthroughs in Cenozoic K–Ar dating (Evernden *et al.*, 1964) and magnetostratigraphy (Opdyke *et al.*, 1979; Lindsay *et al.*, 1984; Opdyke, 1989) that demolished prevailing illusions of synchrony. This perception justifies a rigorous distinction between temporal and stratigraphic terminology, as argued forcefully by Walsh (1998). The complex terminology proposed by this author is, in theory, applicable to all biostratigraphic data, including that of deep-sea planktonic microfossil biostratigraphy. Nevertheless, the common circumstances of "functional isochrony" in this latter body of data make many of the nuanced distinctions created by Walsh (1998) functionally meaningless as well. In striking a balance, then, between a careless indifference to the distinction between time and rock, on the one hand, and an impractical scrupulosity on the other, we accept the opinion of Ager (1981) that in dealing with rock and time, a needlessly complex terminology (or "verbiage" in this author's expression) tends to obscure scientific clarity.

We would go slightly further, and contend that only biostratigraphic levels that have a temporal significance (HOs and LOs that are the stratigraphic evidence of the unique biological events called FADs and LADs) are worthy to be designated in a formal terminology (*contra* Walsh, 1998). In circumstances of palpable diachrony, however, the terminology proposed above can be in agreement with chronostratigraphic time scales only if we use FADs and LADs for the primary event, and recognize regional diachrony in the propagation of the event as evidenced by biostratigraphic LODs and HODs. Thus, the LAD of *Ericsonia formosa* is in Chron C12r (estimated age of 32.8 Ma) whereas its HOD at high latitudes is in Chron C18 at about 39.7 Ma (Berggren *et al.*, 1995c). Similarly, the FAD of *Amaurolithus primus* is near the end of the Sidufjall Event (estimated age of 4.8 Ma) but its LOD in the Mediterranean Basin is within the Nunivak Event and has an estimated age of 4.55 Ma (Berggren *et al.*, 1995a; Sprovieri *et al.*, 2002).

The application of time-significant biostratigraphic boundaries – LOs and HOs – in chronostratigraphy is also complicated by cryptic diachrony, concealed in incomplete samples and missing section. The most important distortion of LOs and HOs results from the truncation of stratigraphic ranges due to unconformities (Fig. 2.1, left of time axis). In such cases, the temporal significance of the biozonal boundaries can be compromised to the point that it loses its meaning. In recognition of the fact that "boundaries" that coincide with unconformable surfaces are not biozonal transitions, these transitions should be called *contacts* to emphasize the artificial superposition of biozones. For example, in section (b) there is a biozonal boundary between Biozones BA and BB but a biozonal contact between Biozones BB and BC (Fig. 2.1). It must be remembered that a normal succession of biozones is no guarantee of complete temporal succession as shown in Figs. 2.1 and 2.2, sections (b) and (c) (see also Aubry, 1993a).

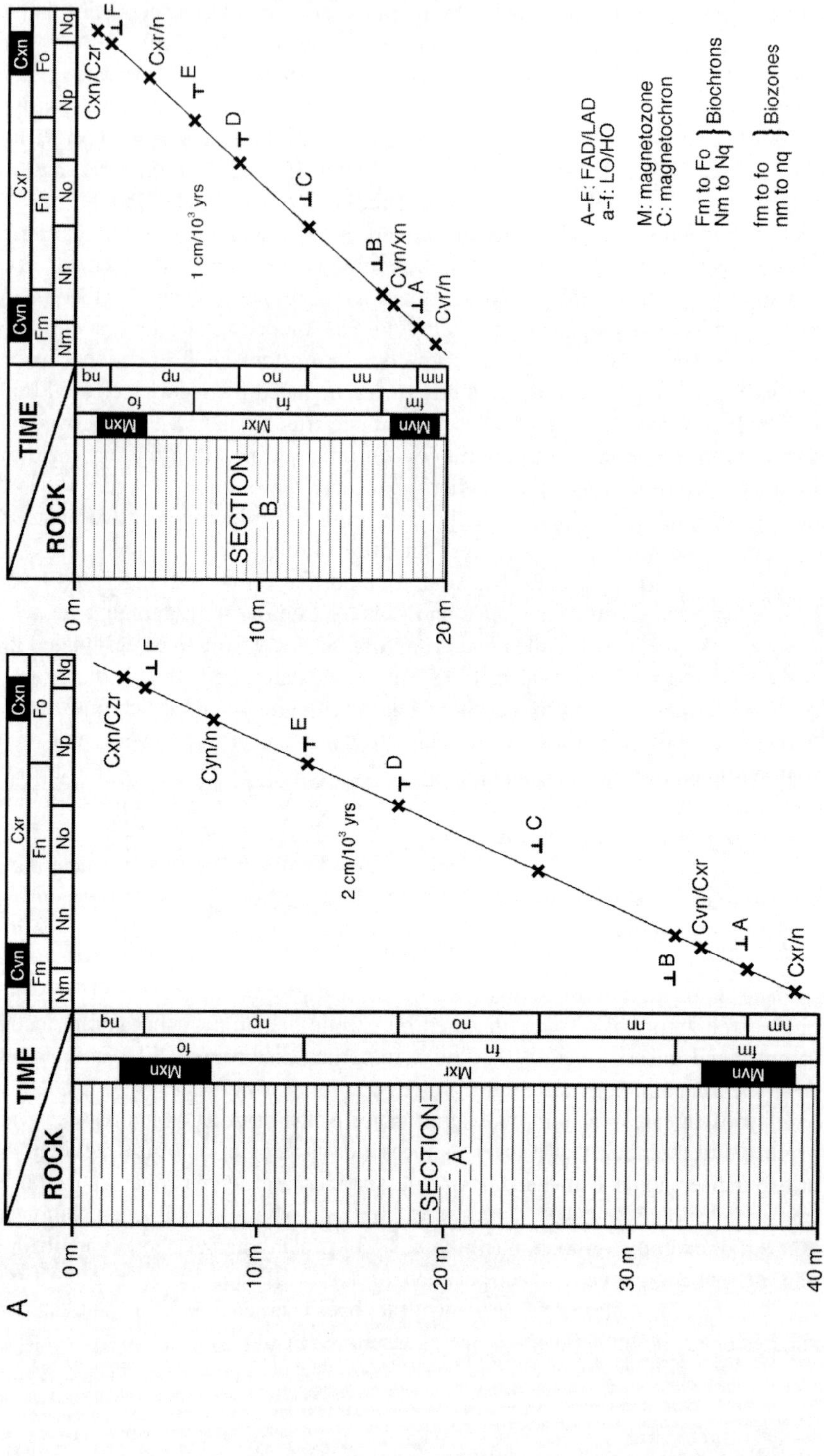
A
ROCK
TIME
SECTION A
SECTION B
0 m
10 m
20 m
30 m
40 m
Cvn
Cxr
Cxn
Fm
Fn
Fo
Nm
Nn
No
Np
Nq
Mvn
Mxr
Mxn
fm
fn
fo
nm
nn
no
np
nq
2 cm/10³ yrs
1 cm/10³ yrs
Cxn/Czr
Cyn/n
Cxr/n
Cvn/Cxr
Cvn/xn
Cvr/n
A–F: FAD/LAD
a–f: LO/HO
M: magnetozone
C: magnetochron
Fm to Fo
Nm to Nq
Biochrons
fm to fo
nm to nq
Biozones

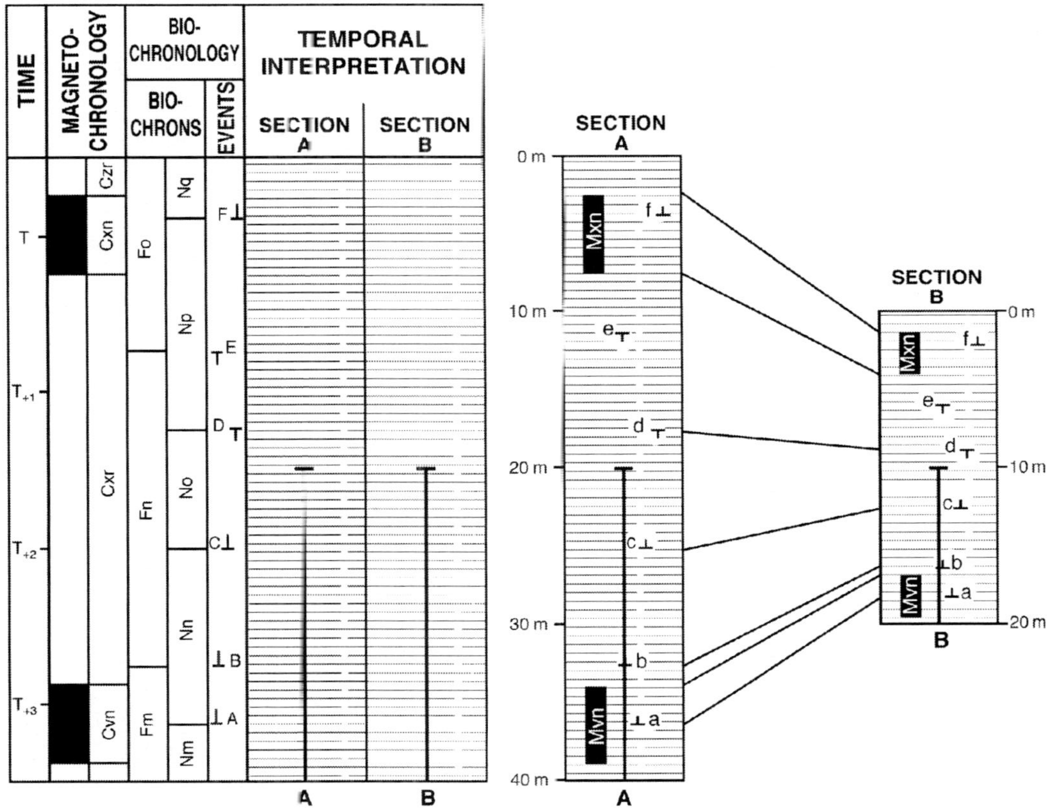


*Figure 2.3A* *Differentiating true paleontologic events from pseudo-events in stratigraphic sections. Sections of different thickness (section A is twice as thick as sections B and C) record a paleontologic event (e.g. the highest occurrence of a fauna as indicated by the bold vertical line) at mid-height in each section (Fig. 2.3A, B, bottom right). Because the position of the HO is the same in the three sections with respect to magnetic reversals (Mxn and Mvn), it is concluded that the HOs of the fauna in the three sections represent the same extinction event. Section (A) is compared with section (B). Both are complete, and represent the entire time span from the beginning of deposition (base of the sections) to the end of deposition (top of the sections), as documented by the temporal interpretation (bottom left) of sedimentation rate curves (top). Thus, the HOs in (A) and (B) record the same faunal extinction event.*

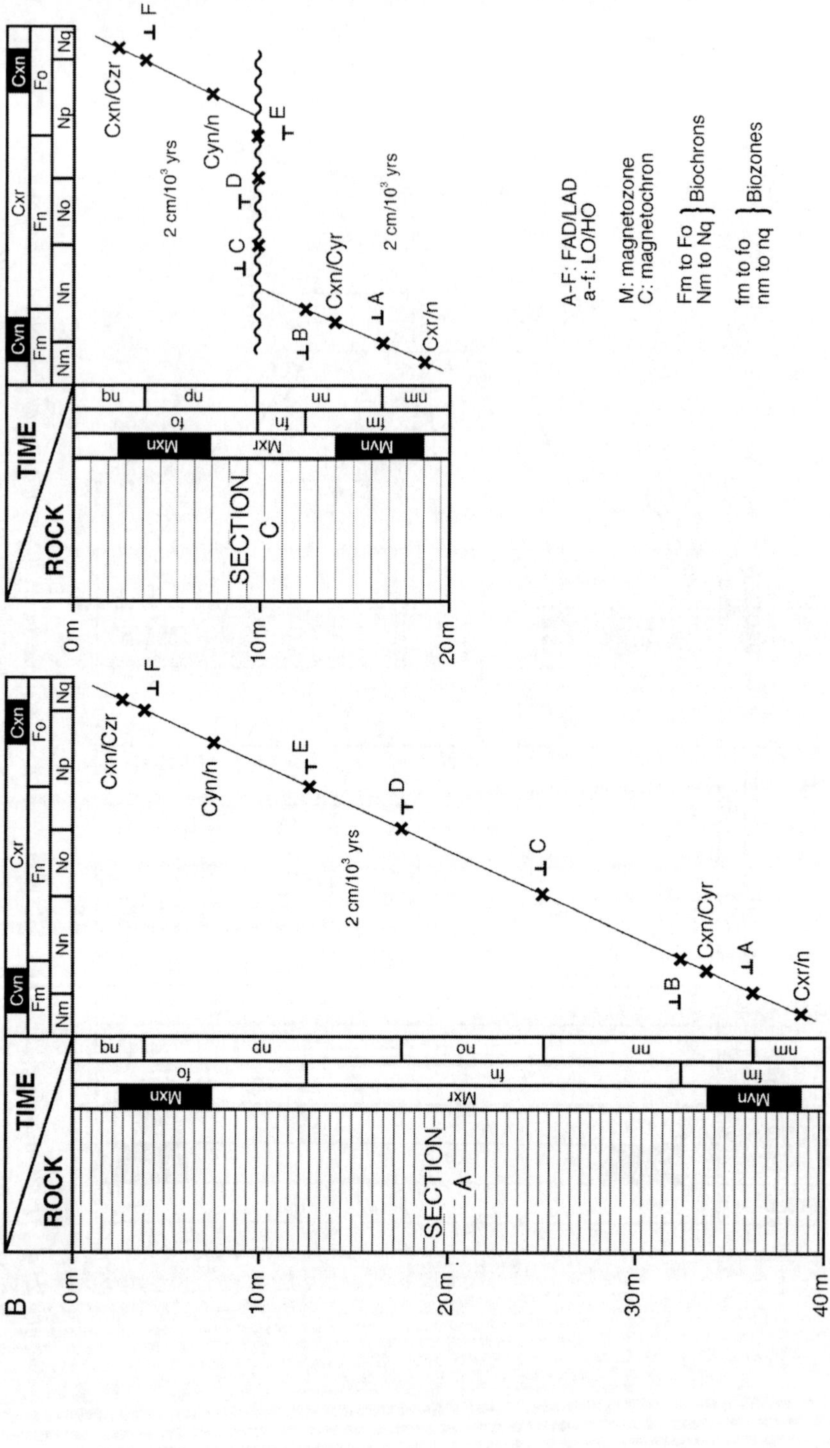
B
ROCK
TIME
SECTION A
SECTION C
0 m
10 m
20 m
30 m
40 m
Mvn
Mxr
Mxn
fm
fn
fo
nm
nn
no
np
nq
Cvn
Cxr
Cxn
Fm
Fn
Fo
Nm
Nn
No
Np
Nq
Cxn/Czr
Cyn/n
Cxn/Cyr
Cxr/n
2 cm/10³ yrs
A–F: FAD/LAD
a–f: LO/HO
M: magnetozone
C: magnetochron
Fm to Fo
Nm to Nq
Biochrons
fm to fo
nm to nq
Biozones

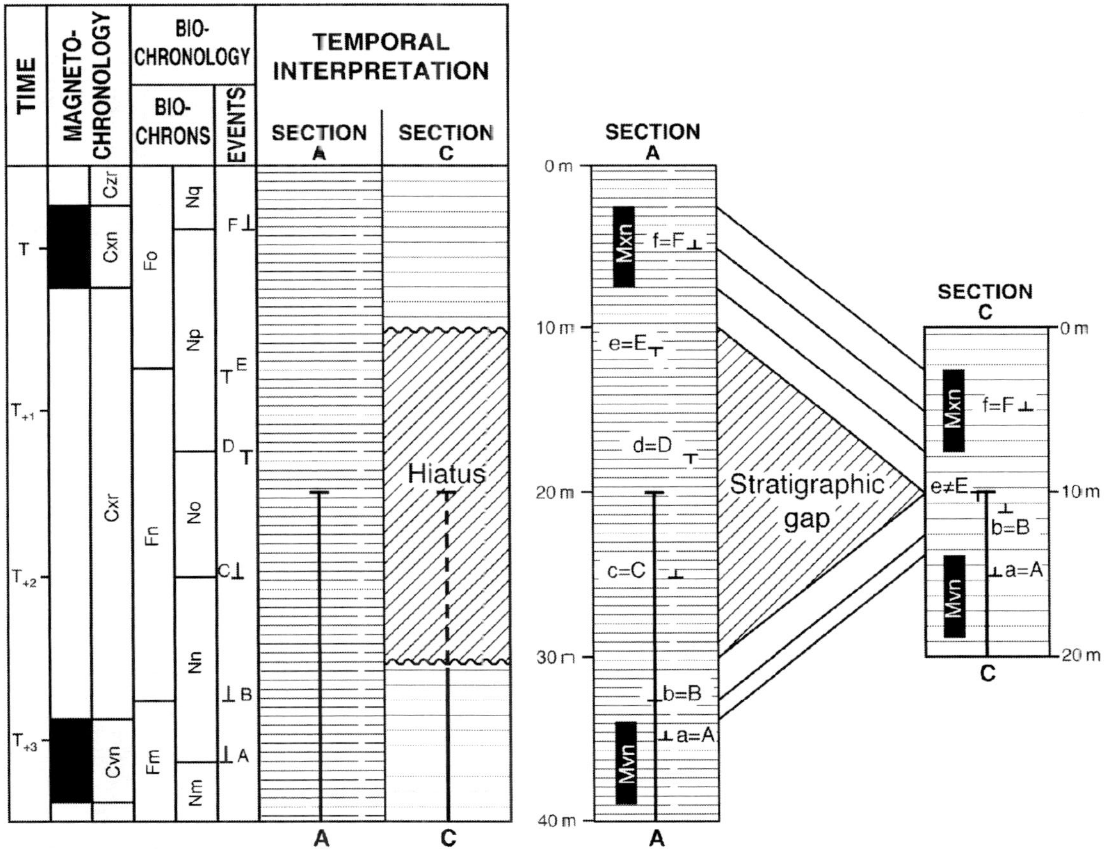


*Figure 2.3B Differentiating true paleontologic events from pseudo-events in stratigraphic sections. Sections of different thickness (section A is twice as thick as sections B and C) record a paleontologic event (e.g. the highest occurrence of a fauna as indicated by the bold vertical line) at mid-height in each section (Fig. 2.3A, B, bottom right). Because the position of the HO is the same in the three sections with respect to magnetic reversals (Mxn and Mvn), it is concluded that the HOs of the fauna in the three sections represent the same extinction event. Section (A) is compared with section (C), which has a substantial stratigraphic gap that removes the upper part of the fauna's true range (top of figure). Through the temporal interpretation (bottom left) it can be shown that whereas section (A) records the true HO event, the highest level of occurrence in (C) is a pseudo-event.*

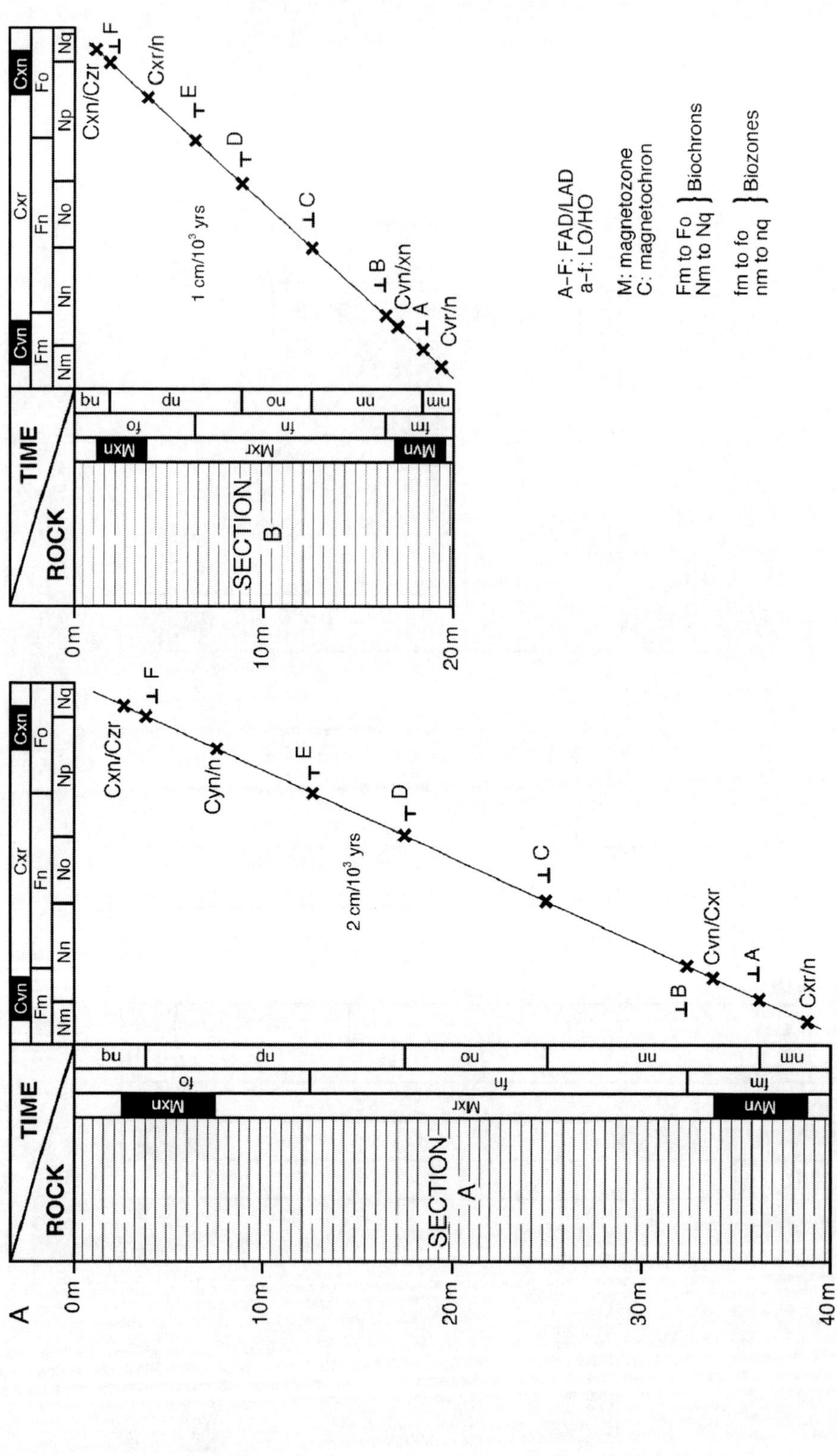
A
ROCK
TIME
SECTION A
SECTION B
0 m
10 m
20 m
30 m
40 m
Mvn
Mxr
Mxn
fm
fn
fo
nm
nn
no
np
nq
Cvn
Cxr
Cxn
Fm
Fn
Fo
Nm
Nn
No
Np
Nq
2 cm/10³ yrs
1 cm/10³ yrs
Cxr/n
Cvn/Cxr
Cyn/n
Cxn/Czr
Cvr/n
Cvn/xn
Cxr/n
A
B
C
D
E
F
A–F: FAD/LAD
a–f: LO/HO
M: magnetozone
C: magnetochron
Fm to Fo
Nm to Nq
Biochrons
fm to fo
nm to nq
Biozones

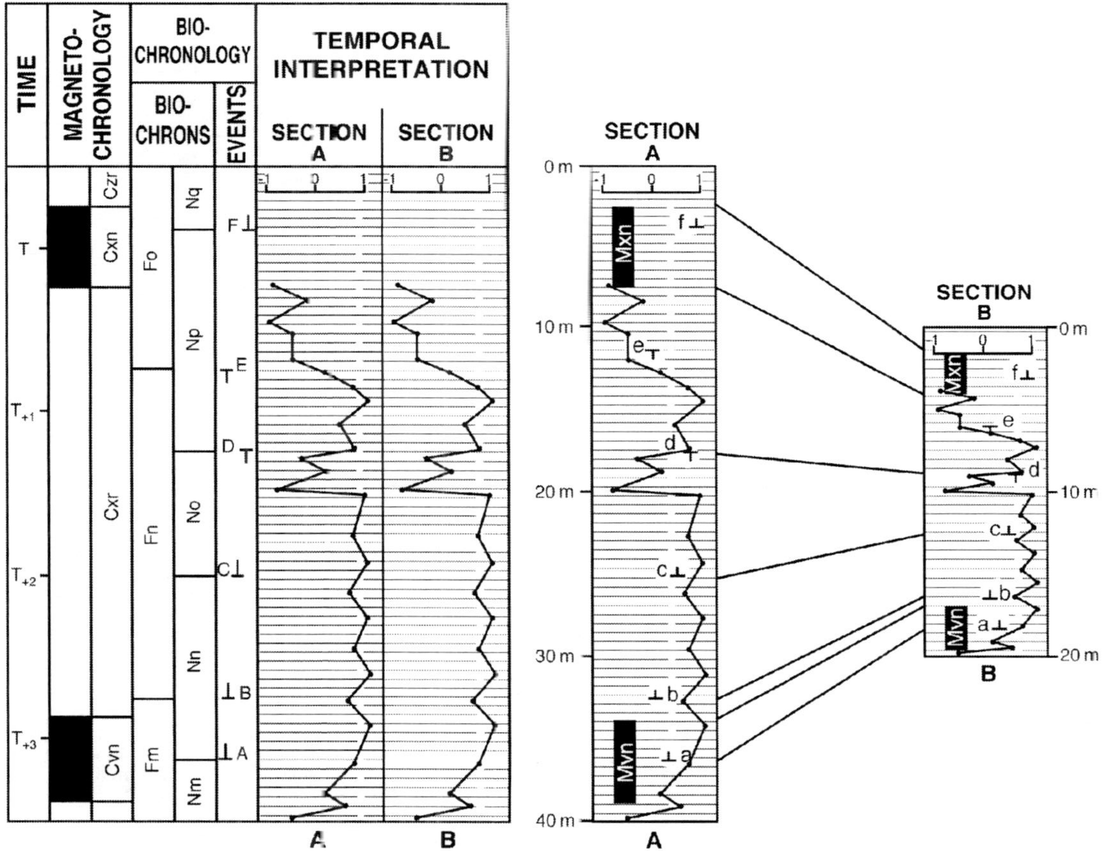


*Figure 2.4A* *Differentiating true isotopic events from pseudo-events in stratigraphic sections. Sections of different thickness (where section A is twice as thick as sections B and C) record an isotopic shift, which falls at mid-height in each section (Fig. 2.4A, B, bottom right). Because the position of the shift is the same in the three sections with respect to magnetic reversals (Mxn and Mvn), it is concluded that the shift in the three sections represent the same isotopic event. Section (A) and section (B) are both complete, and each represents the entire time span from the beginning of deposition (base of the sections) to the end (tops of the sections), as documented by temporal interpretation (bottom left) of sedimentation rate curves (top). Thus both isotopic shifts record the same isotopic event.*

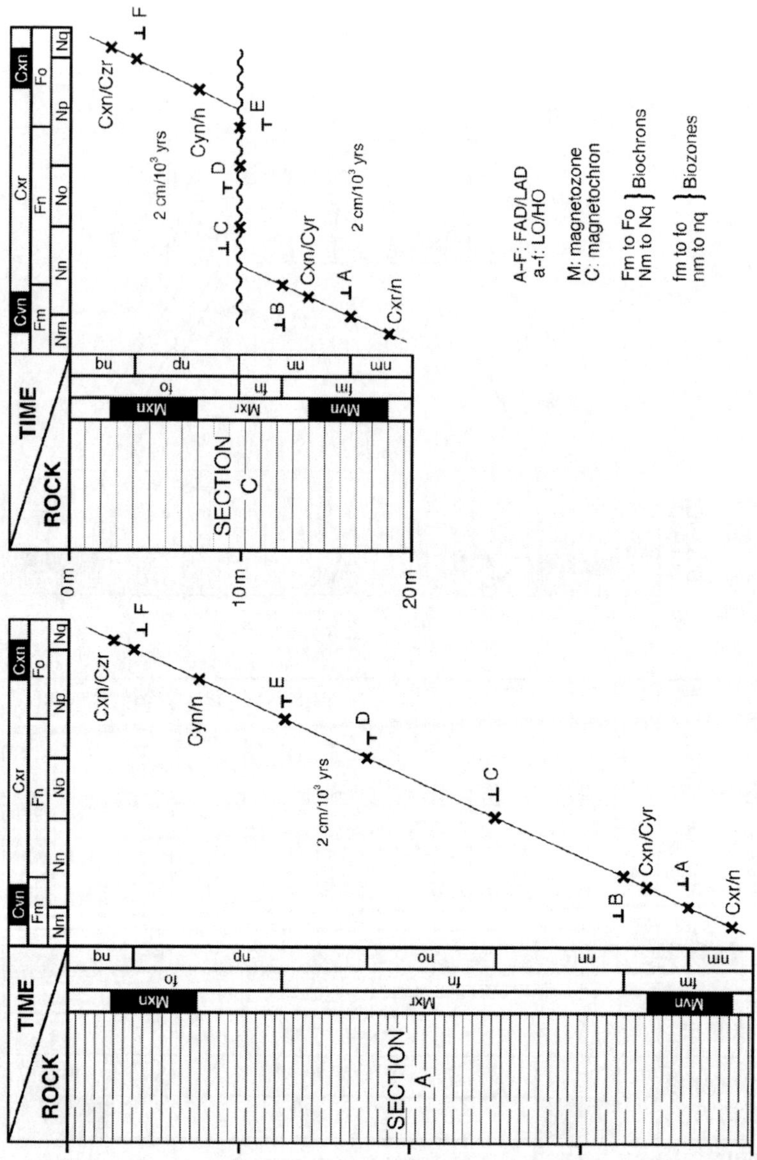
B
ROCK
TIME
SECTION A
SECTION C
0 m
10 m
20 m
30 m
40 m
Mxn
Mxr
Mvn
fo
fn
fm
nq
np
no
nn
nm
Cvn
Cxr
Cxn
Fm
Fn
Fo
Nm
Nn
No
Np
Nq
Cxn/Czr
Cyn/n
Cxn/Cyr
Cxr/n
2 cm/10³ yrs
⊥ A
⊥ B
⊥ C
⊤ D
⊤ E
⊥ F
A–F: FAD/LAD
a–f: LO/HO
M: magnetozone
C: magnetochron
Fm to Fo
Nm to Nq
} Biochrons
fm to fo
nm to nq
} Biozones

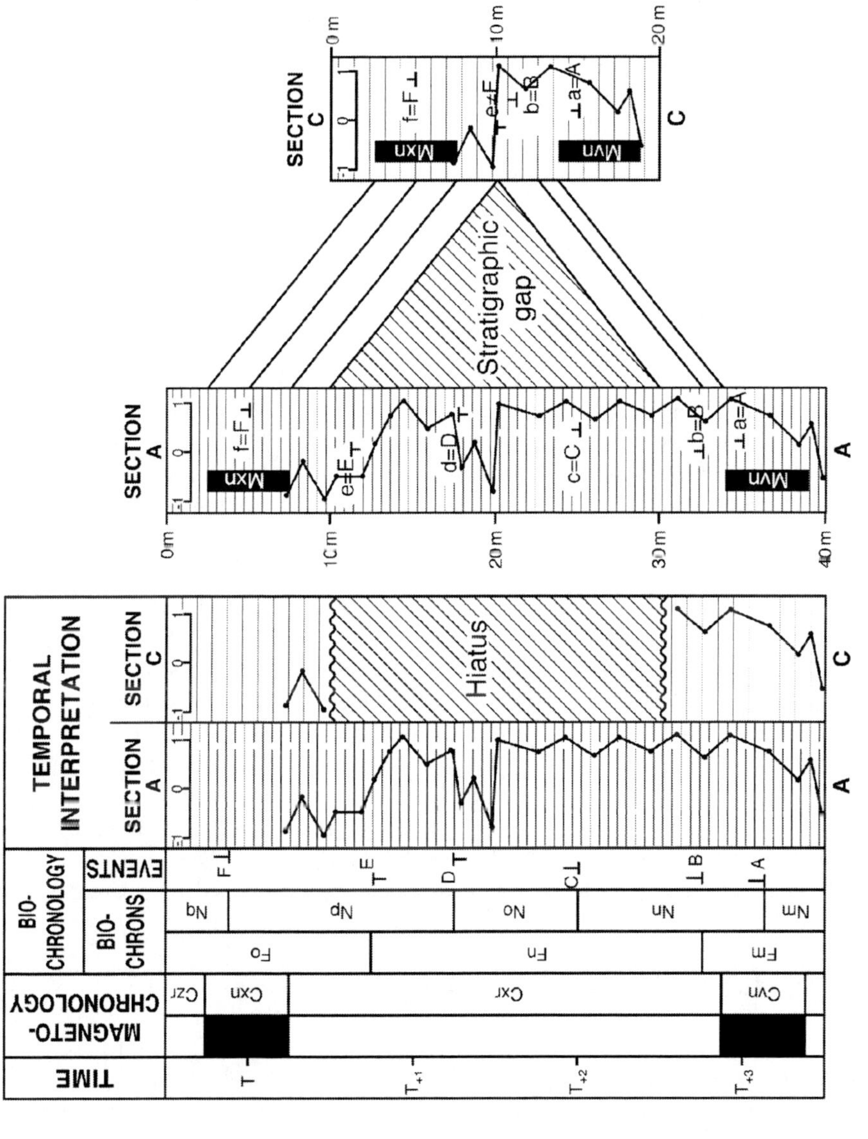


*Figure 2.4B* Differentiating true isotopic events from pseudo-events in stratigraphic sections. Sections of different thickness (where section A is twice as thick as sections B and C) record an isotopic shift, which falls at mid-height in each section (Fig. 2.4A, B, bottom right). Because the position of the shift is the same in the three sections with respect to magnetic reversals (Mxn and Mvn), it is concluded that the shift in the three sections represent the same isotopic event. Section (A) is compared with section (C), which includes a substantial stratigraphic gap at the level of the shift (top of figure). The temporal interpretation (bottom left) shows that whereas section (A) records the true isotopic event, section (C) records a pseudo-event.

These and other sources of error in stratigraphic interpretation and correlation, that result from assuming that a given LO or HO in the strata is the same as the FAD or LAD event which it documents, is discussed at greater length by Aubry (1995, 1998). Unrecognized hiatuses in apparently complete sections are the greatest source of error, far outweighing the consequences of misidentification, poor preservation and imperfect sampling. They not only falsely simplify the record by removing evidence, but also falsely complicate it by creating pseudoevents (Fig. 2.3a, b). If these distortions go unrecognized, they will lead to incorrect timing of geological processes such as paleontological turnover, sedimentological change, geochemical enrichment, and biofacies and lithofacies migrations. The dating of such evidence bears directly on interpretations of change frequency, sedimentation-accumulation rates, and evolutionary tempo. Ultimately, this will result in invalid interpretations of geological history and geologically-calibrated processes.

The geological history of regional and global variations in stable isotope ratios is becoming increasingly important in geochronology, but these observations too can be affected by cryptic defects in the stratigraphic record. We will leave the discussion of cyclostratigraphy to other authors in this work. Here, we note only the growing role of carbon isotope stratigraphy in oceanic–terrestrial correlation, for example the late Paleocene carbon isotope excursion (Kennett and Stott, 1991; Koch *et al.*, 1992, 1995; Stott *et al.*, 1996; Bowen *et al.*, 2001). This geochemical signature reflects changes in the global carbon reservoir that would propagate, in concept, at the present decadal rate of $CO_2$ equilibration (Plattner *et al.*, 2001) – in effect, instantaneously, as far as geochronological controls are concerned. Even so, this inherently isochronous evidence requires the same level of vigilant testing as other tools. Erosional gaps, in particular, can transform the record of isotope ratios, to create the impression of an excursion simply by a jump from one part of the record to another across a hiatus. Such pseudo events (as in biostratigraphy) must be distinguished from real events (Fig. 2.4a, b). On the other hand, a true excursion can be erased, or two excursions can be melded into one where the sedimentary sequence is collapsed by undetected non-deposition. Aubry (1998) and Aubry *et al.* (2000) documented several such false "excursions" in the records published by the DSDP/ODP (Deep-sea Drilling Project/Ocean Drilling Program). Thus, peak-to-peak correlations of isotopic profiles should be considered provisional until actual stratigraphic continuity is demonstrated.

## 2.5 A Case Study: The 1995 Cenozoic IMBS

In the excitement and satisfaction of seeing formerly intractable correlation problems crumble under the application of a newly developed stratigraphic methodology, it is a natural tendency to consider the new method to be superior to all preexisting ones, and to expect that the new method, alone, will bring definitive answers to the problems that it addresses. Where the new methodology can be used to strengthen the capability of other methodologies, even greater expectations are raised. The Integrated Magneto-Biochronologic Scale, or IMBS, of Berggren *et al.* (1995c) was hailed as an exemplary syncretism of the new magnetostratigraphy and the new marine biostratigraphy, both of which emerged from the deep-sea research program. The IMBS brought a strikingly improved level of precision to biostratigraphic correlations, a precision that magnetostratigraphy alone was unable to provide. A similar benefit is apparent from the combination of magnetostratigraphy and isotope stratigraphy (except for Pleistocene sediments), although some biostratigraphic control is also involved.

The integration of the IMBS with rigorously documented stratigraphic data leads to what may be called "temporal analysis" in the evaluation of local stratigraphic sections. With reference to precisely calibrated magneto–biostratigraphic models, it is possible to determine which parts of a stratigraphic section are most complete, what intervals of geological time these represent, where unconformities lie, and what are the ages of unconformable surfaces and the duration of the hiatuses (Aubry, 1995, 1998). Such temporal analysis of sections guarantees that correlations are, as far as possible, truly isochronous, and not simply stratigraphic correlations that are said to be isochronous. In this way we can hope that the evidence of actual events is not compared with pseudoevents, and that the timing of the evidence in the strata is correctly established.

One great benefit from this new advance in methodology was improvement in the various specialized time scales that constitute the Standard Time Scale, and the improvement in the Standard Time Scale in turn flowed back into improved local chronostratigraphy. We see stratigraphy as the physical image of geochronology, whose objective is the construction of time scales and the quantification of historical geology. Time scales represent scenarios of past physical events such as magnetic reversals, isotopic events, tectonic shifts and environmental change, together with events in the evolution of biological life. The time scales are based not only on the evidence preserved in the strata, but also (to an extent that is not widely appreciated) on the positioning of the evidence-bearing strata in composite sections, where different pieces of the story can be compared. It is this latter procedure, that of time correlation of strata, that provides the framework by which the evidence is turned into history.

For most of the pre-Cenozoic time scales (Roberts *et al.*, 1995; Opdyke, 1995; Gradstein *et al.*, 1995; see also Harland *et al.*, 1990), numerical time is derived from wide interpolations between relatively few dated points, and time-correlations are taken directly from the stratigraphic record. In the Cenozoic, as we have have pointed out above, and in the late Jurassic–early Cretaceous interval (Channell *et al.*, 1995), numerical time values are more numerous, and the geochronological significance of stratigraphic correlations can be independently checked; "functional isochroneity" of marine planktonic biostratigraphic correlation, in this context, is closer to true isochroneity due to the quality of external controls. The principal time control for the Cenozoic and the mid-Mesozoic is found in statistical standardization of the paleomagnetic lineations on the slowly-spreading seafloor, which provides a vast measuring tape marked off with more than a hundred reversals of the magnetic poles. This special time scale (actually a chronology, not a chronometer) is referred to as the global polarity time scale or GPTS. With interpolation closely controlled by the standardized seafloor spreading rates, the paleomagnetic time scale is actually calibrated by relatively few numerically dated reference horizons (Cande and Kent, 1992, 1995). The IMBS applies the GPTS to the biostratigraphic record in deep-sea cores, to estimate the ages of paleontologic events such as FADs, LADs, and eventually acmes. Once they have been well characterized, other stratigraphic evidence as well as biostratigraphic and paleoclimatic boundaries can be integrated into the IMBS, such as the singular $\delta^{18}O$ events Ol and Mi (Miller *et al.*, 1991), the terminal-Paleocene $\delta^{13}C$ excursion (Kennett and Stott, 1991), and steps in the strontium isotopic evolution of sea water (Miller *et al.*, 1988; Hodell and Woodruff, 1994). Using paleomagnetic data in facies where biostratigraphic correlations are difficult, the "Vail Curve" of eustatic sealevel changes (Haq *et al.*, 1988; Hardenbol *et al.*, in de Graciansky *et al.*, 1998) can also be tested and more precisely dated and described as part of the unified historical calendar that we call the Standard Time Scale.

Three elements determine the strength of the IMBS: (1) the reliability of the GPTS, or rather its applicability, especially in the "blank zones" between widely-spaced reversals; (2) the acuracy of the radioisotopic and cyclostratigraphic dates used to calibrate the GPTS, and (3) the quality of correlations between the GPTS and the data in deep-sea cores, with reference to both the biostratigraphic record and the paleomagnetic record. The first two elements, and in particular the critical role played by the calibration points, are discussed in Cande and Kent (1995) with regard to the astronomically estimated age of the Thvera calibration tie-point; by Berggren *et al.* (1995c) in a postscript on the reliability of the eight other tie-points in Cande and Kent (1995), and by Berggren and Aubry (1996) and Aubry (1998) with regard to the use of the Paleocene/Eocene Epoch boundary as a tie-point. Although the 1995 IMBS is a noticeable improvement on the 1985 version (Berggren *et al.*, 1985a, b, c), mainly because of an increase in sections with magnetobiostratigraphic data, the dependability of the correlations in the current IMBS is still uneven (see also Berggren *et al.*, 1995c).

The incomparable advantage of deriving a chronology independent of the stratigraphic record is that it frees biostratigraphy from self-referential impotence and allows it to take on true temporal perspective. In this new light, it can be seen that largely unsuspected unconformities riddle the stratigraphic record not only in epicontinental basins, but also in the deep-sea (Aubry, 1995). For instance, in many different coreholes around the world, the interval that includes the lower-middle Eocene (Ypresian–Lutetian) boundary was found to contain hiatuses typically 1 to 4 m.y. in duration, or even more. To go further, from what we know so far it appears that unconformities in the deep-sea are not random, but cluster around certain levels. It may be more precise, however, to state that unconformities involving depositional hiatuses of measurable duration (i.e. duration perceptible to geochronologic resolution: see below) are more common in some stratigraphic intervals than in others. In any case, it seems clear that there are intervals of time, expressed in terms of Cenozoic biochrons (Moore *et al.*, 1978; Aubry, 1993a, b, 1995), that are generally well represented in the stratigraphic record, such as the early Eocene (NP11–NP12), the late Eocene (NP19–20), and the late Miocene (NN11). Other intervals, however, are consistently less complete, such as the Paleocene–Eocene transition (NP9–NP10), the early Eocene–middle

Eocene transition (NP13–NP14a), and the middle–late Miocene transition (NN8–NN9). This implies that there are stratigraphic intervals for which magnetobiostratigraphic correlations are firmly established, due to the wide preservation of paleomagnetic and biostratigraphic data in those intervals, whereas correlations are much less satisfactory for those intervals that are poorly represented. The magnetobiochronological tool is relatively robust, as a result, in lower Eocene and upper Miocene sections, but weaker in the lower part of the upper Miocene. Ideally, in the future the clock-like Milankovich periodicities, documented in cyclostratigraphic analysis, will be available in the older parts of the Cenozoic to further refine the magnetobiostratigraphic correlations, as we have seen for the late Neogene (Shackleton *et al.*, 1995; Hilgen, 1991b; Berggren *et al.*, 1995a; Hilgen *et al.*, 2000; Spovieri *et al.*, 2002). Progress is already being made in establishing a late Paleocene–early Eocene astrochronology (Katz *et al.*, 1999; Norris and Röhl, 1999; Röhl *et al.*, 2000; Cramer, 2001).

Aside from the problems caused by unconformities, there are difficulties in the chronostratigraphy of some specific stratigraphic intervals. For instance, the middle Eocene (NP15 to NP17 zonal interval) magnetobiochronology is generally weak, due to the facts that middle Eocene deep-sea sections worldwide tend to yield neither a good magnetic record nor good microfossil samples because of numerous chert horizons that hinder core recovery. The low quality of the record makes it difficult, of course, to perceive unconformities. Overall, however, magnetobiochronology is probably more reliable for the Paleogene than for the Neogene. This is due in large part to the contrasting magnetic history of the two periods. In the Paleogene, the paleomagnetic model exhibits a relatively simple pattern with few reversals, producing long chrons with few, simple subchrons that are easily identified in stratigraphic sections. In contrast, Neogene magnetic history is one of strikingly more numerous reversals, producing many short chrons and within them multiple brief subchrons, so that it is often difficult to identify one's position in the stratigraphic record. A point in case is the difference between the initial interpretation of the magnetozones in the Neogene of DSDP Site 563 by Miller *et al.* (1985) and its revision by Wright *et al.* (1992). Uncertainty in Neogene paleomagnetic age assignments also likely reflects an insufficient acknowledgement of the incompleteness of the Neogene deep-sea record.

One of the weakest part of the current IMBS is probably the late middle to early late Miocene (NN8–NN9, approximately). Despite several decades of deep-sea drilling and the recovery of cores from numerous thick Miocene sections, the magnetobiostratigraphic correlations in this interval must be characterized as largely uncertain, despite claims to the contrary (Schneider *et al.*, 1997). This situation largely stems from the fact that no core has yet been recovered from a section that yields a good magnetostratigraphic record in direct association with a good biostratigraphic record of *both* planktonic foraminifera *and* calcareous nannofossils. Miller *et al.* (1994) reviewed this problem in the light of direct correlations between magnetostratigraphy (albeit of less than ideal quality) and microfossil biostratigraphy in the middle to upper Miocene section exposed at Buff Bay, eastern Jamaica. This interpretation flatly contradicted the magnetobiostratigraphic correlations that had previously been established in North Atlantic DSDP Sites 558 and 563, and that served in the formulation of the 1985 IMBS (Berggren *et al.*, 1985c): biostratigraphic levels in calcareous nannofossils and planktonic foraminifera that were associated mainly with normal polarity in DSDP Holes 563 and 558, proved to be associated mainly with reversed polarity in the Buff Bay section. To help explain this, Miller *et al.* (1994) pointed to temporal control from carbon-isotope data that suggested that inconsistent relationships between magnetostratigraphy, calcareous nannofossil biozones and those based on planktonic foraminifera, in various Atlantic, Pacific and Indian Oceans cores, as well as the discrepancies with the Buff Bay section, were probably due to marked diachrony of microplankton datums rather than to undeciphered unconformities. This explanation, as suggested earlier, was based on interpretation of a more limited set of examples in the North Atlantic, by Aubry (1993b).

Large and unexplained inconsistencies have been observed in the relationships of magnetic reversals and biostratigraphy in cores from different oceanic basins, or at different latitudes, and even some that are from relatively nearby sites (e.g. Sites 563 and 608: Table 2.1). There are clearly two groupings of these inconsistent relationships: one that is found primarily in North Atlantic sites, the other in tropical coreholes (Table 2.1). Clearly, these inconsistent relationships in compared biohorizons do not reflect a pattern that can be explained by simple latitudinal diachrony in the distribution of oceanic microplankton. Due to the fact that all oceanic microplankton are sensitive to watermass boundaries to a greater or lesser extent, we should observe parallel geographic trends in the diachrony of calcareous nannoplankton taxa and planktonic foraminifera. This, however, is not the case. Whereas diachrony is probably involved to some degree, we suggest that undeciphered unconformities, again, provide the most probable explanation.

In addition to the noted biostratigraphic inconsistencies in, we note that magnetostratigraphic relationships in the interval below the normal polarity interval that is to Chron C5n.2n in the same time frame have been difficult to correlate to the GPTS, in deep-sea drill sites such as Site 845. The following discussion demonstrates how an understanding of stratigraphic incompleteness can help to resolve questions arising from contradictory and incongruous variations in the range limits of planktonic microfossils with respect to magnetostratigraphy.

In one example, the data from DSDP Sites 845 (Raffi *et al.*, 1995) and 608 (Gartner, 1992) strongly suggests that the FAD of *Discoaster hamatus* lies in the mid-part of Chron C5n.2n, and that the LO of *D. hamatus* is therefore not diachronous between the equatorial Pacific and the tropical North Atlantic. In this light, the fact that the recorded LO of *D. hamatus* in Hole 563 apparently occurs at the base of Chron C5n.2n may be explained by the observation of Miller *et al.* (1985), that the normal-polarity interval attributed to this chron in Site 563 can only represent its later part. The inference that the apparent diachrony is due to a hidden unconformity is clearly justified.

In another example of inconsistent dating, Raffi *et al.* (1995) and Schneider *et al.* (1997) proposed that the FAD of *Catinaster coalitus*, based on evidence from the sites 844 and 845, lies in the early part of Chron C5n.2n, while records at other sites suggested an older FAD. In sites 563 and 608, for instance, the LO of *C. coalitus* is

*Table 2.1 Comparison between magneto- and biostratigraphic relationships in sections that have played a critical role in establishing a late middle to early late Miocene biochronology*

| *Source* | *Magneto-stratigraphy* | *Calcareous nannofossils* | *Planktonic foraminifera* | *Correlations* |
|---|---|---|---|---|
| DSDP Site 563 | relatively good; interpretation of the reversal stratigraphy below C5n. 1n is not straightforward (compare with Wright *et al*., 1992) | good (6) | good (6) | LO *N. acostaensis* <1 m below LO *D. hamatus*, at base of C5n. 2n<br><br>N15 < 1 m-thick<br>LO *D. hamatus* slightly above base C5n.2n<br>HO *N. mayeri*, slightly below LO *D. hamatus* at base C5n.2n<br>LO *C. coalitus* ~4 m below LO *D. hamatus*, in Chron C5r<br>LO *C. calycuclus* slightly below LO *D. hamatus* |
| DSDP Site 608 | good in general, but interpretation of the reversals below Chronozone C5n is uncertain (7, p. 39) | good (9, 3) | good (7) | LO *N. acostaensis* ~ base Zone NN8, ~24 m below LO *D. hamatus*<br><br>N15 < 4 m-thick<br>LO *D. hamatus* in mid Chron C5n.2n (3)<br>HO *P. mayeri* 27 m below LO *D. hamatus*<br>LO *C. coalitus* = 14.86 m below LO *D. hamatus*<br>LO *C. coalitus* questionably (7) in Chron C5n.2r<br>LO *C. calyculus* ~14 above LO *D. hamatus* |
| DSDP Site 710 | no data | good (11) | dissolution | LO *D. hamatus* in a reversed polarity interval below C5n.2n (11, but reliability questioned by 10) |

*Table 2.1 Continued*

| *Source* | *Magneto-stratigraphy* | *Calcareous nannofossils* | *Planktonic foraminifera* | *Correlations* |
|---|---|---|---|---|
| | | | | LO *C. calyculus* below Chron C5n.2n (11; but reliability questioned by 10) |
| DSDP Site 714 | no data | good (11) | dissolution | LO *C. calyculus* below LO *D. hamatus* (10) |
| ODP Site 844 | excellent (13) | good (10) | dissolution | LO *D. hamatus* in lower C5n.2n<br>LO *C. calyculus* in lower C5n.2n, slightly below LO *D. hamatus* |
| ODP Hole 845 | excellent (13) | good (10) | dissolution | LO *D. hamatus* in mid C5n.2n<br>LO *C. calyculus* in lower C5n.2n, slightly below LO *D. hamatus* |
| ODP Hole 848 | excellent (13) | good (10) | dissolution | LO *D. hamatus* in mid C5n.2n |
| DSDP Site 289 | no data | good (12) | good (14) | LO *N. acostaensis* in upper Zone NN9, ~33 m above LO *D. hamatus*<br>N15 ~28 m-thick<br>HO *P. mayeri* <4 m above LO *D. hamatus*<br>LO *C. coalitus* = 8 m below LO *D. hamatus*<br>LO *C. calyculus*: 10.9 m above LO *D. hamatus* |
| DSDP Holes 588, 588A | Magnetostratigraphy in Hole 588A unreliable(8). In addition, bottom Hole 588 just in reversed immediately below C5n | good (5) | good; anomalous range of *P. mayeri*, marker of top Zone N14 and regarded by many as synonym *P. siakensis* | LO *N. acostaensis* in upper Zone NN9, 12.5 m above LO *D. hamatus*<br>N15 = 8.5 m-thick (if base taken at HO *P. siakensis*)<br>HO *P. siakensis* 5.2 m above LO *D. hamatus*<br>LO *D. hamatus* in lower third C5n.1n |

*Table 2.1 Continued*

| *Source* | *Magneto-stratigraphy* | *Calcareous nannofossils* | *Planktonic foraminifera* | *Correlations* |
|---|---|---|---|---|
| | | | | LO *C. coalitus* = 5 m below LO *D. hamatus*<br>LO *C. calyculus* = LO *D. hamatus* |
| Buff Bay, Jamaica | highly questionable (8) | good (1) | good (2) | N15/N16 not recovered (unconformity); > 45 m above LO *D. hamatus*<br>N15 > 45 m-thick<br>HO *P. mayeri* < 4 m above LO *D. hamatus*<br>LO *C. coalitus* = 13.4 m below LO *D. hamatus*<br>LO *C. calyculus*: 22.8 m above LO *D. hamatus* |
| Bodjonogoro Well | no data | good (15) | good (15) | LO *N. acostaensis* in upper NN9, 187 m above LO *D. hamatus*<br>N15 = 187 m-thick<br>HO *P. mayeri* = LO *D. hamatus* (but coarse sampling interval)<br>LO *C. coalitus* = 10 m below LO *D. hamatus*<br>LO *C. calyculus* inderterminate: single occurrence 138 m above LO *D. hamatus*. |

N.: Neogloboquadrina; C.: Catinaster; D.: Discoaster. (1) Aubry, 1993b; (2) Berggren, 1993; (3) Gartner, 1992; (4) Kennett and Srinivasan, 1984; (5) Lohman, 1986; (6) Miller et al., 1985; (7) Miller et al., 1991; (8) Miller et al., 1994; (9) Olafsson, 1991; (10) Raffi et al., 1995; (11) Rio et al., 1990; (12) Shafik, 1975; (13) Schneider, 1995; (14) Srinivasan and Kennett, 1981; (15) Bolli, 1966b; W. D. Zachariasse, Utrecht, written communication, 1990; Berggren and Aubry, unpublished data. LO ? lowest; HO ? highest. Text abbreviated for clarity.

identified in an interval with reversed polarity that is older than Chron C5n.2n (Miller *et al.*, 1985; Olafsson, 1991; Clement and Robinson, 1987). There are two possible interpretations of these inconsistent observations: one, that the LO of *C. coalitus* is diachronous between the equatorial Pacific and the North Atlantic. If so, the age of 10.81 Ma estimated by Schneider *et al.* (1997) applies to the LOD of the species in the equatorial Pacific, but does not date its FAD. The other alternative is that there is a narrow gap in the stratigraphy at Sites 844 and 845, so that the lowermost part of the assumed Chron C5n.2n (Schneider, 1995) and the uppermost part of the reversed interval below, in which the first *C. catinaster* might have been found, is absent from the cored sections. This second alternative is controversial, in view of the excellent magnetostratigraphic record of Hole 845 overall, but we note that whereas the match between the magnetic record in Hole 845 and the GPTS (Cande and Kent, 1995) above Chron C5n.2n is nearly perfect, the match between the two below Chron C5n.2n is not as straightforward.

The probability of a disguised hiatus in these Pacific cores is also consistent with the proposal by Raffi *et al.* (1995) and Schneider *et al.* (1997) that calcareous nannofossil Biochron NN8 should be restricted to the earliest part of Chron C5n.2n, based on the discovery of diagnostic taxa from this biozone in Site 845. Backman *et al.* (1990) had earlier proposed that this biochron should be correlated essentially to Chron C5r.2r – that is to say, in that part of the record that may be missing in Site 845 – according to the record at Site 710. This correlation is also suggested in the Buff Bay section (Miller *et al.*, 1994).

Clearly there is a great need to understand the causes of contradictory magnetobiostratigraphic datasets in different sections. Whereas it has been usual to imply diachrony of biostratigraphic events, undeciphered unconformities constitute an equally plausible mechanism for discrepancy. The problem is complex, and involves not only ecological preferences on the one hand and stratigraphic gaps on the other, but also preservational biases. Astrobiochronology offers the means to resolve the conundrums. For instance, at the same time as the HO of *C. miopelagicus* is suspected to be diachronous (Berggren *et al.*, 1995c; Backman and Raffi, 1997), Hilgen *et al.* (2000) remarked on the near synchrony of this event between the Mediterranean and Atlantic Sites 608 and 926.

## 2.6 Conclusions

With the radical improvement in worldwide marine correlations and magnetochronology derived from the first forty years of rotary coredrilling in the deep-ocean basins, considerable advances have been made towards stabilizing and focusing the Cenozoic time scale. Future improvements will come mainly through increasing the number and quality of numerical dates that apply to the new marine chronostratigraphy; already, astrochronology (the historical mirror of cyclostratigraphy) is being widely applied in upgrading the IMBS (Hilgen and Krijgsman, 1999; Hilgen *et al.*, 2000; Cramer, 2001). Uncertainties in the biochronologic framework remain, especially in those intervals that have been difficult to sample (see above). Greater attention to the physical architecture of marine stratigraphy, as well as improving the strategies of correlation, will be needed to reduce the discrepancies that weaken the application of the vast body of knowledge that is now available from the fossil record in the deep-sea.

The development of the integrated magneto-biostratigraphic time scale in the deep-sea record has modified the way we see geological time. Rather than inferring time indirectly and in only approximately quantitatively, through the superpositional logic of stratigraphic succession, the growing accuracy and detail in the geological time scale now permits us to see some (if not yet all) strata as a function of time, instead. Thanks to this, we can treat interpretations of the stratigraphic record with ever greater rigor, leading in turn to new insights that guide further advances in research.

The importance of improving the geological time scale is not just an academic matter. We can be sure that the mechanisms responsible for great globin the past continue to operate today. Depending on their magnitude, such global changes are always detrimental and sometimes catastrophic to the established global ecosystems. That would now include human culture as well. Until the geological record of deep ocean, epicontinental and terrestrial environments can be temporally coordinated to a degree that reveals cause and effect at a scale that is meaningful for environmental scientists and policy makers, however, the practical lessons that we might learn from historical geology will remain a closed book.

## 2.7 Acknowledgements

This contribution is an outgrowth of ideas originally presented in Aubry (1997). We are grateful to W. A. Berggren, D. V. Kent, K. G. Miller, M. Woodburne, W. Zachariasse, F. Steininger and numerous other colleagues for fruitful discussions on the subject of chronostratigraphy and time scales. We thank Jack Cook for preparing the graphics in this paper. Finally thanks to Eduardo for suggesting the topic of this chapter, and for his patience during its preparation.

Chapter 3

# Ecostratigraphy's Basis, using Silurian and Devonian Examples, with Consideration of the Biogeographic Complication

ARTHUR J. BOUCOT
Department of Zoology, Oregon State University, Corvallis, Oregon 97331-2914, USA.



## 3.1 Introduction

Beginning about two-hundred years ago fossils have been used, with increasing stratigraphic precision, to correlate fossiliferous beds near and far. The overall reliability of these fossil-based correlations has been tested again and again with radiometric dating and non-radiometric isotopic techniques, and found to be good. Why is this the case? In order for any dating technique to work there must be time systematic changes affecting one or another variable.

With fossils these time systematic changes are of two types, as discussed previously (Boucot, 1994). These two basic kinds give rise to different types of biostratigraphy. The first, the d'Orbignyan, is based on the time successive series of Phanerozoic, relatively short interval extinctions, each one followed after a brief interval by a short interval of adaptive radiation. D'Orbignyan biostratigraphy recognizes the many extinctions followed by adaptive radiations that characterize the Phanerozoic fossil record. These extinctions and adaptive radiations range from major to minor, a complete spectrum. Each pair of extinctions and subsequent adaptive radiations is separated from the next pair by a lengthy interval of relative evolutionary monotony. This lengthy interval of evolutionary monotony is characterized for each



*E.A.M. Koutsoukos (ed.), Applied Stratigraphy,* 55-71.

community group within the ecological–evolutionary unit and subunit by little or no evidence for phyletic evolutionary change among the abundant to common genera, but some evidence among the uncommon to rare genera in each well studied community group of phyletic (anagenetic) evolution affecting the uncommon to rare genera. Since most of Phanerozoic time is present between the d'Orbignyan events we need a second kind of biostratigraphy for use within each of them.

The second type of time systematic change is the Oppelian which deals with the time intervals between the d'Orbignyan events, and gives rise to Oppelian biostratigraphy. Oppelian biostratigraphy is based on the fact that within each d'Orbignyan unit, which I have termed Ecological–Evolutionary Units and Subunits (Boucot, 1983, 1984a, b), there is within any community type (my term community group, Boucot, 1975) an overlapping time sequence of species within the different genera present in the Community Group, i.e., those uncommon to rare genera showing evidence for phyletic evolution.

Both types ultimately involve evolution. One of them involves the abrupt evolutionary, adaptive radiation changes following extinctions, were termed metacladogenesis (Boucot, 1978) to make it clear that these are relatively major evolutionary changes commonly involving the generation of new families and higher taxa (in contrast to the minor cladogenetic changes within a genus that biogeographically give rise to allopatrically generated species, what I have termed diacladogenesis, 1978). The other type of evolution involves the anagenetic, phyletic evolutionary changes that take place within some of the genera present in community groups during the Oppelian changes occurring in the time intervals between the d'Orbignyan events. These two types of evolutionary change are radically different! The adaptive radiations involve the generation of descendent families with significantly different morphologies, community ecologies, and presumably distinct behaviors and physiologies, i.e., major genomic changes under the control of strong directional selection. In contrast the phyletic changes taking place within some of the uncommon to rare genera characterizing a community group are relatively small morphologic changes, accompanied by no changes in relative abundances of the genera within a community group, and are most easily explained by appealing to strong stabilizing selection and genetic neutralism as the cause of the minor morphological changes.

Examples of such phyletic change in Silurian–Devonian brachiopods I (Boucot, 1997) have worked with include the following: (1) the *Eocoelia* species lineage of the mid to late Llandovery and earlier Wenlock; (2) the many *Stricklandia lens* subspecies to *Costistricklandia* lineage of the early Llandovery to early Wenlock; (3) the *Microcardinalia–Plicostricklandia* lineage of the earlier Llandovery to mid-Wenlock; (4) the *Borealis–Pentamerus–Pentameroides* lineage of the mid to late Llandovery in Eurasia and in the mid-Llandovery to mid-Wenlock of North America; (5) the late Ludlow–Pridoli *Dayia* species lineage of the European Province; (6) the *Howellella vanuxemi–Howellella cyclopterus–Acrospirifer murchisoni* lineage of the Early Devonian Appohimchi Province; (7) the *Dalejina oblata–Discomyorthis* lineage of the Early Devonian Appohimchi Province; (8) the *Glypterina–Ptychopleurella* lineage of the Middle Ordovician through Early Devonian; (9) the *Leptostrophia*

(*Leptostrophiella*)–*Leptostrophia* (*Leptostrophia*)–*Protoleptostrophia* lineage in the later Silurian through Early and Middle Devonian of the Appohimchi Province; (10) the *Nanothyris–Rensselaeria–Etymothyris–Amphigenia* (small followed by large) lineage in the Early and earlier Middle Devonian of the Appohimchi Province.

Additional brachiopod examples include in the Uralian Region the Early Devonian *Karpinskia* species lineage described by Zhivkovich and Chekhovich (1985), Sun and Boucot's (1999) Givetian examples of *Stringocephalus* species level lineages in East Asia and western North America, Rozman's (1999) Silurian species level lineage of *Tuvaella* known in the Mongolo–Okhotsk Region, and the probable species lineage in the European Late Jurassic–Early Cretaceous for the deep-water, "keyhole" brachiopod *Pygope* (Vogel, 1966, 1984). Smith (1984, pp. 115–118) discusses the *Infulaster-Hagenowia* Late Cretaceous echinoid lineage from the chalk of northwestern Europe that is well documented by Gale and Smith (1982). Ward and Blackwelder (1975) document the Miocene to Pliocene *Chesapecten* lineage of eastern North America. Experienced biostratigraphers working up and down the Phanerozoic column will be able to think of additional examples from their own time intervals and groups.

Brett and Baird (1992, 1995; see also Brett *et al.*, 1996) introduced the term "co-ordinated stasis" to emphasize the fact that within all ecological–evolutionary units and subunits the common to abundant genera do not show evidence for directional, morphological, species-level change, i.e., evolution, from the beginning to end of the unit or subunit. These are, of course, the genera and species present in most small to medium size samples, those including no more than a few hundred specimens. When significantly larger collections are made one may recover the uncommon to rare genera that include examples of phyletic evolution, the type referred to in the preceding paragraph. During times of globally high provincialism it is easier to find such examples of phyletic evolution owing to the overall more rapid evolution during such times that correlates well with globally smaller reproductively communicating populations. In their 1995 paper Brett and Baird provided extensive documentation for the behavior of the common to abundant genera and their species within the Silurian–Middle Devonian of the Central Appalachians that conforms to this concept. Boucot (1996, Table 2) listed the globally recognized ecological–evolutionary units with many of their known ecological–evolutionary subunits.

The term ecostratigraphy (Boucot, 1983, 1984a, b) has been employed to describe the evolutionary basis of biostratigraphy. First, organisms on land and sea do not occur in a random manner. Varied assemblages of taxa, genera for example, commonly recur with each other from place to place within any one time interval. In any one time interval there are a varying, but large number of such recurring assemblages. The ecologist commonly terms them communities, although the term guild might be preferable if one subscribes to the idea that a community should include ALL of the organisms from the viral and bacterial up through the tiger sharks and blue whales that co-occur in the ocean, or on land to the large carnivores and herbivores that co-occur with each other. But, since the term community is not commonly used in this all inclusive manner we will continue to employ community rather than guild. Frest *et al.* (1999) made the very helpful suggestion that communities should be named after the numerically dominant taxa, commonly belonging to a single guild and major taxon, whereas the less common taxa commonly belonging to other guilds should be named associations of the numerically dominant taxon community; this procedure enables one, for example, to discuss a group of regularly recurring gastropods occurring in small numbers within a numerically dominant group of brachiopods as a named gastropod association of the numerically dominant brachiopod community, and the same with a regularly recurring, less abundant group of trilobites, etc. This point is important because the associations do not always occur within a single community. This procedure permits a number of associations of numerically minor, distinct taxa to be described from a single community.

## 3.2 Community Frameworks

A community framework is a graphical device that enables one to ascertain whether the community groups present in any one biogeographic unit include all of the expected, potential environments in which different communities might be expected to occur. Boucot (1975), Wang Yu *et al.* (1987) and Boucot and Lawson (1999) provide numerous examples from the Silurian–Devonian.

A community framework has a vertical ordinate on which depth correlated features are arrayed, and a horizontal ordinate on which varying physical features of the environment that are not depth correlated are arrayed (Fig. 3.1 is a specimen). For example, on the vertical ordinate the intertidal and subtidal are noted, the photic and subphotic zones, and the shelf margin equivalent (as indicated by the incoming of the turbiditic, flysch facies). On the horizontal ordinate are noted such features as rough water, quiet water, and normal turbulence, hypersalinity, normal salinity and hyposalinity, sediment grain size, anoxic, low oxygen dysaerobic and normal oxygen and so forth (see Fig. 3.1 and the examples cited). By plotting on a community framework all of the communities present in any one time interval and area one may see at a glance which environments remain unaccounted for as well as lateral relations among communities present. Most of this information will be derived from level bottom communities because the reef complex of communities has been little studied ecologically.

By taking advantage of the community framework approach one may immediately estimate whether or not the overall taxic diversity, high to low, recorded for any area is based on an ecologically sound, comprehensive sample, or not. This is potentially important in assessing diversity changes through time and also for biogeographic analysis. For example, Figure 3.1 indicates that turbulent condition, shallower photic zone (Benthic Assemblages 1 and 2) communities have not been recognized in the Pridoli of the North Atlantic Province, North Atlantic Region at this time (Boucot, 1999a), although they may well be present elsewhere within this time interval and biogeographic unit.

Until now, only marine benthos have been considered. The nektonic and planktonic organisms, the conodonts, graptolites, chitinozoans and acritarchs to name the most abundant and widespread within the Silurian–Devonian, also must be considered

<table>
<tr><th>Benthic Assemblage</th><th colspan="2">Turbulent Conditions</th><th colspan="5">Normal Current Activity</th><th colspan="2">Quiet Conditions</th></tr>
<tr><td>1</td><td></td><td>Thrombolite-Algae-Metazoan Community</td><td>Nuculoid Bivalve Dominated Communities</td><td>Rhynchonellid Community</td><td>Orbiculoid-Linguloid Community</td><td colspan="2">Dizygopleura punctella-Kloedenia normalis-Herrmannina sp. Community</td><td colspan="2">Leperditiid Community</td></tr>
<tr><td rowspan="2">2</td><td rowspan="2"></td><td rowspan="2">Thrombolite Community</td><td colspan="5">Welleria-Dizygopleura Community</td><td rowspan="2">Protathyris Community</td><td rowspan="2">Howollolla-Protathyris Community</td></tr>
<tr><td colspan="5">Zygobeyrichia-Bolbiprimitia Community</td></tr>
<tr><td>3</td><td colspan="2">Amsdenina Community</td><td rowspan="3">Zoophycos Community</td><td colspan="4">Eccentricosta Community</td><td colspan="2" rowspan="3">Atrypa "reticularis" Community</td></tr>
<tr><td>4</td><td colspan="2"></td><td rowspan="2">Dicoelosia-Skenidioides Community</td><td>Sponge Mound Community</td><td></td><td></td></tr>
<tr><td>5</td><td colspan="2"></td><td>Clorinda Community</td><td>Clorinda-Cyrtia Community</td><td>Navispira Community</td></tr>
</table>

*Figure 3.1 Community framework for the Pridoli, latest Silurian, North American Province, North Atlantic Region showing the absence of rough water, turbulent environment communities for deeper water environments (from Boucot, 1999a).*

and their ecologies integrated with that of the benthos. Chen Xu (in Mu En-Zhi *et al.*, 1986) has provided an excellent example, using earlier Silurian graptolite evidence. It is clear that much of the nektonic and planktonic biota is depth stratified. This results in some forms, the very shallow water forms, being associated with the very shallow water marine benthos. In somewhat deeper water the shallow water plankton and nektonic forms are associated with more offshore marine benthos. The relative abundance of these somewhat deeper water forms increases relative to accompanying biota until their maximum depth limit is reached, after which they show no increase in relative abundance. Likewise, for the still deeper water, still more offshore forms occur. One can, therefore, combine the benthic, nektonic and planktonic organisms in a reasonable manner and use all three to achieve a still more refined biostratigraphy.

One can achieve additional biostratigraphic resolution by combining the biostratigraphic results from a single community group, within a specific ecological–evolutionary unit, with the biostratigraphic results from an adjoining community group. Using this Oppelian approach one may achieve the maximum biostratigraphic resolution possible for time intervals within any one ecological–evolutionary unit. Of course, within a particular sedimentary basin, one may achieve additional biostratigraphic resolution by paying attention to community changes that are controlled by physical events within that basin, but almost all of these will be purely local rather than widespread on a global basis, as is the case with some sequence stratigraphic events (Brett, 1998).

### 3.2.1 Community Framework Example

Wang Yu *et al.* (1987, fig. 10; our Fig. 3.5) show how a community framework analysis may be used to better understand environmental conditions within a basin, the South China Yangtze Platform in this case. It is clear (Fig. 3.2) that although there are rich and varied shelly faunas, low diversity to medium and high diversity representing a number of Benthic Assemblage 1 through 3 environments plus a few from Benthic Assemblage 4 and inner 5, there is overall no representation of the rich Benthic Assemblage 5 shelly benthos so widespread in other warm water biogeographic regions (Boucot and Lawson, 1999) that are mostly included within the *Dicoelosia–Skenidioides* Community Group. The total absence of this important, geographically widespread benthic community group from the Yangtze Platform is best explained as owing to anoxic, bottom conditions at Benthic Assemblage 5 depths because of the presence of deeper water graptolitic faunas (Mu En-Zhi *et al.*, 1986) seaward of the shallower water shelly faunas, i.e., pelagic graptolitic fauna thrived in the shallower and deeper parts of the water column above the deeper water, anoxic bottom waters.

Community frameworks may be similarly used to tell at a glance which environments are present in a particular basin or biogeographic unit. This relieves the worker of trying to recall all of the complex environmental details in areas with which he is not intimately familiar. Basin analysis possibilities are significantly enhanced by such an analysis in which one can consider both physical and biological indicators of environment. Keep in mind that benthic organisms are commonly far more sensitive indicators of bottom conditions on and within the sediment than are conventional logging and petrographic parameters. Organisms are sensitive to such things as thixotropic and dilatant sediments, as well as turbidity conditions, that defy the sedimentary analyst.

## 3.3 The Biogeographic Complication

Until now we have discussed things as though all community groups occur everywhere in the world, i.e., a 100% cosmopolitan world with no global climatic gradient or longitudinal barriers to reproductive communication. Such a situation would materially simplify correlation by means of fossils, just as would the possibility that both marine and nonmarine biotas occurred together everywhere, or that every possible environmental permutation occurred everywhere. Unfortunately, the world is not like this today, nor is there any evidence from the past to suggest that it ever has been. There has always been a a global climatic gradient (Boucot *et al.*, in press, lays out the global Cambrian through Miocene climatic gradient changes), which has sporadically fluctuated from high to low, intervals with continental glaciation at one or both poles, and intervals with many longitudinal barriers to reproductive communication as contrasted with few such intervals. Biogeography at lower latitudes is essentially a history of the coming and going of longitudinal barriers to reproductive communication; relatively cosmopolitan conditions and relatively provincial conditions.

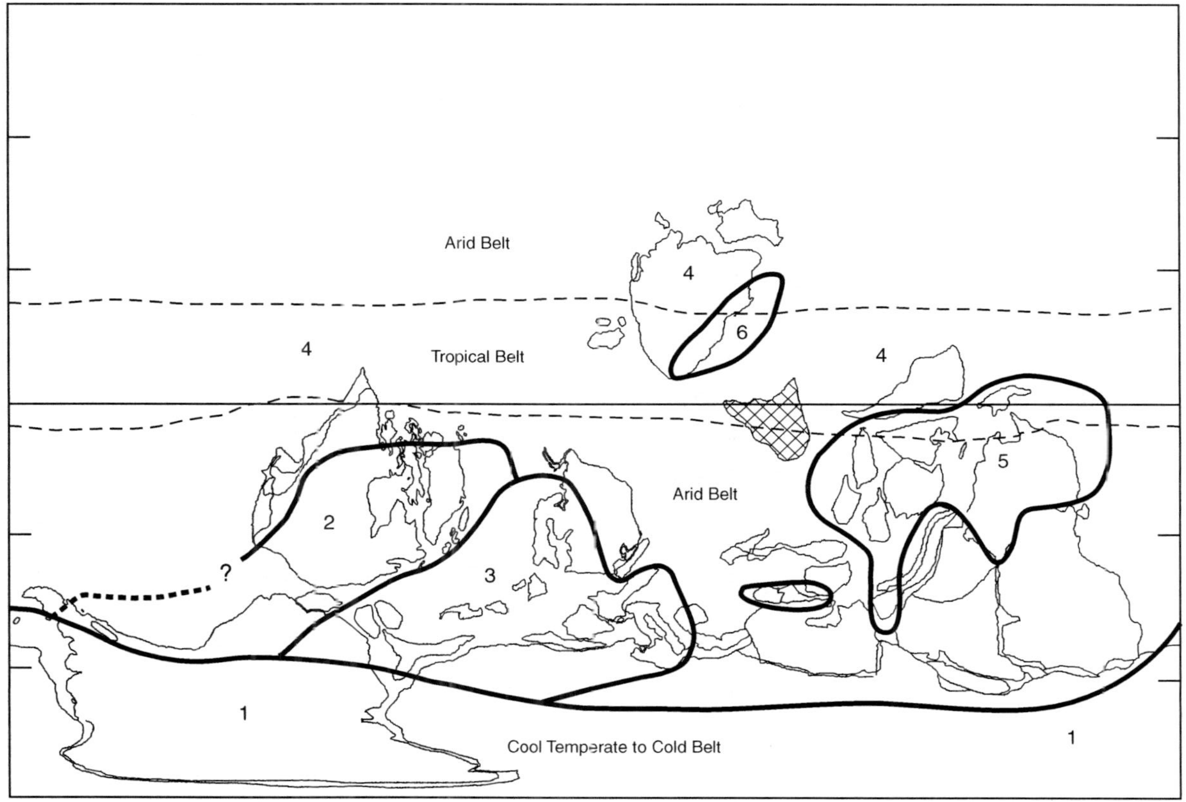


*Figure 3.2* Silurian paleogeography with biogeographic units outlined: 1. Afro-South American Subrealm; 2. North American Province; 3. European Province; 4. Uralian–Cordilleran Region undivided; 5. Sino-Australian Province; 6. Mongolo–Okhotsk Province; 7. Iran (from Boucot and Blodgett, 2001).

Correlation between different biogeographic entities, just as between differing ecological entities, depends on the recognition of either localities with a mixing in specific beds or interbedding of biotas containing key taxa belonging to the biogeographic entities on either side. Additional to these correlation possibilities is the presence of key taxa that bridge the biogeographic boundaries. For example, the correlation of the earlier Devonian neritic faunas of the Malvinokaffric Realm (see Boucot and Racheboeuf, 1993, for discussion of this realm) with those present in the extra-Malvinokaffric realms for many years could only be done by a rough stage of evolution type argument since there are few shared genera, whereas the more recent development of a higher land plant spore biostratigraphy usable in both biogeographic entities makes for far better correlation precision.

A global climatic gradient, with cooler or even cold conditions at higher latitudes, guarantees that there will be biotas adapted to these conditions at higher latitudes. These higher latitude biotas are invariably lower in diversity at every level from class to species, far fewer taxa, just as at present (Cox and Moore, 2000), with the Holocene of course being characterized by a very high global climatic gradient, probably the highest in the Phanerozoic. Boucot (1975) provides data from the Silurian–Devonian, and it is common knowledge that the high latitude faunas of the Cambro–Ordovician are lower diversity during the Middle Cambrian through Late Ordovician, not to mention the well known, high latitude, low diversity Permo-Carboniferous *Eurydesma* faunas; the situations during the Mesozoic and Cenozoic are little different. The reasons for lower diversity at higher latitudes are presumably complex, and may change over time with different causal factors changing their relative importance. High seasonality, with stronger temperature fluctuations will typify times of high global climatic gradient. Other factors include marked seasonal changes in light intensity and duration, more seasonal reproduction, and many other factors. These factors are not easy to distinguish very far back in time, except for light seasonality being fixed, because of changing global climatic gradients through time.

Needless to say, an adaptive radiation of cool to cold climate biotas at high latitudes insures that there will always be correlation problems and uncertainties vis-à-vis lower latitude, warmer climate biotas. Within the Silurian–Devonian one needs only think of the lower level of correlation precision between biotas of the cool climate Malvinokaffric Realm and the warmer climate extra-Malvinokaffric Realm until the end of the earlier Middle Devonian, Eifelian, when the global climatic gradient lowers abruptly with consequent large scale extinction of cool to cold climate biotas (Boucot, 1975, 1988).

For the geologist the cool to cold regions are characterized, additional to their endemic faunas, by widespread siliciclastic lithologies, with glittering, unweathered mica flakes being widespread in the Middle Cambrian through Ordovician and within the Silurian–earlier Middle Devonian, and an absence of limestone, dolomite, redbeds, evaporites, reefs, lateritic deposits and other warm climate indicators. The reverse is the case for the extra-Atlantic Realm and extra-Malvinokaffric Realm lithologies.

At lower latitudes, where warm conditions dominate in the neritic environments (we know nothing of bathyal or abyssal circumstances during the Silurian–Devonian), and presumably at lower elevations on land (our knowledge of the freshwater and terrestrial biotas during the Silurian–Devonian is relatively limited, especially regarding elevation correlated distinctions), biogeographic distinctions arise from the insertion and removal of longitudinal barriers to reproductive communication through time. One thinks here of the later Cenozoic examples, such as the Isthmus of Panama, the Red Sea barrier, as land barriers for neritic organisms, as well as of the barrier erected for most organisms with short planktonic larval stages by the distance from the central to the eastern Pacific (no inter-island transport possible!; such land barriers for the marine organisms are, of course, "bridges" for terrestrial organisms). In the earlier Paleozoic and even mid-Paleozoic our knowledge of contemporary paleogeography is too poor for us to be able to recognize the longitudinal barriers responsible for differing levels of low latitude provincialism. However, the presence of reasonably high levels of low latitude provincialism for some time intervals guarantees that such barriers must have been present. Their location can be postulated using the presence of boundaries between biogeographic units as the clues.

As examples of these relations one can consider the biogeography of the Silurian and Devonian, with a few words devoted to the latest, Hirnantian part of the Ordovician. Overall from the Hirnantian through the Famennian, end of the Devonian, we see a global climatic gradient going from high in the Hirnantian to very low in the Famennian.

The Hirnantian is characterized by widespread, Southern Hemisphere continental glaciation, while the Frasnian and Famennian are characterized by a very low global climatic gradient with no evidence for cold climates. However, during the Famennian there is a brief very latest, Strunian interval with continental glaciation in some parts of South America (Streel *et al.*, 2000). Between the Hirnantian and the Famennian there is a marked decline in the global climatic gradient.

### 3.3.1 Biogeographic Nomenclature

Modern biogeographers commonly employ, in descending hierarchical order, the terms Realm, Region, Province and Subprovince as biogeographic units. Some (Westermann, 2000) drop the term "Region" in favor of Subrealm. Some employ the term Superrealm as well. No matter what hierarchy one prefers the critical thing is to recognize that there are differing levels of endemism that give rise to biogeographic provincialism. It is also well to keep in mind that one does not always encounter a gradual transition through all units of the hierarchy, i.e., one may go from realm to province in some cases without evidence for any regions. Also keep in mind that not all paleontologists in the past have employed the biogeographic hierarchy. Some have referred to all of their biogeographic units as "provinces," which results in some confusion in that units now referred to as realms or regions were earlier referred to as provinces.

## 3.4 Silurian–Devonian Biogeography

Silurian–Devonian biogeographic patterns were established after the demise of widespread continental glaciation. After extinction of the cold water fauna of the latest Ordovician, Hirnantian, there is an extended interval characterized by a moderately high global climatic gradient, Early Silurian (Llandovery) time with evidence for continental glaciation in the Amazonian region and possibly within the Paraná Basin to the south, of Brazil, but nowhere else (Boucot *et al.*, in press). A geographically extensive area at high southern latitudes from the Early Silurian through the earlier Middle Devonian (Eifelian) is characterized by a cool climate benthic marine fauna. The global climatic gradient abruptly lowers near the end of the Eifelian, the Kačak Event (Boucot, 1988), following which there is no evidence for cool climate marine faunas through the Famennian (including the Strunian as well).

From the Llandovery through the Eifelian there are varying levels of marine benthic provincialism at lower latitude, warm regions, with the highest level present in the later part of the Early Devonian (Pragian–Emsian). Following the Kačak Event, latest Eifelian, there is a low but not lowest level of provincialism during the later Middle Devonian (Givetian), followed by relatively cosmopolitan conditions in the Late Devonian, Frasnian, Famennian; this appears to be true at the generic level, but may not be at the specific level which has not been carefully investigated.

The causes for the lower latitude provincialism are currently unknown, but presumably are due to the insertion of barriers to reproductive communication at lower latitudes. Unfortunately our knowledge of Silurian–Devonian paleogeography is still too rudimentary to permit any definitive recognition of the physical barriers to reproductive communication in the neritic environments.

It is worth emphasizing that for every time interval, including the Silurian-Devonian, certain organisms are favored for biogeographic purposes. These favored organisms commonly happen to be those which have been most subject to extended, globally extensive taxonomic treatment. For example, the study of modern marine biogeography (Ekman, 1953) is dominated by the echinoids because of the extensive work on this group earlier in the 20th Century by Mortensen. Other marine groups, such as the molluscs play a subordinate role, probably owing to the absence of a "Mortensen" devoted to their study globally. On land the situation is little different, with birds, mammals plus a few insect and flowering plant groups playing the dominant role, while many other potentially important groups such as mites and nematodes playing only a minor role owing to lack of taxonomic attention globally.

In the Silurian–Devonian the biogeographically key role today is played by the articulate brachiopods. Other taxa, including the gastropods, rugose corals, trilobites, ostracodes and pelmatozoans play only supporting roles mostly owing to the lack of really comprehensive global attention to their taxonomy–morphology and distribution; in the future this situation may well change. At present, one can say that the non-brachiopod groups just cited are permissively in agreement with the biogeographic conclusions arrived at with the aid of the brachiopods. Groups such as the relatively rapidly evolving conodonts, graptolites and goniatites during this interval are vital for biostratigraphic purposes but owing to their relatively broad geographic distribution during certain time intervals are of limited use for biogeography, i.e., relatively cosmopolitan organisms by their very nature are of limited value for biogeography, although biostratigraphically valuable if rapidly evolving.

### 3.4.1 Silurian Biogeography

Silurian biogeography from beginning to end is dominated by a cool to cold Southern Hemisphere Malvinokaffric Realm, Afro-South American Subrealm contrasting with lower latitude relatively warm units (Fig. 3.2). The Afro-South American Subrealm is relatively uniform biotically, characterized by a very low diversity fauna (Boucot, 1975; Boucot and Blodgett, 2001), with the brachiopods *Clarkeia* and *Heterorthella* being relatively abundant and widespread. This subrealm has not been subdivided biogeographically. Many brachiopod orders, superfamilies and families characteristic of lower latitude warm regions are absent. This is also true of many other phyla and classes that are widespread in warmer regions, including such things as varied rugose corals, calcareous algae, stromatoporoids, varied bryozoans, most shelled cephalopods, and so forth, whereas conularids and hyolithids tend to be relatively common.

The Afro-South American Subrealm is recognized in the southern two-thirds of South America from the Amazon Basin and southern Peru south into northwestern Argentina (Salta and Jujuy Provinces) and into the western part of the Paraná Basin in Paraguay. It may be present in Ghana, based on a microfossil occurrence and does occur in Guinea (Racheboeuf and Villeneuve, 1992). The South African Disa Siltstone was formerly considered to be lowest Silurian, but is now assigned to the Hirnantian (Boucot, 1999b) with the presence of *Heterorthella* indicating ancestral relations. Areas of boundary mixing with extra-Afro-South American Subrealm faunas are known in the Argentine Precordillera and Puna.

Lower latitude warm regions of the Silurian (Fig 3.2) are assigned to the North Silurian Realm (Boucot, 1975, 1990), which is divided into North Atlantic and Uralian–Cordilleran regions. The North Atlantic Region is divided into North American and European provinces (the latter includes the Merida Andes of central Venezuela (Boucot *et al.*, 1972) and northeastern Tamaulipas, Mexico (Boucot *et al.*, 1997)). The Uralian-Cordilleran Region is partially subdivided (Rong Jia-yu *et al.*, 1995) into Sino-Australian and Mongolo–Okhotsk provinces. They also suggested (their fig. 4) that eastern Laurentia might represent an additional province in contrast to western and Arctic North America, the Uralian belt to which the Pre-Carboniferous of the Carnic Alps and Karawanken belong, and possibly parts of Kazakhstan. The platform Silurian of Iran may represent another unit featuring relatively abundant, endemic *Xerxespirifer*. All of this suggests that there was a moderately high level of warm region provincialism during the later Silurian. Provincialism for the Early Silurian, the Llandovery, is not nearly as well understood; it might be more cosmopolitan but it might be equally provincial since adequate data for a decision are currently unavailable.

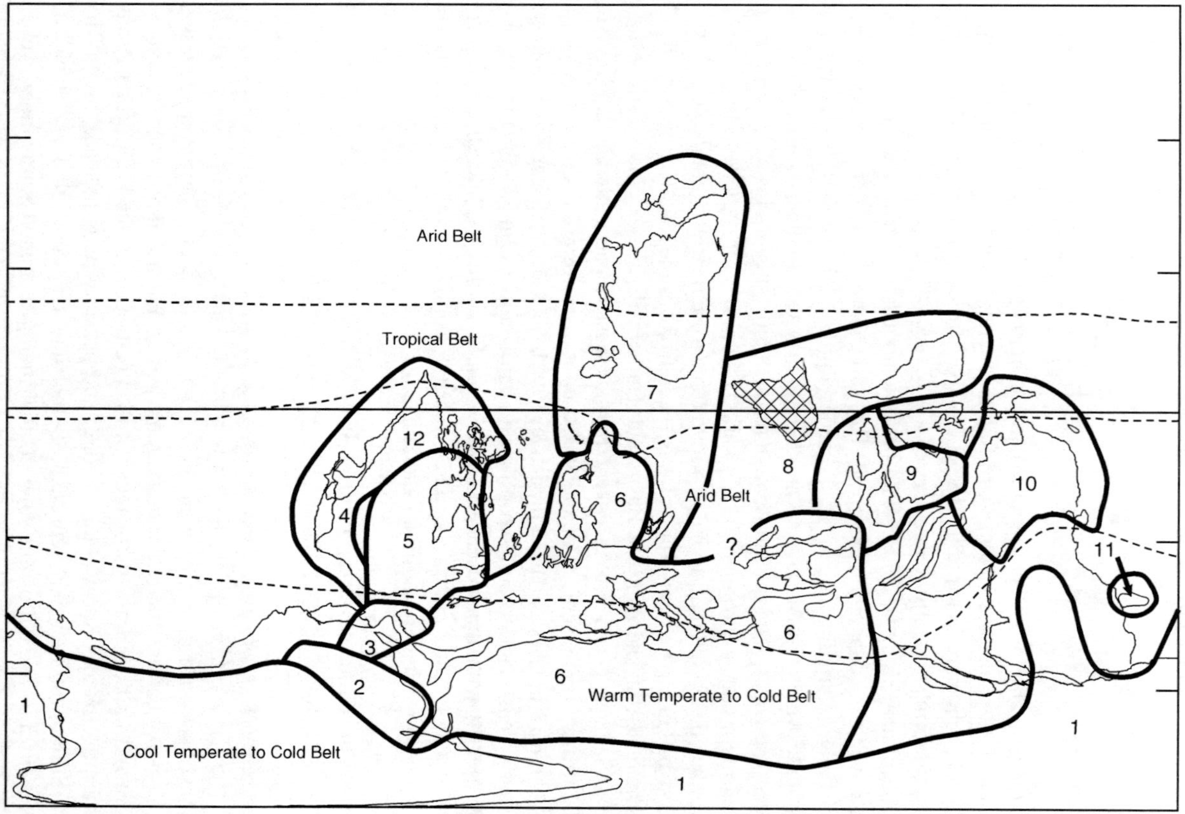


*Figure 3.3* *Early Devonian paleogeography with biogeographic units outlined: 1. Malvinokaffric Realm; 2. Amazon Subprovince; 3. Colombian Subprovince; 4. Nevadan Subprovince; 5. Appohimchi Subprovince; 6. Rhenish–Bohemian Region; 7. Uralian Region; 8. Balkhash–Mongolo–Okhotsk Region; 9. South China Region; 10. Tasman Region; 11. New Zealand Region; 12. Cordilleran Region (from Boucot and Blodgett, 2001).*

So little is known of nonmarine organisms in the Silurian that no attempt is possible to characterize it biogeographically; there are only a few vertebrate, ostracode and higher land plant localities known. This situation might change when more nonmarine localities are recognized.

### 3.4.2 Devonian Biogeography

Devonian provincialism continues to increase gradually from that present in the later Silurian, with highest levels achieved during the later Pragian–Emsian interval, following which there is a steady decrease into the relatively cosmopolitan Late Devonian. All of this, of course, refers to level bottom benthic provincialism because, as with the Silurian, there has been little effort to analyse reef biogeography. In principle reef biogeography should exhibit higher levels of provincialism than the level bottom owing to the far more scattered nature of the reef occurrences.

During the Devonian (Figs. 3.3, 3.4) there is, as remarked earlier, a significant decrease in the global climatic gradient and the cool climate Southern Hemisphere Malvinokaffric Realm disappears near the end of the Eifelian in the Kačak Event, although holdovers, such as a few scattered Malvinokaffric Realm trilobites do persist into the Givetian and possibly even into the Late Devonian of Brazil. In the Early Devonian through later Middle Devonian the warm water regions are divided into a New World Eastern Americas Realm and an Old World Realm present in the lower latitude parts of the Old World plus parts of western and Arctic North America.

There are enough data from the nonmarine Devonian to begin to characterize levels of provincialism. What we know of the provincialism of the higher land plants, the tracheophytes (vascular land plants), appears to agree very well with that for the marine benthos, although why this should be is currently unclear. The same is true for the vertebrates, both jawed and jawless, although here the possibility for marine larval stages might provide an explanation, i.e., many may have been anadromous although this is still unproved.

There are some significant biostratigraphic changes within the Malvinokaffric Realm that feature the earlier Pragian or Emsian appearance of the endemic, large terebratuloid brachiopods (for example, mutationellinids such as *Scaphiocoelia* and *Pleurothyrella*) followed by their disappearance in the Eifelian. Overall, the Malvinokaffric Realm is characterized by the relatively abundant, probably eurytopic brachiopod *Australocoelia*, absent however in Antarctica, and the large spiriferid *Antarctospirifer*. The Malvinokaffric Realm brachiopods for the most part are easily derived from Eastern Americas Realm antecedents, although this is not the case for the highly endemic trilobites where an Old World Realm origin is most likely. Varied nuculoid and some other bivalves are widely distributed within the Malvinokaffric Realm but their global biogeographic relations remain poorly understood. The few gastropods within the realm are relatively cosmopolitan types and the single rugose coral is known elsewhere only in the Eastern Americas Realm. The Malvinokaffric Realm faunas are recognized in the southern two-thirds of South America

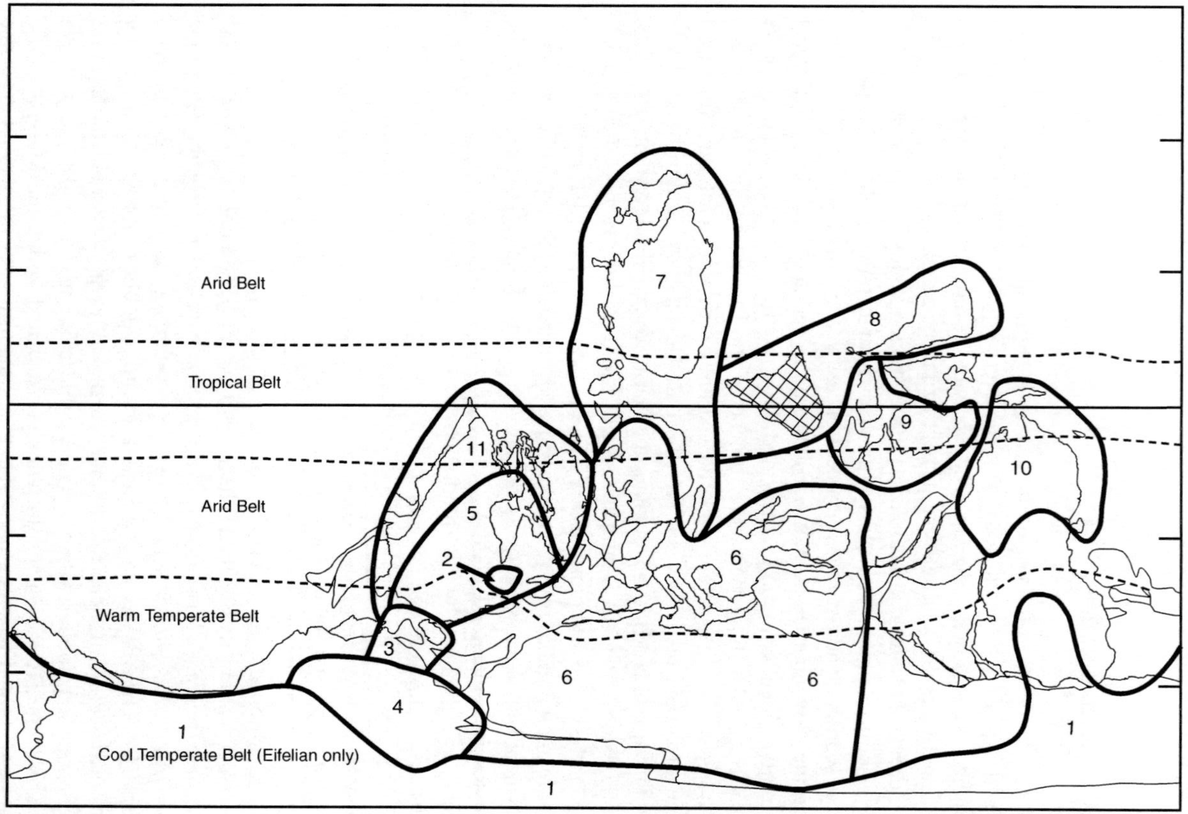


*Figure 3.4* *Earlier Middle Devonian (Eifelian) paleogeography with biogeographic units outlined: 1. Malvinokaffric Realm; 2. Michigan Basin-Hudson Bay Subprovince; 3. Colombian Subprovince; 4. Amazon Subprovince; 5. Appohimchi Subprovince; 6. Rhenish–Bohemian Region; 7. Uralian Region; 8. Balkhash–Mongolo–Okhotsk Region; 9. South China Region; 10. Tasman Region; 11. Cordilleran Region (from Boucot and Blodgett, 2001).*

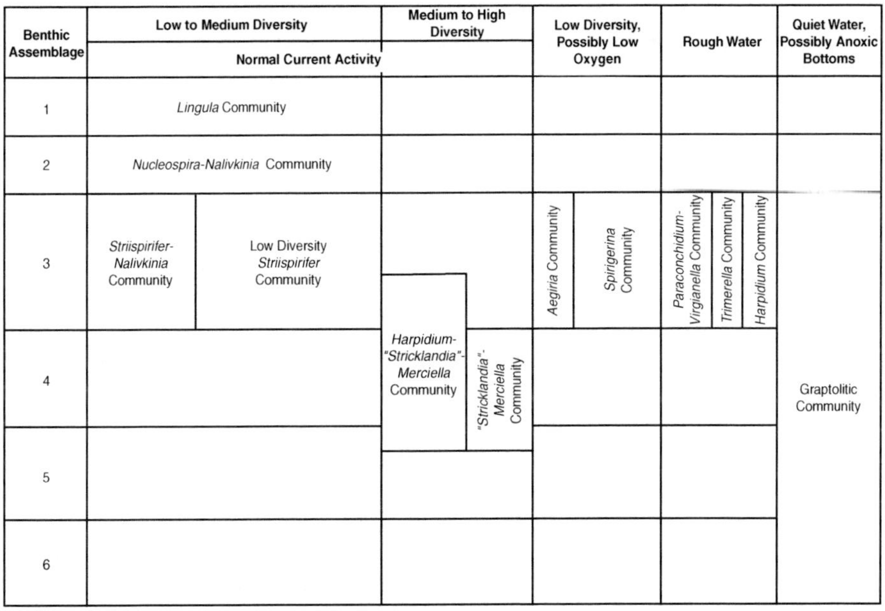


*Figure 3.5 Community framework for the South China earlier Silurian. Note the absence of benthic units from Benthic Assemblage 5 and most of 4 (from Wang Yu et al., 1987, fig. 10).*

(Figs. 3.3, 3.4), from southern Peru and adjacent Bolivia southerly through northwestern Argentina (Salta and Jujuy Provinces), the Paraná Basin in Brazil and Paraguay, Uruguay, the Sierras Australes south of Buenos Aires, Ghana, Guinea (Racheboeuf and Villeneuve, 1989), South Africa, the Falkland Islands and Antarctica. A few Malvinokaffric Realm brachiopod genera are known as boundary mixing region items in New Zealand and Tasmania, as are Malvinokaffric trilobites in similar mixing areas of northern Brazil (Parnaíba and Amazon Basins).

During the Pragian–Emsian interval, the Devonian interval of highest provincialism, within the Eastern Americas Realm one can recognize Colombian, Appohimchi, Amazon and Nevadan subprovinces (Fig. 3.3). The Colombian Subprovince is based on the presence of somewhat different stratigraphic ranges for some of its genera as contrasted with the Appohimchi, the Amazon Subprovince by a mixing of Malvinokaffric Realm endemic trilobite genera with Eastern Americas Realm brachiopod genera, i.e., biogeographic boundary region mixing, and the Nevadan Subprovince by a variety of species distinct from those present elsewhere within the realm. Note that there is no evidence within the Eastern Americas Realm for region level units, i.e., provincialism drops from the realm to the subprovince level without intermediate units as long as one pays attention to relative levels of generic provincialism. During the Eifelian there is evidence within the Appohimchi Subprovince for reef complexes, as is also true within the Michigan Basin Subprovince, the latter not being recognized during the Early Devonian. Reefs have not been recognized within the

Amazon and Colombian subprovinces, although they might be present in some of the poorly studied, massive dolomites present in the Nevada Subprovince region. Overall evidence suggests that the Eastern Americas Realm probably represents a warm temperate rather than a subtropical or tropical climate. The absence of abundant, widespread calcareous algae is one symptom, as well as gradational relations with the cool climate Malvinokaffric Realm on the one hand, and with the warmer Old World Realm in Nevada on the other hand.

The Old World Realm, probably represents a subtropical to tropical climatic regime with lateritic products such as bauxite known from some localities. This realm has been divided into Rhenish–Bohemian, Uralian, Cordilleran, Tasman, New Zealand, and South China regions plus a possible Siberian region. Blodgett *et al.* (2001) further divide the Cordilleran Region into Nevada and Western Canada provinces, with the Western Canada Province extending from at least northern British Columbia to the boundary region between Alaska and the Yukon. Herrera and Racheboeuf (1997) make a strong case for considering the Early Devonian brachiopods of the Argentine Precordillera a zone of mixing between the Malvinokaffric Realm and either or both the Old World and Eastern Americas realms. Their conclusion about a zone of mixing is in agreement with a somewhat similar but more poorly dated occurrence in northern Chile (Boucot *et al.*, 1995).

A major biogeographic discordance occurs in North America in the late Eifelian as far east as eastern New York and adjacent Pennsylvania where the Stony Hollow Member of the Union Springs formation (Ver Straeten *et al.*, 1994), the basal unit of the Hamilton Group, of latest Eifelian age yields a typical Old World rather than Eastern Americas Realm fauna (Fig. 3.4). This anomaly can best be explained by appealing to the short term presence of a warm water gateway that diverted Old World Realm waters briefly into eastern Laurentia.

The Appohimchi Subprovince, the most thoroughly described part of the Eastern Americas Realm is characterized by a large number of endemic brachiopods as well as some endemic trilobites, rugose corals and gastropods.

## 3.5 Conclusions

Ecostratigraphy enables one to use a combined community ecologic and biogeographic analysis to provide a far more detailed, evolutionarily based biostratigraphy. It makes clear the lower level correlation precision all too commonly present between biotas from different biogeographic and ecologic units. It capitalizes on the improved correlation precision made possible by combining data from adjoining, either intertonguing or mixed, ecologic and biogeographic units. It emphasizes the importance of careful taxonomic work on the phyletically evolving species within generic lineages, typically the uncommon to rare genera within any community group. It emphasizes the critical importance of working with large fossil collections that have a much better chance of recovering the uncommon to rare genera, the rarefaction phenomenon.

## 3.6 Acknowledgements

I am grateful to José Henrique Gonçalves de Melo (Petrobras-Cenpes, Rio de Janeiro), for his careful review of the manuscript, to Carlton Brett (University of Cincinnati) for careful review and the addition of useful information, as well as to Robert Blodgett, U.S. Geological Survey, Anchorage, Alaska, for the same.

Chapter 4

# Devonian Palynostratigraphy in Western Gondwana

STANISLAS LOBOZIAK[1] (deceased), JOSÉ HENRIQUE G. MELO[2] and MAURICE STREEL[3]

[1] *Ex* Université des Sciences et Technologies de Lille (U.S.T.L.), Sciences de la Terre, Villeneuve d'Ascq, France.

[2] PETROBRAS-CENPES, Cidade Universitária, Quadra 7, Ilha do Fundão, 21949-900 Rio de Janeiro, RJ, Brazil.

[3] Université de Liège, Services associés de Paléontologie, Paléobotanique et Paléopalynologie, Sart-Tilman, Bât. B 18, B-4000 Liège, Belgique.

---





*E.A.M. Koutsoukos (ed.), Applied Stratigraphy,* 73-99.

## 4.1 Introduction

At its origin palynology was a branch of Botany, but ever since it became also an important speciality of Micropaleontology. Its applications are numerous in varied scientific and economic fields.

In the scope of geosciences, palynology studies the whole of the organic matter present in sedimentary rocks. In addition to the amorphous organic matter, it also focuses on the structured elements concentrated by means of chemical processes that eliminate the mineral components of the sediment. These structured elements fall in two major categories according to their provenance:

– those of nonmarine origin, as for example cuticle and wood fragments, cryptospores, spores and pollen grains, freshwater phytoplankton, etc.; and
– those of marine origin, like microphytoplankton elements (acritarchs, prasinophytes and related forms, dinoflagellates) and chitinozoans.

The diversity of the biological material seen with the optical microscope, as well as its diagenetic state, give varied and accurate information on the age, depositional setting and thermal evolution of sedimentary rocks. The use of palynology in stratigraphy is its most common geological application. However, palynology can also provide valuable information concerning several other fields of geology, such as paleogeography, paleoecology, paleoclimatology, tectonics, eustasy, evolution of the organic matter during diagenesis, and hydrocarbon generation.

## 4.2 Principles and Methods of Study

The foundation of any palynological research is technique, which on its turn is based on varied procedures, the methods used depending on the results required and available resources.

All sediments that contain organic matter are liable to provide some palynological residue. Fine grained clastics (silts, clays), such as light grey to black, not too diagenesed or oxidized pelites and lutites, are potentially the most palyniferous sediments, whereas coarser-grained rocks and even pelites with red, green or yellow colors are generally unsuitable.

Laboratorial procedures include three phases: a mechanical phase, a chemical phase, and residue mounting. The steps to be followed are not the same in all cases, but instead depend on the composition of sediments and the palynological material that has to be analysed. For sediments in general there are standard methods to which laboratories adapt their own processing techniques. A detailed description of the varied procedures is found in Wood *et al.* (1996), and thus, only a brief overview of the main laboratorial steps is given next.

### 4.2.1 The Mechanical Process

Contamination during sampling of sedimentary rocks is a frequent problem in palynology. Therefore as concerns outcrop samples, trenching to expose a fresh surface is recommended. Of all types of samples recovered from wells, cuttings are the least desirable because of caving problems but, unfortunately, they often represent the only available source of palynological information in the oil industry. So they must be carefully cleaned and sieved so as to minimize contamination risks. Conventional and sidewall cores are preferred. However, because they are often penetrated by or coated with drilling mud they are not immune to contamination, and therefore washing is usually required in order to eliminate the mud cake. Because laboratorial contaminations are also frequent, all equipment must be carefully washed and cleaned before every new batch of samples is processed.

Washed rock samples are first crushed and sifted. The amount of rock to be submitted to chemical treatment depends on its colour, which is often a function of its organic matter content, as well as on the lithology itself.

### 4.2.2 The Chemical Process

The aim of the chemical procedure is extraction and concentration of the organic residue from sedimentary rocks. The general procedure for processing paleopalynological samples and eliminating the mineral matter involves a routine series of inorganic acid (HCl–HF–HCl) attacks followed by density separation and organic matter oxidation.

Numerous processes have been proposed in the literature for the extraction of the organic matter, but none have thus far solved the problem of completely separating the organic and mineral parts. The most important difficulty lies mainly on the size of particles of the mineral residue which are insoluble in HF. These particles, the chemical composition of which is not yet accurately known, are usually very small sized. They are agglomerated into clumps with colloidal aspect and are more abundant in residues from rocks with higher percentage of clay minerals. Several chemical agents have been used in attemps to eliminate these undesirable materials: washing the residue with diluted and hot $HNO_3$ immediately after the treatment with HF (Reissinger, 1939), which is the most classic method; the use of oxidizing agents before and during the HF treatment (de Jekhowsky, 1959); the ultrasonic treatment (Funkhouser and Evitt, 1959; Dumait, 1962; Alpern, 1963) or the use of a «Millipore» filtration instrument with a 15 μm mesh sieve (Streel, 1965). Microwave processing machines have also been developed (Shane in McGregor *et al.*, 1996), the heating of palynological samples by microwaves reducing the time and the volume of rock and chemicals needed for the acid treatment. The combined use of nitric acid and the microwave technique enhances the oxidation of superfluous kerogen without affecting the palynomorph concentration in palynological preparations (Jones, 1994).

### 4.2.3 Mounting

Once concentrated, the organic matter intended for optical microscope studies is pipetted and dispersed on a transparent slip cover, gently heatened for drying, and then fixed on a glass slide by means of a mounting media with appropriated optical properties so that palynomorphs can be examined under transmitted light. Homogeneous slides are obtained by double mounting with a dispersing agent like Hydroxyethylcellulose subsequently covered with Euparal.

Palynologists have adopted a variety of mounting media. Some also commonly used are glycerine jelly, Canada balsam, silicone oil and elvacite. They differ in durability, optical properties and their effect on palynomorphs. None is superior in all respects.

Organic residues can be stored dry or in a liquid, as for example ethanol, methanol or a mixture of water and phenol. All residues must be chemically neutralized because residual chemicals may cause decay of palynomorphs, or crystals may form in the final mount.

### 4.2.4 Microscopy

Bright-field microscopy with or without phase interference or contrast is the commonly preferred method for palynomorph examination. Incident light fluorescence microscopy is also a common procedure. It provides data that allow palynologists to reach a variety of interpretations, mainly in palynofacies studies. The use of the transmission electron microscopy (T.E.M.), and particularly scanning electron microscopy (S.E.M.), has also become popular. Both techniques represent excellent auxiliary tools to the light microscopy, chiefly in research concerned with taxonomic and ultrastructural aspects of palynomorphs, because they allow elucidation of external features at very high magnifications combined with the examination of internal structures. However, due to the higher operational costs and lengthier procedures involved, electron microscopy studies are seldom considered in routine palynology of the oil industry.

## 4.3 The Main Devonian Palynomorph Groups and Brief Comments on Their Morphology

Here we will only take into account the palynomorph groups most used in palynological studies of Devonian sediments, i.e. acritarchs and related morphotypes, chitinozoans and miospores. Cryptospores, on their turn, have their main application in Ordovician–Silurian biostratigraphy, and for this reason they will not be included in this discussion.

### 4.3.1 Chitinozoans

Chitinozoans are an extinct group of organic-walled microfossils only known *in situ* from marine sedimentary rocks of Ordovician through Devonian age. Chitinozoans are characterized by vesicles of radial symmetry and varied shape (e.g. bottle-, enrlenmeyer- and urn-shaped, globose, subcylindrical, etc.). Vesicles in general are within the 50–2000 μm size range but most of them do not exceed 250 μm in length. They occur singly, in linear or coiled chains, or still in cocoon-like aggregates and radial clusters. The vesicle wall can be glabrous or variably ornamented with rugulae, foveae, granules, crests, felt- or web-like coverings, spines and longer processes which may be simple, ramified, coalescent, spongy, etc. Distinctive external structures can be developed, such as a marginal carina or cingulum, diverse apical (basal) structures related to intervesicle linkage in a chain, or a membranous sleeve extending well beyond the vesicle base. Membranous envelopes or wing-like expansions around the vesicle also characterize some chitinozoan taxa. The type and arrangement of ornamentation, the external structures, the overall vesicle shape, and the closing mechanism of the aperture (either an external circular lid, the operculum, or an internal tubular plug, the prosome), altogether play a major role in the modern chitinozoan classification.

Chitinozoans were probably first illustrated by Hinde (1882), who found them on the bedding planes of shale samples taken from the early Wenlock Högklint Beds at Visby, Gotland Island, Sweden (Y. Grahn, writ. com., 10 April 2001). However, it was only half a century later that they were finally named by Eisenack (1931) after their supposed zoological affinity and the chitin-like composition of the vesicles, which appear characteristically dark in most light microscope preparations. The most influential publications on those microfossils are mainly dated from the mid-1950's onwards, when their outstanding biostratigraphic significance was first demonstrated by the French in the course of oil exploration campaigns in North Africa. The application of S.E.M. analyses to chitinozoan studies has become common practice mainly during the past three decades. It has proved truly revolutionary to the taxonomic classification of this fossil group, because it permits the accurate observation of critical morphological features that could not be sufficiently characterized in conventional (light microscope) investigations.

Chitinozoans have already been tentatively assigned in the literature to such disctinct groups as protists, protozoans, metazoans, and fungi. The problem of the biological affinity of chitinozoans remains still unsolved. However, the dominant opinion amongst specialists (Paris, 1981; Grahn, 1981; Miller, 1996; Paris *et al.*, 1999) has been to regard them as eggs or egg cases of some unknown group of extinct, possibly soft-bodied marine metazoans (Grahn's 1981 enigmatic "chitinozoophorans"). Because chitinozoans cannot be compared with certainty to any known group of extant or fossil taxa, their classification is largely artificial and relies only on observable morphological and structural features of the vesicle.

Several authors laid the foundation for chitinozoan morphological terminology and classification. Morphological terms used nowadays in the description of chitinozoans derive mainly from Combaz *et al.*'s (1967) terminology, which has undergone further

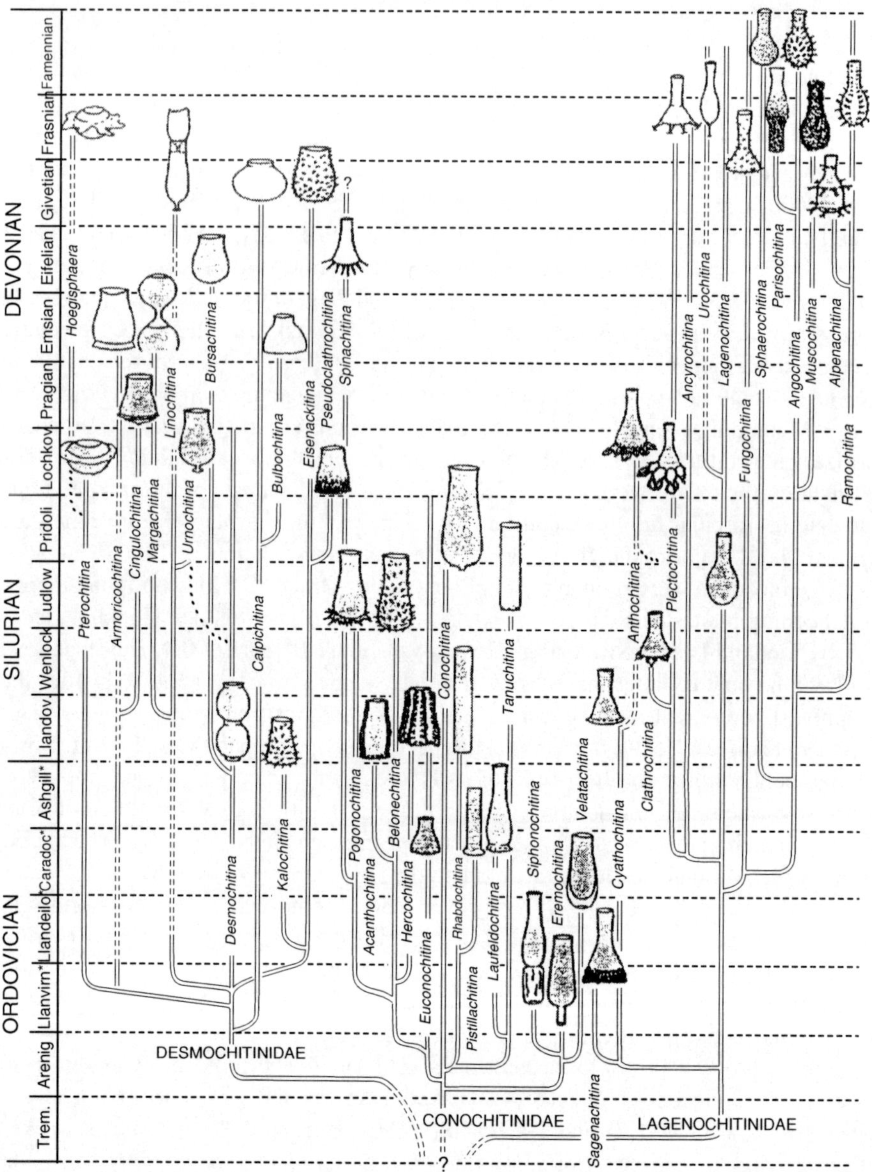


*Figure 4.1 Phylogeny of selected chitinozoan genera, after Paris (1981, 1990), F. Paris apud Miller (1996), and Paris et al. (1999), modified with newer South American data. * – Ordovician stages currently in revision.*

adaptations by Laufeld (1974) and Paris (1981), and lately also by Miller (1996) and Paris *et al.* (1999). The suprageneric classification of chitinozoans into Orders, Families and even Subfamilies, first proposed by Eisenack (1931), is regaining growing acceptance in the recent literature (e.g. Paris, 1981; Achab *et al.*, 1993;

Miller, 1996; Paris *et al.*, 1999) in place of the simpler but less useful classification system based on the alphabetical listing of genera (Laufeld, 1967, 1974; Jenkins, 1970). The newest scheme of chitinozoan classification, by Paris *et al.* (1999), complies with the ICZN rules and is based on diagnostic morphological features ranked after statistical and evolutionary criteria. A tentative phylogeny of selected chitinozoan genera is shown in Figure 4.1.

### 4.3.2 Organic-Walled Microphytoplankton

Devonian organic-walled marine microphytoplankton consists essentially of acritarchs and allied morphotypes, such as prasinophytes. Acritarchs are organic-walled microfossils of varied morphology and poorly known, probably heterogeneous biological affinities (Servais *et al.*, 1997), which cannot be assigned to any definite group of organisms. Most of them – though not all – probably represent the resting cysts of varied marine phytoplankton groups. Prasinophytes, on their turn, represent a well-defined class of cyst-forming, planktonic green algae (Chlorophyta). Many forms previously characterized as herkomorph, pteromorph and sphaeromorph acritarchs are now regarded as fossilized prasinophyte cysts or phycomata (Wicander, 1974; Tappan, 1980; Wicander and Wood, 1981; Playford and McGregor, 1993; Strother, 1996). In addition to acritarchs and prasinophytes, other "acritarch-related" groups (both algal and non-algal in origin) also make up a significant proportion of Devonian microphytoplankton assemblages (Le Hérissé *et al.*, 2000). Some "colonial" forms usually associated with acritarchs – such as the Western Gondwanan genera *Proteolobus* and *Petrovina*, of Late Devonian age – have been regarded as coenobial chlorococcalean algae (Oliveira and Burjack, 1996; Wood, 1997).

Devonian microphytoplankton morphology is very diverse. Most specimens consist of a single, hollow vesicle that may bear outgrowths (unbranched or ramified processes) as well as planar membranes that form septa, muri, wings, or a velum. The processes and vesicle wall can be further ornamented with sculptural elements such as grana, knobs, small spines, rugulae, striae, etc.

The vesicle symmetry, size and shape, the excystment method, the wall structure, and the shape of processes as well their distribution, density and mode of attachment on the vesicle, constitute the main descriptive characters of acritarch/prasinophyte morphology.

A generalized successive appearance of the main types of excystment opening, plus some selected vesicle outlines showing a variety of ornamentation, are illustrated in Figure 4.2.

### 4.3.3 Miospores

Miospores are mostly issued from vascular plants (tracheophytes). In the fossil record they are only represented by their external envelope, the exine. This bears diagnostic

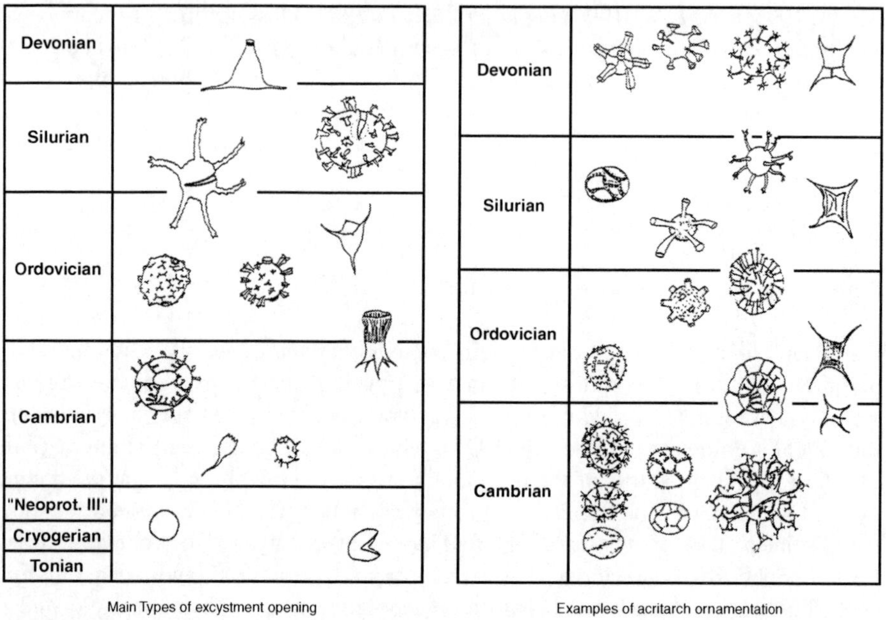


*Figure 4.2 Generalized succession of appearance of main types of excystment opening and ornamentation variety in selected Paleozoic acritarchs, after Martin (1993).*

structures and/or sculptural elements that became more and more elaborated along geological time.

Miospores originated in the Ordovician Period (Steemans, 1999) when they developed, in one of their poles, the first undoubted trilete mark which is the trace of a dissociating tetragonal tetrad. Haptotypic (i.e., linked to the tetrad status) proximo-subequatorial features are first noticed in the earliest retusoid miospores, which are often recovered from coeval strata. Distal thickening of the exine (the patina) was soon differentiated in that same period. Proximal radial muri and interradial papillae as well as diverse sculptural patterns (verrucate, foveolate, apiculate, reticulate, and rarely biform) appear during the Silurian Period. Zonate two-layered forms, i.e., miospores bearing an equatorial extension of the external layer of the exine (the sexine), first occur in the beginning of the Devonian Period to soon give rise to an explosive phase of miospore diversification. Indeed, by the end of the Early Devonian they had developed most of their Devonian characteristics (Alpern and Streel, 1972; Richardson and McGregor, 1986). Camerate miospores, i.e., miospores having a cavity between the sexine and the inner layer (the nexine), evolved from two-layered zonate miospores and display almost all kinds of exinal sculpture, including typical biform ornaments and bifurcated spines. The average spore size had increased from less than 25 μm (the size of extant bryophyte spores) to 50–150 μm (the size of extant fern spores), occasionally even reaching megaspore size (>200 μm) possibly in relation to incipient plant heterospory. During Middle and Late Devonian times miospore size changed again in two steps. A dramatic increase started little before the Emsian/Eifelian transition, with species (many of them camerate) often surpassing the 200 μm size limit, probably related now to the development of true heterospory. Close to the Frasnian/Famennian boundary most of the miospores returned to the 50–150 μm size range, possibly in response to new strategies in land plant reproduction such as the differentiation of the seed and prepollen. The first occurrence of endoreticulate miospores near that stage boundary could reflect such a change. Therefore, miospores encountered in the Middle and Late Devonian may include not only isospores (from homosporeous plants), microspores (from heterosporeous plants), and prepollen (from possible gymnosperm ancestors), but also megaspores below the 200 μm size limit. The most characteristic features of the miospore evolution are shown in Figure 4.3.

## 4.4 Palynostratigraphy

Biostratigraphy and correlation of sedimentary rocks are the primary applications of palynology in geological sciences. Because palynomorphs are commonly recovered from both marine and nonmarine sediments, they have the potential for correlating deposits of continental and transitional settings with those of marine origin. This is particularly true in the case of miospores, which are often transported far away from the mother plant by running water and wind and dispersed over a variety of sedimentary environments related to distinct biofacies.

### 4.4.1 Types of Biostratigraphic Units

Various types of biostratigraphic units and elements that are considered in their descriptions have been used in the palynological literature. Among them, three major types of biozones are roughly recognized, in terms of the International Stratigraphic Guide (ISG) and the North American Commission on Stratigraphic Nomenclature (NACSN), sometimes with significant differences in interpretation. They are:

- the range zones, on their turn comprising: (a) interval zones which are strata between the successive stratigraphically highest and/or lowest occurrences of index taxa, and/or (b) Oppel Zones which, according to ISG, represent the known range of one single taxon or group of selected taxa or, according to Hedberg (1976, p. 58), may be defined as an association or aggregation of selected taxa of restricted and largely concurrent ranges, chosen as indicative of approximate contemporaneity;
- the assemblages zones, which include strata characterized by an association of three or more distinctive taxa; and
- the acme or abundance zones, composed of strata characterized by maxima (epiboles) in the relative abundance of one or more taxa.

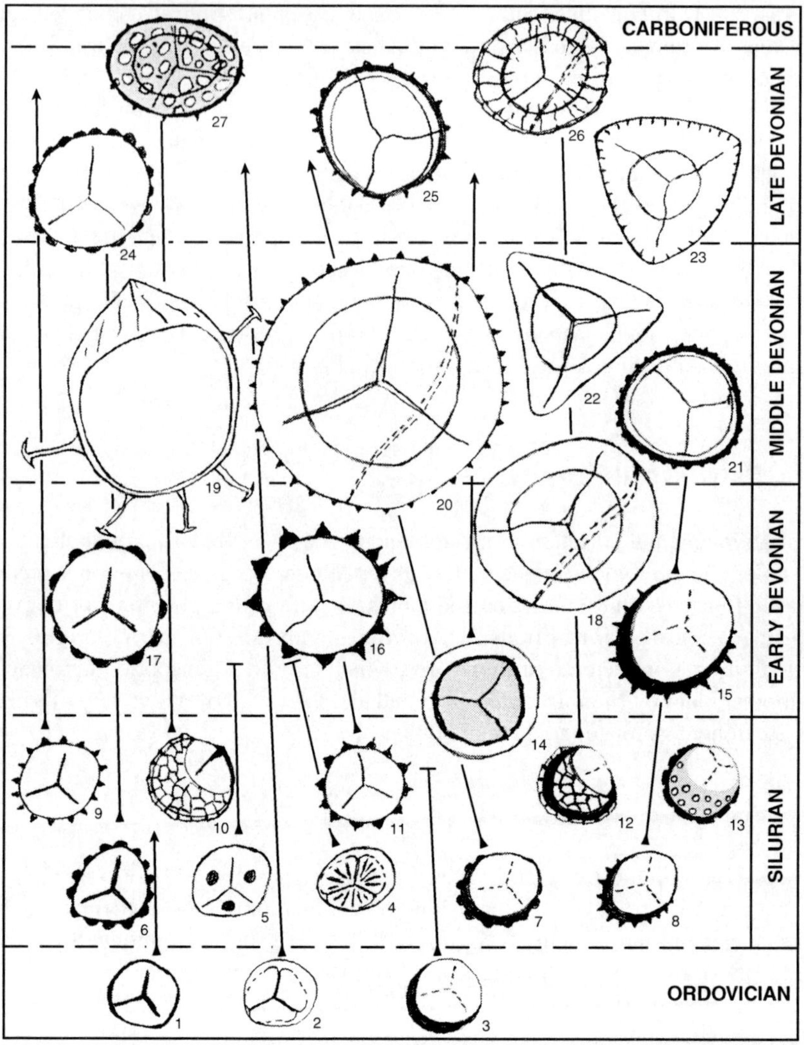


*Figure 4.3* Morphological trends of miospores from Ordovician to basal Carboniferous: *1: the first undoubted trilete mark at one pole; 2: haptotypic (linked to the tetrad status) proximo-subequatorial features (retusoid miospores); 3: distal thickening of the exine (patina); 4: proximal radial muri; 5: interradial papillae; 6, 17 and 24: verrucate sculpture; 7: patinate verrucate sculpture; 8 and 15: patinate apiculate sculpture; 9: apiculate sculpture; 10: reticulate sculpture; 11 and 16: biform sculptural elements; 12: patinate reticulate sculpture; 13: patinate foveolate sculpture; 14: equatorial extension of an external layer of the exine, the sexine (acamerate); 18: two-layered zonate, camerate (cavity between the sexine and the inner layer, the nexine); 19: two-layered azonate with bifurcated spines; 20, 21 and 25: two-layered zonate, camerate, with apiculate sculpture; 22: two-layered zonate, acamerate; 23: two-layered zonate, camerate, endoreticulate; 26: three-layered zonate, camerate; 27: two-layered zonate, camerate, with apiculate, reticulate sculpture.*

As concerns the Devonian miospore stratigraphy two zonal concepts were adopted. One of them is the assemblage zone concept, mainly used by Richardson and McGregor (1986) in their miospore zonation of the Old Red Sandstone Continent and adjacent regions. These authors described assemblage zones defined by a combination of different criteria, such as the co-occurrence of characteristic taxa, the first appearance of selected species, and the appearance of selected morphological features. This concept was proposed as the only one that would allow correlation on a global scale. Indeed, the Old Red Sandstone Continent and adjacent regions obviously comprise different Devonian phytogeographic provinces, and therefore, in order to accommodate a maximum of available data it was necessary to apply such a comprehensive zonal concept. Note that in Eastern Europe Russian authors still favor a concept of Assemblage Zone based on, and generally named after, the acme of characteristic species, whereas the true inceptions of taxa are seldom taken into account. Such kind of biozonation is still largely applied to a Devonian phytogeographic area in the eastern part of northern Euramerica.

The other concept is the interval zone adopted by Streel *et al.* (1987) for the type marine Devonian strata of the Ardenne-Rhenish regions in Western Europe. Streel *et al.*'s miospore zonal scheme comprises a series of interval zones entirely erected within the limits of one single phytogeographic province, to the contrary of Richardson and McGregor assemblage zones. The advantage of the interval zone concept is that it allows unequivocal correlation with interval zones based on varied fossil groups.

### 4.4.2 Calibration with Other Reference Units

The main purpose of the biostratigraphic tool, aside from paleoenvironmental evaluations, is to date sediments and correlate them accurately throughout long distances. To make such purpose feasible, the relationships between the proposed biozones and the litho- and chronostratigraphic units, as well as their relation with biozones based on other fossil groups, should be established as precisely as possible. So the interval zones are especially suitable for such calibration.

An example is given in Streel and Loboziak (1996), who propose a table of correlation for the Middle and Upper Devonian of the Ardenne-Rhenish regions, in which the first occurrences of thirty miospore key species are calibrated according to the standard conodont zonation. Of course, the quality of these correlations, named the Quality Index of Correlation, depends not only on the distance in time and space between each miospore *datum* and the nearest faunal *datum*, but also on the kind of stratigraphic unit that is being used.

## 4.5 Paleoenvironment and Paleogeography

The occurrence, composition and state of preservation of the organic matter in sediments yield information on its transport to the depositional site, on the paleoenvironment in which it has been deposited, and on its alteration once deposited. For instance, it is a well-established fact in the palynological study of marine sedimentary rocks that the proportion of land-derived organic debris (land plant cuticles, wood, and miospores) tends to increase inshoreward. The opposite trend is often characterized by the predominance of amorphous (unstructured) organic matter, of probable algal origin, and marine palynomorphs over scarcer terrestrial organic matter. Post-depositional erosion or mass transportation of palyniferous sediments may cause palynomorphs to be reworked into deposits of different geological ages (most often within stratigraphically younger sediments). Reworked palynomorphs, which are considered as mere contamination by some biostratigraphers, may however help to detect tectonic uplifts and eustatic falls (Streel and Bless, 1980; Eshet *et al.*, 1988; Di Pasquo and Azcuy, 1997). Moreover, they provide a key to the source area of the sediments into which they have been recycled. Therefore, reworked palynomorphs are regarded as valuable tools for the reconstruction of important geological events.

When the quantitative analysis of nonreworked miospore assemblages is possible, it permits the recognition of different paleoenvironmental settings of the nonmarine realm. Downstream (coastal lowland) "coal" swamp and swamp margin environments are directly controlled by short-term sea-level changes: any sea-level highstand will induce a high fresh-water table in those environments and therefore increase their importance by means of a higher proportion of swamp plant miospores in neighboring seas, highstands of sea level corresponding to the extension of coastal lowland swamps. The upstream (upland) "coal" swamp and upstream swamp margin environments are not directly affected by the sea-level fluctuation. Their retreat or expansion are first controlled by dry versus wet climates. Wet climates produce flooding episodes that, in turn, carry more upland miospores into the marine basin.

Studies carried out in North Africa and Brazil (Loboziak *et al.*, 1988; Streel *et al.*, 1988; Loboziak and Streel, 1989) have demonstrated the potential of recognizing Western European defined miospore zones, particularly those from the Ardenne-Rhenish regions (Streel *et al.*, 1987), in the regional Devonian of Western Gondwana. A quantitative approach based on the relative abundance of Devonian miospore taxa, first inside Western Gondwana (Loboziak *et al.*, 1989), then between the latter and Southern Euramerica (Streel *et al.*, 1990), indicates a remarkable uniformity of land plant vegetation, and therefore of climate, from paleotropical to paleopolar realms during at least the Givetian and Frasnian (Streel *et al.*, 2000b).

The aforementioned miospore-based studies suggest that Southern Euramerica and Western Gondwana constituted a single major phytogeographic province during the Middle and Late Devonian. Although the biostratigraphic scheme developed in the Ardenne-Rhenish region can be applied to practically the whole of this phytogeographic province, it does not mean that the flora was homogeneous over such a huge area. For example, some species were restricted to Western Gondwana during the Middle and early Late Devonian (Loboziak and Streel, 1995a), where they can represent a significant part of the miospore assemblages. This becomes particularly evident if the percentages computed for each sample take into account the number of specimens rather than the number of species. Two kinds of climatic restrictions are evoked to explain this lack of homogeneity: (a) macroclimatic, probably latitudinal, differences determining subprovinces, and (b) microclimatic, mainly altitudinal, variations affecting plant communities in local areas.

In the marine realm, according to Molyneux (in Molyneux *et al.*, 1996, p. 515–516), the distribution of microphytoplankton is conditioned in a general way by two major environmental trends. These, on their turn, dictate ultimately several interrelatd physical, chemical and biotic controls such as: nutrient, light and oxygen supply, predation, competition, and seawater depth, salinity, pH, temperature, hydrodynamics, and turbidity, amongst others. Although primarily envisaged to explain the distribution pattern of Paleozoic microphytoplankton, Molyneux's trends have also a bearing on the co-occurrence of other marine palynomorph groups with more problematic paleoecology, such as chitinozoans (see below). These trends are:

– the relative distance from shoreline (Molyneux's inshore–offshore trend): by and large, the abundance, diversity and morphological complexity of microphytoplankton assemblages in shelf paleoenvironments tend to increase offshorewards, with less diverse, more eurytopic (sphaeromorph-dominated) assemblages predominating in the shallower marine settings. However, this general tendency may be variably affected by environmental factors determined by coastal physiography, seabottom relief, marine currents, etc.; and
– the paleolatitudinal trend: this leaves a less evident imprint on the composition of ancient microphytoplankton assemblages than do paleoecological factors related to shoreline distance. Latitudinal controls are influenced ultimately by climate and paleogeography. They may be detected in the form of acritarch/prasinophyte provinces, which are ideally distributed as latitude-parallel belts, best documented for the Ordovician and Silurian, but also recognizable in the Devonian (Le Hérissé *et al.*, 1997a). Some authors (Wood, 1984; Vavrdová *et al.*, 1993; Vanguestaine, 1986; Vavrdová and Isaacson, 1996, 1999, 2000) record different microphytoplankton taxa characteristic of late Middle to Late Devonian assemblages from Western Gondwana and Euramerica. Wood's (1984) Late Devonian *Umbellasphaeridium saharicum* Bioprovince is believed to characterize high latitude marine environments of the Southern Hemisphere supposedly influenced by a cold water subpolar gyre (Isaacson and Díaz Martínez, 1995). References to additional works concerned with Devonian acritarch paleobiogeography are provided by Le Hérissé *et al.* (2000, p. 196–197).

As discussed by Paris (1996, p. 544–545), chitinozoans too, with their supposedly planktonic mode of dispersion, were sensitive to environmental parameters related to both the inshore–offshore and paleolatitudinal trends. This can be inferred from variations in the taxonomic composition of chitinozoan assemblages (diversity), the absolute abundance of chitinozoans (expressed in counts of specimens per gram of rock), and their relative frequency in the total palynological residue (ratio of chitinozoans to other palynomorphs and land plant clasts). Wicander and Wood (1997) combined chitinozoan diversity trends with microphytoplankton and miospore quantitative data in order to estimate paleoshoreline fluctuations related to transgressive/regressive cycles in the Middle Devonian of Iowa, U.S.A. Evidence of faunal differentiation and endemism at genus and species levels is recorded by Paris *et al.* (2000) in Devonian chitinozoan assemblages from separate areas like Australia, Brazil and Ghana.

## 4.6 Devonian Palynostratigraphic Schemes in Western Gondwana

### 4.6.1 Chitinozoans

Chitinozoans altogether span from Early Ordovician to Late Devonian and have been reported from all continents. Many chitinozoan species and even some genera have restricted stratigraphic ranges and relatively widespread geographic distribution, which makes them excellent biostratigraphic tools.

Devonian chitinozoans have been reported and described from a variety of Western Gondwanan regions, such as Brazil, Argentina and the Malvinas Islands, Bolivia, Uruguay, Paraguay, Ghana, and North Africa (see main references in Grahn, 1992; Wood, 1994; Paris, 1996; and Paris *et al.*, 2000). However, relatively few works have focused on the determination of accurate stratigraphic ranges and the erection of regional biozones based partly or entirely on chitinozoans in Western Gondwana. Of these, the main ones refer to Brazilian basins, Uruguay, eastern Paraguay and northwestern Argentina (Lange, 1967a–b; Quadros, 1982, 1988; Volkheimer *et al.*, 1986; Ottone, 1996; Grahn, 1997, 1998a–b, 1999a–b; Grahn *et al.*, 1997, 2000, 2001, 2002 and 2003; Grahn and Melo, 2002, 2003 and in press, a–b; Grahn, 2003a–b; Azevedo-Soares and Grahn, in press); Bolivia (Limachi *et al.*, 1996; Dufka in Vavrdová *et al.*, 1996; Grahn, 2002); Gondwanan portions of SW Europe (Cramer, 1964; Díez and Cramer, 1978; Paris, 1981); Ghana in West Africa (Anan-Yorke, 1974); and Algerian and Libyan basins of North Africa (Taugourdeau and Jekhowsky, 1960; Jardiné and Yapaudjian, 1968; Massa and Moreau-Benoit, 1976; Boumendjel, 1987; Boumendjel *et al.*, 1988; Paris in Streel *et al.*, 1988).

Most of the chitinozoan datings and biozonations published before the late 1990's for northern Brazilian basins and Ghana are now in need of revision following the latest advances of Devonian miospore biostratigraphy in the Solimões, Amazon and Parnaíba Basins of Brazil. Grahn (1992) summarized the stratigraphic distribution of Devonian chitinozoans in Brazil, but without designating formal biozones, and some of his age conclusions also require update because based on now obsolescent databases. Paris (1996, text-fig. 8) presented a range chart of selected Devonian chitinozoan species integrating data from Gondwanan regions and other parts of the world. Paris *et al.*

(2000) were the first to present a provisional global biozonation of Devonian chitinozoans including eighteen interval range zones and one total range biozone, based on chitinozoan data from all continents. Unfortunately, however, limitations to the worldwide application of these biozones still persist, mainly posed by such constraints as chitinozoan endemism, scanty biostratigraphic data for the Emsian, Frasnian and Famennian stages, and insufficient information from several regions of the world. On the other hand, ongoing and future studies will surely bring about further improvements to the biostratigraphy of Devonian chitinozoans in the coming years. For example, a relatively large number of species (mostly yet undescribed, in addition to those newly erected by Grahn, 2002, Grahn and Melo, 2002 and in press, a–b) occur in late Middle to Late Devonian strata of Bolivia and northern Brazil. Although possibly endemic to Western Gondwanan areas, some of these forms could eventually add considerable refinement to the regional Eifelian–Famennian chitinozoan biostratigraphy.

### 4.6.2 Organic-Walled Microphytoplankton

Considerable advances have been made in recent years as concerns the taxonomy of Devonian organic-walled microphytoplankton (chiefly acritarchs and prasinophytes). A relatively large amount of such taxonomic works dealing with Western Gondwanan assemblages has accumulated during the past four decades, particularly as concerns SW Europe, North Africa, and South America. On the other hand, continued effort has to be concentrated on the biostratigraphy of Devonian microphytoplankton taxa as well as their paleoecology (including a better understanding of the main factors that control their environmental distribution).

Unfortunately, many of the papers dealing with Devonian acritarchs and related forms provide insufficient information on their accurate age and stratigraphic distribution. Relatively few, too widely separated contemporary acritarch/prasinophyte successions have been studied in detail thus far, so hampering the erection of a widely accepted global Devonian microplankton biozonation. Nevertheless, as more Devonian sections are continuously studied from a variety of locations and paleoenvironments, it becomes evident that, beside the many taxa with restricted geographic distribution or too extended stratigraphic ranges, there are also a number of cosmopolitan or semi-cosmopolitan species with short stratigraphic ranges that are useful for long-distance correlations. This is demonstrated in range charts compiled by Le Hérissé (in Molyneux *et al.* 1996, text-fig. 10) and Le Hérissé *et al.* (2000, tab. 1) for selected Devonian species recorded in various continents, although some of the ranges presented need revision because based on partly obsolete bibliographic databases for Western Gondwana. Several of these taxa can be used for intercontinental correlations despite some degree of provincialism in their geographic distribution.

In Western Gondwana, the most significant biostratigraphic accounts derive from North Africa (Algerian Sahara, western and eastern Libya) and northern Brazil (Solimões, Amazon and Parnaíba Basins). Discrete published results also concern Iran

(Hashemi and Playford, 1998); Spain (Cramer, 1964); Ghana (Anan-Yorke, 1974); and southern Brazil (Paraná Basin), Paraguay, Uruguay, Bolivia, and Argentina in South America (Daemon *et al.*, 1967; Pöthe de Baldis, 1974, 1978, 1979; Barreda, 1986; Oliveira *et al.*, 1995; Ottone, 1996; Ottone and Rossello, 1996; Vavrdová and Isaacson, 1996; Vavrdová *et al.*, 1993, 1996; Le Hérissé *et al.*, 1997b; Oliveira, 1997; Quadros, 1999). In addition, some regional palynozones for the Devonian of Bolivia have been erected on the basis of mixed and poorly documented phytoplankton/miospore data (Lobo-Boneta, 1975; Lobo-Boneta *et al.*, 1977; Suárez-Soruco and Lobo-Boneta, 1983; Limachi *et al.*, 1996).

Jardiné (1972, text-figs. 1, 2) presented integrated range charts of selected Devonian microphytoplankton species from SW Europe, the Algerian Sahara, and South America. Still regarding the Algerian Sahara an acritarch/prasinophyte range chart and a regional biozonation are also given in Jardiné *et al.* (1974, fig. 2), based on results from the Grand Erg Occidental (Magloire, 1968; Lanzoni and Magloire, 1969) and the Illizi (Fort Polignac) Basin (Jardiné and Yapaudjian, 1968). The range chart shows the distribution of selected 160 microphytoplankton species and a palynostratigraphic scheme alongside a composite lithostratigraphic column.

More recent data from the Illizi Basin concern the Emsian to latest Famennian («Strunian») stratigraphic interval (Coquel and Latrèche, 1989; Moreau-Benoit *et al.*, 1993). Several phytoplankton assemblages are listed but no biozonation is proposed, all datings being based on the associated miospore results.

In regard to the Hammadah Basin of western Libya (also named Rhadamès or Ghadamis Basin in the previous literature), Moreau-Benoit (1984) gives a list of taxa identified in Givetian and latest Famennian («Strunian») samples from three wells and their stratigraphic distribution. Some of the most significant taxa are described and illustrated. In Cyrenaica (northeast Libya) organic-walled microphytoplankton assemblages have been recognized from samples assigned to the Givetian–Famennian stratigraphic interval of several exploratory wells. The most significant forms are illustrated in Paris *et al.* (1985). The stratigraphic and paleogeographic implications of these assemblages are discussed in a second paper (Vanguestaine in Streel *et al.*, 1988). A list of taxa identified in the uppermost Devonian of the Mourzouk Basin is provided by Vavrdová (1987).

In northern Brazil two main biozonations have been proposed. They concern the Parnaíba (formerly Maranhão) Basin (Brito, 1971 partly modified by Quadros, 1982) and the Solimões Basin (Quadros, 1988), both of which require update in terms of their taxonomic bases and age implications. In addition, the Late Devonian microphytoplankton biostratigraphy of the Amazon Basin is currently under review (Le Hérissé, 2001; A. Le Hérissé, unpubl. data). As a whole, the organic-walled microphytoplankton record in those three basins concerns the stratigraphic interval currently dated by miospores from possibly the latest Early Devonian (late Emsian) to the end-Devonian (Loboziak and Melo, 2000, text-fig. 2: A, B). Riegel's (1996) post-Devonian "phytoplankton blackout", recognized through an abrupt decrease in the relative abundance and diversity of Carboniferous assemblages all over the world (Le Hérissé *et al.*, 2000, p. 201), can be also identified right above the Famennian/Tournaisian boundary in northern Brazilian basins despite the maintenance of widespread marine settings in those areas (Melo and Loboziak, 2003).

### 4.6.3 Miospores

Publications on the stratigraphic distribution of Devonian miospores actually started with the classic work of Naumova (1953) on the Russian platform, in the northeast part of the Euramerican Continent. Ever since, additional Devonian sections have been zoned by various authors, mainly within Euramerica but also elsewhere around the world.

In respect of the Devonian of Western Gondwana, the earliest and most significant miospore-based syntheses were established in the late 60's and 70's. They concern mainly the Illizi (Fort-Polignac) Basin of the Algerian Sahara (Jardiné and Yapaudjian, 1968), the Hammadah Basin of western Libya (Massa and Moreau-Benoit, 1976), and the Brazilian Amazon, Parnaíba and Paraná Basins (Daemon *et al.*, 1967; Daemon and Contreiras, 1971; Daemon 1974, 1976). Other relevant Devonian miospore investigations concern Iran (Coquel *et al.*, 1977), Paraguay (Menendez and Pöthe de Baldis, 1967), the Cordillera Oriental of Bolivia (McGregor, 1984), northernmost Bolivia (Vavrdová *et al.*, 1993, 1996), and the Subandean regions of southern Bolivia and northwestern Argentina (Pérez-Leyton, 1991; Ottone, 1996; Melo, 2000).

In the past few decades numerous studies have been carried out in North Africa, Brazil and Saudi Arabia. As shown in Figure 4.4, most of these investigations are relatively recent, and several were made by the same group of authors.

#### *4.6.3.1 Miospore Events*

Figure 4.5 depicts the most significant Devonian miospore events recorded in Western Gondwana (Loboziak, 1999; Loboziak and Melo, 2000, 2002; Melo and Loboziak, 2001, 2003). The events are dated by correlation with faunally calibrated interval zones of the Ardenne-Rhenish regions (Streel *et al.*, 1987) and the assemblage zones of the Old Red Sandstone Continent and adjacent regions (Richardson and McGregor, 1986). With the exception of some stratigraphically useful species that are restricted to Western Gondwana, most of the selected taxa are common to both Southern Euramerica and Western Gondwana.

During most of the Early Devonian, i.e. from Lochkovian to the early late Emsian, miospores were diverse and rather small-sized. Richardson and McGregor (1986) report prominent flimsy zonate taxa from the base of their *E. micrornatus – S. newportensis* Assemblage Zone within the early Gedinnian (Lochkovian). In the Ardenne-Rhenish region (Streel *et al.*, 1987; Steemans, 1989) the zonates, represented by the species *Cirratriradites diaphanus*, are noted for the first time in the equivalent MN Oppel Zone, more precisely in the upper part of the phylogenic zone *E. micrornatus* var. *sinuosus* (Si). Small non-spinose zonates have been recorded in various miospore assemblages of the Hammadah, Illizi and Al Kufrah Basins of North Africa, as well as in Bolivia and the Brazilian Amazon and Paraná basins. Based on independent chitinozoan–acritarch datings of regional Upper Silurian–Lower Devonian miospore successions in the Hammadah Basin, western Libya, Rubinstein and Steemans (2002)

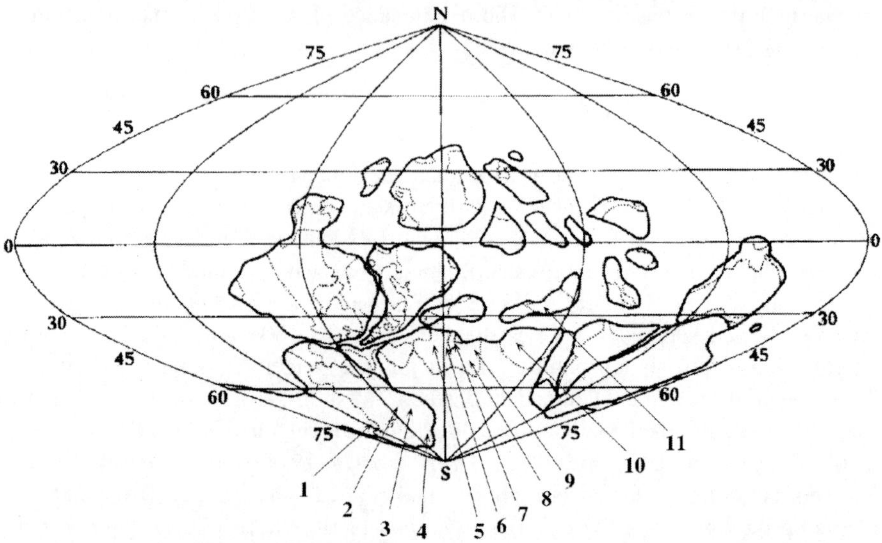


**1.** - Cordillera Oriental and Subandean regions, Bolivia (McGregor, 1984; Perez-Leyton, 1991; Melo, 2000).

**2.** - Amazon Basin, northern Brazil (Daemon and Contreiras, 1971; Daemon, 1974, 1976; Loboziak *et al.* 1991, 1993, 1996, 1997a, b; Melo and Loboziak, 2001, 2003).

**3.** - Parnaíba Basin, northern Brazil (Daemon, 1974, 1976; Loboziak *et al.* 1992b, 1993; Rodrigues *et al.*, 1995; Loboziak *et al.*, 2000b; Grahn *et al.*, 2001).

**4.** - Paraná Basin, southern and western-central Brazil (Daemon *et al.*, 1967; Daemon, 1976; Burjack *et al.*, 1987; Loboziak *et al.*, 1988; 1995; Dino and Rodrigues, 1995).

**5.** - Grand Erg occidental, Algerian Sahara (Lanzoni and Magloire, 1969).

**6.** - Illizi (Fort-Polignac), Algerian Sahara (Jardiné and Yapaudjian, 1968; Attar *et al.*, 1980; Boumendjel *et al.*, 1988).

**7.** - Hammadah (Rhadamès-Ghadamis) Basin, Tunisia-Libya (Massa and Moreau-Benoit,1976; Massa *et al.*, 1980; Loboziak and Streel, 1989; Loboziak *et al.*, 1992a; Rubinstein and Steemans, 2002).

**8.** - Al Kufrah Basin, Libya (Grignani *et al.*, 1991).

**9.** - Cyrenaica, Libya (Streel *et al.*, 1988).

**10.** - Northern and Central Saudi Arabia (Loboziak and Streel, 1995b; Steemans, 1995; Loboziak, 2000).

**11.** - Eastern Elburz, Iran (Coquel *et al.*, 1977).

*Figure 4.4 Selected occurrences of Devonian palynofloras in Western Gondwana and respective data sources. Paleogeographic reconstruction after Heckel and Witzke (1979).*

now consider that the earliest zonates there, including such species as *Breconisporites simplex* and *Perotrilites laevigatus*, could be as old as Late Silurian (Pridoli).

Another significant feature in Lower Devonian palynofloras is the first occurrence of the index reticulate species *Dictyotriletes emsiensis*. This species is an eponym of Richardson and McGregor's (1986) *D. emsiensis–V. polygonalis* Assemblage Zone, which is an equivalent of the *V. polygonalis–D. wetteldorfensis* (PoW) Oppel Zone defined in the Ardenne (Streel *et al.*, 1987). In this latter region, the first occurrence of *Dictyotriletes emsiensis* (Streel *et al.*, 1987; Steemans, 1989) characterizes the base of the *D. emsiensis* (E) Interval Zone, the youngest subdivision of the *B. breconensis–E. zavallatus* (BZ) Oppel Zone which underlies the PoW Biozone. According to

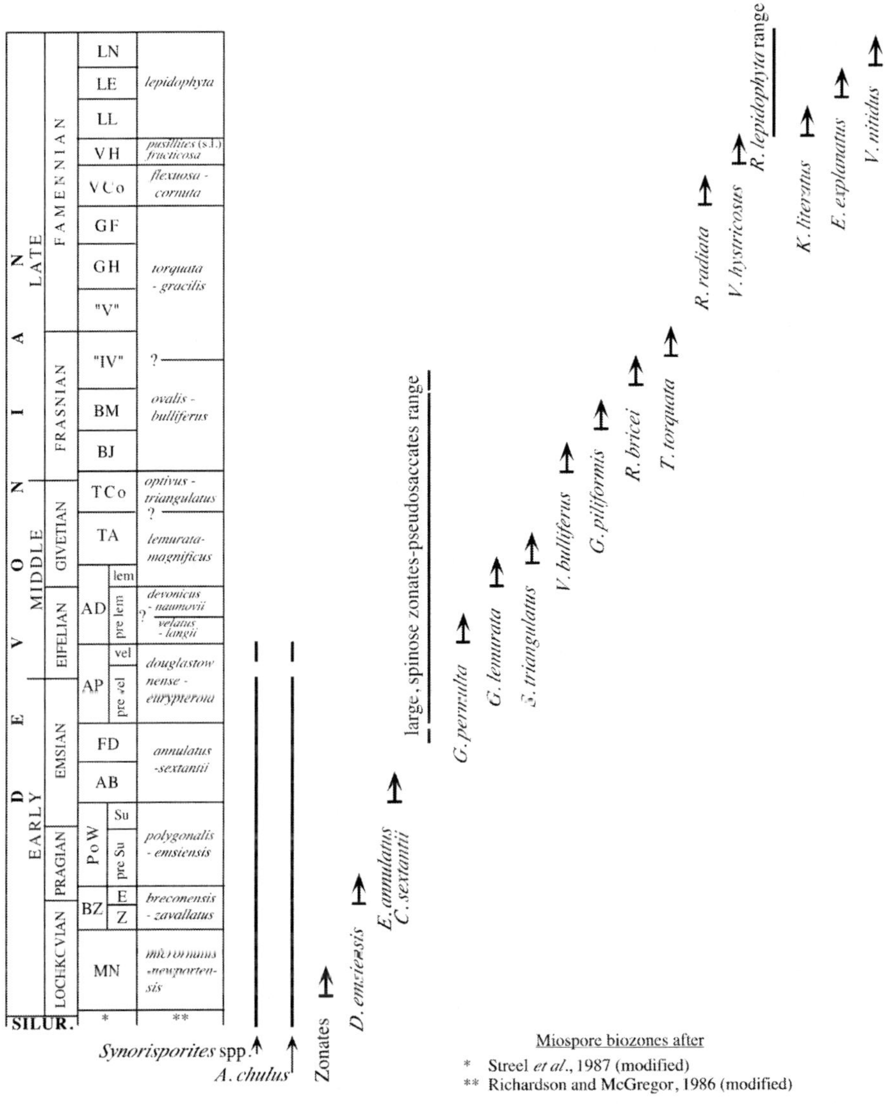


*Figure 4.5 Main Devonian miospore events in Western Gondwana, after Loboziak (1999), Loboziak and Melo (2000, 2002), and Melo and Loboziak (2001, 2003), modified.*

Streel *et al.* (2000a, p. 18, text-fig. 8), the base of the E Biozone is close to (but below) the lower limit of the Pragian stage. However, detailed studies of Lower Devonian palynofloras in northern Brazilian basins (Rubinstein *et al.*, in press) now suggest that the stratigraphic inception of reticulates belonging to the so-called *Dictyotriletes emsiensis* Morphon could be as old as early late Lochkovian (Si Zone-equivalent), at least in some of the Western Gondwanan regions where the same bio-event has been recognized (North Africa, Saudi Arabia, Brazil and Bolivia).

The next miospore event is the first occurrence of *Camarozonotriletes sextantii* and *Emphanisporites annulatus*, a well-known radially-ribbed species with a distal concentric ring. The inception of these two taxa defines the base of the *E. annulatus - C. sextantii* Assemblage Zone of Richardson and McGregor's zonation, somewhere above the base of Emsian, as now redefined (Yolkin *et al.*, 1997) by the appearance of the conodont *P. kitabicus*. According to Streel *et al.* (1987) the *E. annulatus–C. sextantii* Assemblage Zone corresponds to the *E. annulatus–B. bellatulus* (AB) and *E. foveolatus–V. dubia* (FD) Oppel Zones of the Ardenne-Rhenish regions. *Camarozonotriletes sextantii* is present in North Africa and Saudi Arabia but has not yet been recorded in Brazil.

*Emphanisporites annulatus* is an Emsian–Givetian species with world-wide distribution. It has been recorded in various Western Gondwanan regions, such as North Africa, Saudi Arabia, Bolivia, and Brazil. Its accurate stratigraphic inception in Brazilian basins is still uncertain, because early Emsian strata are either missing or poorly characterized palynologically there. This is particularly true in the Amazon and Parnaíba Basins, whereas in the Paraná Basin the species was used by Daemon *et al.* (1967), under the designation *Emphanisporites* cf. *erraticus*, to characterize their D2b and D3 Intervals, attributed by those authors to the Emsian–Eifelian.

A renewal of the palynoflora occurred in the late Emsian, mainly due to a dramatic increase in miospore size. This event is well documented by the appearance of large, prominently spinose zonates and pseudosaccates belonging to the *Samarisporites/ Grandispora* complex and the disappearance of several smaller forms, particularly reticulate elements well known in the Lower Devonian.

The large spinose zonates–pseudosaccates first occur and proliferate in the transition between the *E. annulatus–C. sextantii* and the overlying *G. douglastownense–A. eurypterota* Assemblage Zones of Richardson and McGregor (1986). This same interval of latest Emsian age correlates with the transition between the *E. foveolatus–V. dubia* (FD) and the succeeding *A. apiculatus–G. protea* (AP) Oppel Zones of the Ardenne-Rhenish regions (Streel *et al.*, 1987).

In Western Gondwana this group of miospores is abundant throughout the Eifelian, attains its epibole in the Givetian and becomes gradually rarer in the Frasnian. Some specimens, mainly belonging to *Grandispora incognita*, still persist into palynophase «IV», which is the highest miospore subdivision of the Frasnian according to Streel *et al.*'s (1987) zonal scheme. Therefore, assemblages of large zonates-pseudosaccates are useful in identifying the late Emsian–early late Frasnian interval.

Several successive first occurrences of characterisitic or index species defining miospore events are recorded within the *in situ* range of the large zonates–pseudosaccates in Western Gondwana. In ascending order they are:

– the first occurrence of several endemic zonates and pseudosaccates in Western Gondwanan areas, such as *Grandispora permulta*, within a stratigraphic interval equivalent to the Ardenne-Rhenish AP/ *A. acanthomammillatus–D. devonicus* (AD) transition of late early Eifelian age;

- the appearance of *Geminospora lemurata*, a species widely dispersed all over the world, at the base of the *G. lemurata* (lem) Interval Zone, very close to, but above, the base of the Givetian;
- the appearance of *Samarisporites triangulatus* and allied congeneric forms at the base of the *S. triangulatus–A. ancyrea* (TA) Oppel Zone of late early Givetian age;
- the first occurrence, in the early Frasnian, of miospore taxa bearing tabulate sculpture: *Verrucosisporites bulliferus* at the base of the *V. bulliferus–C. jekhovskyi* (BJ) Oppel Zone, and *Geminospora piliformis*, an endemic Western Gondwanan species, within a stratigraphic interval equivalent to the BJ/*V. bulliferus – L. media* (BM) zonal transition of the Ardenne-Rhenish zonation; and
- the first occurrence of *Rugospora bricei*, a small-sized pseudosaccate species with fine rugulate sculpture at the base of the palynophase «IV» of late Frasnian age.

A return to palynofloras dominated by smaller-sized miospores is documented close to, but below, the Frasnian/Famennian boundary. From this level to the upper Famennian, miospore assemblages in the Ardenne-Rhenish regions include various small-sized, spinose pseudosaccate species of the genus *Grandispora*. The successive first occurrences of these species characterize the bases of most of the miospore zone subdivisions from palynophase "IVc" through the *A. verrucosa–V. hystricosus* (VH) Oppel Zone.

Unfortunately, the lower and middle Famennian interval has not been sufficiently documented in North Africa and the Middle East. Results from the Hammadah Basin (Massa and Moreau-Benoit, 1976) and the Al Kufrah Basin (Grignani *et al.*, 1991) are disputable because of the absence of diagnostic Famennian miospore species amidst the described assemblages and the lack of any faunal control. By contrast, a late Famennian (Fa2c) palynoflora was recorded in the Illizi Basin (Boumendjel *et al.*, 1988) and Cyrenaica (Paris *et al.*, 1985; Streel *et al.*, 1988).

In the northern Brazilian basins index species of latest Frasnian to middle Famennian Ardenne-Rhenish palynozones are rare or unknown, even along potentially coeval, carefully investigated rock sections. Possible causes of this problem may include lithologies unsuitable for miospore preservation, biostratigraphic gaps related to sedimentary condensation, or the scarcity of contemporary vegetation cover in those high latitude areas due to climatic constraints. Therefore no Western European-defined miospore zones within this stratigraphic interval have been individualized on an indisputable basis. Sedimentary sections of same age are seemingly absent in the Paraná Basin.

The successive inceptions of at least three reliable diagnostic species can be taken into account within the latest Frasnian–late Famennian of Western Gondwana areas. These are:

- *Teichertospora torquata*, an eponym of the *T. torquata–G. gracilis* Assemblage Zone of the Old Red Sandstone Continent biozonation not yet recorded in Ardenne-Rhenish regions;

- *Rugospora radiata*, an eponym of the *R. flexuosa–G. cornuta* Assemblage Zone of the Old Red Sandstone Continent and which first occurs at the base of the *D. versabilis–G. cornuta* (VCo) Oppel Zone in Ardenne-Rhenish regions; and
- *Vallatisporites hystricosus*, the first occurrence of which characterizes the base of the *V. pusillites* (*sensu lato*)–*A. fructicosa* Assemblage Zone in the Old Red Sandstone Continent (Richardson and Ahmed, 1988) and of the *A. verrucosa–V. hystricosus* (VH) Zone in the Ardenne-Rhenish regions (Maziane *et al.*, 1999).

The youngest Devonian characteristic miospore event is the first occurrence, at the base of the latest Famennian or «Strunian», of *Retispora lepidophyta*, a well-known cosmopolitan species which permits accurate chronocorrelation all over the world. Its total range within the «Strunian» is subdivided into three biozones, each of these being characterized by the first occurrence of a diagnostic species, viz.: *Knoxisporites literatus* (LL), *Indotriradites explanatus* (LE) and *Verrucosisporites nitidus* (LN).

The «Strunian» interval has been palynologically investigated in various areas of North Africa (Lanzoni and Magloire, 1969; Attar *et al.*, 1980; Massa *et al.*, 1980; Grignani *et al.*, 1991) and the Middle East (Coquel *et al.*, 1977) giving rise to different regional biozonations based on miospore assemblages. Their comparison with the biozones defined in northern Brazilian basins (Loboziak *et al.*, 2000a) suggests that the lower part of the *Retispora lepidophyta* total range may be missing in these regions, although an assignment to the LL Zone has been proposed by Streel *et al.* (1988) for several samples from well A1-37 in Cyrenaica.

In northern Brazilian basins most of the latest Famennian miospore assemblages contain *Indotriradites explanatus* and a few other characteristic species, mainly *Vallatisporites verrucosus* and *Vallatisporites vallatus*, which first occur only in the upper subdivisions of the *Retispora lepidophyta* range. Therefore, the LL Zone could not be identified in any of the investigated sections and may be entirely absent in northern Brazil. This indicates a biostratigraphic, and probably also lithological, gap between at least the VH Zone (or still, older parts of the underlying VCo Zone) and the base of the section containing *Retispora lepidophyta*, which is further characterized by massive reworking of Middle Devonian and Frasnian palynomorphs. Due to the scarcity, or even local absence, of *Verrucosisporites nitidus* in most of the studied sections the *Retispora lepidophyta* range cannot be accurately subdivided. Therefore a comprehensive LE–LN zonal attribution is preferably proposed for this entire interval.

In the subsurface of the Paraná Basin, Loboziak *et al.* (1995a) reported a LE-LN palynoflora from diamictites. This may document the first *in situ* occurrence of «Strunian» miospores in that basin. Diamictites and other glacio-marine sediments with same age are widespread in northern Brazilian basins (Loboziak *et al.*, 1992b, 1993, 1995b, 1996, 1997b, 2000b; Grahn *et al.*, 2001; Streel *et al.*, 2000b, 2001; Melo and Loboziak, 2001, 2003) and also occur in Bolivia, where they have been dated by means of their miospore and acritarch content (Pérez-Leyton, 1991; Vavrdová *et al.*, 1991, 1993, 1996; Díaz *et al.*, 1993a–b, 1999).

## 4.7 Applications of Devonian Palynology to Petroleum Geology: Some Western Gondwanan Examples From South America

Palynology constitutes a powerful auxiliary tool to hydrocarbon exploration in a variety of ways. Of these the most important concerns biostratigraphy, involving primarily the palynomorph-based subdivision, dating, and correlation of sedimentary rocks. However, palynology can also provide critical information on the thermal evolution of organic matter and hydrocarbon generation, the reconstruction of sedimentary environments, and the geological history of sedimentary basins. Most of these matters have already been sufficiently covered in recent works (Tyson, 1995; Batten, 1996a, b; McGregor *et al.*, 1996), and so, only three examples concerning the palynostratigraphic application of Devonian organic-walled microfossils to the hydrocarbon exploration in sedimentary basins of South America are provided next.

### 4.7.1 Subsurface Correlation of Hydrocarbon Reservoirs in Southern Subandean Bolivia

In structurally complicated Paleozoic basins, complex faulting, folding and deformation may affect inextricably siliciclastic sequences characterized by monotonous, recurrent pelite/sandstone successions. Under these circumstances, seismic resolution becomes very limited, and palynostratigraphy eventually provides a most reliable tool for the recognition and tracing of exploratory target intervals from one well to another. This was the case in an area of southern Subandean Bolivia drilled for gas by a pool of oil companies including PETROBRAS. There, two sandy intervals with very similar lithology but quite distinct ages (and palynological content) – ascribed to the Devonian Santa Rosa and Huamampampa formations – were easily confused with each other on seismic sections. Only one of them included the main productive reservoir, and poorly understood lateral shifts of shaley–sandy lithofacies further complicated the issue. Fortunately the biozonal succession in a few critical wells was soon elucidated, based on accurate palynological correlation with carefully analysed, structurally less complicated reference sections in the nearby Devonian outcrop belt. The palynological signature and age of the main rock units – including the target interval – could be then established (Melo, 2000), and improved correlations provided a basis for new interpretations of regional facies relationships in the light of sequence stratigraphy concepts (Miranda *et al.*, 2000; Souza Cruz *et al.*, 2000). Palynological support has proved crucial to structural and stratigraphic interpretations in the area, and therefore to drilling decisions involving several millions of dollars.

### 4.7.2 Integrated Palynological–Geochemical–Clay Mineral Characterization of Potential Source Rocks in the Devonian of Parnaíba Basin, Northern Brazil

High-resolution palynostratigraphy can be coupled with analytical techniques such as organic geochemistry and clay mineralogy to achieve a detailed characterization and

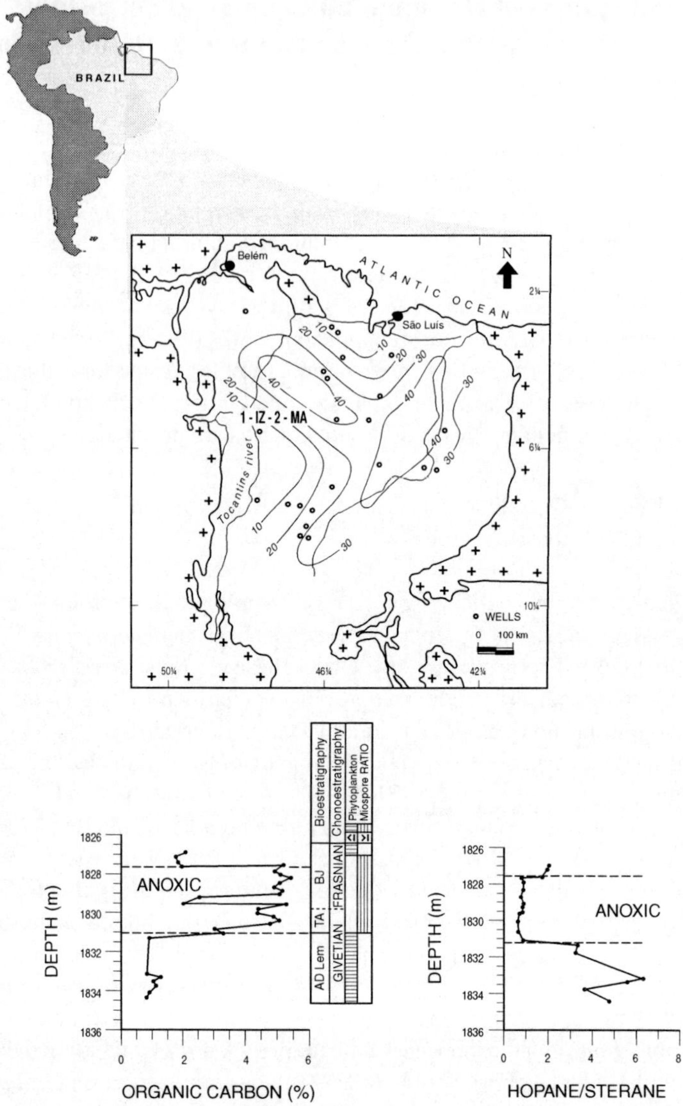


*Figure 4.6 Integrated miospore biostratigraphy, geochemistry, and clay mineralogy data from Devonian potential hydrocarbon source rocks in well 1-IZ-2-MA, Parnaíba Basin, northern Brazil, after Rodrigues et al. (1995), modified. The Parnaíba Basin map shows the location of the studied well and isolith curves (in meters) for the Frasnian radioactive shale interval.*

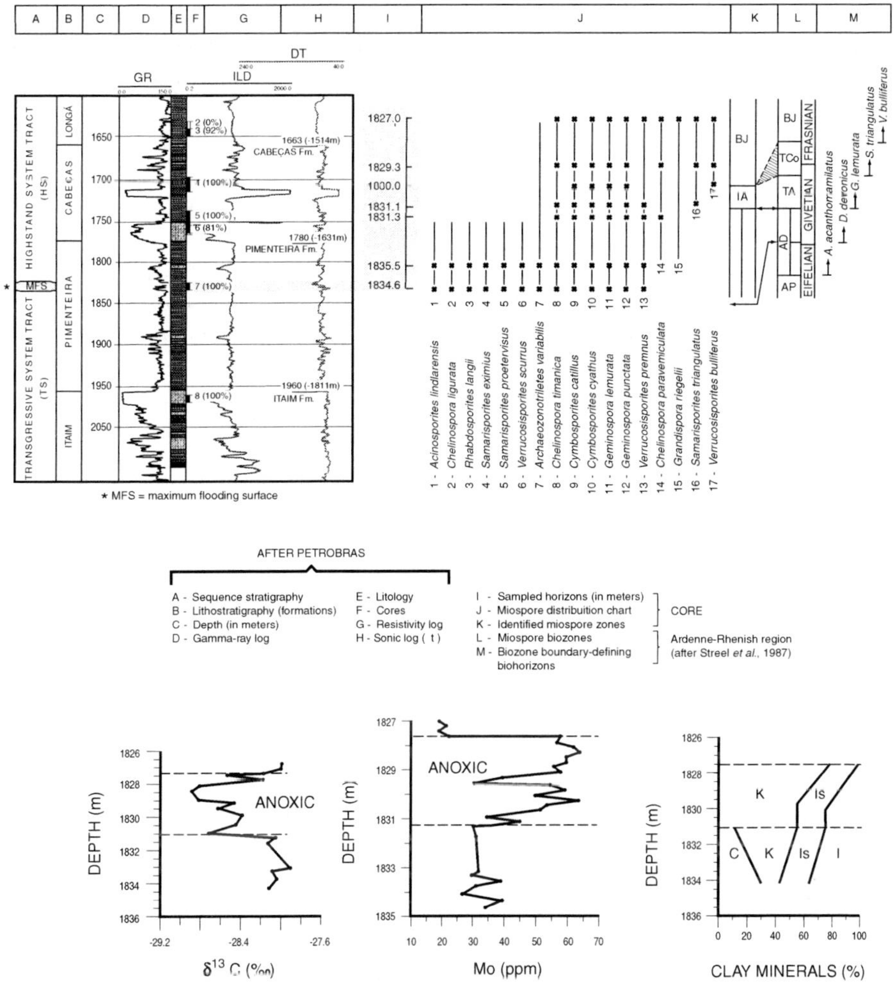


*Figure 4.6 Continued*

subdivision of potential hydrocarbon source rocks. This is well exemplified by Rodrigues *et al.*'s (1995) study of Late Devonian black shales belonging to the Pimenteira Formation (Parnaíba Basin, northern Brazil). Based on a carefully sampled core section from well 1-IZ-2-MA (Imperatriz nr.2), those authors integrated miospore biostratigraphy with varied geochemical bulk analyses (total organic carbon, total sulphur, hydrogen index), biomarker and carbon isotopic analyses, trace element data and clay mineralogy, in order to characterize the early stages of the Frasnian anoxic event in the western part of the Parnaíba Basin (Figure 4.6). Bands of sedimentary condensation, originated under distal marine settings, and detected palynologically in the form of apparent biostratigraphic gaps ("missing" miospore biozones) and high microphytoplankton concentrations, were shown to coincide with maximum flooding surfaces and increased anoxia based on organic geochemical data. Such integrated high-resolution studies are valuable in the accurate delimitation of intervals with highest hydrocarbon potential within lithologically uniform black shale sections.

### 4.7.3 Palynological Dating of Frasnian–Famennian Anoxia and Glacio-Eustatic Events in the Amazon Basin of Northern Brazil, and their Bearing on Basin Modelling

In higher-paleolatitude Devonian basins such as Brazilian ones, traditional high-resolution biostratigraphic tools like conodonts are very scarce or absent due to climatic constraints. Under these circumstances, the recognition of suitable palynozones (such as Western European-defined miospore zones, which are conodont-calibrated in the type marine Devonian of Ardenne-Rhenish areas) provides the best way to estimate the duration of regional sedimentary cycles. These data are critical to calculations of sediment accumulation rates and other basin modelling procedures followed in the oil industry. An example from the Upper Devonian of northern Brazil is given next.

The Late Devonian Barreirinha Formation of the Amazon Basin is an essentially pelitic unit which comprises a moderately thick (less than 100 m in average), lower section of highly radioactive, well-laminated, organic-rich black shales, and a usually thicker upper section of far less radioactive and fissile, mainly dark grey shales. The lower Barreirinha sub-unit represents the main potential hydrocarbon source rock of the Amazon Basin. The formation is overlain by siltstones, sandstones and diamictites belonging to the Curiri Formation (although some diamictites have also been partlly attributed in older PETROBRAS reports to the succeeding Oriximiná Formation). Altogether, the Barreirinha and Curiri Formations depict a broad transgressive–regressive cycle in the Upper Devonian of the Amazon Basin. According to earlier palynological accounts (Daemon and Contreiras, 1971; Daemon, 1974), the two units were loosely correlated with the entire Frasnian and Famennian respectively, and no significant stratigraphic gaps were recognized between or inside them. In terms of the newest Late Devonian time-scale based on conodont biochronology (Sandberg *et al.*, 1997 *apud* Streel *et al.*, 2000b, text-fig. 35), this could imply supposedly continuous sedimentation along a time-span of up to 15 Ma, with no additional information on sedimentary rate variations.

The latest miospore-based assessments of the Barreirinha and Curiri Formations (Loboziak *et al.*, 1996, 1997a–b; Loboziak and Melo, 2000; Melo and Loboziak, 2001, 2003) now permit more improved estimates of the relative duration and intensity of Late Devonian sedimentary pulses in the Amazon Basin. The lower Barreirinha unit is demonstrated to correspond to a sustained anoxic phase characterized by very diluted, distal marine, organic-rich clay sedimentation, punctuated with several non-depositional (condensed sedimentation) episodes of variable duration. If Loboziak *et al.*'s (1997a) correlation of the entire lower Barreirinha with parts of the TA/BJ – GF miospore zonal range of Western Europe is accepted, and using Streel *et al.*'s (2000b, text-fig. 35) equivalence between the Late Devonian miospore and conodont zonal schemes as a standard, then a time-span of up to 10 Ma may have elapsed between the onset of the lower Barreirinha sedimentation and that of the upper Barreirinha unit. Both this latter and the overlying lower Curiri unit *sensu* Loboziak *et al.* (1997a–b) (i.e., the diamictite-free part of the formation that also includes the *Protosalvinia* Zone) display a regressive, roughly shallowing-upward character (Melo *et al.*, 1996; Melo and Loboziak, 2003). Although invariably thicker than the lower Barreirinha, the upper Barreirinha–lower Curiri package corresponds to only the VCo–VH miospore zones of Western Europe, thus pointing out to a ca. 3 Ma phase of highly intensified marine sedimentation. A minimum stratigraphic gap of only about 0.5 Ma (corresponding to the regional absence of the LL miospore Zone in the Amazon Basin) separates the top of that section from overlying glaciogenic strata of the upper Curiri unit. This latter is usually characterized by the presence of diamictites, extensive reworking of pre-"Strunian" palynomorphs, and attains considerable thickness along the basin margins (often within the 75–200 m range, occasionally even more). The upper Curiri unit records massive, glacially driven erosion and resedimentation throughout the basin during a time-span no longer than ca. 1.0 Ma, which is the combined duration of the LE and LN miospore Zones (Streel *et al.*, 2000b, text-fig. 35). Its top is characterized by a renewed flooding of the basin very close to the Devonian/Carboniferous boundary (the LN/VI–HD transition, after Melo and Loboziak, 2003).

## 4.8 Acknowledgements

The authors are indebted to Dr. Alain Le Hérissé (Université de Bretagne Occidentale, Brest, France) and Dr. Yngve Grahn (Universidade do Estado do Rio de Janeiro – UERJ, Rio de Janeiro, Brazil) for the critical review of the manuscript and their many invaluable suggestions which led to a much improved work.

Chapter 5

# Carboniferous and Permian Palynostratigraphy

GEOFFREY PLAYFORD[1] and RODOLFO DINO[2]

[1] Department of Earth Sciences, The University of Queensland, Brisbane 4072, Australia.

[2] PETROBRAS-CENPES, Cidade Universitária, Quadra 7, Ilha do Fundão, 21949-900 Rio de Janeiro, Brazil; Universidade do Estado do Rio de Janeiro (UERJ), Faculdade de Geologia/DEPA, Maracanã, 20559-013 Rio de Janeiro, Brazil.



## 5.1 Introduction

From an historical perspective, early investigations (conducted during the late 19$^{th}$ and early to mid 20$^{th}$ centuries) of the spore–pollen contents of upper Paleozoic terrestrial sedimentary strata can be viewed as having provided the impetus, indeed the essential foundations, for paleopalynological studies of an extensive range of pre-Quaternary sequences. This, in turn, has led to the burgeoning status appertaining to paleopalynology, especially from the mid 1940's onwards. Accordingly, in concert with significant improvements in optical microscopes and the development of electron microscopy, paleopalynology has become widely acknowledged as a vital and uniquely important micropaleontological discipline. Its major applications, of particular importance in



*E.A.M. Koutsoukos (ed.), Applied Stratigraphy,* 101-121.

geological exploration, are undoubtedly in biostratigraphic zonation of Phanerozoic and, to a lesser extent, Proterozoic successions of both terrestrial and nearshore marine origin. Moreover, in the case of the facies-independent spores and pollen grains – whose distribution effectively transgresses otherwise impeditive nonmarine/marine biofacies boundaries – precise chronostratigraphic correlations can be established between Silurian and younger sequences deposited under such disparate circumstances.

Upper Paleozoic palynological investigations were initially centered in Western Europe, most notably in the Upper Carboniferous (Pennsylvanian) coal-bearing sequences of Britain and Germany (see historical reviews by Smith and Butterworth, 1967; Jansonius and McGregor, 1996; Owens, 1996). Similar groundbreaking studies were conducted on coeval coals of North America, as summarized by, *inter alia*, Phillips *et al.* (1973) and Peppers (1996). These North American researches in early decades of the $20^{th}$ century can be viewed as culminating in the publication by Schopf *et al.* (1944), which contains the first comprehensive taxonomic treatise on Carboniferous spores (and spores generally). In the Gondwana realm of the southern hemisphere and peninsular India, pioneering spore–pollen investigations commenced somewhat later, and focused more broadly on Pennsylvanian-basal Permian "pre-glacial" and glacigene strata and on stratigraphically succeeding Permian coals and associated deposits; e.g. in Australia (Dulhunty, 1945, 1946; Balme, 1952; Balme and Hennelly, 1955, 1956a, b), India (Virkki, 1946; Bharadwaj, 1962), southern and central Africa (Hart, 1960, 1965a, b), Antarctica (Balme and Playford, 1967), and South America (Sommer, 1953; Menéndez, 1965; Daemon, 1966; Azcuy, 1975).

From the 1960's onwards, spores and pollen grains have become increasingly utilized as a prime or ancillary basis for local, regional, and even international chronostratigraphic correlations of Carboniferous and Permian sequences. The essential focus of the present account is, thus, palynostratigraphic. However, it would be remiss not to cite here other important applications of upper Paleozoic palynological analyses; among them, paleoecological and paleogeographic syntheses, paleofloristic reconstructions, source-rock analyses, thermal maturation histories, and provenance studies.

## 5.2 Carboniferous and Permian Palynomorphs: A Synopsis

Preserved, often very profusely, in Carboniferous–Permian sequences that may otherwise appear virtually unfossiliferous, palynomorphs are represented chiefly by an extensive array of small spores and pollen grains (miospores); to a lesser extent by megaspores and microphytoplankton (acritarchs and prasinophytes); and, very rarely, by scolecodonts and chitinozoans. These organic-walled, essentially acid-resistant microfossils can be extracted readily, for detailed study and identification, by means of relatively simple, physico-chemical laboratory procedures (e.g. Phipps and Playford, 1984; Traverse, 1988; Wood *et al.*, 1996). Routine examination of slide-mounted, concentrated palynological residues containing the smaller palynomorphs (miospores, microphytoplankton) is normally and satisfactorily conducted by means of a high-powered stereo-binocular microscope equipped with transmitted light source. Larger, semi-opaque palynomorphs (megaspores, chitinozoans, scolecodonts) are normally studied in the first instance by relatively low powered light microscopy using incident light. For investigation of the often intricate detail of palynomorph morphology, which may have taxonomic importance or implications, scanning- and transmission-electron microscopy can very usefully augment light-microscopic observations.

### 5.2.1 Miospores

As defined by Guennel (1952, pp. 9-10), the term miospore incorporates "all fossil spores and spore-like bodies smaller than 0.20 mm, including homospores (= isospores), true microspores, small megaspores, pollen grains, and pre-pollen". Guennel acknowledged the "arbitrary" nature of this grouping, in particular the <200 μm size limitation. Nonetheless, as noted by Traverse (1988, p. 59), the term is reasonably appropriate for accommodation of functional microspores as distinct from functional megaspores, although some of the latter are <200 μm and some pollen grains may (uncommonly) exceed 200 μm.

#### *5.2.1.1 Small Spores*

In the absence of a definitive term for spores, as distinct from pollen grains, that fit into the miospore size category, the term "small spores" as used by some previous authors (e.g. Kosanke, 1969) is adopted herein. It should be noted, however, that there seems to be a tendency among a few paleopalynologists to apply the term miospore exclusively to such small spores (i.e. to the exclusion of pollen); this, of course, is contrary to Guennel's (1952) original inclusive definition.

Small spores are represented abundantly in upper Paleozoic sequences. They are overwhelmingly predominant components of Mississippian and older Paleozoic terrestrial palynofloras, but are joined by increasing representation of pollen grains, including so-called prepollen (Traverse, 1988, pp. 56, 110), in the later Carboniferous and Permian. The spores were produced by a considerable range of lower vascular plants (seedless metaphytes), chiefly lycopods, "horsetails" (sphenophytes), and ferns. Of these, the first two groups attained evolutionary peaks – in respect of numbers, diversity, and stature (spectacular arborescence in some cases) – during the Carboniferous, when they were conspicuous components of forest and "coal swamp" communities, especially in the Euramerican region.

Because the botanical affinities of these plant microfossils are often very difficult to ascertain in other than very generalized terms, they are named and treated taxonomically on the basis of morphological features that are, in the main, readily discernible and definable via conventional light microscopy. Smith and Butterworth (1967), Traverse (1988), and Playford and Dettmann (1996) provided synopses of spore-morphological attributes and outlined the currently accepted "turmal" classificatory scheme, which is acknowledged as being inherently artificial but may, in part, reflect genuine botanical affiliations.

Upper Paleozoic assemblages of small spores comprise mainly radiosymmetric trilete forms (with a three-rayed tetrad mark functioning as the germinal aperture), but bilaterally symmetrical, bean-shaped spores bearing a single-rayed (monolete) tetrad mark are often plentiful in Pennsylvanian and Permian deposits. The layering, structures, and sculpture of the spore wall (exine) vary considerably, with the result that a very substantial number of form genera and form species have been defined. Many of the latter are short-ranged and hence important stratigraphically. In addition to simple spores with essentially unmodified, single-layered (acavate) exines, morphologically more complex forms may exhibit two-layered exines (as in cavate/pseudosaccate spores), equatorial structures such as cingula, auriculae, or coronae, and a range of sculptural projections or indentations (see, for example, representative photomicrographs in such publications as Smith and Butterworth, 1967; Hart, 1969; Kosanke, 1969; Clayton *et al.*, 1977; Utting, 1996, pls. 1, 2).

*5.2.1.2 Pollen Grains*

Prior to the inception of the angiosperms (true flowering plants) at the beginning of the Cretaceous, producers of pollen grains were exclusively gymnosperms, the more primitive category of seed plants that had their origins in the Devonian. During the late Paleozoic, such gymnospermous groups as seed ferns (pteridosperms), cordaitaleans, and primitive coniferaleans became increasingly important elements of the land vegetation and hence significant contributors to terrestrial palynofloras. The morphological characteristics of Carboniferous–Permian pollen grains, constituting the basis of their form taxonomy and classification, are diverse, as will be evident from illustrations in, for example, Traverse (1988, Fig. 9.6), Owens (1996, pl. 2), and Utting (1996, pls. 1, 2).

In gross terms, many of these gymnospermous pollen grains possess a prominent central body (corpus) which is enveloped by a relatively thin, more or less inflated, bladder-like membrane (saccus) in the case of monosaccate forms. Or the corpus may be flanked by a pair of sacci (one at each "end" of the corpus), as in typical bisaccate pollen grains. Relative sizes of corpus and saccus or sacci are varied among different form categories and this is manifested especially at form-generic level. Moreover, the corpus itself may be striate or taeniate, with relatively thick, parallel, band-like strips of exine separated by thinner clefts of intervening exine. In what appears to be a more or less globally synchronous series of palyno-events during the Pennsylvanian through Permian interval, monosaccates became subordinate to non-taeniate bisaccates, and these in turn to taeniate bisaccates, with the latter attaining conspicuous dominance in the Permian especially in Gondwana (and persisting, albeit declining progressively, through the Triassic). The other principal (and non-saccate) group of pollen grains represented in late Paleozoic palynofloras are more simply constructed ellipsoidal unilayered forms. These are mostly monocolpates in that they characteristically feature a distinct longitudinal germinal aperture termed a colpus (or sulcus).

### 5.2.2 Megaspores

Knowledge of the megaspore fossil record has been derived largely from coals and associated richly carbonaceous deposits, because of their relative abundance therein. Megaspores, being relatively large palynomorphs, have a much lesser dispersive capability than miospores, tending to be restricted mostly to depositional environments proximate to growth sites of their parental plants and hence to essentially autochthonous, organic-rich rocks. However, they do occur, albeit sporadically, in somewhat less carbonaceous, clastic deposits. These latter are normally processed in the laboratory primarily for their miospore content by methods that are not conducive to recovery of other than fairly low numbers of fragmented megaspores. Successful retrieval of intact megaspores from their host rocks obviously involves application of more delicate preparation techniques, including careful sieving, as described by Traverse (1988, pp. 473–475) and Scott and Hemsley (1996, p. 629). Similar precautionary measures are necessary for recovery of chitinozoans (see 5.2.5, below).

Megaspores are and were (from Early Devonian times) produced during the life cycles of heterosporous, lower vascular plants (pteridophytes); viz., some lycopods and ferns, also extinct progymnosperms. In contrast to homosporous pteridophytes which shed only one kind of spore (homospores, a.k.a. isospores, classified paleopalynologically as miospores), the heterosporous pteridophytes produce, on an individual plant, both microspores (of the miospore category) and the larger megaspores (mostly $>200$ μm in diameter). Progressing from the Devonian initiation of heterospory, the heyday of heterosporous pteridophytes – in particular the lycopods – was undoubtedly during the Carboniferous. Thereafter they declined quantitatively and qualitatively, although megaspores do occur, sometimes plentifully, in younger deposits. In present-day floras, megaspores are produced freely by only a few lycopods and ferns.

The morphology of megaspores, like that of small spores, is considerably varied and similar morphological terms (and form-classificatory tenets) are applicable to fossil representatives of both groups. However, some megaspores do exhibit certain structures not encountered among their smaller relatives, necessitating the use of additional terms. Moreover, in conjunction with observations from optical and scanning electron microscopes, transmission electron microscopy is being applied increasingly for determination of exine ultrastructure, which in turn provides insights into botanical affiliations (Scott and Hemsley, 1996).

### 5.2.3 Organic-Walled Microphytoplankton

Conventionally included in this broad palynomorph category are acritarchs, prasinophyte phycomata, and dinoflagellates ("dinocysts"). The last-named are abundant and diverse protists, mainly though not exclusively marine. Because they are essentially post-Paleozoic, flourishing from the Jurassic onwards, they are outside the scope of the present account. On the other hand, acritarchs and prasinophytes do merit consideration here because of their presence in Paleozoic deposits.

*5.2.3.1 Acritarchs*

The Acritarcha was established by Evitt (1963, p. 300) as an informal taxonomic "group" consisting of a heterogeneous, polyphyletic, "catch-all" conglomeration of cysts or cyst-like microfossils whose natural affinities are uncertain or, at best, equivocal. Evitt recommended that they be treated taxonomically in accordance with the International Code of Botanical Nomenclature, and this has been universally accepted. Clearly, the majority of these enigmatic microfossils are of eukaryotic–algal origin (Martin, 1993), probably representing the resting cysts of various algal groups. Acritarchs range from the Proterozoic through to the present day, and although predominantly marine, many brackish- and fresh-water forms have been reported. They reached their acme, in terms of abundance and diversity, in early to mid Paleozoic times, but declined sharply from the latest Devonian onwards and are relatively uncommon post-Permian.

Most acritarchs are less than 100 μm in diameter (though some attain several hundred micrometers). They possess a highly resistant wall (eilyma) composed of a sporopolleninous organic compound, and consequently are extracted readily from sedimentary rocks by applying essentially the same techniques as are used for miospore preparations. Being mainly marine in occurrence, they commonly co-occur with miospores in nearshore marine deposits, but predominate over the latter progressively offshore. The basic morphology of acritarchs is relatively simple: a hollow central body or vesicle that comes in various shapes (and sizes) and comprises a one- or, less commonly, two-layered, transparent or translucent wall. Projecting from the eilyma, itself smooth or finely sculptured, there may be hollow or solid processes of diverse shapes, sizes, and distribution, that may (if hollow) communicate with the vesicle interior. The eilyma may possess a consistent slit or other type of "excystment" opening. These and other form features of acritarchs – as discussed and illustrated by, *inter alia*, Martin (1993) and Strother (1996) – provide an objective basis for the recognition of form genera and form species (e.g. Fensome *et al.*, 1990, 1991). A suprageneric (*ipso facto*, phenetic) classification of acritarchs was proposed by Downie *et al.* (1963) but is not widely adopted. Taxonomic accounts normally incorporate photomicrographs from both optical and scanning electron microscopy.

*5.2.3.2 Prasinophytes*

Prasinophytes are generally regarded as primitive unicellular green algae. Some authors have classed them – paleopalynologically – as acritarchs (e.g. Playford and Dring, 1981; Knoll, 1996, p. 73), but features such as the microstructure and surface modifications of the wall of these cyst-like, organic-walled microfossils suggest, though not always unequivocally, that they represent phycomata of prasinophycean green algae (Colbath and Grenfell, 1995). Thus interpreted, prasinophytes extend from the Proterozoic through all of Phanerozoic time, during which they have evidently been relatively stable in terms of their fundamental morphology and their successful adaptation to a range of aquatic (marine, brackish, fresh-water) environments (Guy-Ohlson, 1996). In palynological residues from marine Paleozoic rocks, prasinophyte phycomata are often reasonably plentiful but are generally subordinate, quantitatively and qualitatively, to acritarchs.

### 5.2.4 Scolecodonts

Scolecodonts are the disjunct, fossilized, pseudochitinous mouthparts of marine segmented worms (polychaetous annelids) exhibiting considerable morphological variation. They occur, sometimes quite profusely, in association with other organic-walled microfossils (acritarchs, prasinophytes, chitinozoans) in palynological residues extracted from marine Paleozoic rocks. Scolecodonts have received comparatively little detailed study, not only because of their generally lesser abundance than their palynological associates, but also because of inherent taxonomic problems (single vs. multielement) not unlike those that have plagued the studies of conodonts in past decades. Scolecodonts range from the Early Ordovician and were most common and diverse in Paleozoic times. They offer some potential, as yet unrealized but possibly limited, in biostratigraphy. Known environmental sensitivities of modern polychaetes suggest that they may ultimately prove efficacious as paleoecological indicators (Szaniawski, 1996).

### 5.2.5 Chitinozoans

Like scolecodonts, chitinozoans are relatively large, morphologically varied, and exclusively marine palynomorphs, with a durable organic wall composed of pseudochitin. As with megaspores and scolecodonts, the successful procurement of chitinozoans from enclosing rock necessitates application of more delicate laboratory procedures (e.g. Traverse, 1988, pp. 473-475; Miller, 1996, pp. 329–330) than those employed for retrieval of smaller palynomorphs.

The vesicle (a.k.a. test) of chitinozoans is typically bottle- or urn-shaped, sometimes tubular, with considerable size and sculptural variation; most are *ca.* 150–300 μm in length. In an authoritative account, Miller (1996) concluded that, although found either singly or aggregated, all chitinozoans probably developed originally as colonial groupings (e.g. Paris and Nõlvak, 1999, fig. 2) and were probably mostly planktic. A continuing and seemingly intractable problem pertains to the biological affiliations of chitinozoans. Miller (1996, pp. 315–318) and Paris and Nõlvak (1999, pp. 318–320) have furnished useful commentaries on this intriguing topic. Paris *et al.* (1999) inclined towards Grahn's (1981) hypothesis that chitinozoans are the eggs of soft-bodied metazoans (so-called chitinozoopharans).

Chitinozoans are studied routinely with reflected light binocular microscopy and, more definitively, using scanning electron microscopy (see, for example, Miller, 1996, pls. 1–5; Paris, 1996, pls. 1–3).

Some authors consider that chitinozoans range from Cambrian (even Precambrian) through Permian (e.g. Traverse, 1988, p. 5; but cf. his fig. 1.2). Miller (1996), Paris (1996), and Paris *et al.* (1999) are, however, in unison that bona fide, in-situ chitinozoans are unauthenticated from either pre-Ordovician or post-Devonian rocks. In any event, reports of chitinozoans from Carboniferous and Permian rocks are so rare that, even if in situ, they have no or negligible biostratigraphic importance; and some, at least, have evidently been recycled.

## 5.3 Stratigraphic Applications

From the above discussion, it will be evident that small spores and pollen grains (i.e. miospores), representing the land flora, are the palynomorphs that have greatest significance in Carboniferous–Permian stratigraphy. They form the basis of numerous zonation schemes that are, with varying degrees of precision, applicable to dating and correlation of terrestrial and also marine successions in many parts of the world. Indeed, as mentioned previously herein, the burgeoning development of paleopalynology, in particular palynostratigraphy, over the past five or six decades owes much to the early investigations of spore–pollen palynofloras retrieved in multifarious profusion from upper Paleozoic strata.

The practical effectiveness of fossils in biostratigraphic correlation depends on the formulation of clearly defined biostratigraphic units, known collectively as biozones (or palynozones where based on palynomorphs). A review of Carboniferous–Permian palynological literature – or at least of those works having some stratigraphic emphasis or aspiration – shows that a diversity of criteria have been utilized in the definition of miospore-based biostratigraphic units, which moreover are sometimes informally labelled (i.e. with numbers and/or letters). While conformity with international codes of stratigraphic (including biostratigraphic) nomenclature (Hedberg, 1976; Salvador, 1994) is clearly desirable, the efficacy of those palynozonations which fail to so conform is not, in general, significantly compromised. Many palynozones fall into the category of assemblage zones; others are essentially interval zones, concurrent range zones, or taxon range zones. FADs and LADs (first- and last-appearance datums of selected taxa) provide important boundary criteria for most zonal units (e.g. Peppers, 1996; Price, 1997). Some palynozones are based (desirably) on specified reference sections; others are not clearly or objectively designated in that respect.

Dating of palynozones with respect to the relative geological timescale obviously depends on reliable correlation with marine sequences bearing age-definitive faunas. Most palynozones embody largely terrestrial stratigraphic intervals (i.e. essentially devoid of marine fossils), thus posing considerable hindrances to their precise dating. However, the zones may be traceable laterally into reliably dated marine sections; moreover, marine intercalations and radiometrically datable igneous rocks within predominantly terrestrial palyniferous sections often provide vital chronostratigraphic information or tie-points, as do magnetostratigraphic data from both sedimentary and igneous rocks.

Stratotypes for stages of the Carboniferous and Permian systems are based classically in Europe, and their defining biostratigraphic criteria (fundamentally, but not exclusively, ammonoids, conodonts, and fusulinid Foraminifera) are subjects of ongoing revisionary scrutiny and refinement (e.g. Paproth and Streel, 1984; Lane and Ziegler, 1985; Brenckle and Manger, 1991; Jin Yugan *et al.*, 1997). The reliability of biostratigraphic correlation with the stratotypes, and hence age determination in terms of the international timescale, depends on many separate and interrelated factors, including fossil content, paleogeography, paleoecology, and faunal and floral provincialism. For the Devonian, long-distance correlations based on both marine faunal and terrestrial floral data are more readily achievable than is the case with the later Paleozoic. For instance, from a purely palynological perspective, McGregor and Playford (1993) have demonstrated the applicability of miospore–zonal correlations between Devonian successions of the Old Red Sandstone Continent (Northern Euramerica) and of Australia (Eastern Gondwana); this provides confirmation of previously reported intimations of floristic cosmopolitanism during much of the Devonian period. Particularly striking, as discussed by many authors (more recently, Playford and McGregor, 1993; Streel and Loboziak, 1996), is the fact that the Devonian–Carboniferous boundary is recognizable globally by the exiting of a distinctive and seemingly omnipresent, late Famennian miospore species, *Retispora lepidophyta* (Kedo) Playford, 1976. In the later Paleozoic – more particularly during the later Carboniferous and Permian – regionalism among global floras developed progressively (e.g. Chaloner and Meyen, 1973; Chaloner and Lacey, 1973; Meyen, 1987; Chaloner and Creber, 1988), and is most conspicuously manifest by the Permian development of the highly distinctive *Glossopteris* flora throughout and confined to Gondwana. The inevitable consequence of this is that long-distance palynostratigraphic correlations, particularly Gondwanan–Euramerican, become increasingly less feasible from Mississippian through Permian time.

Ensuing paragraphs are intended to provide a brief and fairly selective (for space limitations, non-comprehensive) coverage of the stratigraphic applications of Carboniferous and Permian miospore assemblages, and with reference to some of the more notable examples of such applications from various parts of the world.

### 5.3.1 Carboniferous Palynostratigraphy

As noted by Higgs (1996, p. 553), the Carboniferous has received the most extensive (and intensive) palynological study of all the Paleozoic systems. This applies most cogently to the British Isles and to mainland Western and central Europe, with particular but by no means exclusive emphasis on the coal-bearing Upper Carboniferous (Pennsylvanian).

Western European Carboniferous miospore assemblages were illustrated and classified into 12 biozones by Clayton *et al.* (1977); these collectively span the upper Visean to the base of the Autunian. Although subsequent studies have produced some modifications to the zonal scheme, it remains an important palynostratigraphic standard for that region and facilitates external correlation, particularly with Pennsylvanian successions in other parts of Euramerica and, to a lesser extent, even beyond that province (e.g. Owens *et al.*, 1978, 1989; Owens, 1996). Clayton (1996) focused on the recognition (initially by Sullivan, 1965, 1967) of palynologically expressed provinces in the Mississippian, especially in the later (Visean) part of the epoch, and discussed the consequences of these for long-distance correlations. Moreover, he presented an updated miospore zonation scheme (Clayton, 1996; Fig. 5.1 herein) for the Tournaisian through lower Namurian succession of Western Europe. Earlier, Riley (1993) had usefully established precise chronostratigraphic

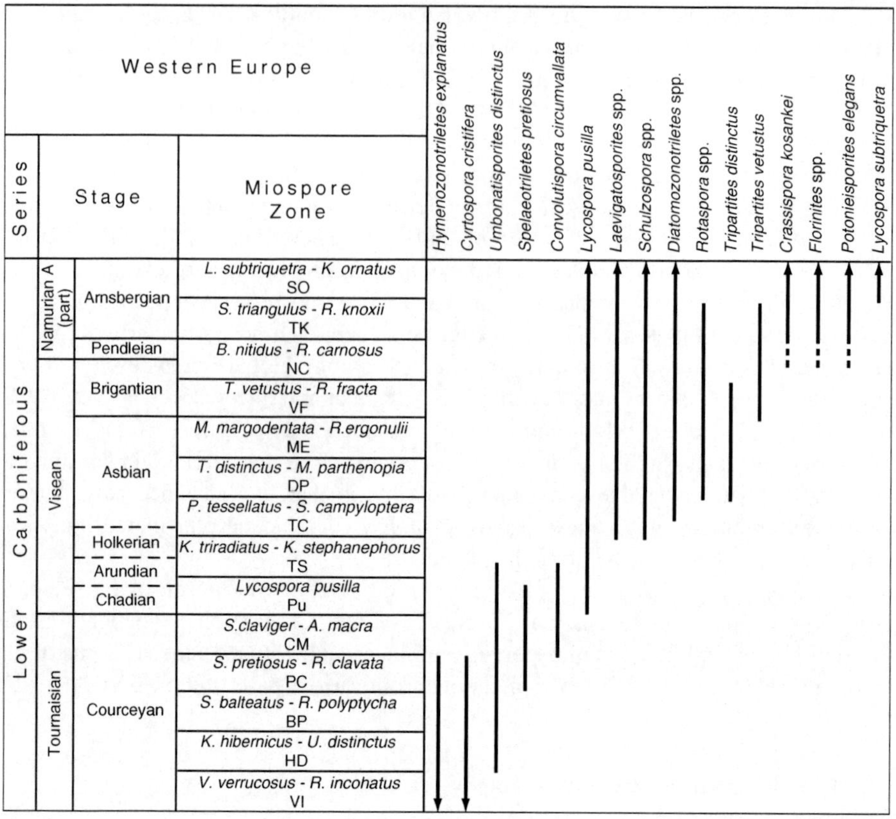


*Figure 5.1 Miospore zonation of the Western European Lower Carboniferous (Mississippian), with ranges of selected diagnostic taxa. Adapted from Clayton (1996, text-fig. 2).*

correlations among British Mississippian zonations based on various marine faunal groups and on miospores.

Many Carboniferous miospore zonations have been promulgated in various parts of the world, some extending into the Permian. They have proven stratigraphically effective in local and regional contexts; e.g. in Russia (Byvscheva, 1974; Peterson, 1999), eastern Canada (Utting *et al.*, 1989), U.S.A. (Ravn, 1986), North Africa (Coquel *et al.*, 1988; Loboziak and Clayton, 1988), Saudi Arabia (Clayton, 1995; Clayton *et al.*, 2000), Australia (Kemp *et al.*, 1977; Playford, 1985; Jones and Truswell, 1992), China (Gao Lianda, 1985), and South America (Playford and Dino, 2000b; Césari and Gutiérrez, 2001). Long-distance chronostratigraphic correlation (e.g. from Gondwana to northern hemisphere regions) based on ranges of certain species is practicable to a limited extent for the Mississippian (e.g. Playford, 1978, 1991). However, with the advent of increasing phytogeographic differentiation of land floras and hence declining miospore cosmopolitanism, international correlations or comparisons on a palynofloral basis becomes much less feasible as the remaining Carboniferous and the Permian are ascended.

A particularly impressive example of extended or interregional palynostratigraphic correlation – in this case embracing Midcontinent U.S.A., Western Europe, and the Donetz Basin of the former U.S.S.R – was furnished by Peppers (1996). While the principal focus of Peppers's research was on the palynology of Pennsylvanian coal-bearing sequences of the Illinois Basin and of other U.S. basins, he was able to detail similarities between his U.S. (Illinois Basin-based) miospore zonation and those established in successions of various parts of the European domain. The basis of his zonal correlations, including vertical ranges of important miospore species in individual successions, is graphically depicted in Peppers (1996, pls. 1, 2), and a synoptic version of his inferred trans-Euramerican chronostratigraphic correlations is presented herein (Fig. 5.2).

### 5.3.2 Permian Palynostratigraphy

Precise palynological demarcation of the Carboniferous–Permian and Permian–Triassic boundaries remains elusive in many parts of the world (e.g. Helby, 1966; Balme, 1970, cf. Foster, 1979, 1982; Balme, 1980a, b; Piasecki, 1984; Gomankov, 1992; Foster *et al.*, 1998; Playford and Dino, 2000b). This uncertainty stems from several factors, including unfossiliferous or non-palyniferous deposits in critical stratigraphic intervals, and, more fundamentally, controversy and equivocation regarding faunal criteria for diagnosing the systemic boundaries in different marine realms and provinces (see Utting and Piasecki, 1995, pp. 239–241). In any case, as discussed below, Permian palynofloras are likewise strongly influenced by regional and global phytogeographic variations. Notwithstanding problems of pinpointing the lower and upper limits of the Permian system, its miospore floras are unmistakably distinctive, in gross character as well as in detail, from those hosted by Carboniferous and Triassic strata. Palynological zonation of Permian successions, with (preferably) or without independent marine–biostratigraphic age constraints, is extensively utilized in sedimentary basin analyses worldwide.

Major alterations in the composition and diversity of land floras around the end of the Carboniferous were triggered by pronounced geological and climatic changes. These changes, of global extent, are manifest paleofloristically by increasing qualitative and quantitative representation of primitive conifers, pteridosperms, and other gymnosperms and by concomitant decline of some cryptogam groups, most notably extinction of the arborescent lycopods that were so prevalent in coal-forming swamp forests of the Pennsylvanian. Moreover, the latest Pennsylvanian–Early Permian was a time of accelerated phytogeographic differentiation. This resulted in the establishment of several distinct, paleolatitudinal, megafloral provinces (which themselves exhibited changing geographic configurations during the Permian: Chaloner and Meyen, 1973; Chaloner and Lacey, 1973; Chaloner and Creber, 1988). In Laurasia, the northerly

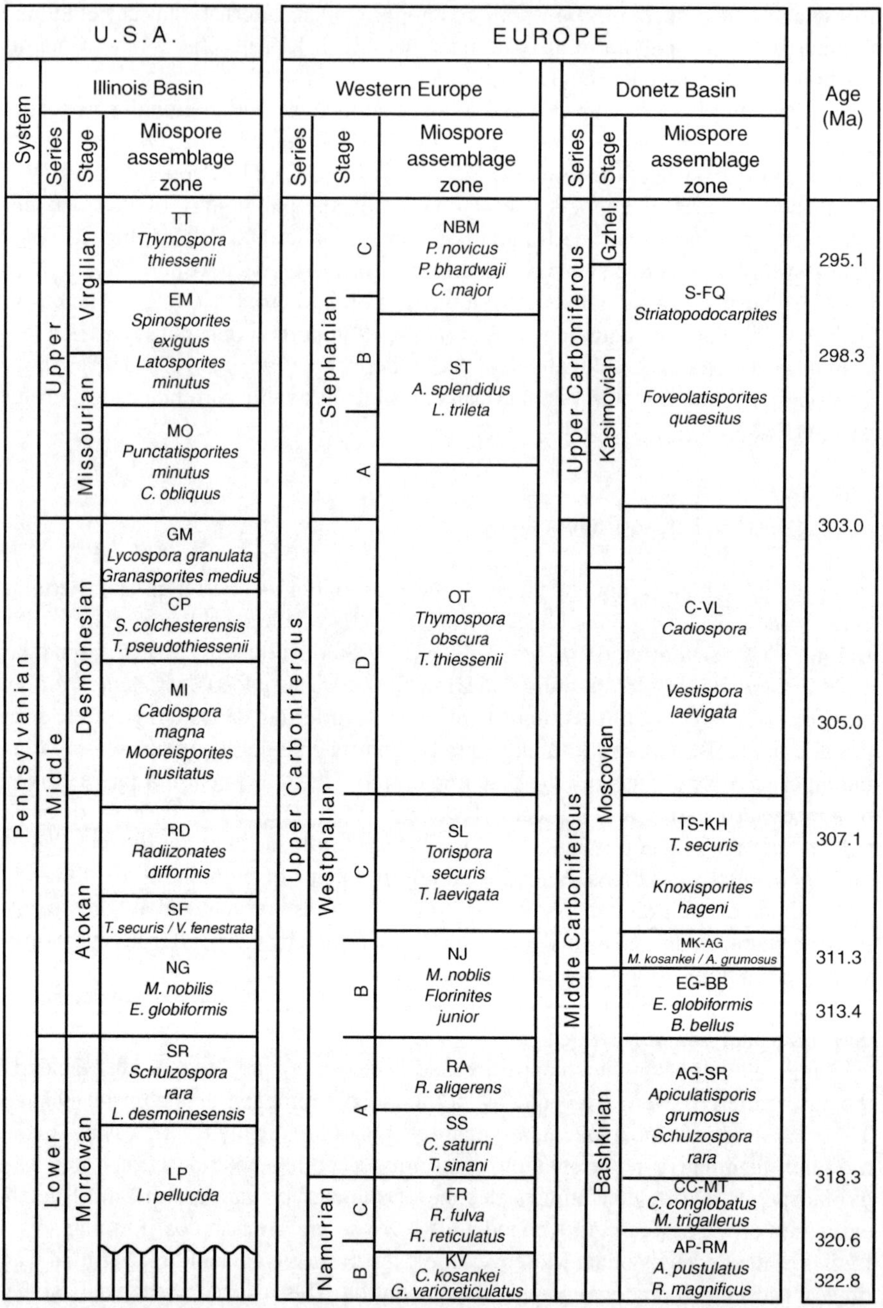


*Figure 5.2 Pennsylvanian (Upper Carboniferous) trans-Euramerican chronostratigraphic correlations based on miospore assemblage zones recognized in Midcontinent U.S.A. (Illinois Basin), Western Europe (principally Britain), and Russia (Donetz Basin). Adapted from Peppers (1996, pl. 1).*

Angara province became well-differentiated (with Siberian, Pechoran, Far-Eastern, and Sub-Angaran subdivisions discernible in the later Permian), as did the equatorial Cathaysia and Euramerica (Atlantic and North American) floral provinces. Largely confined to the vastness of the Gondwana southerly supercontinent is the Gondwana province, which is characterized especially by a highly distinctive pteridospermous (glossopterid-dominated) flora. This floral association became well-established following the waning of the Pennsylvanian–earliest Permian glaciation and proliferated through subsequent Permian time. The temporal and spatial distributions of late Paleozoic floral provinces are summarized by Utting and Piasecki (1995, figs. 1, 2).

Obviously, the conspicuous changes in the terrestrial megafloras - Carboniferous vis-à-vis Permian – are reflected in the miospore palynofloras, i.e. in the assemblages of spore–pollen propagules that are preserved in such abundance and diversity in Permian sedimentary rocks. Furthermore, as would be anticipated, the phytogeographic provinces cited above are mirrored by the distinctive character of the palynofloras preserved in those respective provinces, but with transitional assemblages recorded in some regions, e.g. in North Africa, where the Euramerica and Gondwana provinces are conterminous. Indeed, as noted by Utting and Piasecki (1995, p. 256), miospore floras from otherwise unfossiliferous Permian sequences often facilitate attribution of the latter to a particular floral province, assist in delineating provincial boundaries, and indicate variations in the paleogeographic spread of provinces through Permian time. However, Balme (1970, p. 431) has cautioned that palyno-phytogeographic inferences based on identification of other than well-characterized, morphologically distinctive miospore species may prove erroneous or misleading. He recommended specifically that most form genera should be disregarded in this particular context.

Not surprisingly, therefore, Permian palynostratigraphic correlations are essentially restricted paleogeographically, i.e. provincially bound (Utting and Piasecki, 1995; Warrington, 1996). However, in a broader, virtually global perspective, one can observe, *inter alia*, the proliferation of saccate pollen grains, especially taeniate bisaccates, in part presumably reflecting homeomorphic phenomena involving different parental plant groups in the various provinces (Foster, 1978, 1979; Gomankov *et al.*, 1998). Studies of the miospore palynology of Permian successions have been prosecuted, albeit in varying detail, in all of the aforementioned provinces, usually with no more than passing reference to any associated megaflora and directed primarily towards stratigraphic goals. The publications by Utting and Piasecki (1995) and Warrington (1996) together constitute an impressive compendium of Permian miospore research. Even if sufficient space were available, it would obviously be a redundancy to duplicate here their comprehensive reviews. Quite clearly, as with the Carboniferous and other systems but certainly exemplified by Permian researches, meticulous taxonomic analyses of miospores really does produce impressive stratigraphic-correlative outcomes.

While it may appear somewhat invidious to focus on a particular publication to the exclusion of others warranting similar notice, the work of Backhouse (1991) on the Permian palynostratigraphy of the nonmarine, coal-bearing Collie Basin (southwestern Australia) can justifiably be singled out as a notable example of the application of Permian spores–pollen in local, regional, and intraprovincial (in this case intra-Gondwanan) correlation. An essential ingredient of Backhouse's detailed palynostratigraphic synthesis was his own meticulous and well-illustrated taxonomic analysis (Backhouse, 1991, pp. 257–306, pls. I–XXIV), well-supported by previous Australian Permian systematic–palynological studies (principally Balme and Hennelly, 1955, 1956a, b; Foster, 1975, 1979; Rigby and Hekel, 1977). Backhouse was thus able to formulate a sequence of ten successive miospore zones (mainly referred to as "palynostratigraphic units") embracing the largely subsurface (cored and open-cut) basinal succession (glacigene Stockton Formation and overlying Collie Coal Measures). The zone boundaries were diagnosed from FADs of selected miospore species, which were adopted as zone eponyms; however, the zonal characterizations incorporated the ranges and quantitative representations of many

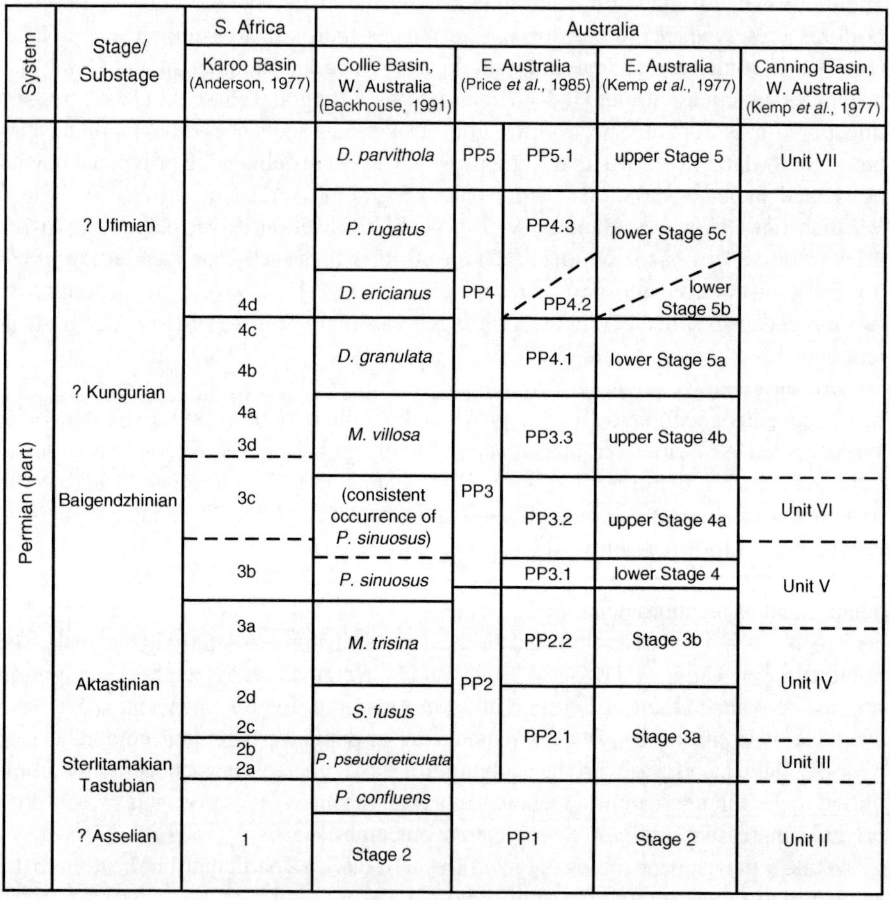


*Figure 5.3 Inferred correlation of Permian palynozonations established in Australia and the Karoo Basin of South Africa. Adapted from Backhouse (1991, fig. 10).*

additional species (Backhouse, 1991, pp. 245–251; figs. 7, 8, 11). The palynozonation facilitated correlation (a) within and among the three Collie sub-basins; (b) with upper Paleozoic sequences of other palynologically documented Australian basins (see also Backhouse, 1993, 1998; Eyles *et al.*, 2002); and (c), elsewhere in southern Gondwana, specifically with South Africa's northern Karoo Basin (see Backhouse, 1991, figs. 9, 10; Fig. 5.3 herein). The age-range of the Collie Basin sequence was adduced as latest Carboniferous/Asselian through early Late Permian, mainly on the grounds of limited marine faunal evidence from correlative strata in other Western Australian basins (Backhouse, 1991, pp. 256–257). Dating of individual zones in terms of the Russian (Urals) Permian standard was (and remains) understandably tentative, as implied by the absence of stage boundaries in the left-hand column of Figure 5.3.

## 5.4 Pennsylvanian–Permian Palynostratigraphy of the Amazonas Basin: A Case History

The intracratonic Amazonas Basin extends over a substantial area – some 500,000 km$^2$ – of northern Brazil. As the basin lies largely within the Amazonas rainforest, outcrops are uncommon or inaccessible, so that knowledge of its stratigraphy and structure has accrued largely from drilling and geophysical operations. The youngest part of the largely Paleozoic sedimentary succession, termed the Tapajós Group or Carboniferous–Permian Supersequence, usually exceeds 2500 m in thickness and comprises four conformable formations defined from cored successions in numerous exploratory wells drilled by Petróleo Brasileiro S.A. (Petrobras). Given that the Tapajós strata are essentially available only as small samples (cores, cuttings) and that they are the products of neritic to continental sedimentation, it follows that palynology has, since the very early phases of basin exploration, provided the major tool for dating and correlating the strata.

The initial studies, produced as unpublished internal reports by Petrobras palynologists, dealt with samples from upper Paleozoic sections encountered in particular boreholes in the Amazonas Basin and in the contiguous Solimões Basin (i.e. the western sector of the Amazonas Basin *sensu lato*). These reports included identifications of selected spores and pollen grains to generic level or, where deemed feasible, to specific level. Descriptions and illustrations were normally not included; and the species, assigned generically, were often given informal alphabetical and/or numerical designations. This company work culminated in the publication by Daemon and Contreiras (1971) which effectively encompassed the entire stratigraphic column of Silurian through Quaternary. They subdivided the upper Paleozoic succession into eight palynostratigraphic units, of which the upper four (termed, in ascending order, Intervals XIII–XVI) embraced the Pennsylvanian–Permian of the basin. These zonal intervals can be regarded generally as assemblage zones, each of them being defined on an association of spore–pollen forms, some restricted to the particular zone and some entering or exiting therein. Vertical distributions of the zone-defining taxa were depicted by Daemon and Contreiras (1971) in a substantial range-chart, but neither descriptions nor illustrations of the palynomorphs were provided. Correlations between the well sections were effected by application of the zonal scheme through much of the Amazonas Basin and, extending westwards, into the Solimões and Acre Basins. As Playford and Dino (2000b, p. 123) commented, although one can easily point to deficiencies in the taxonomic and stratigraphic basis of Daemon and Contreiras's scheme, it must be acknowledged as a significant, ground-breaking achievement.

The most recent and most comprehensive account of the palynology and palynostratigraphy of the upper Paleozoic sequence (Tapajós Group) in the Amazonas Basin consists of a two-part monograph by Playford and Dino (2000a, b). Their study was based on over 440 core samples collected from 26 wells distributed throughout the basin and covering the group's four formations (in ascending order, Monte Alegre, Itaituba, Nova Olinda, and Andirá Formations). Because no single well section included the entire Tapajós sequence, a composite palynostratigraphic reference section was assembled from six wells (Playford and Dino, 2000a, text-fig. 6). This served as an objective basis for palynozonal subdivision of the Tapajós Group *in toto* (Fig. 5.4) and facilitated correlation with and among the numerous other palyniferous well sections. The palynozonation – comprising seven stratigraphically successive assemblage zones – was underpinned by detailed taxonomic analysis of the palynofloras, including documentation and illustration (with optical and scanning electron micrographs) of some 90 species of spores and pollen grains, certain examples of which are figured herein (Figs. 5.5, 5.6). Detailing the occurrence of all the species identified in the palyniferous samples (e.g. Playford and Dino, 2000b, text-fig. 1) provided the database necessary for establishing the stratigraphic ranges of the taxa within the Amazonas Basin, thus facilitating their biostratigraphic application. Accordingly, Playford and Dino's (2000b, pp. 119–123, text-fig. 2) seven spore–pollen assemblage zones were defined on such criteria as constant or characteristic species associations, FADs and LADs, and relative abundances of particular taxa (Fig. 5.4). The zonal nomenclature followed the recommended and widely adopted practice of using a binomial name eponymously, based on a species confined to the zone or one that is particularly plentiful or otherwise especially characteristic of the zonal interval. For example, the eponym of the *Raistrickia cephalata* Zone is a distinctively sculptured species of trilete spores (see Playford and Dino, 2000a, pp. 18–19, pl. 3, figs. 8–15, pl. 4, figs. 10–12, text-fig. 6; Fig. 5.4 herein) that is restricted to and relatively common within the zone, with well-defined FAD and LAD demarcating the zone's lower and upper limits respectively. The zonation scheme was calibrated chronologically as precisely as possible (Fig. 5.4), mainly with reference to available independent (mostly marine-faunal) evidence. The zonation thus serves the need for precise biostratigraphic correlation throughout the basin-wide extent of the Tapajós Group, and exemplifies the value of detailed palynological analyses in petroleum exploration.

Crucial to the scope of the Playford and Dino (2000a, b) research were the numerous (>200) wells drilled in the basin from which abundant core material was available. Not all of the samples proved palyniferous, and those that were contained plant microfossils in varying concentrations and states of preservation. Such variation notwithstanding, the zonation proved serviceable even in sections that, for instance, had been subjected to thermal metamorphism (which tends to adversely affect

| BIOZONATION | | LITHOSTRATIGRAPHIC OCCURRENCE | INFERRED AGE |
|---|---|---|---|
| PALYNOZONES | SUMMARY CHARACTERISTICS | | |
| *Tornopollenites toreutos* | FAD of *Tornopollenites toreutos*, *Verrucosisporites insuetus*, *Thymospora obscura*, *Laevigatosporites minor*; taeniate bisaccates predominant throughout. | middle and upper Andirá Formation | Late Permian |
| *Vittatina costabilis* | FAD of *Vittatina* spp. (*costabilis*, *vittifera*, *saccata*, *subsaccata*); *Hamiapollenites andiraensis*, *H. fusiformis*, *Pakhapites* spp., *Corisaccites alutas*, *Lueckisporites virkkiae* introduced; exit of *Illinites unicus*, but other taeniate bisaccates predominant (particularly in upper part of zone). | upper Nova Olinda and lower Andirá Formations | Early/Late Permian (Sakmarian to Kazanian) |
| *Raistrickia cephalata* | Delimited by LAD & FAD of *Raistrickia cephalata*, *Peppersitesellipticus*, etc.; characteristic association of latter spp. with *Cycadopites* cf. *follicularis*, *Striatosporites heyleri*, *Apiculatasporites daemonii* & taeniate bisaccates; LAD of *Striomonosaccites incrassatus*, *A. daemonii*, among others. | middle Nova Olinda Formation | Westphalian D |
| *Striatosporites heyleri* | FAD of *Striatosporites heyleri*, *Apiculatasporites daemonii*; continuing *Illinites unicus*; increased diversity of taeniate & non-taeniate bisaccates; decrease in *Spelaeotriletes triangulus/arenaceus*; LAD of *Mabuitasaccites crucistriatus*, *Crucisaccites* cf. *latisulcatus*. | lower Nova Olinda Formation | Westphalian C-D |
| *Illinites unicus* | *Illinites unicus* very abundant; associated with *Spelaeotriletestriangulus/arenaceus*, *Endosporites globiformis* & zonate–cingulate spores (*Vallatisporites*, *Cristatisporites*); FAD of *Barakarites rotatus*, *V. arcuatus*; slight increase of taeniate bisaccates. | upper Itaituba Formation | Westphalian C |
| *Striomonosaccites incrassatus* | FAD of *Illinites unicus*, *Striomonosaccites incrassatus*, *Mabuitasaccites crucistriatus*, *Meristocorpus explicatus*; abundant *Spelaeotriletes triangulus/arenaceus*; bilateral monosaccates (excluding taeniate forms) infrequent; taeniate bisaccates inconspicuous, but taeniate monosaccates common. LAD of *Costatascyclus crenatus*, *Potonieisporites seorsus*. | lower to middle Itaituba Formation | Westphalian C |
| *Spelaeotriletes triangulus* | *Spelaeotriletes triangulus/arenaceus* abundant; *Lophotriletes lentiginosus*, *Waltzispora polita*, *Potonieisporites marleniae* diagnostic; monosaccate pollen (*Potonieisporites*, *Caheniasaccites*, *Costatascyclus*, *Plicatipollenites*) strongly represented; taeniate pollen & monolete spores infrequent. | Monte Alegre and basal Itaituba Formations | Westphalian A-B |

*Figure 5.4* Pennsylvanian–Permian palynozonation, Tapajós Group, Amazonas Basin, northern Brazil. Adapted from Playford and Dino (2000b).

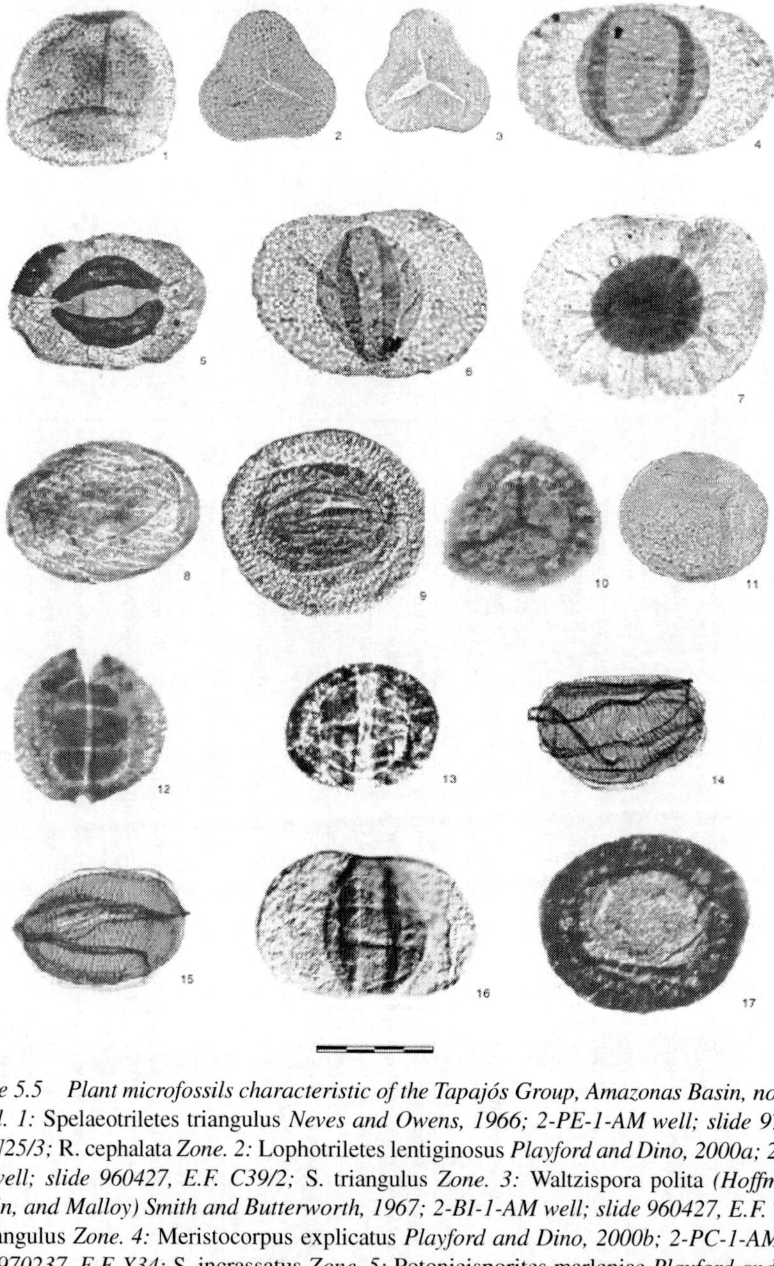

*Figure 5.5 Plant microfossils characteristic of the Tapajós Group, Amazonas Basin, northern Brazil. 1:* Spelaeotriletes triangulus *Neves and Owens, 1966; 2-PE-1-AM well; slide 970221, E.F. N25/3;* R. cephalata *Zone. 2:* Lophotriletes lentiginosus *Playford and Dino, 2000a; 2-BI-1-AM well; slide 960427, E.F. C39/2;* S. triangulus *Zone. 3:* Waltzispora polita *(Hoffmeister, Staplin, and Malloy) Smith and Butterworth, 1967; 2-BI-1-AM well; slide 960427, E.F. T51/4;* S. triangulus *Zone. 4:* Meristocorpus explicatus *Playford and Dino, 2000b; 2-PC-1-AM well; slide 970237, E.F. X34;* S. incrassatus *Zone. 5:* Potonieisporites marleniae *Playford and Dino, 2000a; 1-AM-2-AM well; slide 970176, E.F. V41/1;* S. triangulus *Zone. 6:* Potonieisporites seorsus *Playford and Dino, 2000a; 2-BI-1-AM well; slide 960427/43, E.F. N44/4;* S. triangulus *Zone. 7:* Costatascyclus crenatus *Felix and Burbridge emend. Urban, 1971; 2-PC-1-AM well; slide 970237, E.F. J32/2;* S. incrassatus *Zone. 8:* Mabuitasaccites crucistriatus *(Ybert) Playford*

palynomorph preservation). The taxonomic documentation, based upon detailed systematic descriptions and photomicrographic illustrations of the diversity of spores and pollen grains, was judged an essential prerequisite for their biostratigraphic utilization; i.e. for the definition of their component palynozones. This underlines the fact that, as with any paleontological research having primarily stratigraphic objectives, a rigorous taxonomic approach normally yields appreciable dividends in terms of stratigraphic-correlative efficacy. Moreover, in this particular case history, it proved feasible to extend the palynostratigraphic correlation beyond the Amazonas Basin, albeit with lesser precision, to other palynologically documented regions of the Gondwana supercontinent (Playford and Dino, 2000b, text-figs. 3, 4).

## 5.5 Acknowledgements

The management of Petrobras-Cenpes (Rio de Janeiro), together with the Fundação de Amparo à Pesquisa do Estado do Rio de Janeiro (FAPERJ; Project no. E-26/170.127/2001), have provided much-appreciated support of this and cognate research by the authors. Sincere thanks are expressed to Dr. L. Antonioli for assistance in technical matters.

*Figure 5.5 Continued*

*and Dino, 2000b; 1-TR-1-AM well; slide 970299, E.F. M48/4;* S. incrassatus *Zone. 9:* Striomonosaccites incrassatus *Playford and Dino, 2000b; 2-PC-1-AM well; slide 970237/5, E.F. K40/2;* S. incrassatus *Zone. 10:* Vallatisporites arcuatus *(Marques-Toigo) Archangelsky and Gamerro, 1979; 2-IP-1-AM well; slide 9706103, E.F. M47/3;* I. unicus *Zone. 11:* Apiculatasporites daemonii *Playford and Dino, 2000a; 2-PE-1-AM well; slide 9706190A/20, E.F. K37/2;* S. heyleri *Zone. 12:* Illinites unicus *Kosanke, 1950; 9-FZ-2-AM well; slide 9705473, E.F. M23/2;* I. unicus *Zone. 13:* Illinites unicus *Kosanke, 1950; 2-PE-1-AM well; slide 9706190, E.F. E27;* R. cephalata *Zone. 14:* Striatosporites heyleri *(Doubinger) Playford and Dino, 2000a; 2-PE-1-AM well; slide 970221, E.F. K54/4;* R. cephalata *Zone. 15:* Striatosporites heyleri *(Doubinger) Playford and Dino, 2000a; 2-PE-1-AM well; slide 9706190, E.F. L36/3;* R. cephalata *Zone. 16:* Protohaploxypinus hartii *Foster, 1979; 2-PE-1-AM well; slide 9706190, E.F. A23/3;* R. cephalata *Zone. 17:* Barakarites rotatus *(Balme and Hennelly) Bharadwaj and Tiwari, 1964; 1-FZ-1-AM well; slide 970473, E.F. L56/2;* I. unicus *Zone. Scale-bar unit = 30 μm.*

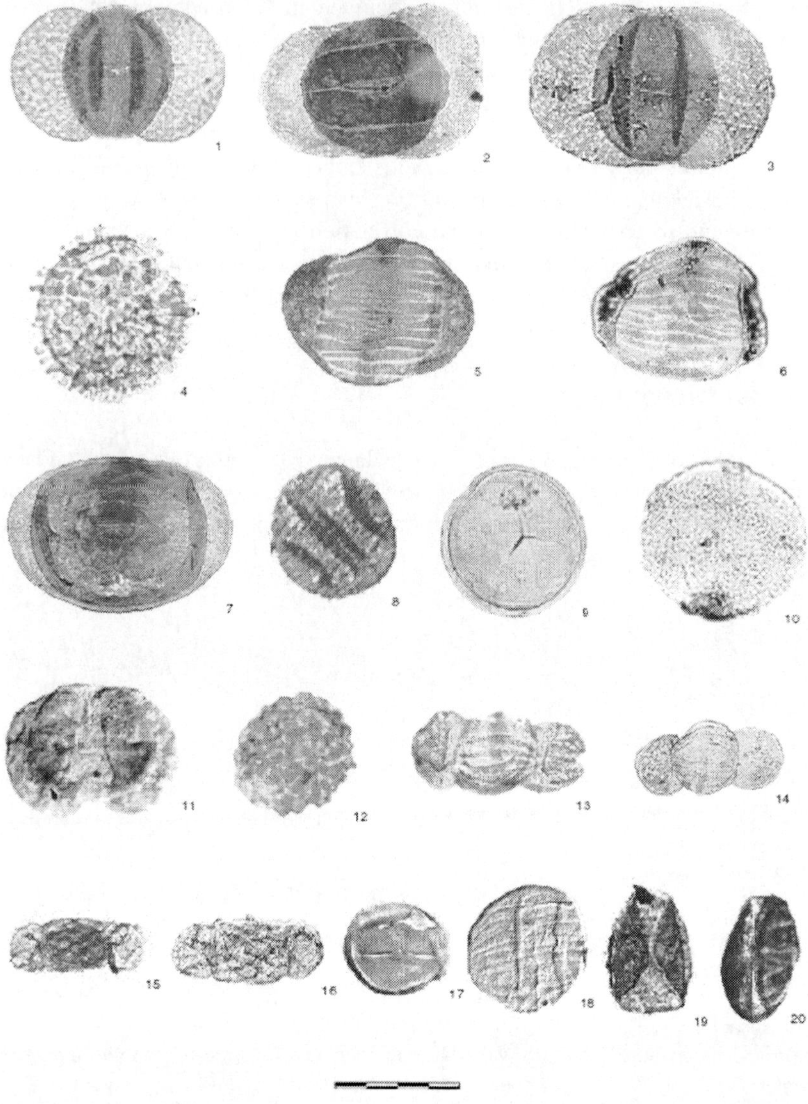

*Figure 5.6 Plant microfossils characteristic of the Tapajós Group, Amazonas Basin, northern Brazil. 1:* Limitisporites scitulus *Playford and Dino, 2000b; 2-PC-1-AM well; slide 970237/41, E.F. N35/1;* S. incrassatus *Zone. 2:* Lunatisporites onerosus *Playford and Dino, 2000b; 2-PE-1-AM well; slide 970221A/67, E.F. K39/2A;* R. cephalata *Zone. 3:* Potonieisporites pyriferus *Playford and Dino, 2000b; 2-PE-1-AM well; slide 970221A/3, E.F. K40/3;* R. cephalata *Zone. 4:* Raistrickia cephalata *Bharadwaj, Kar, and Navale, 1976; 2-PE-1-AM well; slide 9706190, E.F. R34/3;* R. cephalata *Zone. 5:* Vittatina saccata *(Hart) Playford and Dino, 2000b; 9-FZ-28-AM well; slide 9705908, E.F. F26/4;* T. toreutos *Zone. 6:* Vittatina subsaccata *Samoilovich, 1953; 9-FZ-2-AM well; slide 9705473, E.F. X36/3;* V. costabilis *Zone. 7:* Peppersites ellipticus

*Figure 5.6 Continued*

*Ravn, 1979; 2-PE-1-AM well; slide 9706190A, E.F. B56/1;* R. cephalata *Zone. 8:* Vittatina costabilis *Wilson, 1962; 9-FZ-28-AM well; slide 9705915, E.F. G57/1;* V. costabilis *Zone. 9:* Punctatisporites *sp.; 9-FZ-28-AM well; slide 9705908, E.F. L52;* T. toreutos *Zone. 10:* Verrucosisporites insuetus *Playford and Dino, 2000a; 9-FZ-28-AM well; slide 9705908, E.F. N43/4;* T. toreutos *Zone. 11:* Lueckisporites virkkiae *Potonié and Klaus emend. Clarke, 1965; 9-FZ-28-AM well; slide 9705916, E.F. K50;* V. costabilis *Zone. 12:* Thymospora obscura *(Kosanke) Wilson and Venkatachala, 1963; 9-FZ-28-AM well; slide 9705915, E.F. L63/4;* V. costabilis *Zone. 13:* Hamiapollenites fusiformis *Marques-Toigo emend. Archangelsky and Gamerro, 1979; 9-FZ-28-AM well; slide 9705899, E.F. X48/2;* T. toreutos *Zone. 14:* Hamiapollenites andiraensis *Playford and Dino, 2000b; 9-FZ-28-AM well; slide 9705916/3, E.F. L40;* V. costabilis *Zone. 15:* Tornopollenites toreutos *Morgan, 1972; 9-FZ-28-AM well; slide 9705915, E.F. M41/4;* V. costabilis *Zone. 16:* Tornopollenites toreutos *Morgan, 1972; 9-FZ-28-AM well; slide 9705915, E.F. A61/3;* V. costabilis *Zone. 17:* Laevigatosporites minor *Loose, 1934; 9-FZ-28-AM well; slide 9705911, E.F. O53/4;* V. costabilis *Zone. 18:* Vittatina costabilis *Wilson, 1962; 9-FZ-28-AM well; slide 9705908, E.F. N43/1;* T. toreutos *Zone. 19:* Pakhapites fusus *(Bose and Kar) Menéndez, 1971; 9-FZ-28-AM well; slide 9705915, E.F. V50/1;* V. costabilis *Zone. 20:* Pakhapites fusus *(Bose and Kar) Menéndez, 1971; 9-FZ-2-AM well; slide 9705473, E.F. H30/4;* V. costabilis *Zone. Scale-bar unit = 30 μm.*

Chapter 6

# Biostratigraphy of the Non-Marine Triassic: Is a Global Correlation Based on Tetrapod Faunas Possible?

CESAR L. SCHULTZ
Universidade Federal do Rio Grande do Sul (UFRGS), Instituto de Geociências, Caixa Postal 15001, 91501-570 Porto Alegre, RS, Brazil.



## 6.1 Introduction

The Triassic Period, as indicated by its name, is characterized, in its type-section (in Germany), by a succession of three series: the Bunter (= Lower Triassic), the Muschelkalk (= Middle Triassic) and the Keuper (= Upper Triassic), corresponding to the Early, Middle and Late Triassic epochs. Besides, the system/period as a whole have been divided into five to seven stages/ages (Fig. 6.1). Despite the dominance of continental deposition, the biostratigraphy and biochronology of these ages are based totally on marine invertebrates, mainly ammonoids and conodonts (e.g. Dagys and Weitschat, 1993; Gradstein *et al.*, 1995), coming from complete marine sessions from which other parameters as thickness, depositional rate, sea-level changes, magnetometric and radiometric data were also considered to establish the standard ages (Stipanicic, 2002). Nevertheless, neither the boundaries among these divisions nor the GSSPs (**G**lobal boundary **S**tratotype **S**ection and **P**oints), for each one are yet consensual [IUGS/ICS Subcommission on Triassic Stratigraphy (STS), Business Meeting, Halle/Saale, Germany, 1998]. Even in marine deposits, it is difficult to get good associations of biostratigraphic



*E.A.M. Koutsoukos (ed.), Applied Stratigraphy,* 123-145.

and geochronologic data. According to Gradstein *et al.* (1995), within the Triassic, only the Ladinian/Anisian Boundary can be considered an important anchor point due to the presence of tuff layers in the basal part of the lowermost Ladinian *Nevadites* ammonite Zone, which indicates an age around 232 m.y. for this boundary.

However, ammonoids and conodonts are useless for correlation of Triassic continental strata, and the extensive prevalence of "red beds" among these layers precludes the use of spores and palynomorphs for this purpose. Notwithstanding, most of the papers dealing about stratigraphy and biostratigraphy of non-marine Triassic sequences adopt the same division defined to the standard marine sequence, sometimes in a quite speculative way.

To by-pass this problem, the use of fossil tetrapods has been suggested as the key for correlation among non-marine Triassic sediments (e.g. Bonaparte, 1973; Cox, 1973; Romer, 1966, 1975; Cooper, 1982; Ochev and Shishkin, 1989; Shishkin *et al.*, 1995; Shubin and Sues, 1991; Lucas, 1990, 1993, 1998, 2001; Lucas *et al.*, 1998; Lucas and Hancox, 2001; Lucas and Heckert, 2002; Hunt and Lucas, 1991). This tool is particularly reliable for this period since Triassic tetrapod faunas – and also the floras – show one of the highest grades of cosmopolitanism in the history of life on Earth.

According to Shubin and Sues (1991) the Triassic is unique because it is the only period of tetrapod history during the entire length of which a single landmass existed. Few (if any) physical barriers for biotic interchange among continental tetrapods existed during this period, generating unparalleled potential for global biotic interchange. This paleogeographic condition, added to a relative climatic uniformity, favoured the dispersion and the cosmopolitanism of the floras and tetrapod faunas.

Triassic tetrapod-bearing layers occur in South America, Antarctica, Africa, Australia, Asia, Europe and North America (Fig. 6.2), and major faunal successions

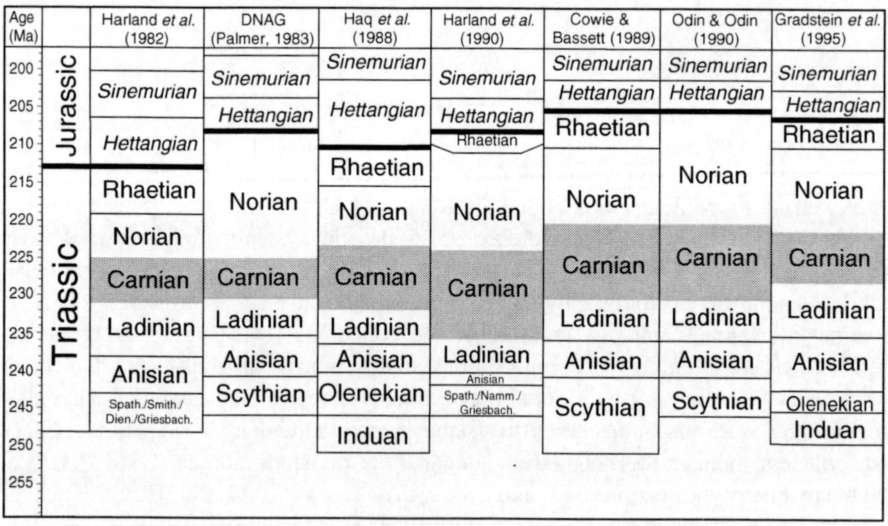


*Figure 6.1 Some time-scales for the Triassic (modified from Gradstein et al., 1995).*

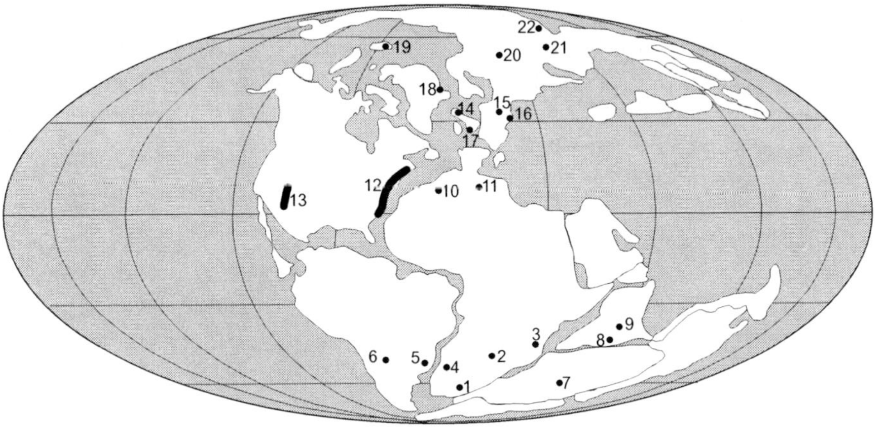


*Figure 6.2* *Map of Triassic Pangea showing locations of principal vertebrate fossil assemblages: 1 = Karoo basin, South Africa; 2 = Zambia; 3 = Tanzania; 4 = Namibia; 5 = Parana basin, Brazil; 6 = Argentina; 7 = Transantarctic Mountains, Antarctica; 8 = Pranhita-Godavari Valley, India; 9 = Damodar, India; 10 = Essaouira basin, Morocco; 11 = Ilizi basin, Algeria; 12 = Newark Supergroup basins, eastern USA-Canada; 13 = Moenkopi and Chinle basins, western USA; 14 = Scotland; 15 = Germanic basins; 16 = northern Italy; 17 = Devon, UK; 18 = eastern Greenland; 19 = Svalbard; 20 = Russian Urals; 21 = Junggur basin, China; 22 = Ordos basin, China. (Modified from Lucas, 1998).*

and evolutionary trends among the fossil groups can be easily detected by correlation among them.

Notwithstanding the progress reached until now, the possibility of good correlation among non-marine Triassic strata and even the establishment of a time scale based on their tetrapod faunas do not resolve entirely the problem, since it is still necessary to link these data with those of marine Triassic, not only regarding to the limits between the stages/ages but specially to the definitions of the P–T and T–J boundaries, both controversial even among marine paleontologists. To aim this goal, it will be necessary to improve and enchain not only stratigraphic and paleontologic data, but also radiometric and magnetometric data (or even other geological techniques that could be useful for this purpose), whenever it is possible.

## 6.2 The Triassic: General Aspects

During the Triassic Period, the continents formed a united landmass known as the Pangea supercontinent, introduced by Wegener (1924). Pangea began to assemble during the end of the Carboniferous, with the collision of Gondwana and Laurasia and reached its maximum development in Triassic with the addition of Kazakhstan, Siberia, parts of China and southeastern Asia (Ziegler *et al.*, 1983; Veevers, 1991, 1994). During the Triassic, Pangea was distributed symmetrically in relation to the equator, forming a continuously exposed landmass from about 85° N to 90° S (Ziegler *et al.*, 1983). According to Stanley (1989), the Triassic was defined by Friedrich August von Alberti in 1834, in Germany, as a system bearing a unique fauna, bounded by the Permian extinction below and by another extinction above. The name (originally "Trias") refers to the occurrence, in that country, of a succession of three distinctive stratigraphic units: a marine sequence, the Muschelkalk (or "mussel limestone") sandwiched between two predominantly non-marine sequences, the Bunter deposits beneath it, and the Keuper deposits above. Nevertheless, this definition is not appropriate for all sequences deposited during Triassic time. Through this period, sea level was relatively low (Vail *et al.*, 1977) and the sedimentary deposits around the Pangea were dominantly continental, with restricted marine sediments occurring only at the western shoreline of ancestral North and South America, in the east of Asia and along the northern and southern coastline of the Tethys (Tucker and Benton, 1982) (Fig. 6.3).

According to Embry (1988), nine eustatic sea-level cycles of third-order, and four of second order, can be recognized for this period, although the Triassic has been divided in seven stages, on basis on marine faunas. Embry (1988) stressed that worldwide stratigraphic, sedimentologic, and paleontologic data for the Triassic are relatively sparse in comparison to those for the Jurassic, Cretaceous and Tertiary, so that his attempt at a Triassic eustatic sea-level curve could be changed with the addition of more data.

Regarding to the Triassic climate, the absence of glacial deposits (tillites) added to the generalized occurrence of plants and evidence of seasonal changing climatic conditions in latitudes up to 60°, suggests that global climate was much hotter than nowadays, with high average annual temperatures even in polar latitudes (Holz and Scherer, 2000). To many authors (e.g. Robinson, 1973; Hallam, 1985; Parrish *et al.*, 1986) an essentially arid climate is accepted to the Triassic. Indeed, Frakes (1979) considered the climate of the Middle Triassic was possibly the most arid in the history of the Earth.

Besides the hotter climate, the latitudinal belts of humidity and aridity were completely different from those existing nowadays. According to Hallam (1985), Pangea was characterlized by a wide-spaced climatic zoning, with three climatic zones: one eminently humid at high latitudes, a seasonally humid zone at middle latitudes, and an arid zone at the lowest latitudes. These climatic zones are supported by quantitative atmospheric modelling of the Triassic (e.g. Wilson *et al.*, 1994).

This climatic arrangement, despite the great cosmopolitanism observed in the flora and fauna, suggests by itself the possibility of some degree of endemism between north and south portions of Pangea, and it really occurs, which difficult the establishment of global tetrapod biozones.

## 6.3 Triassic Tetrapod Biostratigraphy and Biochronology – A Retrospective

According to Kitching (1995), one of the first attempts at dividing a Triassic sequence into paleontological zones based on tetrapods, was made by Seeley (1892) in the

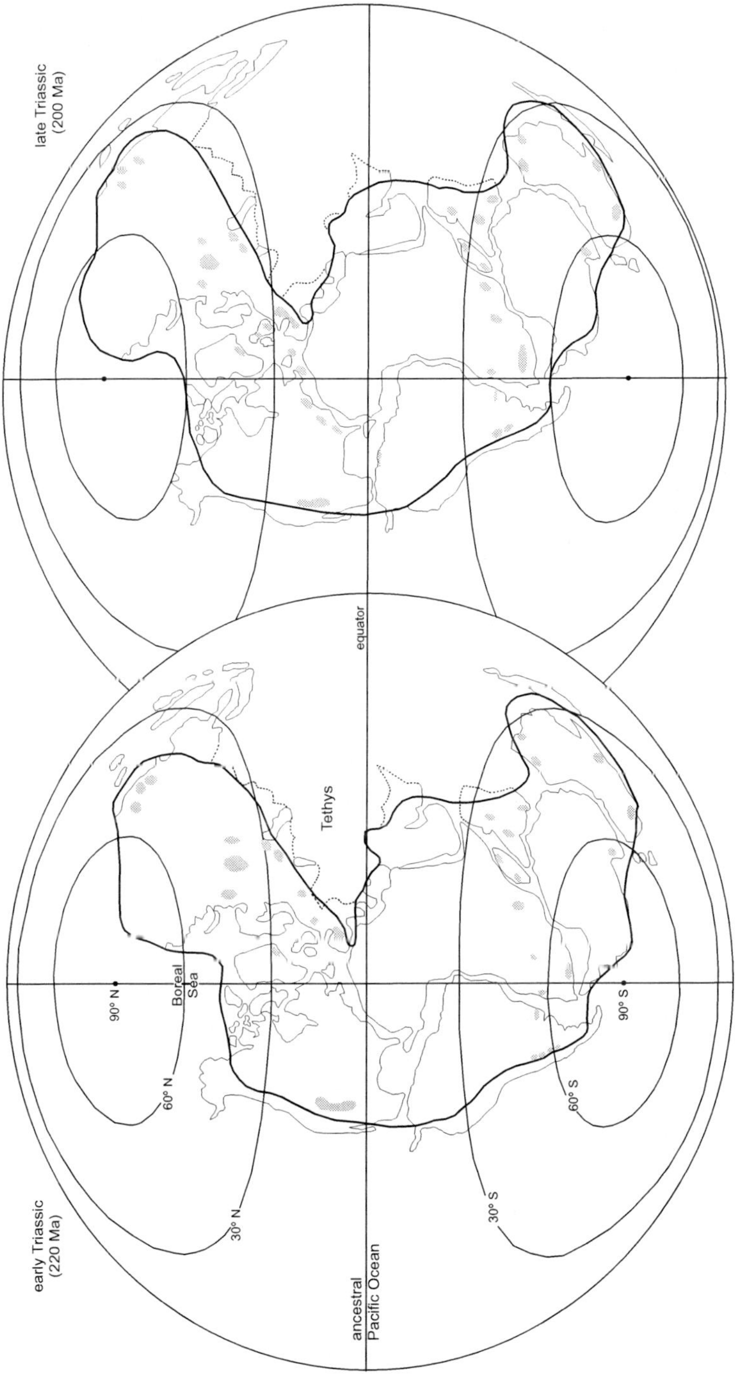


*Figure 6.3* *Coastline of the Tethys in the Triassic (modified from Tucker and Benton, 1982).*

Karoo Basin of South Africa. Some years later, Broom (1906, 1907, 1909) divided the Beaufort into six zones, three of them in the Lower/Middle Triassic, characterized, from the lower to the upper one, by the presence of the *genera Lystrosaurus, Procolophon* and *Cynognathus*. Dutoit (1918, *apud* Groenwald and Kitching, 1995) remarked on the unique use of the *Lystrosaurus* Zone for regional mapping purposes. However, the *Lystrosaurus* and *Cynognathus* Zones have been used by paleontologists world-wide and the identification of biozones based on these genera was accepted for many years in South America, Antarctica, India, Russia and China.

Romer (1966, 1975) proposed a threefold division of Triassic tetrapod assemblages on a global scale. At the base was the **Zone A (Predominantly Therapsid)** dominated by archosauriform diapsids and dicynodonts, which was best documented from the Lower Triassic portion of the Beaufort Group of South Africa. The **Zone B (Intermediate Assemblage)** was dominated by traversodont cynodonts and rhynchosaurian reptiles and was better known from the Middle Triassic of Argentina, Brazil, and Tanzania. The Upper Triassic **Zone C (Predominantly Archosaur)**, by its turn, was dominated by a great diversity of archosaurian reptiles, especially dinosaurs. According to Shubin and Sues (1991), Romer and other authors (e.g. Colbert, 1984) used these assemblages to hypothesize that the Triassic was a time of competitive replacement of nonmammalian synapsids by diapsids. Other authors (e.g. Anderson and Cruickshank, 1978; Tucker and Benton, 1982; Ochev and Shishkin, 1989) also adopted such kind of triple division to the non-marine Triassic, but not exactly the same proposed by Romer, 1966, 1975. Indeed, Tucker and Benton (1982) associated the occurrence of these three major vertebrate faunas with the presence of three successive different floras ("transitional", *Dicroidium* and Conifer). Besides, each association of fauna and flora was found in a different sedimentary facies association (Fig. 6.4): (A) fluvio-lacustrine sandstones and mudrocks with coal seams and abundant plant material; (B) fluvio-lacustrine sandstones and mudrocks with rare coals, some red beds and calcretes, occasional gypsum and plant material common; (C) fluvio-lacustrine red beds with calcrete, playa mudrocks, gypsum and halite, and gypsum deposits and aeolian sandstones.

According to Tucker and Benton (1982), these successive changes of faunas, floras and sedimentary facies reflect climatic changes leading to increasing aridity towards the end of Triassic, affecting particularly Gondwana and Laurasia. In turn, these climatic changes resulted from plate motions of Pangea northwards. In fact, this paper constituted the first attempt to integrate the biostratigraphy of the non-marine Triassic as a whole with stratigraphic and paleoclimatic data.

However, Holz and Scherer (2000), also using sedimentologic and paleontologic evidences, agree that paleoclimatic changes occurred along the Triassic, but following a global trend towards a humid – rather than arid – paleoclimate at the end of the period. According to the same authors, sedimentary evidences for this hypothesis are the substitution, during the Carnian/Norian, of red beds bearing gypsum by paralic sediments with coal and also by the presence of kaolinite in continental and marine sediments.

Regarding to the causes of the faunal successions, Benton (1983, 1986, 1989, 1994) instead the competitive replacement hypothesized by Romer (1966, 1975) and others, proposes an opportunistic ecological replacement model.

Thus, the division of the Triassic into three series/epochs seems to be clearly supported, regarding both the marine and non-marine record, but it does not mean that the boundaries among the divisions are directly correlated, once that it is not common that index fossils from each one of these environments occur together. This problem just increases as detailed as a global correlation is intended.

Several other authors proposed biostratigraphic zonations for the non-marine Triassic using different criteria (Fig. 6.5). Cooper (1982) presented a more detailed proposal of a global tetrapod biostratigraphy using mainly dicynodonts. He established six zones for the Triassic, but considered the *Lystrosaurus* zone as Permian rather than Early Triassic as most authors.

Ochev and Shishkin (1989) based on the tetrapod faunas coming from triassic sediments of Cis-Uralian region (East European Platform), divide the Triassic also in three successive global epoches referred to as *Proterosuchian* (Early Triassic), *Kannemeyeroid* (Middle Triassic) and *Dinosaur* (Upper Triassic). The Proterosuchian time includes the *Neorachitome Fauna* (Induan-Lower Olenekian) and the *Parotosuchus Fauna* (Upper Olenekian = Spathian). In the Kannemeyeroid Epoch, the *Eryosuchus* and *Mastodonsaurus* Faunas correspond to the Muschelkalk and Lettenkohle of the germanic basin, respectively. The proterosuchian epoch is marked by the dominance of the primitive thecodonts over other reptiles; the succeeding kannemeyeroid epoch demonstrates the radiation of anomodonts and the appearance of the pseudosuchians, the gomphodonts and the advanced bauriamorphs. The assemblages of the proterosuchian epoch are of particular biostratigraphic importance because they constitute the only regional sequence so far known among the Triassic tetrapod faunas that permits a direct comparison with the marine sections (due to the presence of common labyrinthodont genera).

Better results regarding to more detailed subdivisions of continental Triassic have been obtained from provincial studies. Bonaparte (1966a, b, 1973, 1982) proposed "reptile ages" for the Triassic of Argentina. This author characterized the faunal associations taking into account the evolutive stage of determinate *taxa*, its presence or absence and the dominance of some groups in the faunas, as well as its possible relationship with other gondwanic paleofaunas (Stipanicic and Riccardi, 2002). Bonaparte (1973) present a general chronological scheme to the Argentinean and Brazilian Triassic in which a series of *Reptile/Ages* were established based on **Local-Faunas** [*sensu* Wilson, 1950 (*in* Simpson, 1971), i.e. "geo-biotic units constituted by the totality of the species collected in some important outcrops or group of outcrops, composed by organisms that lived in the same geologic time and in the same area but are not representative of the totality of an extensive litho or chronostratigraphic unit"]. These Reptile/ages were: **Puestoviejense Age** (Eotriassic) including the Local-Faunas of the Puesto Viejo and Rio Mendoza Formations of Argentina; **Lower Chañarense Age** (Early Middle Triassic) including the Local-Fauna of the Los Chañares

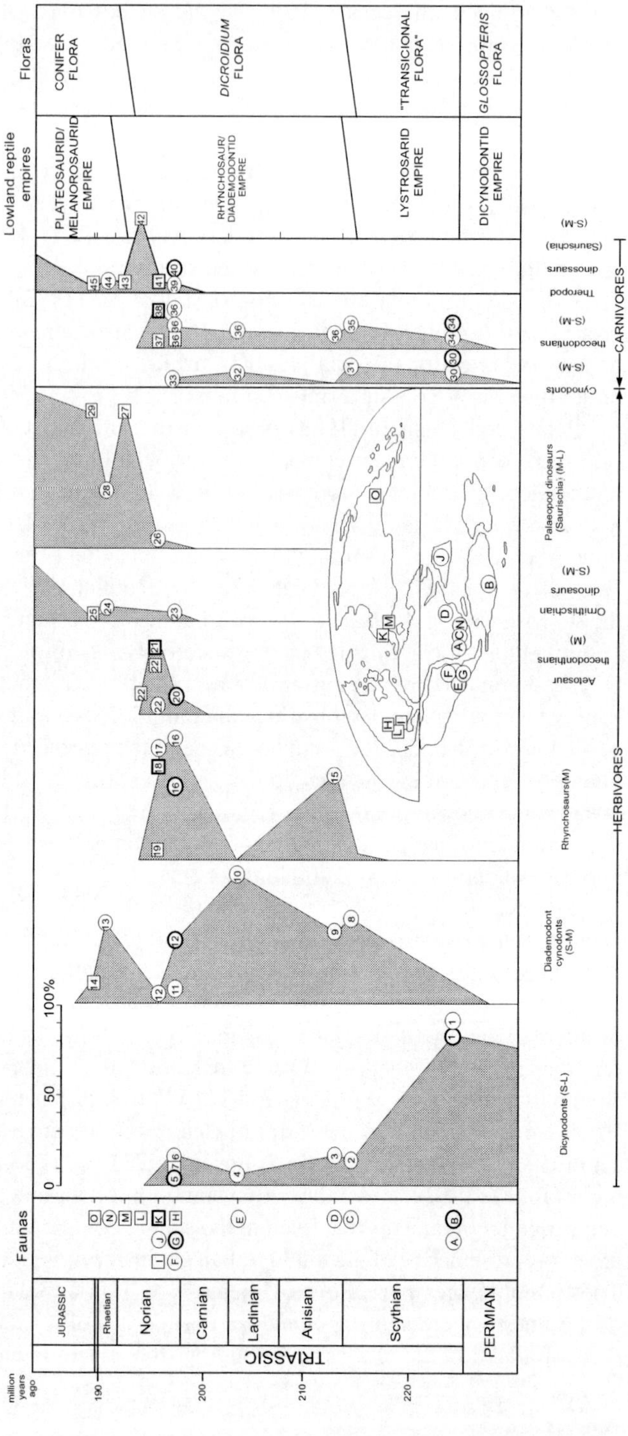
Floras
CONIFER FLORA
DICROIDIUM FLORA
"TRANSICIONAL FLORA"
GLOSSOPTERIS FLORA
Lowland reptile empires
PLATEOSAURID/ MELANOROSAURID EMPIRE
RHYNCHOSAUR/ DIADEMODONTID EMPIRE
LYSTROSARID EMPIRE
DICYNODONTID EMPIRE
Faunas
million years ago
JURASSIC
Rhaetian
Norian
Carnian
Ladinian
Anisian
Scythian
TRIASSIC
PERMIAN
Dicynodonts (S-L)
Diademodont cynodonts (S-M)
Rhynchosaurs(M)
Aetosaur thecodontians (M)
Ornithischian dinosaurs (S-M)
Palaeopod dinosaurs (Saurischia) (M-L)
Cynodonts (S-M)
thecodontians (S-M)
Theropod dinosaurs (Saurischia) (S-M)
HERBIVORES
CARNIVORES

*Figure 6.4* *Changes in abundance of major terrestrial reptile groups during the Triassic. Histogram percentage scale indicated. Broad adaptative zones of the animals are shown: S = small (length <1 m); M = medium (length 1–2 m); L = large (length >2 m). Circles = Southern Hemisphere Faunas. Squares = Northern Hemisphere Faunas. Heavy circles or squares = two different faunas coming from the same stratigraphic horizon in the same hemisphere.* ***Major Faunas****: A =* Lystrosaurus *Zone, South Africa; B = Fremouw Formation, Antarctica; C =* Cynognathus *Zone, South Africa; Manda Formation, Tanzania; E = Chañares Formation, Argentina; F = Santa Maria Formation, Brazil; G = Ischigualasto Formation, Argentina; H = Popo Agie Formation, USA; I = Dockum Formation, USA; J = Maleri Formation, India; K = Lossiemouth Formation, Scotland; L = Upper Chinle Formation, USA; M = Knollenmergel, Germany; N = Elliot Formation, South Africa; O = Lower Lufeng Series, China.* **Major Genera**: *1* = Lystrosaurus; *2* = Kannemeyeria; *3* = Tetragonias; *4* = Dinodontosaurus; 5 = Ischigualastia; *6* = Stahleckeria; *7* = Placerias; *8* = Diademodon; *9* = Scalenodon; *10* = Massetognathus; *11* = Traversodon; *12* = Exaeretodon; *13* = Tritylodon; *14* = Bienotherium; *15* = Stenaulorhynchus; *16* = Scaphonyx; *17* = Paradapedon; *18* = Hyperodapedon; *19* = *Rhynchosaur (unnamed)*; *20* = Aetosauroides; *21* = Stagonolepis; *22* = Typothorax; *23* = Pisanosaurus; *24* = Fabrosaurus, Heterodontosaurus; *25* = Tatisaurus; *26* = *unnamed anchisaur*; *27* = Plateosaurus; *28* = Thecodontosaurus; *29* = Lufengosaurus; *30* = Thrinaxodon; *31* = Cynognathus; *32* = Probelesodon; *33* = Belesodon; *34* = Chasmatosaurus; *35* = Erythrosuchus, Euparkeria; *36* = *various thecodontians*; *37* = Poposaurus; *38* = Ornithosuchus; *39* = Staurikosaurus; *40* = Herrerasaurus, Ischisaurus; *41* = Saltopus; *42* = Coelophysis; *43* = Halticosaurus; *44* = Syntarsus; *45* = Lokousaurus. *(Modified from Tucker and Benton, 1982).*

| PERIOD | | Bonaparte (1966a) (Argentina) | Romer (1975) | Cooper (1982) | Ochev & Shishkin (1989) | Lucas (1993) (China) | Lucas & Hunt (1993) [W. USA] | Huber *et al.* (1993) [E. N. AM.] |
|---|---|---|---|---|---|---|---|---|
| TRIASSIC | LATE | | C | *Plateosaurus* zone | dinosaurian epoch | | Apachean LVF | Cliftonian LVF |
| | | Coloradian | | *Placerias* zone | | | Revueltian LVF | Neshanician LVF |
| | | | | | | | Adamanian LVF | Conewagian LVF |
| | | Ischigualastian | | *Stahleckeria* zone | | | Otischalkian LVF | Sanfordian LVF |
| | MIDDLE | Chanarian | B | *Dinodontosaurus* zone | kannemeyeroidean epoch | | | |
| | | | | *Tetragonias* zone | | Ningwuan LVF | | Economian LVF |
| | EARLY | Puestoviejan | A | *Kannemeyeria* zone | proterosuchian epoch | Ordosian LVF | | |
| | | | | | | Fuguan LVF | | |
| | | | | | | Jimsarian LVF | | |

*Figure 6.5 Some proposed biostratigraphic zonations for the Triassic (modified from Lucas, 1998).*

Formation of Argentina; **Upper Chañarense Age** (Late Middle Triassic), including the Local-Fauna of the Santa Maria Formation of Brazil; **Ischigualastense Age** (Early Neotriassic), including the tetrapod association of the Ischigualasto Formation of Argentina; and **Coloradense Age** (Late Neotriassic) including the Local-Faunas of the Los Colorados Formation and Laguna Colorada Formation (*El Tranquilo* Local-Fauna), both in Argentina.

In Brazil, a biostratigraphic zonation for the Triassic, based on the scheme of Bonaparte (1973) was elaborated by Barberena (1977) and later enhanced by Barberena *et al.* (1985a, b, 1991). In these last papers, the authors identified, to the south Brazilian Permo-Triassic package, seven Local-Faunas, five into the Triassic. However, although the use of Local-Faunas could bring good results in some cases, it is not proper to large-scale correlations, as will be discussed ahead, justly due to its local reach.

Just aiming such a larger-scale approach, Lucas (1993) proposed, for the Chinese Early-Middle Triassic, a succession of paleofaunistic associations called Land-Vertebrates Faunachrons (LVFs). The beginning of each LVF is defined by the First Appearance Datum (FAD) of a widespread tetrapod genus. Following the same method, Lucas and Hunt (1993) proposed Late Triassic LVFs based on the Chinle Group tetrapod record from the western United States, while Huber *et al.* (1993) and Lucas *et al.* (1998) proposed Middle-Late Triassic LVFs based on the Newark Supergroup record of eastern North America.

In eastern Europa, Sennikov (1995) introduced a new approach to the study of triassic biotas. He defined a succession of terrestrial and freshwater tetrapod communities, reconstructed as food-webs (Figs. 6a,b).

This author identified an Early and Mid-Triassic thecodontian-dicynodontian terrestrial community, with evolution from Early Triassic lystrosaurids to Mid Triassic kannemeyerids, and a temnospondyl aquatic community. The change of tetrapod faunas at the Permo-Triassic boundary was characterized by (1) replacement of therapsids by archosaurs, and (2) replacement of the top carnivores and wholesale changes in the food web structure.

Barberena *et al.* (1993), Scherer (1994) and Schultz *et al.* (1994, 2000) returned to the perspective of a formal biostratigraphy to the Middle-Upper Triassic of South Brazil (Fig. 6.7), through the individualization of two Cenozones, the Therapsid Cenozone, below, and the Rhynchosaur Cenozone, above, followed by a third informal unit, the "*Jachaleria* Level", characterized by the presence of this tuskless dicynodont. The Therapsid Cenozone was considered as Middle Triassic (probably "Ladinian") due to the dominance of therapsids, while the Rhynchosaur Cenozone, by the same reason, was attributed to Late Triassic (probably "Carnian") following the models used for other authors (e.g. Tucker and Benton, 1982). The *Jachaleria* Level, by its turn, was considered as "Norian" by correlation with Argentina, where the same genus occurs at the basal layers of the Los Colorados Formation, which were considered as having this age (e.g. Stipanicic and Marsicano, 2002). Radioisotopic dating of the basis of the underlying Ischigualasto Formation (227.8 $\pm$ 0.3 million years, Rogers *et al.*, 1993) indicates an early to middle Carnian age for this unit (and, by faunal correlation, for the Rhynchosaur Cenozone of Brazil). However, Rogers *et al.*, 1993

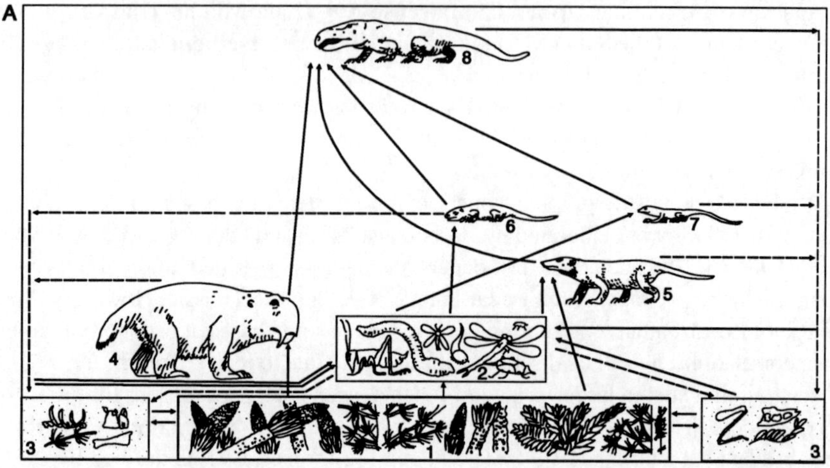


*Figure 6.6A Reconstructed food-web for the terrestrial component of the Lower Vetluga Community (Early Triassic, Induan) of eastern Europe. Lines with arrows indicate the movement of energy through the community: solid lines show feeding pathways, and dashed lines show decay pathways. 1 = plants; 2 = invertebrates; 3 = plant and animal detritus; 4 =* Lystrosaurus*; 5 =* Scalopognathus*; 6 = Procolophonid* Phaanthosaurus*; 7 =* Blomosaurus*;* *8 =* Vonhuenia *(modified from Sennikov, 1995).*

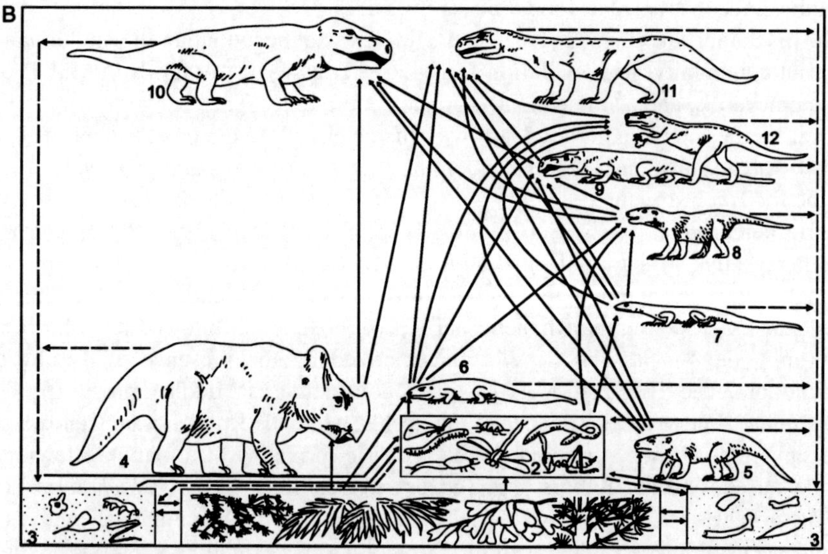


*Figure 6.6B Reconstructed food-web for the terrestrial component of the Donguz Community (Mid Triassic, Anisian/Ladinian) of eastern Europe. Lines with arrows indicate the movement of energy through the community: solid lines show feeding pathways, and dashed lines show decay pathways. 1 = plants; 2 = invertebrates; 3 = plant and animal detritus; 4 =* Rhadiodromus; *5 =* Antecosuchus; *6 = Procolophonid* Kapes; *7 = prolacertiform; 8 =* Dongusaurus; *9 =* Sarmatosuchus; *10 =* Erythrosuchus; *11 =* Dongusuchus; *12 =* Dorosuchus *(modified from Sennikov, 1995).*

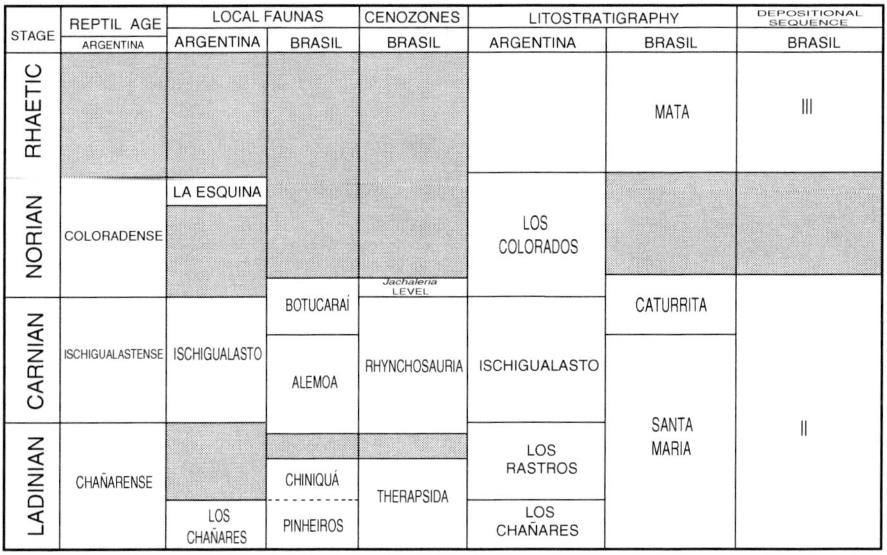


*Figure 6.7 Lithostratigraphic and biostratigraphic correlation between Brazil and Argentina (modified from Schultz et al., 2000).*

estimate the depositional time-range of the Ischigualasto formation between 1–4 million/years, indicating that both the basis of the overlying Los Colorados Formation in Argentina and the *Jachaleria* Level in Brazil would be still Carnian in age (rather than Norian as suggested also by Cox, 1991 and Battail, 1993). The upper layers of Los Colorados bear a quite distinct fauna (La Esquina), dominated by prosauropod dinosaurs, to which is attributed a Norian Age by most authors (e.g. Bonaparte, 1973; Tucker and Benton, 1982; Lucas and Hancox (2001)). Whatever be the precise time-range of the Ischigualasto Formation, it is undoubted inserted into Carnian, which confirms that the dominance of hyperodapedontidae rhynchosaurs in the Triassic tetrapod faunas occurred during this age.

Lucas and Huber (1998) reviewed global Late Triassic tetrapod biochronology and suggested the broad utility of the Chinle Group tetrapod biochronology proposed by Lucas and Hunt (1993). Following this idea, Lucas (1998) established (Fig. 6.8) eight temporally successive assemblage zones of tetrapod (amphibian and reptile) fossils, which provide the basis for dividing Triassic time in eight **L**and-**V**ertebrate **F**aunachrons (LVF).

Neveling *et al.* (1999) re-studied the interval separating the *Lystrosaurus* and the *Cynognathus* assemblage zones (respectively in the Katberg and Burgersdorp Formations), in South Africa, which was believed to be barren of tetrapod fossils, and recognized distinct fossil tetrapods assemblages in that level, named "impoverished zone." Based on these new findings, the authors established a new *Procolophon* Zone (not that of Broom, 1906) and divided the overlying *Cynognathus* Zone into Sub-zones "A" and "B" (Fig. 6.9). According to these authors, the presence of the amphibian

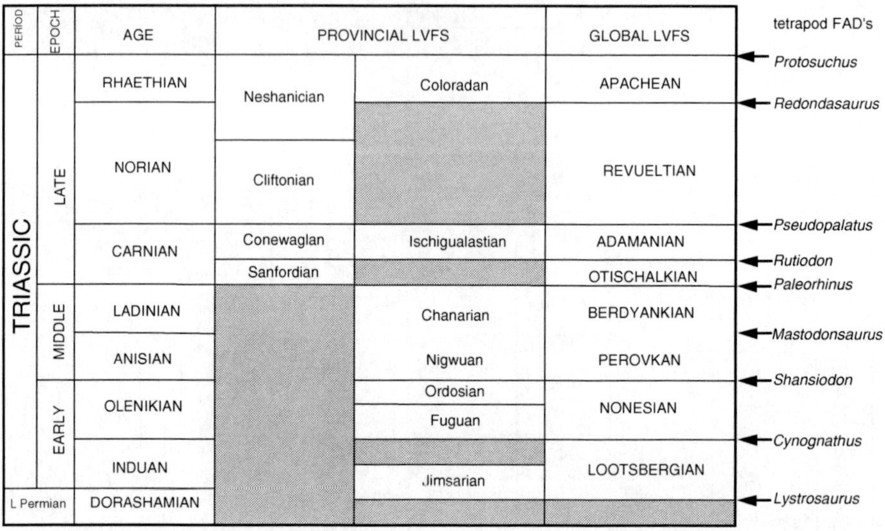


*Figure 6.8 Triassic provincial and global LVFs, their correlation to SGCS (standard chronostratigraphic scale) and the tetrapod FADs that define their boundaries (modified from Lucas, 1998).*

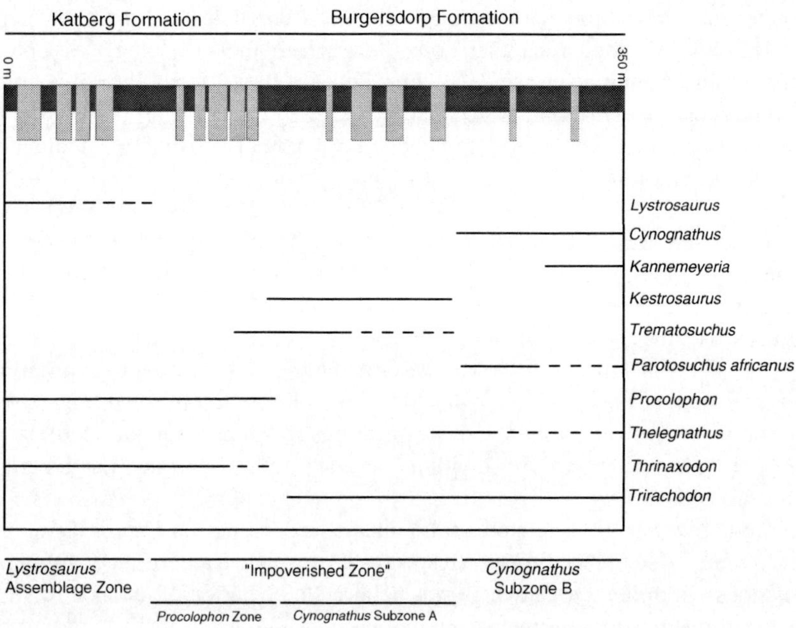


*Figure 6.9 Biostratigraphic faunal distribution of the tetrapod fossils from the Lower Triassic of South Africa (modified from Neveling et al., 1999).*

*Trematosuchus* in the upper reaches of the Katberg Formation (*Procolophon* Zone) is strongly indicative of an Upper Olenekian age for these strata. Thus, the Lootsbergian LVF from Lucas (1998) could be potentially divided in two portions, based on Procolophonids.

Lucas and Hancox (2001) considered the younger Triassic nonmarine strata in southern Africa, assigned to the Molteno and Elliot formations, as problematic for correlation. The only tetrapod fossils from the Molteno Formation are footprints, and the presence of dinosaur tracks suggest late Carnian as a maximum age for the upper part of the formation, because the oldest records of dinosaur fossils are Otischalkian (=late Carnian: Lucas, 1998). The lower Elliot, by its turn, contains a tetrapod fossil assemblage dominated by prosauropod dinosaurs (the *Euskelosaurus* range zone, Kitching and Raath, 1984). On basis on the presence of a traversodontid cynodont, Gauffre (1993) argued for a lower Carnian age for the Lower Elliot, as well as for both Ischigualasto and Santa Maria Formations, but Lucas and Hancox (2001) points out three problems in this assertion: (1) traversodontids are known through the Rhaetian; (2) the Ischigualasto and upper Santa Maria assemblages are late Carnian; and (3) no *taxa* are shared between Ischigualasto-Santa Maria and the lower Elliot, so their correlation lacks any basis. The dominance of prosauropod dinosaurs in the Lower Elliot leads Lucas and Hancox (2001) to assign it a Norian age because: (1) the only other prosauropod-dominate Late Triassic faunas – that of the German Knollenmergel and that of the Los Colorados Formation in Argentina – are Norian; (2) *Euskelosaurus* may be an ecological vicar of *Plateosaurus*; (3) its footprint ichnofauna is closest to Norian ichnofaunas; and (4) Lower Jurassic strata directly overlie the lower Elliot.

However, even Lucas and Hancox (2001) admit that none of these arguments is incontrovertible. We agree that a correlation between the Lower Elliot and Los Colorados-Knollenmergel assemblages, based on the dominance of prosauropod dinosaurs, is more probable than between Ischigualasto-Santa Maria and Lower Elliot faunas, but it doesn't means that the correlated Los Colorados-Knollenmergel-Lower Elliot faunas are necessarily Norian, once that Ischigualasto-Santa Maria assemblages are not undoubted Late Carnian for the reasons just discussed.

The middle and upper Elliot, by its turn, contains a more diverse tetrapod assemblage defined as the "*Massospondylus* range zone" including, besides the prosauropod dinosaur *Massospondylus*, a coelurosaur and several ornithischians, sphenosuchian archosaurs, a proganochelyid turtle, a tritylodontid and other cynodonts and also mammals (Lucas and Hancox, 2001). For these authors, the age of the *Massospondylus* range zone is unclear and they conservatively assign it an Early Jurassic age.

In Brazil, a new tetrapod fauna, dominated by non-mammalian cynodonts, was recently presented by Abdala *et al.* (2001) for the Middle-Upper Triassic. This new assemblage is the only within the Santa Maria Formation characterized by the prevalence of cynodonts and absence of both dicynodonts and rhynchosaurs, suggesting it represents a temporal interval not previously recognized for the Brazilian Triassic (Fig. 6.10). However, the geographic occurrence of this fauna is very restricted (as pointed out by Lucas, 2001) suggesting, at first, it would be more prudent to consider it as a Local-Fauna rather than a Zone. Nevertheless, the hypothesis of the occurrence of a distinct faunal assemblage for this interval fits into the faunal succession observed by Tucker and Benton (1982) (Fig. 6.4), i.e. a peak of cynodont-dominated faunas just before the explosion of the hyperodapedontidae rhynchosaurs around the world.

Besides, a new Triassic fauna, also dominated by cynodonts, has been discovered in Madagascar (Flynn *et al.*, 1999, 2000). This fauna, which has so far yielded procolophonids, rhynchosaurs, dinosaurs, sphenodontians, dicynodonts and mainly cynodonts, is regarded provisionally to be late "Ladinian"/early "Carnian" in age. The described Malagasy traversodontids show notable similarities to some of the traversodontid types from the Santa Cruz do Sul fauna (Abdala *et al.*, 2001). These similarities, even when restricted just to cynodonts, let Abdala *et al.* (2001) infer, together with the arguments above discussed, a late Ladinian (but not necessarily early Carnian) age for the Santa Cruz Fauna. Finally, the age inferred for this biozone fits to the limit between the Berdyankian and Otischalkian LVFs from Lucas (1998), who considered (see discussion ahead) that an other LVF could be potentially be recognized for that interval, although data to do this were insufficient at that moment.

Lucas (2001) reviewed the Brazilian Triassic biostratigraphy and concluded that the stratigraphic data (or lack thereof) and the composition of the tetrapod assemblages of the Santa Maria Formation support only the recognition of two, temporally successive tetrapod faunas separated by a substantial hiatus (Fig. 6.10). According to the same author, recent attempts to refine this correlation by recognizing more subdivisions (e.g. Schultz, 1995; Abdala *et al.*, 2001) are laudable, but they lack support from lithostratigraphic or biostratigraphic data. To get a detailed lithostratigraphic framework of the Brazilian Triassic tetrapod localities may be impossible, given the heavily vegetated landscape of Rio Grande do Sul, where outcrops are intermittent and few exposures encompass more than several meters of stratigraphic section.

From that, Lucas (2001) erected a *Dinodontosaurus* Assemblage Zone (= Berdyankian, Lucas, 1998) and a *Hyperodapedon* Assemblage Zone (=Otischalkian, Lucas, 1998). Besides, in this paper, the author proposes a synonymy for some *taxa* (*Ischigualastia* = *Stahleckeria* in the *Dinodontosaurus* AZ and *Jachaleria* = *Ischigualastia* in the *Hyperodapedon* AZ) so reinforcing the existence of only two biozones. However, neither the proposed synonymy nor the inclusion of the fauna from the upper portion of Santa Maria and that from the Caturrita Formation in a single biostratigraphic unit are accepted by Brazilian paleontologists (see discussion ahead).

Lucas and Heckert (2002) based on recent revision of some South American and Malagasy rhynchosaurs (Contreras, 1999; Langer and Schultz, 2000; Langer *et al.*, 2000a, b), which redefined most of them as belonging to the genus *Hyperodapedon*, as well as in a rhynchosaur recently documented from the Upper Triassic Popo Agie Formation of Wyoming and also attributed to the same genus, define a *Hyperodapedon* biochron of late to latest Carnian (Otischalkian-Adamanian) age for strata in North America, Scotland, India, Zimbabwe, Tanzania, Madagascar, Argentina, and Brazil (Fig. 6.11).

| LITHOSTRATIGRAPHY | Barberena (1977) | Barberena *et al.* (1985a) | Barberena *et al.* (1985b) | Schultz *et al.* (1995) | Schultz (1995) | Abdala *et al.* (2001) | Lucas (2001) | |
|---|---|---|---|---|---|---|---|---|
| Caturrita Fm. | Rhyncho-cephalia Assemblage Zone | *Scaphonyx* Zone | Botucaraí local fauna | *Jachaleria* interval | Association 7 | *Jachaleria* interval | *Hyperodapedon* Assemblage Zone | Adamanian (Ischigualastian) *Latest Carnian* |
| Santa Maria (upper) | | | | Rhynchosaur Cenozone | Association 6 | Rhynchosaur Biozone | | |
| | | | | | Association 5 | | | |
| | | | Alemoa local fauna | | Association 4 | | | Otischalkian *Late Carnian* |
| Santa Maria (lower) | Therapsida Assemblage Zone | *Dinodon-tosaurus* Zone | | | | Traversodontid Biozone | *Dinodon-tosaurus* Assemblage Zone | Berdyankian (Chañarian) *Ladinian + Early Carnian* |
| | | | Chiniquá lf | Therapsid Cenozone | Association 3 | *Dinodon-tosaurus* Biozone | | |
| | | | Pinheiros lf | | Association 2 | | | |
| | | | | | | | | Perovkan |

*Figure 6.10* *Schemes of Triassic tetrapod biostratigraphy of the Santa Maria and Caturrita formations (modified from Lucas, 2001).*

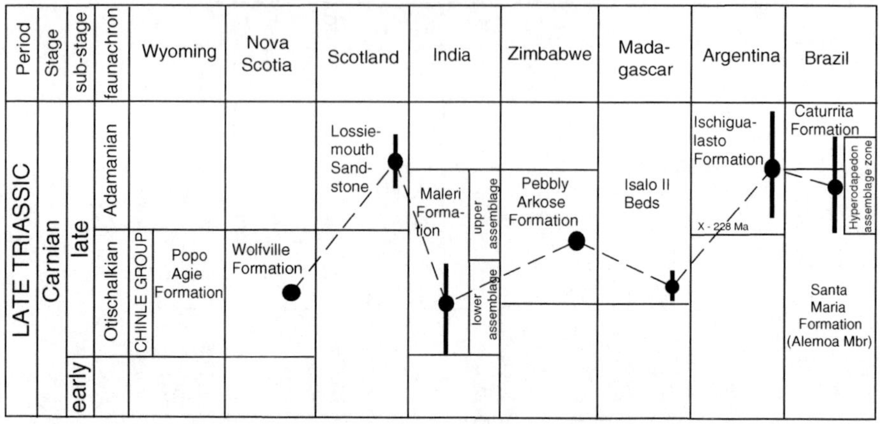


*Figure 6.11 Global correlation of* Hyperodapedon *localities (modified from Lucas and Heckert, 2002).*

## 6.4 Discussion

Anderson (1980) claims that reliable correlations depend on reliable biostratigraphy based on comprehensive sampling and sound taxonomy. Regarding to these aspects, non-marine sequences will always be less reliable than marine ones due the non-continuity of the basins, the lateral faciologic variation of the sequences and the scarcity of the samples, which leads to a high number of doubtful taxonomic interpretations. Most of the authors cited above perform faunal correlations at the level of genera. However, a "*genus*" is an arbitrary taxonomic unit whose monophyly can only be confirmed after a phylogenetical analysis, and it was not made for many *taxa* used in the correlations.

Shubin and Sues (1991) stressed that the age of early Mesozoic continental strata can only rarely be refined beyond the stage level. Intercontinental correlations generally prove difficult because of the depositional heterogeneity of the standard Triassic sequence in the Germanic basin of central Europe. From this sequence only the Upper Triassic (Keuper) produced significant assemblages of terrestrial plants and tetrapods. Besides, as cited by Ochev and Shishkin (1989), the assemblages of the Early Triassic regional sequence constitute the only so far known among the Triassic tetrapod faunas that permits a direct comparison with the marine sections, due to the presence of common labyrinthodont genera. According to the same authors, tetrapod faunas provide the best means for correlation of triassic fossil-bearing continental deposits. However, in biostratigraphic studies, the purpose cannot be restricted to a simple co-ordinating the particular faunas into one or another sort of the correlation charts as it often occurs. The data obtained in such a way should be analyzed in order to reveal the most universal biotic replacements which could serve as an integral framework for comparing the regional faunal changes.

### 6.4.1 The Question of Cross-Correlation Between Marine and Non-Marine Timescales to Triassic

Lucas (1998) believes that the Triassic LVFs may provide a framework for the correlation of Triassic non-marine deposits with a temporal resolution comparable to the seven Triassic Stages/Ages of the standard chronostratigraphic scale (SGCS). Nevertheless, although it constitutes the finest degree of resolution for a global tetrapod-based biochronocorrelation available until the moment, some aspects of this scheme must be viewed with caution.

The main question, as in all proposals of such kind, is the link between the non-marine ages (LVFs or any other) and the SGCS (based on marine strata and fossils), specially regarding to Middle and Late Triassic ages (see Benton, 1986, for further discussion). In Lucas' (1998) scheme we find:

- **Lootsbergian LVF** (non-marine) X **Induan** (SGCS): cross-correlation provided by the occurrence of characteristic Lootsbergian temnospondyls in ammonite-bearing Induan strata of the Wordy Creek Formation in Eastern Greenland. However, it is uncertain if Lootsbergian correlates to the end of the Induan (Lucas, 1998). This is the ideal situation, i.e. the presence of shared fossils in marine and non-marine correlated strata.
- **Nonesian LVF X Olenekian:** the occurrence of the temnospondyl *Parotosuchus* in marine Olenekian strata of the Mangyshlak Peninsula in western Kazakstan and the occurrence of *Aphanerama* or *Parotosuchus* records in Svalbard, Germany and/or North America support this correlation.
- **Perovkan LVF X Anisian:** Direct correlation can be made by the presence of marine facies containing Anisian conodonts and tetrapods in the Lower Röt (Upper Buntsandstein), Germany.
- **Berdyankian LVF X (Ladinian + Early Carnian):** there is no direct cross-correlation to the standard chronostratigraphic scale. According to Lucas (1998) it is possible that there is a need for another LVF between the Berdyankian and Perovkan, but it could not be defined and characterized at that time due to inadequate data. In the same way, there is potential to recognize an LVF between the Berdyankian and Otischalkian (base of Late Carnian), although data to accomplish this are still insufficient.
- **Otischalkian LVF X Late Carnian:** correlation supported by the presence of *Paleorhinus* and *Metoposaurus* records in marine strata in Austria, palinostratigraphy and magnetostratigraphy. The tetrapod *taxa* that support this correlation, however, are not undoubted. Fara and Hungerbühler (2000) and Hungerbühler (2000), for example, demonstrate that the monophyletic status of the phytosaur *Paleorhinus* (the guide-fossil for the Otischalkian LVF) is very doubteous. For Lucas

(1998) the Otischalkian is not represented in South America. However, South American paleontologists (e.g. Schultz *et al.*, 2000; Abdala *et al.*, 2001) consider both Ischigualasto Formation of Argentina and the upper layers of the Santa Maria Formation as, at least, beginning in Early Carnian, once that the radiometric age obtained by Rogers *et al.* (1993) in Argentina match the base of the Carnian in most of the timescales (e.g. Ross *et al.*, 1994; Gradstein *et al.*, 1995; Gradstein and Ogg, 1996). Lucas (1998), however, adopts the paleomagnetic data and the time-scale of Kent *et al.*, (1995), in which that age fits in the late Carnian. It just demonstrates that, even the existence of absolute ages in some strata not necessarily solves such kind of controversies regarding to biochronology.

– **Adamanian LVF X Latest Carnian:** the correlation is based on palynostratigraphy and magnetostratigraphy (using also the data of Kent *et al.*, 1995, among others). The *Hyperodapedon* biochron proposed by Lucas and Heckert (2002) to late/latest Carnian (Otischalkian-Adamanian) provides a biochronological resolution coarser than that achieved using other *taxa*, principally phytosaurs and aetosaurs, used in Lucas' (1998) scheme, and it does not solve the problem of the direct cross-correlation to the standard chronostratigraphic scale. According to Lucas (2001) the faunas of the upper layers of the Santa Maria Formation and those of the overlying Caturrita Formation belonged both to the *Hyperodapedon* Assemblage Zone, correlated to the Ischigualastense Reptile-Age (=Adamanian, Lucas, 1998). The main faunistic argument for this purpose would be the presence of the rhynchosaur *Hyperodapedon* in both Brazilian formations, notwithstanding all the rest of the *taxa* is different in each unit. However, the rhynchosaur of the Caturrita Formation (*"Scaphonyx" sulcognathus*) is a hyperodapedontinae, but is not *Hyperodapedon* at all (Langer and Schultz, 2000).
– **Revueltian LVF X Norian:** Palynostratigraphy, magnetostratigraphy and sequence stratigraphy suggests the type Revueltian assemblage is of Norian Age (Lucas, 1998). Revueltian correlates approximately with the entire Norian, however, whether or not the beginning and end of the Revueltian and Norian are exact equivalents is unclear. There is no direct correlation to the marine sequences of the SGCS.
– **Apachean LVF X Latest Norian:** According to Lucas (1998), the Apachean is the most difficult Triassic LVF to correlate globally. This almost certainly reflects a provincialization of the global tetrapod fauna. There is no direct cross-correlation to the standard chronostratigraphic scale.

So, proposals of global tetrapod based time-scales as that of Lucas (1998) are laudable and must be enhanced, but still must overcame some problems, specially taking into account that: "the use of fossils in the establishment of a relative time-scale includes three important points: first, *the reliability of the taxonomic identification of the fossils and the correlation among the biotas* (which is particularly important to fragmentary records); second, *the reliability of the ages attributed to the fossil-bearing strata* (fundamental to correlate non-continuous units), and third, *the faciologic control regarding to the fossil-bearing levels* (which interferes directly in the correlation and is very important in continental environments). So, to pay no attention to anyone of these points can lead any kind of biostratigraphic hypothesis to become highly speculative" (Marsicano, 2002).

In the Triassic, due to its particular paleogeographic and paleoclimatic conditions, these problems used to be still more difficult to solve than in other periods. Most of the continental area of Pangea was subject to erosion, with deposition taking place in restricted areas, mainly in narrow rift and some epicontinental basins. Triassic successions are rarely complete, commonly showing depositional hiatuses that hinder the correlation among the different basins (Embry, 1988). Besides, the tetrapod fossil record is often fragmentary, making difficult precise taxonomic identification and its correlation.

### 6.4.2 Problems Regarding Correlations Among Continental Basins

It is easy to realize that correlation between non-marine sediments reveals more obstacles than in marine ones. Marine sedimentation is controlled by sea-level changes, but what about continental deposition? Continental tetrapod-bearing layers are mainly of fluvial origin. Some authors (e.g. Shanley and Mccabe, 1994; Miall, 1996) advocate that tectonics is the main factor that controls the accommodation of alluvial sedimentation. Others (Wright and Marriot, 1993; Koss *et al.*, 1994; Olsen *et al.*, 1995; Emery and Myers, 1996) stressed the role of eustasy in controlling the creation of space for fluvial deposition. Anyway, the concept that base level variations control accommodation (the basis of sequence stratigraphy) is surely applicable to fluvial strata. However, the use of sequence stratigraphy to continental systems is still controversial due to the presence of two conflicting paradigms (Miall and Miall, 2001): the first one – the "Global eustasy paradigm" – claims that eustatic variations are global and, for this reason, their signs would be registered somehow in the sedimentary basins of the whole planet, and it could be traced from one to another; on the other hand, the "Complexity paradigm" states that there are many factors which affect the depositional architecture of a basin, precluding long-distance correlations.

### 6.4.3 The Role of Taphonomy in the Samples Used to Correlation

Vertebrate taphonomists must also take into account that vertical distribution and time-averaging may introduce strong preservation biases in terrestrial taphocenosis preserved in fluvial facies (e.g. Behrensmeyer and Hook, 1992). This bias is controlled by the depositional style of the channel and the overbank deposits. Smith (1993) demonstrated that bone weathering increases and bone density decreases with distance along a channel. Thus, the proximity of the burial site to the main channel and the frequency and intensity of the floods that bury the bones in the floodplain control the taphonomic signature of the vertebrate fossils.

Willis and Behrensmeyer (1994) state that the preserved succession is not a result of a gradual aggradational sedimentation of the overbank fines, but rather a product of sedimentation followed by long periods of non-sedimentation and soil formation. In some cases, the fossil record would be biased by the alternation of habitats formed by phases of aggradation and phases of fluvial incision separated in time for a lot of years. Lag deposits of reworked bones, representing time averaged taphocenoses, could be formed during the phases of fluvial incision. For example, Smith (1993) discusses the time resolution regarded to the fluvial facies of the Hoedemaker Member of the Late Permian Teekloof Formation of South Africa. The bones found at the floodplain indicates a floodplain accretion rate of 5.5 mm/y, while the pedogenetic features of the paleosoils, in turn, indicate periods of about 10,000 years of almost non-deposition, with an accretion rate of 0.035 mm/y.

Thus, it is clear that the preservation of terrestrial vertebrate fossils (even bones, tracks or footprints) depends on the facies type of the fluvial system, i.e. the development of channel and floodplain facies. The development of these facies, in turn, is controlled by variation of base level, so that one can conclude that the taphonomic preservation, as well as the distribution of terrestrial body and trace fossils is a function of base level change. In this context, their preservation, as well as their first and last occurrences within a given stratigraphic unit may not be biologically, but stratigraphically controlled, in a manner to what Holland (1995a,b) demonstrated for the distribution of marine fossils. Thus, biostratigraphic problems, concerning both the temporal resolution and interbasinal correlation, may reflect the lack of integration between studies concerning fluvial sequence stratigraphy and taphonomy. Most of the biostratigraphers, working with terrestrial faunas, use lithostratigraphic columns as a geologic basis for their work, and take first and last appearance of certain *taxa* as horizons for correlation. In this context, biostratigraphic mismatches may reflect different sedimentation regimes. Taking this into account, sequence boundaries and key flooding surfaces may be traceable laterally throughout the marine to non-marine succession and will provide a tool for correlation between terrestrial and marine guide fossils (see Holz and Simões, in this book, Chapter 12).

Meanwhile, the potential – and the limitations – of a biostratigraphy and biochronology for the non-marine Triassic based on tetrapods must be stressed. New findings of Triassic tetrapods continuously provide new data to improve, as demonstrated, this biostratigraphic framework. On the other hand, variations in the faunal content from a place to another can have different explanations, including time span, paleoecology (specially endemism related to climatic latitudinal zonation) or sampling. Indeed, the role of the time-averaging in such kind of continental fossil-bearing sequences is not yet fully evaluated, and any attempt to establish a biostratigraphy for the non-marine Triassic based on tetrapods must be based on a detailed taphonomic work.

### 6.4.4 Can Radiometric and Paleomagnetic Data Solve the Problem?

In order to reinforce the link between marine and non-marine time-scales, it would be necessary, in addition to the continuous improvement of paleontologic and stratigraphic correlation, to obtain more radiometric and paleomagnetic data, mainly from tetrapod-bearing strata. The age obtained by Rogers *et al.* (1993) at the basal layers of the Ischigualasto Formation, Argentina, is the only radiometric datum coming from tetrapod-bearing non-marine Triassic sequences available until today. Additional data can be obtained, for example, from some other Argentinean Middle-Upper Triassic sedimentary sequences, which are interbedded with volcanoclastic layers (Stipanicic and Marsicano, 2002). This work is just in development in Argentina and Brazil and new data are expected for the next years. Regarding to paleomagnetic data, Vizan (*apud* Stipanicic and Marsicano, 2002) advertises: "among the first-order problems which until now persists universally in the scope of paleomagnetism, is the uncertainty about the pre-Jurassic configuration of Pangea. The knowledge about the position of the plates during Triassic is still controversial and it is difficult accept any model as definitive." Nevertheless, it constitutes an important tool to enhance the geochronological framework of the Triassic, and its use must be disseminated, mainly on non-marine sequences, where such kind of data is practically absent.

## 6.5 Conclusions

The Triassic was the only period of tetrapod history during the entire length of which a single landmass existed, and this paleogeographic condition favoured the dispersion and the cosmopolitanism of the floras and tetrapod faunas. In turn, Triassic successions are rarely complete, commonly showing depositional hiatuses that hinder the correlation among the different basins. Besides, endemism of land vertebrate assemblages may also be due to facies, sampling and taphonomic biases. Notwithstanding, a global correlation among non-marine Triassic strata and the establishment of a time scale based on their tetrapod faunas is quite possible, but it is still necessary more evidences to link these data with those of the marine sequences.

To complete this goal, the continuity of the fieldwork on Triassic strata shall probably provide new radiometric and paleomagnetic data, both from marine and continental strata, which today are still very scarce. Besides, fieldwork in non-marine sequences will provide also new findings of tetrapods, which can continuously improve this tetrapod-based biostratigraphic framework if they are properly studied and compared. However, the most essential thing is that fieldwork on continental tetrapod-bearing Triassic beds must be based in a systematic taphonomic analysis and on the application of the concepts of sequence stratigraphy for continental strata.

## 6.6 Acknowledgements

I thank Rogério Roque Rubert and Paula Dentzien Dias, for the drawing of the figures; Michael Holz and Claiton Scherer, for their comments and suggestions; Maurice E. Tucker for permission to use to use some figures; the reviewers Michael J. Benton (University of Bristol), Ernesto L.C. Lavina (Unisinos, São Leopoldo, Brazil), and José Fernando Bonaparte (Visiting Researcher CNPq/UFRGS); and the Editor, for not letting me give up.

Chapter 7

# The K–T Boundary

EDUARDO A. M. KOUTSOUKOS
PETROBRAS-CENPES, Cidade Universitária, Quadra 7, Ilha do Fundão, 21949-900
Rio de Janeiro, RJ, Brazil.



## 7.1 Introduction

Chronostratigraphic boundaries have great importance in the geosciences. These occur in both marine and continental realms, provide key markers for geological correlations, and for attaining more refined paleogeographic and biogeographic reconstructions. Stage boundaries are commonly characterized by significant geological and paleobiological events which leave a distinct signature in the stratigraphic record.

The Cretaceous-Tertiary (K–T) boundary marks the end of the Mesozoic Era, about 65 million years ago, when widespread oceanic basins had global circulation patterns somewhat similar to present day (e.g. Gordon, 1973; Berggren and Hollister, 1974; Hancock and Kauffman, 1979; Haq *et al.*, 1987, 1988). For the last two decades of the 20th Century there has been much multidisciplinary research on more than 200 K–T boundary sections worldwide, with about two-thirds being outcrops (Lima, 1999, and references therein), from oceanic to marginal basins, interior seaways and continental sites. This makes the K–T transition the best documented ever in published geoscience literature.

This paper summaries these results and in so doing shows the strengths and weakness of the different approaches that have been used.



*E.A.M. Koutsoukos (ed.), Applied Stratigraphy,* 147-161.

## 7.2 Nature of the Stratigraphic Record

### 7.2.1 Stratigraphic Completeness and Sampling Bias

The K–T boundary is globally recognized by an event which involved the extinction of around 64 to 85% of all species on Earth (e.g. Sepkoski, 1990, 1994; MacLeod *et al.*, 1997), including organisms in both marine and terrestrial environments. This makes it the second largest mass extinction in Earth's history (the first is the Permian-Triassic boundary, when nearly 94% of all species disappeared). Major issues include:

- the precise positioning of the K–T boundary and the completeness of the stratigraphic record (e.g. Dingus, 1984; Benton, 1989; MacLeod and Keller, 1991; MacLeod, 1996a,b),
- the pace and magnitude of the extinction process (i.e. Was it gradual or catastrophic? Why is there a clear discrepancy between the timing and intensity of the mass extinction at high and low latitudes?),
- survivorship and origination rates per fossil group in the aftermath of the event (recovery interval),
- the nature of the boundary event itself (see below);
- and, ultimately, evaluate the ecosystem dynamics in response to major environmental changes.

To assess the completeness of a stratigraphic boundary a detailed, high-resolution, biostratigraphic study has to be undertaken, and further complemented with a thorough sedimentological evaluation and supporting stratigraphic analyses, such as of stable isotopes, cyclostratigraphy, etc. These combined analyses would have the obvious advantage of highlighting any minor change in sediment accumulation rate and mark the presence of possible non-depositional hiatuses in the succession, which are of difficult perception after single visual analyses.

A less than rigorous sampling method (e.g. Signor and Lipps, 1982) and the vagaries of biogeographically restricted/stenotopic species occurrences (e.g. Wignall and Benton, 1999), greatly decreases the chances of finding rare diagnostic taxa. Coupled with failure to identify properly the presence of event deposits (such as hardgrounds and bioturbated layers, gravity flows and turbidites, where reworking and fossil mixing commonly take place), condensed sections, and discontinuities in sedimentary successions, these have often lead to erroneous interpretations in estimating patterns and rates of biotic extinction, survivorship and speciation across the K–T boundary transition (e.g. Smit, 1982; Argast *et al.*, 1987; Rigby Jr., 1987; Keller, 1988, 1989a,b, 1993; Eaton *et al.*, 1989; Liu and Olsson, 1992; Olsson and Liu, 1993; Huber, 1996; MacLeod and Huber, 1996; Wignal and Benton, 1999). In addition, common reworking of uppermost Cretaceous fossils into lower Danian beds further contributes to the difficulties in precisely positioning the boundary (e.g. Florentin *et al.*, 1991; Montgomery *et al.*, 1992; Zachos *et al.*, 1992; Olsson and Liu, 1993; Huber, 1996; Huber *et al.*, 2002).

### 7.2.2 The K–T Boundary Stratotype at El Kef, Tunisia

The GSSP stratotype of the K–T boundary has been designated by the International Commission on Stratigraphy (ICS) at the El Kef outcrop section in northern Tunisia, which has a continuous and exceptionally thick calcareous pelagic sedimentary sequence, with no unconformities, hardgrounds, or stratigraphic gaps, and with excellent preserved micropaleontological, geochemical and lithological marker criteria (Smit, 1982; Cowie *et al.*, 1989; Odin, 1992). The section, deposited at the upper bathyal (estimated paleodepth of 300–500 m; Speijer and Van der Zwaan, 1994, 1996), yields an abundant and well-preserved microfossil record, allowing for the reconstruction of a complete succession of bioevents across the K–T transition.

The main criteria defining the K–T boundary at the El Kef stratotype section are summarized in Remane *et al.* (1999). The boundary is placed at the extinction horizon of nearly all upper Maastrichtian planktonic foraminifera and of many coccolith species, and coincides with a lithological change, from uppermost Cretaceous gray marlstones to a 50-cm-thick dark, organic rich claystone, which records the sequential appearances of the first Tertiary species (Fig. 7.1). Marking the exact positioning of the K–T boundary, at the base of the claystone, lies a 2–3 mm thick red oxidized layer enriched in Ir and Ni-rich spinels (Fig. 7.1). The boundary is also marked by a negative $\delta^{13}C$ shift (e.g. Keller and Lindinger, 1989; Keller *et al.*, 1995; MacLeod, 1996c; Ginsburg *et al.*, 1997; Smit *et al.*, 1997; Robin and Rocchia, 1998) and a $^{87}Sr/^{86}Sr$ maximum anomaly (Vonhof and Smit, 1997), which have been related to drastic global environmental changes induced by the K–T boundary event. However, Ir-enriched clay layers associated with geochemical anomalies are not restricted to the K–T boundary horizon (e.g. at the P$\alpha$/P1a boundary in Poty, Brazil: Albertão *et al.*, 1994; Koutsoukos, 1998; in Mexico, Guatemala, and Haiti: Keller *et al.*, 2003; and at the P1b/P1c boundary in Negev, Israel: Hansen *in* Hansen *et al.*, 2001), and thus cannot be considered alone as a defining K–T boundary criterion.

A turnover of the benthic foraminifera across the boundary transition was recorded by Speijer and Van Der Zwaan (1996). A diverse latest Maastrichtian upper bathyal benthic foraminiferal assemblage was replaced by a Danian low-diversity, middle-outer neritic assemblage, with about 30% of extinction rate.

### 7.2.3 Patterns of Extinction and Speciation

The stratigraphic record is much more complete in marine than in continental successions, where the knowledge of patterns of extinction and speciation becomes often blurred by the scattered fossil recovery. Despite vagaries in fossil occurrence and preservation (e.g. Signor and Lipps' effect, 1982), the well-preserved sedimentary and fossil record documented from several continuous terminal Cretaceous and K–T boundary sections, clearly indicate that: (a) extinctions of some taxonomic groups occurred well before the boundary (e.g. of marine organisms such as inoceramid and rudist bivalves, ichthyosaurs, plesiosaurs, mosasaurs and pliosaurs, and on land, of flying reptiles – pterosaurs, by the mid-Maastrichtian – Table 1), apparently in

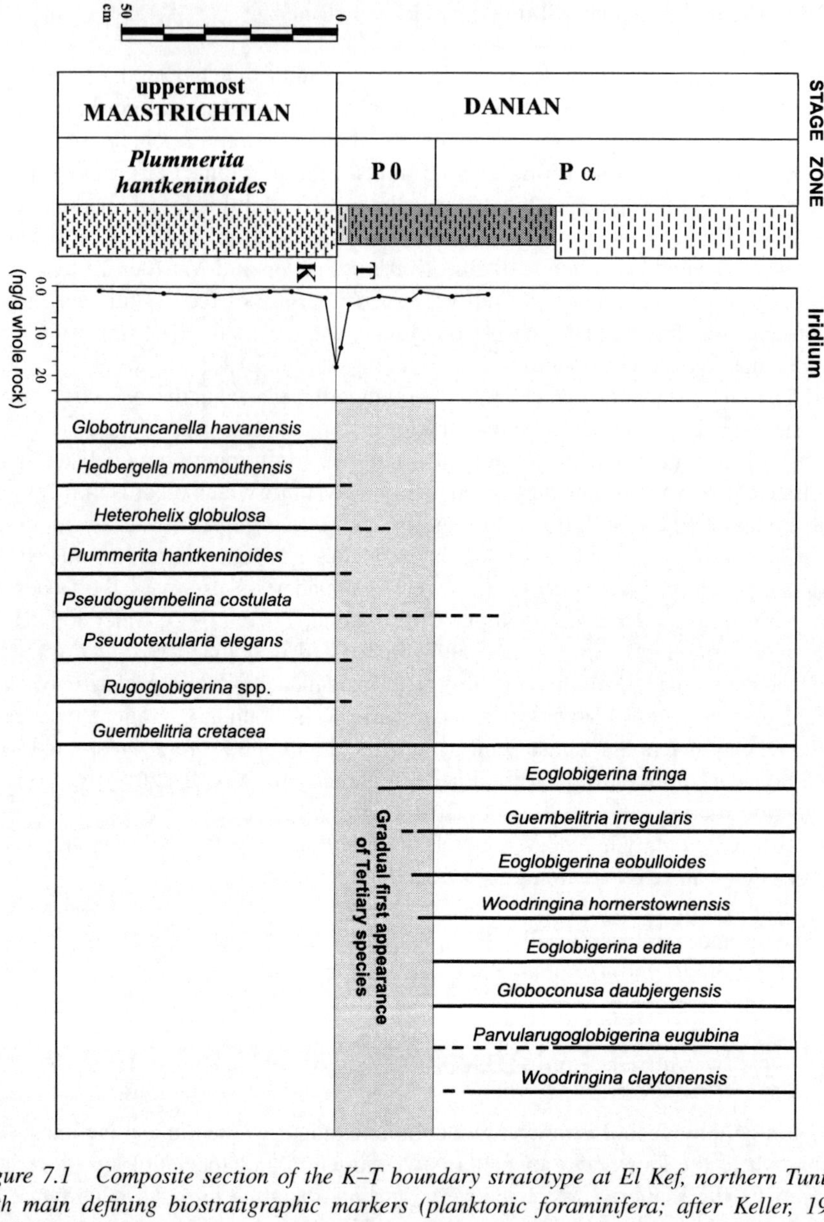


*Figure 7.1 Composite section of the K–T boundary stratotype at El Kef, northern Tunisia, with main defining biostratigraphic markers (planktonic foraminifera; after Keller, 1988; D'Hondt, 1991; Brinkhuis et al., 1994; Keller et al., 1995; MacLeod, 1996c; Smit et al., 1997; Li and Keller, 1998), and Ir profile (Robin et al., 1991; Keller et al., 2003).*

response to slow environmental changes such as the gradual lowering of sea-level (Haq *et al.*, 1987; Stanley, 1984; Hallam, 1987, 1992; Hallam and Wignall, 1999) and cooling temperatures during the early to late Maastrichtian (Haq *et al.*, 1987; Barrera, 1994; Barrera and Savin, 1999; Hallam and Wignall, 1999; Miller *et al.*, 1999; Stoll and Schrag, 2000); (b) there was a catastrophic event at the boundary, which severely affected or eliminated several taxonomic groups in terrestrial and marine ecosystems.

Of the diverse biota which existed at the end Cretaceous, several terrestrial and marine groups perished or suffered great losses at or around the boundary event. The marine ecosystems saw the demise of several taxonomic groups, such as most species of ammonite cephalopods and belemnites. Major extinctions also affected several other groups in the plankton (e.g. foraminifera, calcareous nannofossils), and, to a lesser extenct, in the benthos (e.g. larger benthic foraminifera, and many corals). On the continents, dinosaurs (ornithiscians and non-avian saurichians), marsupial mammals, and many plant species (except the ferns and seed-producing plants; Hickey, 1984; Johnson and Hickey, 1990; Hickey and McSweeney, 1992) became extinct. On the other hand, most placental mammals, birds, crocodilians, turtles, snakes, lizards, amphibians (frogs and salamanders), nautiloid cephalopods, fishes, bryozoans, brachiopods, gastropods, echinoderms, smaller deep-water dwelling benthic foraminifera, radiolarians, dinoflagellates, and diatoms, were little affected by the K–T event (cf. references in Table 7.1).

## 7.3 Causes of the K–T Boundary Event

The nature of the K–T boundary event, and related stratigraphic record and biotic changes (mode and rates of species extinction and radiation) have been subject of much debate and conflicting arguments for the last two decades. Several hypotheses have been proposed trying to explain the mass extinctions at the end Cretaceous, but the most commonly advocated causes are global sea-level changes (e.g. Stanley, 1984; Hallam, 1987), catastrophic volcanism (e.g. McLean, 1985; Rampino and Stothers, 1988; Courtillot, 1990a,b; Courtillot *et al.*, 1988, 1996; Glasby and Kunzendorf, 1996), and extraterrestrial bolide (asteroid or comet) impacts (e.g. Alvarez *et al.*, 1980, 1984a,b, 1992; Alvarez and Muller, 1984; Rampino and Stothers, 1984; Schwartz and James, 1984; Alvarez and Asaro, 1990; Hildebrand and Boynton, 1990; McLaren and Goodfellow, 1990; Smit, 1990; Sigurdsson *et al.*, 1991; Florentin *et al.*, 1991; Shartpton *et al.*, 1992, 1993; Ivany and Salawitch, 1993). These three processes share in common the likely effects of a drastic change in global climate, which would have disrupted global ecosystem dynamics and triggered a sharp increase of extinction rates in both continental and marine realms. It has also been shown that species of high-latitude cooler regions seem to have had greater survivorship rates than the ones thriving along warm low latitudes (e.g. non-marine ostracodes in northern Alaska: Brouwers and De Deckker, 1996; bivalve and gastropods of the Arctic Ocean: Marincovitch Jr., 1996), suggesting that, whatever was the nature of the event, it somewhat favored species adapted to cold-water conditions and/or had a lesser impact on high latitudes.

*Table 7.1 Overview of selected references for biotic extinction and recovery patterns across the K–T boundary*

| MARINE ECOSYSTEMS | |
|---|---|
| **ZOOPLANKTON** | **PHYTOPLANKTON** |
| **Planktonic Foraminifera** | **Calcareous Nannofossils** |
| Smit (1982) | Bramlette and Martini (1964) |
| Keller (1988, 1989a,b) | Perch-Nielsen (1979, 1982) |
| Liu and Olsson (1992) | Percival and Fischer (1977) |
| Olsson and Liu (1993) | Monechi (1979) |
| Keller *et al.* (1993) | Thierstein (1981) |
| Huber (1996) | Perch-Nielsen *et al.* (1982) |
| Koutsoukos (1996) | Henriksson (1996) |
| MacLeod (1996) | Pospichal (1994) |
| MacLeod *et al.* (1997) | Gardin and Monechi (1998) |
| | Mai (1999) |
| **MICROBENTHOS** | **Dinoflagellates** |
| **Benthic Foraminifera** | Willams and Bujak (1985) |
| Widmark and Malmgren (1992) | Brinkhuis and Zachariasse (1988) |
| Coccioni and Galeotti (1994, 1998) | Moshkovitz and Habib (1993) |
| Kuhnt and Kaminski (1993, 1996) | Elliot *et al.* (1994) |
| Speijer and Van der Zwaan (1996) | Askin and Jacobson (1996) |
| Widmark (1997) | Schioler *et al.* (1997) |
| Culver (2003) | Brinkhuis *et al.* (1998) |
| **Ostracodes** | |
| Coles (1990) | |
| Watley (1990) | |
| Fauth (2000) | |

*Table 7.1 Continued*

| | |
|---|---|
| **INVERTEBRATES** | **CONTINENTAL ECOSYSTEMS** |
| **Mollusks** | **Land Vertebrates** |
| Kauffman (1984)<br>Ward *et al.* (1986)<br>Kennedy (1989, 1993)<br>Raup and Jablonski (1993)<br>Sepkoski (1990)<br>Jablonski and Raup (1995)<br>Johnson and Kauffman (1996)<br>Marshall and Ward (1996)<br>Jablonski (1998)<br>Goolaerts *et al.* (2004) | Axelrod (1984)<br>Russell (1984, 1996)<br>Bonaparte *et al.* (1987)<br>Brouwers *et al.* (1987)<br>Lucas *et al.* (1987)<br>Sloan *et al.* (1986)<br>Sloan (1987)<br>Unwin (1987)<br>Charig (1989)<br>Benton (1989, 1992)<br>Archibald and Brayant (1990)<br>Sheehan *et al.* (1991)<br>Archibald (1991, 1992)<br>Zikui *et al.* (1991)<br>Horner *et al.* (1992)<br>Sheehan and Fastovsky (1992)<br>Clemens and Nelms (1993)<br>Cutler and Behrensmeyer (1996)<br>Kardong (1997)<br>MacLeod *et al.* (1997)<br>Hurlbert and Archibald (1995)<br>Zweers *et al.* (1997) |
| **Brachiopods** | **Land Plants** |
| Surlyk and Johansen (1984) | Orth *et al.*(1981)<br>Hickey (1984)<br>Tschudy *et al.* (1984)<br>Nichols *et al.* (1986)<br>Wolfe and Upchurch (1986)<br>Spicer (1989)<br>Johnson and Hickey (1990)<br>Wolfe (1990, 1991, 1992)<br>Nichols (1992)<br>Hickey and McWeeney (1992) |
| **Bryozoans** | |
| Håkansson and Thomsen (1979,1999) | |
| **Echinoderms** | |
| Jefferey (1997)<br>Smith and Jeffery (1998, 2000) | |
| **Corals** | |
| Kauffman (1984) | |
| **VERTEBRATES** | |
| Case (1995)<br>Hoganson *et al.* (1997) | |

Sites worldwide have recorded a major eustatic sea-level fall to have occurred throughout the late Maastrichtian (Stanley, 1984; Hallam, 1987, 1992), which would have contributed to the progressive exposure and disappearance of vast areas in epicontinental seas, such as in the western interior seaway of North America (Williams and Stelck, 1975).

An intense episode of widespread continental flood basalt volcanism, particularly associated with the Deccan Traps in the Indian Ocean and elsewhere (McLean, 1985; Courtillot, 1990a; Courtillot *et al.*, 1988, 1996), would have additionally contributed to global environmental instability by increasing greenhouse-inducing atmospheric $CO_2$ levels and, consequently, induced extinctions. Radiometric dating of the lava pile succession has suggested controversial age assignments, but most studies suggest a short period (of less than 2 Ma) for the main eruptions beginning at or slightly before the K–T boundary (Wignall, 2001).

The asteroid or comet impact hypothesis was proposed by Alvarez *et al.* (1980), based on unusually high concentrations of the Platinum group element Iridium, a heavy metal very sparsely found on Earth's crust but enriched in meteorites, which was detected in the thin boundary clay layer of a K–T section near Gubbio, northern Italy. This suggested an extraterrestrial origin for the Ir-enriched layer, which would have been deposited globally as dust impact debris of the vaporized bolide. Subsequently, overwhelming supporting evidence to the impact theory came from several studies of different K–T boundary sections around the world, which also reported the Ir anomaly (e.g. Alvarez *et al.*, 1984a,b; Schmitz *et al.*, 1992; Bruns *et al.*, 1997; Norris *et al.*, 1999), occurrence of impact-shocked quartz grains (Bohor *et al.*, 1984, 1987; Izett, 1987; Bohor, 1990), microtektites (Hansen *et al.*, 1986; Izett *et al.*, 1991; Sigurdsson *et al.*, 1991; Smit *et al.*, 1992; Koeberl, 1993; Norris *et al.*, 1999), Ni-rich spinel, a mineral formed by fusion and oxidation in the atmosphere of meteoritic material (Kyte and Smit, 1986; Robin *et al.*, 1991; Robin and Rocchia, 1998), soot, produced from impact-induced forest wildfires (Wolbach *et al.*, 1985, 1990), impact-generated tsunami deposits (e.g. Bourgeois *et al.*, 1988; Hildebrand and Boynton, 1990; Smit *et al.*, 1992), and the c. 65 Ma-old (crater melt rock dated as $64.98 \pm 0.05$ Ma) giant impact crater at Chicxulub on the Yucatán Peninsula (e.g. Hildebrand *et al.*, 1991, 1995; Swisher III *et al.*, 1992; Koeberl, 1993; Kring, 1993; Kring and Boynton, 1993; Sharpton *et al.*, 1992, 1993; Bralower *et al.*, 1998).

The detailed and integrated stratigraphic study of several K–T boundary sections has shown compelling evidence of a complex sequence of events across the boundary transition. The occurrence of possible impact-derived material (tsunami beds, anomalous enrichments of Ir and other siderophile elements, microspherules, and shocked quartz) in deposits yielding *in situ* early Danian microfossil assemblages (e.g. Brazos River, Texas: Montgomery *et al.*, 1992; Coxquihui, Veracruz, east central Mexico, and Bochil, Chiapas, southern Mexico: Keller *et al.*, 2003; Beloc, Haiti: Florentin *et al.*, 1991, Stinnesbeck *et al.*, 1999, Keller *et al.*, 2003; Poty, Pernambuco, NE Brazil: Koutsoukos, 1998), suggests possible multiple closely-spaced impact events, spanning about 1 Ma or less (multiple events theory: e.g. Hut *et al.*, 1987; Schmitt, 1989).

Reports of a latest Maastrichtian global cooling and sea-level fall followed by a warming and sea-level rise trend (Haq *et al.*, 1987; Barrera and Savin, 1999; Clarke and Jenkyns, 1999; Hallam and Wignall, 1999), coupled with environmental deterioration/turnover (Officer *et al.*, 1987; McCartney *et al.*, 1990), have also been argued as evidence of pre-boundary, gradual (non-catastrophic) changes. Many species are known to have become extinct earlier or had a greater diversity decline in the high southern latitudes (e.g. earlier disappearance of inoceramid bivalves and belemnites, and major diversity drop of ammonite faunas in the Antarctica: Zinsmeister and Feldmann, 1996), which would support the hypothesis of an end Cretaceous global cooling constraining the thriving of these species in cooler high latitude regions.

Hence, the controversy is far from settled. It seems likely, however, that there was not a single cause to account for all the end Cretaceous mass extinctions, but likely a complex interplay of somewhat coeval and climatically-related global phenomena coupled with and/or induced by possible multiple impact events which took place across the K–T transition.

## 7.4 Case Study: The K–T Boundary at Poty, Ne Brazil: Event Stratigraphy

The stratigraphic, biotic and environmental changes recorded in a K–T boundary section near Recife (the Poty section), in Pernambuco, northeastern Brazil (Figs. 7.2, 7.3), illustrates some of the problems of studying the K–T transition in low-latitude South Atlantic regions.

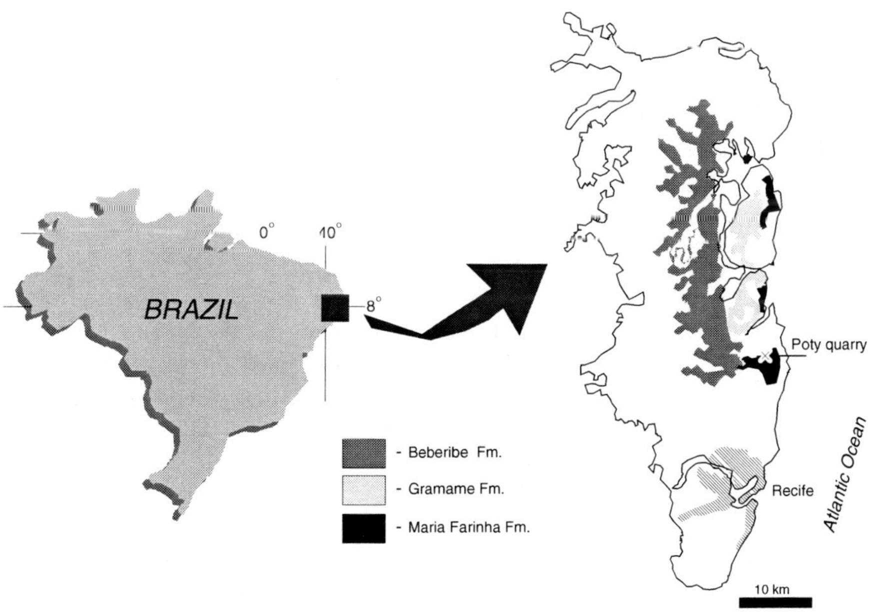


*Figure 7.2 Location map of the K–T boundary Poty section, in northeastern Brazil.*

*Figure 7.3* *The Poty quarry section, showing the positioning of the K–T boundary – see colour version of this figure in Appendix.*

## 7.4.1 The Stratigraphic Record and Depositional Dynamics

Closely spaced samples were collected in the Poty quarry from measured outcrop sections and from a core (Poty #1) drilled in June 1992 in the SE part of the quarry. Samples were processed for foraminiferal analysis using standard techniques, with a minimum of 300 specimens collected from the size fraction >63 μm. The planktonic foraminiferal zonation of the Poty section (Fig. 7.4), based on outcrop and core samples, and discussions on the nature of the event beds across the K–T boundary transition have been presented in Albertão *et al.* (1994) and Koutsoukos (1996a,b, and 1998).

### *7.4.1.1 Maastrichtian*

At the Poty quarry the uppermost Maastrichtian outcropping section (*Plummerita hantkeninoides* Zone; Figs. 7.3, 7.4), about 10 m thick, is represented by alternating beds of carbonate mudstones and marlstones of the Gramame Formation. The succession was deposited in an upper bathyal environment (estimated depth around 300 to 400 m). This is suggested by the abundance of deep-water dwelling, high-trochospiral planktonic foraminifera, such as large specimens of *Contusotruncana*. There are also benthic forms commonly found in upper bathyal environments, such as *Coryphostoma midwayensis, Cibicides hedbergi, Pyramidina rudita, Fursenkoina* sp., *Guttulina adhaerens, Nonionella cretacea, Neoflabelina* ex gr. *pilulifera, N. rugosa, Vaginulinopsis midwayana, Orthokarstenia whitei* and *Siphogenerinoides bramlettei* (Fig. 7.4). In addition, nearly all known latest Maastrichtian planktonic foraminifera (except *Abathomphalus*) occur in these beds (fig. 2 of Koutsoukos, 1996b).

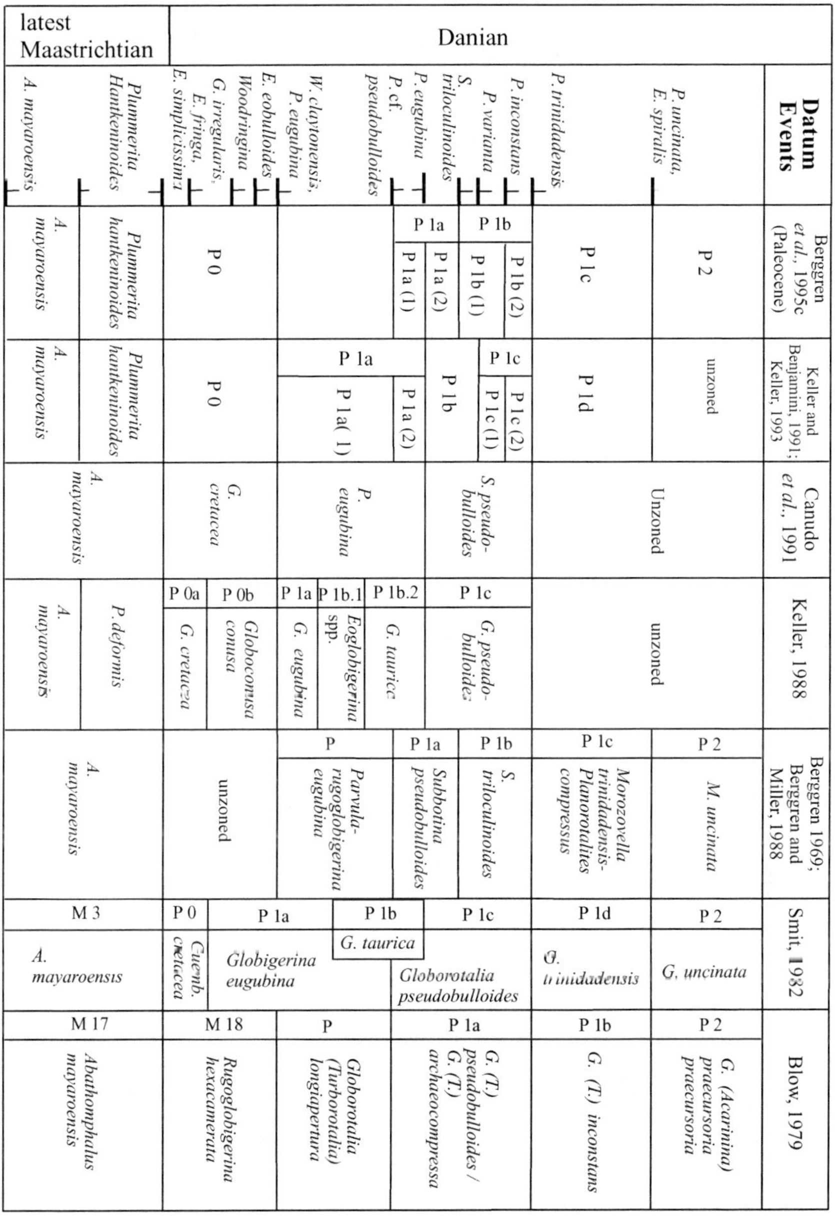


*Figure 7.4 Integrated foraminiferal zonal schemes. The datum level sequence in the left hand column is inferred after MacLeod and Keller (1991), Liu and Olsson (1992), Olsson et al. (1992), Keller (1993), Berggren et al. (1995c), and Koutsoukos (1996a, b).*

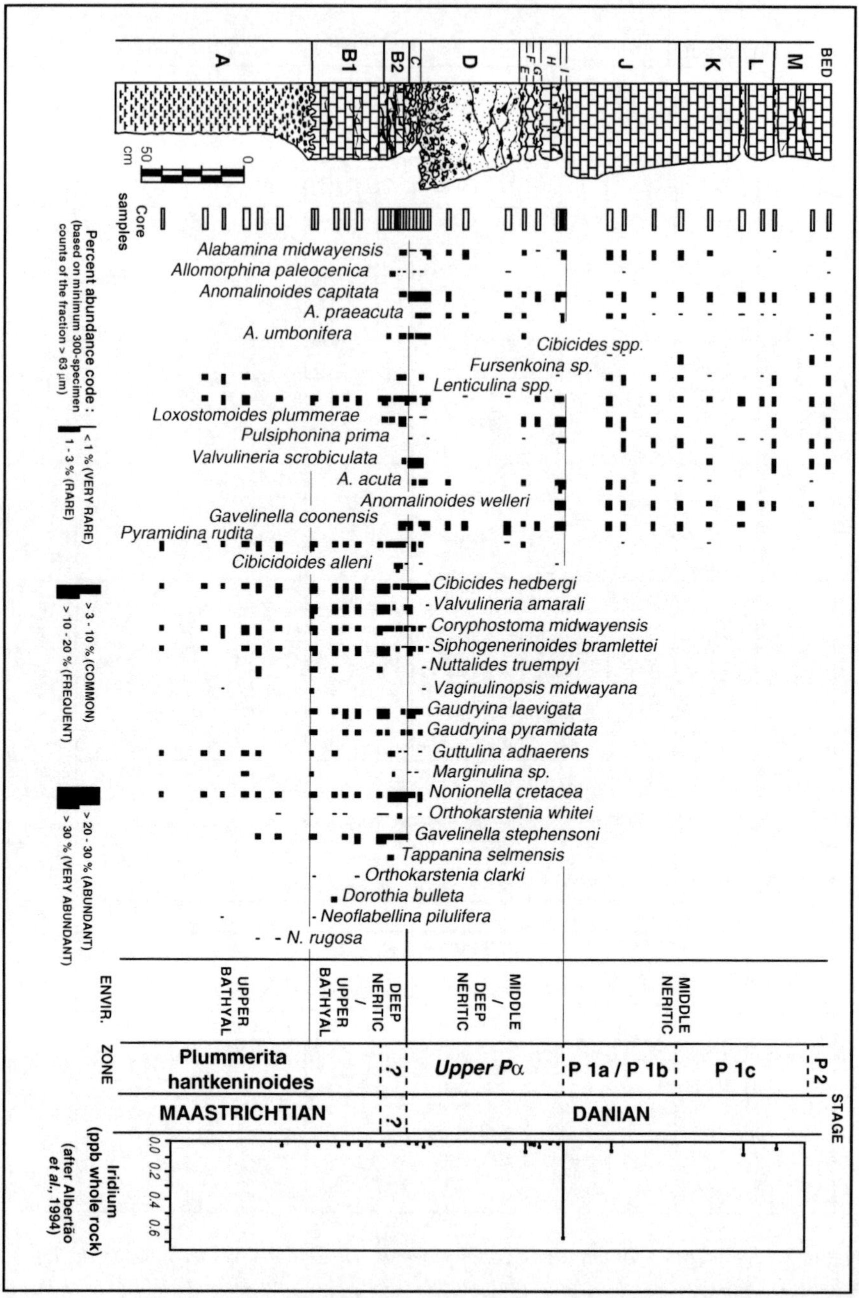


*Figure 7.5 Lithostratigraphy and distribution of selected benthic foraminifera across the K/T boundary section in the Poty quarry (samples from core Poty #1).*

*Figure 7.6 Detail of beds across the K–T boundary at the Poty quarry. Beds A to B, uppermost Maastrichtian: marlstone (A) and mudstone/wackestone (B) deposits in upper bathyal to deep neritic environments. Beds C to I, lowermost Danian: marly limestone breccia (C), graded bioclastic packstone (D), and interbedded fine-grained limestones and marlstones (E–I); probable early Danian impact-generated tsunami deposits in a middle to deep neritic environment – see colour version of this figure in Appendix.*

A non-graded, nodular carbonate mudstone/wackestone (bed B, base of the Maria Farinha Formation; Figs. 7.4, 7.5), about 50 cm thick, overlies the marlstones. It contains the first occurrences of *Gavelinella stephensoni, Orthokarstenia clarki, O. parva, Nuttallides truempyi, Valvulineria amarali, Gaudryina laevigata, G. pyramidata* and *Dorothia bulleta*. This bed appears to be a slump or mud-flow deposit, which originated at or near the neritic/bathyal transition, containing mixed deep neritic and upper bathyal microfossils and, more rarely, upper bathyal taxa from underlying beds. Alternatively, this deposit could have accumulated during a sea-level low and the microfossil mixing caused by the extensive bioturbation observed in these beds (*Chondrites* and *Thalassinoides* burrow systems). The depositional setting for this layer appears to have been the same as that of the underlying marlstones, or slightly shallower water (deep neritic-upper bathyal, c. 150–250 m). A sharp erosional surface marks the K–T boundary at the top of this bed, and a significant sea-level fall, probably in the order of between 100 and 300 m, is inferred across the K–T transition (Figs. 7.4, 7.5). A sea-level fall of similar magnitude was recorded at the El Kef K–T boundary stratotype section by Speijer and Van der Zwaan (1994, 1996), who inferred a shallowing from an upper bathyal environment in the latest Maastrichtian to middle-outer neritic conditions in the earliest Danian.

*7.4.1.2 Danian*

A thinly bedded (5.5 cm-thick) marly limestone breccia (bed C; Figs. 7.4, 7.5), rests upon the K–T boundary, and is interpreted as the initial deposit of an impact-triggered tsunami event (surge of the wave). This is overlain by a 50 cm-thick graded bioclastic packstone (bed D), interpreted as to represent deposition during the return of the main wave, and by a 15–20 cm thick succession of alternating beds of fine-grained limestones and marlstones (beds E to I), which probably represent the deposits of multiple attenuated secondary tsunami waves. These beds record the first occurrences of *Allomorphina paleocenica, Nonionella ovata, N. soldadoensis, Tappanina selmensis, Loxostomoides plummerae, Pulsiphonina prima, Anomalinoides acuta, A. capitata, A. praeacuta, Valvulineria scrobiculata, Alabamina midwayensis, Gavelinella coonensis* and *Cibicidoides alleni*, and are interpreted to have been deposited in a middle to deep neritic environment (around 50 to 150 m depth). The foraminiferal assemblages are representative of the continental shelf "Midway-type fauna" of Berggren and Aubert (1975).

These beds yield abundant reworked Cretaceous microfossils and rare early Danian specimens (foraminifera and ostracodes; upper part of the P$\alpha$ foraminiferal zone; Koutsoukos, 1998, Fauth, 2000). Apparently the P0 and the lower P$\alpha$ zones are missing, probably eroded and/or mixed within the lower Danian beds. Probable impact-related ejecta material, such as altered microspherules and shattered shock-metamorphozed quartz grains, occur scattered throughout these lower Danian beds.

A thin hemipelagic claystone layer (layer I), about 1–3 cm, lies on top of the event sequence and is marked by an Ir anomaly, which is up to 69 times the background level (Albertão *et al.*, 1994). It probably represents the condensed fall-out material of an impact event in the early Danian. This bed contains abundant early Danian planktonic foraminifera of the upper P$\alpha$ Zone, and authochthonous benthic foraminifera (Figs. 7.4, 7.6) which indicate deposition in a middle to deep neritic setting for the event beds. Its top marks the boundary between the Paleocene P$\alpha$ and P1a foraminiferal zones.

It has been argued that the Danian microfossils present in beds C to I are due to bioturbation mixing, as similarly reported by Scasso *et al.* (in press) from a K–T boundary section in the Neuquén basin. This may truly occurs in boundary sections where intense bioturbation is detectable. However, several aspects of the Poty section stratigraphic record give support to the Danian dating of the beds: (1) though bioturbation is present, and particularly conspicuous at the the top of layer H (which marks the end of the interpreted tsunami event and, as thus, the return to normal depositional conditions), it is rare to absent in the underlying beds G to C, the base of which lie down about 75 cm below layer I; (2) the first occurrence of the Danian microfossils occurs consistently at the same horizon in closely-spaced samples collected from randomly selected sections in the outcropping area, and also in core Poty #1; (3) Danian microfossils which are known to be extinct at the top of the P$\alpha$ foraminiferal Zone, such as *Eoglobigerina fringa, Guembelitria irregularis* and *Parvularugoglobigerina eugubina* (the total-range zonal marker), are exclusively recorded within layers C to I, and do not occur above (the upper limit marks the P$\alpha$/P1a zonal boundary), and thus can not be interpreted as contaminated by bioturbation; (4) in addition to that, thin sections of the sample with the Ir anomaly from layer I (Fig. 7.7) show only well-developed diagnostic Danian microfossils, and no Cretaceous forms, which gives further support to the Danian dating of the event.

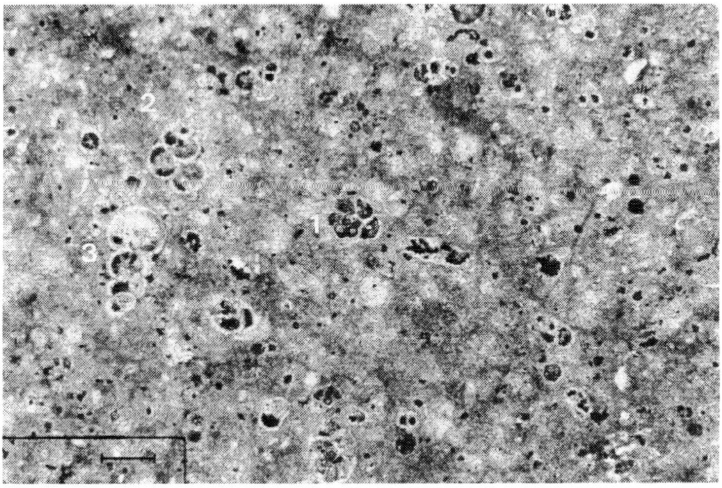


*Figure 7.7 Thin-section micrograph of the marlstone sample with the Ir anomaly from layer I, lower Danian, Poty quarry. Planktonic foraminifera of the Danian upper Pα Zone: 1.* Parvularugoglogigerina eugubina (longiapertura morphotype)*; 2.* Woodringina claytonensis; *3.* W. hornestownensis. *Scale bar = 100 μm – see colour version of this figure in Appendix.* Photo by Gilberto A. Albertão, reproduced by courtesy.

The uppermost Danian beds exposed represent deposition in progressively shallower neritic environments, with stronger influence of storm events. The benthic foraminifera decline sharply in diversity and are characterized by assemblages dominated by anomalinids with subordinate vaginulinids (Fig. 7.5).

To conclude, the Poty section presents evidence (i) of a sea-level fall, in the magnitude of 100 to 300 m, at the K–T boundary in low-latitude South Atlantic regions; and (ii) of possible impact-triggered tsunami deposits in the early Danian, which supports the hypothesis of a sequence of multiple closely-spaced impact events around the K–T boundary transition.

## 7.5 Acknowledgements

I thank Paul Potter (University of Cincinnati) and Peter Szatmari (Petrobras-Cenpes, Rio de Janeiro) for reviewing the manuscript and for their helpful comments. This work is part of a research conducted at the Geologisch-Paläontologisches Institut of the University of Heidelberg supported by a research fellowship from the Alexander von Humboldt Foundation, which is very gratefully acknowledged.

Part III

# The Search for Clues: Analyzing and Sequencing the Record

Chapter 8

# Chemostratigraphy

RENÉ RODRIGUES
Universidade do Estado do Rio de Janeiro (UERJ), Faculdade de Geologia, Rua São Francisco Xavier 524, Bloco A, Sala 2016, Maracanã, 20559-900 Rio de Janeiro, RJ, Brazil.



## 8.1 Introduction

*Chemical stratigraphy* or *chemostratigraphy* involves the application of organic and inorganic geochemical data to characterize and correlate strata. Sequences can be subdivided into stratigraphic units with diagnostic geochemical signatures. These signatures potentially provide a means of defining and correlating sedimentary units over wide areas. Such sort of studies is specially important in areas in which traditional methods of correlation, as biostratigraphy and wirelog signatures, have proven to be inadequate for precise correlations. This is specially true in cases of poorly fossiliferous or non-fossiliferous sequences, in areas with very high sedimentation rates in which biostratigraphic resolution is poor, and in many thick monotonous marine shales or carbonates in which the wireline log signatures are monotonous and fail to provide accurate correlation.



*E.A.M. Koutsoukos (ed.), Applied Stratigraphy,* 165-178.

*Figure 8.1 Location map of studied Brazilian sedimentary basins.*

This chapter is not intended to discuss exhaustively all geochemical parameters that could be used in *chemostratigraphy*, but rather to present and discuss those used in applied stratigraphic interpretations from selected Brazilian basins (Fig. 8.1).

## 8.2 Organic Matter

Sedimentary organic matter provides a variety of indicators that can be used to reconstruct the history of geological records of ancient marine and continental environments. The organic matter concentration in rocks depends on its production and preservation in the depositional environment.

The major source of organic matter entombed in the sediments is the organic-walled phytoplankton that thrived in the photic zone of the aquatic environment. Land plants can also be important additional contributors to the organic matter preserved in the sediments. As a consequence, determination of these different types of organic matter is important for interpretation of the paleoenvironmental conditions. The phytoplankton records the primary productivity of the pelagic environment, while the terrestrial plants reflect the local land organic productivity and transport to an aquatic depositional environment.

Organic matter, being a reduced form of carbon, is unstable under aerobic depositional conditions. Under normal oxygenated water conditions its degradation is relatively rapid during its transport from the water surface to the sediment–water interface, and continues further within the bioturbated layer of the sediment. Usually, only a very small percentage of the primarily produced organic matter survives the remineralization process and is preserved in the sediments (e.g. Wakeham *et al.*, 1980; Emerson and Hedges, 1988). Normally these depositional conditions predominate in the sedimentary record and are characterized by stratigraphic intervals with low organic carbon contents, as normally found in Tertiary shale sections of the Brazilian marginal basins (see case studies of the Cenozoic). On the other hand, it is also important to identify the organic matter alteration, taking into account that degradation processes modify the original composition and concentration of the organic matter, despite the relatively short time involved from the primary organic matter production to its final burial in the sediments. The more labile fractions of organic matter (lipids and proteins) are selectively degraded, creating compositional changes as organic particles sink. This means that algal organic matter degrades faster than land-derived (terrestrial) organic matter. Detailed organic petrographic studies and rock-eval pyrolysis allow us to make such differentiations (Fig. 8.2). Despite this fact, the sedimentary organic matter retains considerable information about its source and depositional environment.

During times of strong stratification of water column and/or of high primary production, dissolved oxygen is used before it can be replenished, and anoxic bottom waters are developed. Organic matter preservation appears to be enhanced under these conditions (e.g. Demaison and Moore, 1980). Pedersen and Calvert (1990) postulate that increased inflow of organic matter coupled with high productivity can produce organic carbon-rich sediments in oxygenated water columns. However, it is hard to explain the very high concentrations of organic carbon ($>10\%$) associated with very high hydrogen indices found in some black shales (like those shown in the upper high part of Fig. 8.2), without invoking increased primary productivity and inflow of organic matter, combined with improved organic matter preservation and limited clastic dilution. These depositional conditions are rare in the sedimentary record, and are normally associated with global dysoxic–anoxic events (e.g. Schlanger and Jenkyns, 1976; Ulmishek and Klemme, 1992). These stratigraphic intervals are easily identified by anomalous high organic carbon contents, high hydrogen index values, and high concentrations of sulfur-related chemical elements (mostly Mo, V, Cu, Pb; see Figs. 8.3, 8.4)

## 8.3 Major, Minor and Trace Elements

In a marine environment subject to siliciclastic deposition under low oxygen bottom water conditions, the main factor which controls pyrite formation is the availability of iron and sulfur ions. Under normal salinity conditions, a reducing marine environment

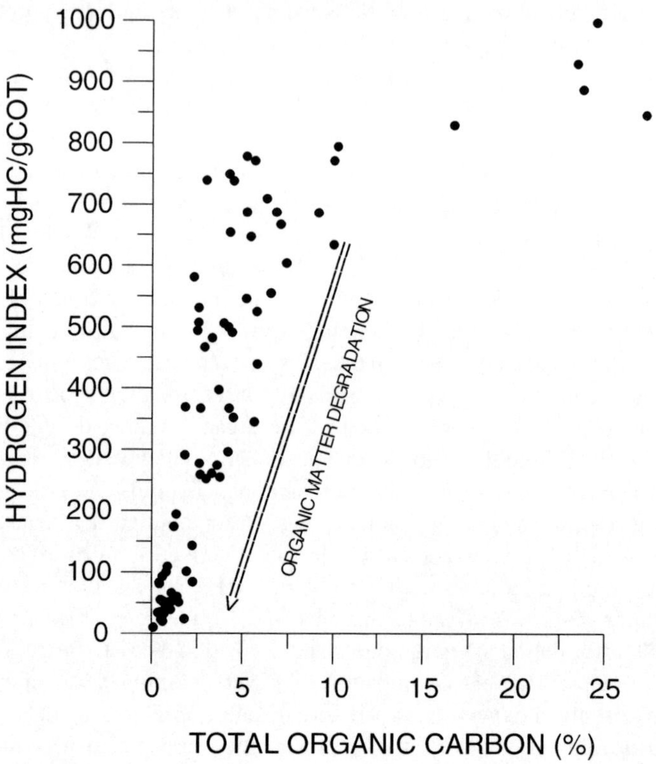


*Figure 8.2 Total organic carbon x hydrogen index, immature section of the Codó Formation, Lower Cretaceous, Parnaíba Basin, northeast Brazil.*

usually has an excess of sulfur ions released by the reduction of sulphates through anaerobic bacterial activity. In these settings, the main constraint to pyrite formation would be the free amount of reactive iron ions (Raiswell and Berner, 1985). These environmental conditions can be identified by the high direct correlation of organic carbon, sulfur and iron.

To some extent, it is a well-known fact that certain metals tend to concentrate in the form of sulfide and organometallic compounds within autochthonous organic matter deposited in a reducing environment (e.g. Le Riche, 1959; Kulbicki and Rumeau, 1967). Trace elements, as U, Mo, V, Cu, and Ni, are statistically associated with both organic carbon and sulfur. However, frequently U, Mo and V show greater covariance with organic carbon than with sulfur. Benthic flux measurements using bell jars revealed a significant release of some of these metals from the sediment during oxic bottom water conditions, while these same metals were fixed in the sediment during times of dysoxic-anoxic bottom water (e.g. Westerlund *et al.*, 1986). This study focus mostly on Mo, which is among the most diagnostic element for sediments deposited under dysoxic–anoxic bottom water conditions.

In contrast, Mn generally keeps in solution in the water column under anoxic conditions, and precipitates in the form of oxides and carbonate under less reducing conditions (Frakes and Bolton, 1984).

The detrital clastic sources for other major, minor, and trace elements can be inferred by the relationship among them and concentrations of Al, an abundant lithogenic element. Some of the elements that show high correlation with Al include Ti, Mg, Na, K, Ga, La, and Nb.

Barium is considered to be a marker of organic productivity and biogenic sediments (e.g. Dymond, 1981; Papavassilou and Cosgrove, 1982; Schmitz, 1987). However, data shown in the case-studies (see Fig. 8.3) indicate that the Ba concentrations does not correlate with organic carbon, but with Al. This suggests that Ba was supplied to the basin together with terrestrial detrital influx, mostly in clay minerals.

## 8.4 Carbon and Oxygen Isotopic Composition of Carbonate

### 8.4.1 Oxygen Isotopes

The primary variables which determine the oxygen isotope signal in marine carbonates are water temperature and the $^{18}O/^{16}O$ composition of the sea water. The isotopic composition of oxygen in a solid phase is not the same as that in water when the two are in equilibrium. There is a differential partition in the solid and liquid phases, with $^{18}O$ being enriched in the solid phase. This enrichment in $^{18}O$ during carbonate formation is temperature-dependent: at higher temperatures, more carbonate ions with $^{18}O$ atoms could remain in solution, and therefore the carbonates precipitated under these temperature conditions should be less enriched in $^{18}O$ than those formed under lower temperatures (e.g. Urey, 1947; Emiliani, 1955). For a constant oxygen isotopic composition of the sea water, the $^{18}O/^{16}O$ ratio in carbonates deposited in isotopic equilibrium will decrease by 0.23‰ with each degree centigrade increase of temperature (e.g. Epstein *et al.*, 1953; Emiliani, 1955). Thus, with this findings it was possible to estimate at which water temperature the carbonate precipitation took place (e.g. Epstein *et al.*, 1951, 1953; Craig, 1965; Erez and Luz, 1983).

The most important process that can produces modification in the oxygen isotope composition of the sea water is evaporation. This process concentrates the heavy $^{18}O$ isotope in the liquid phase, which so becomes relatively enriched in $^{18}O$ as compared to the vapor phase that becomes enriched in the light $^{16}O$ isotope. During glaciations, water vapor in the clouds precipitates $^{16}O$-rich water on ice caps in the form of snow, leaving the ocean water enriched in $^{18}O$. This means that changes in oxygen isotope ratios in carbonates during ice ages are primarily an ice volume signal, with a minor effect due to temperature (e.g. Shackleton, 1967).

The above relationships have provided the basis for numerous studies of past climate changes, continental glaciations and changes in marine current systems. Furthermore, by studying the isotopic compositional changes in both benthic and planktonic foraminifera it is possible to investigate the structure between surface and deep water and their evolution in the geological history.

### 8.4.2 Carbon Isotopes

The variables affecting the carbon isotope composition of carbonates are still not completely understood. Primary $\delta^{13}C$ values in marine carbonates are considered to be a function of the $^{13}C/^{12}C$ of dissolved inorganic carbon ($\Sigma CO_2$) in surface water, since there is little fractionation associated with carbonate precipitation.

Significant changes in carbon isotope ratios reflect changes in carbon cycling of the ocean, particularly between reservoirs of very different isotopic compositions. The two most important metabolic processes affecting the $\delta^{13}C$ of $\Sigma CO_2$ are respiration and photosynthesis. Phytoplankton photosynthesis strongly discriminates against $^{13}C$, preferentially incorporating $^{12}C$ in organic matter formed by this process. Thus, phytoplankton productivity or bacterial oxidation of organic matter in the water column will significantly affect the $\delta^{13}C$ values of the $\Sigma CO_2$ in the surface water. As organic matter is enriched in the light isotope $^{12}C$, its removal by burial from oxidative recycling renders the ocean richer in $^{13}C$, and therefore increases the $\delta^{13}C$ values of carbonates precipitated under these conditions (Scholle and Arthur, 1980). This water conditions normally characterize dysoxic–anoxic events as will be discussed later.

It is a well known fact that the isotope composition of planktonic foraminifera tests is enriched in $^{13}C$ (heavy) and in $^{16}O$ (light) compared to that in benthic foraminifera, which is lighter in carbon isotopes and heavier in oxygen isotopes. Thus, by utilizing isotopic compositional changes in both benthic and planktonic foraminiferal tests it is possible to investigate how the structure between surface and deep water responded to the influence of changing sources of deep waters and the process of their formation, as well as the cycling of nutrients during their transit around the ocean basins (Kennett and Stott, 1990).

During extinction phenomena, as that of the Cretaceous–Tertiary (K–T) boundary, the ocean would have no gradient of carbon isotope from top to bottom, because the water surface would no longer be depleted in $^{12}C$ due to photosynthesis (reduction of primary productivity). Carbonates deposited at that time should have a more negative $\delta^{13}C$ signature as compared to values from the underlying and overlying carbonates, which would explain the negative $\delta^{13}C$ excursion at the K–T boundary (e.g. Hsü and Mckenzie, 1985).

During diagenesis carbon isotopes are also fractionated by bacterial reactions, forming $CO_2$ rich in $^{12}C$ through organic matter oxidation or sulfate reduction, and $CO_2$ enriched in $^{13}C$ due to methanogenesis. If both oxygen and carbon isotopes become simultaneously negative, this could mean fresh-water diagenesis, and therefore a possible sequence boundary (see case studies of the Cenozoic).

## 8.5 Brazilian Case Studies

Whole rock isotope analyses are generally considered to have little stratigraphic value. Nevertheless, despite the presence of some diagenetic overprint, the carbon and oxygen isotope records resemble those derived from careful analyses of individual foraminifera species. Moreover, the general trend is not masked when viewed within the framework of sequence stratigraphy. Also, strong diagenetic overprints may be related to sea-level lowstands that correlate well with sequence stratigraphic boundaries. Based on these findings it is intended to demonstrate here the usefulness of whole-rock sample analyses applied to chemostratigraphic studies, in particular those based on ditch-samples, which are the commonest sort of rock samples available from oil-well drillings. All examples presented are from Brazilian sedimentary basins (Fig. 8.1), involving intervals ranging in age from the Devonian to Cenozoic.

### 8.5.1 Methods

All samples were hand-picked and split into two parts, for organic and inorganic geochemistry investigations. For organic geochemistry studies, one aliquot was ground in rotary mill and, after dissolution of carbonates by hydrochloric acid, analyzed for organic carbon using a Leco SC-444 apparatus. Rock-Eval analyses were performed using a Delsi II instrument, according to the procedures described by Espitalié *et al.* (1977). Major and trace element determinations have been obtained by X-ray fluorescence in a Phillips PW-1480 equipment. For carbon and oxygen isotope analyses, carbonate powder was reacted with phosphoric acid, and the measurements made according to standard techniques on a Finnigan MAT 252 mass spectrometer with an on line carbonate preparation device (Kiel device). The $\delta^{13}C$ and $\delta^{18}O$ values are reported in relation to the PDB standard. All analyses were carried out at the Petrobras Research Center in Rio de Janeiro.

### 8.5.2 The Paleozoic

The Frasnian dysoxic–anoxic event is relatively well-known global event (e.g. Ulmishek and Klemme, 1992), and is widely represented in Paleozoic basins of north and northeast Brazil (Rodrigues, 1995, 2001).

In the Parnaíba Basin (northeast Brazil, Fig. 8.1), this radioactive shale interval is best developed in the central and northwestern parts of the basin, where it may reach thickness up to 40 m. These marine strata represent the maximum flooding event of the Devonian sequence, and are easily recognized in well logs by their higher radioactivity, sonic transit time and resistivity, and by their low density values. Such features are caused by the high concentration of organic matter in relation to that of the surrounding rocks (Fertl *et al.*, 1986). Detailed geochemical and miospore-based biostratigraphic investigation of this event were carried out by Rodrigues *et al.* (1995).

The dysoxic–anoxic interval is made up of fissile, dark shales, with organic carbon content ranging from 4.0 to 5.0%, thus contrasting with the lower concentrations of intervals immediately below and above, 1.0% and 2.0%, respectively (Fig. 8.3). The lower lithological contact (with partially bioturbated shale and very fine-grained sandstone) is abrupt, and suggests quick replacement of oxic bottom conditions by dysoxic–anoxic ones, possibly in response to sudden sea-level rises and a consequent decrease in bottom water circulation. Higher up in the studied section, the disappearance of fine lamination and the increased silt content of the shales could imply a

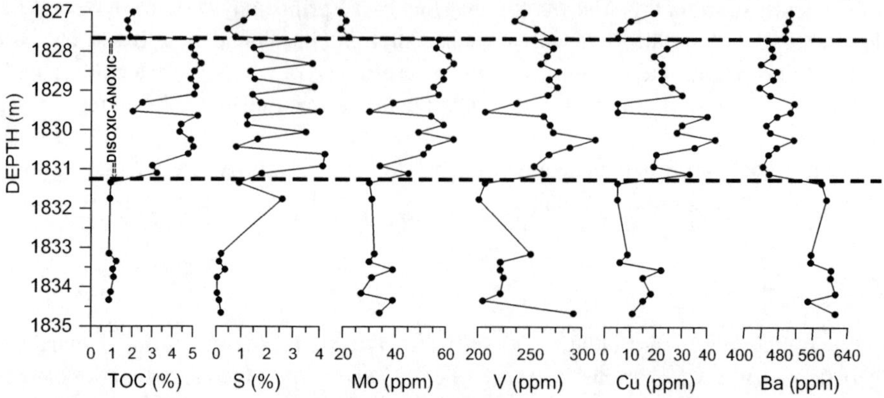


*Figure 8.3 Frasnian dysoxic–anoxic event as shown by the total organic carbon (TOC), S, Mo, V, Cu and Ba distribution pattern, Parnaíba Basin, northeast Brazil.*

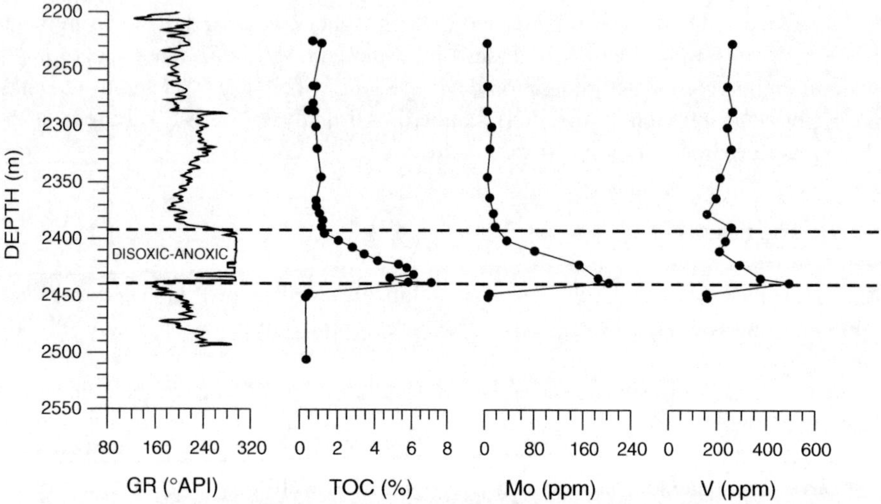


*Figure 8.4 Frasnian dysoxic–anoxic event as shown by the gamma ray log (GR), total organic carbon (TOC), Mo and V distribution pattern, Amazon Basin, north Brazil.*

higher sedimentation rate, as well as progradation in a highstand system tract with a relative increase in bottom water oxygenation.

The distribution pattern of sulphur, Mo, V, Cu, Pb and hydrogen index, similar to that of organic carbon, points out to the reduction of sulphate through anaerobic bacterial activity, and a better preservation of organic matter under dysoxic–anoxic bottom water conditions, during deposition of the interval with higher organic carbon content (Fig. 8.3).

In the Amazonas Basin, located to west of Parnaíba Basin (Fig. 8.1), this dysoxic–anoxic event is also very well documented (Rodrigues, 1973; Trigüis and Rodrigues, 2000; Rodrigues, 2001), and comprise a more than 40 meters thick section of black shales (Fig. 8.4). Ranging in age from Frasnian to Famenian, it represent a distal condensate section of Parnaíba Basin, where in the same age interval was deposited the dysoxic–anoxic black shales (maximum flooding) and the progadational interbeds of shales, siltstones and sandstones (highstand system tract) of the Pimenteira Formation.

### 8.5.3 The Mesozoic

The Brazilian marginal basins are directly related to the rifting of the African and South American plates. Thick sequences of lacustrine and fluvial deposits were accumulated during the Early Cretaceous rifting phase. Following this, tectonic activity has been restricted to subsidence (Estrella *et al.*, 1984) while intermittent marine transgressions took place in the lower Aptian. The sedimentary record of these marine transgressions is characterized by the presence of chemical fossils derived from marine algae and carbonates with higher $\delta^{13}C$ values, as compared to the underlying sequences that present biomarkers and carbon isotopic signatures consistent with freshwater to saline lacustrine environments (e.g. Takaki and Rodrigues, 1984; Mello *et al.*, 1988; Rodrigues, 1995). The isotopic signatures (Fig. 8.5) can be correlated in several wells in the Espirito Santo and Campos Basins, and probably correspond to the same isotope events observed by Menegatti *et al.* (1998) in the lower Aptian, *Globigerinelloids blowi* foraminiferal zone of the Roter Sattel (Swiss Pre-Alps) and Cismon (southern Alps of northern Italy) sections.

The global distribution of bituminous black shales within the Aptian and Albian stages and at the Cenomanian–Turonian boundary has led to the interpretation that these deposits result from global Oceanic Anoxic Events (Schlanger and Jenkyns, 1976). Widespread deposition of marine organic rich shales took place about the Aptian–Albian transition, mostly in northeast and equatorial Brazilian marginal basins. The black shale interval yields high total organic carbon content (TOC up to 25%) and high hydrogen index (HI up to 980 mgHC/gTOC) (Fig. 8.6).

At the Cenomanian–Turonian boundary, even more widespread deposition of marine bituminous shale occurred in the equatorial through southeast Brazilian offshore marginal basins. Co-variation of the high organic carbon content in pelagic rocks with positive shifts of carbon isotope values within the corresponding shallow carbonate facies points out to closely related phenomena, caused by the same dysoxic-anoxic event. In order to test the applicability of the $\delta^{13}C$ shifts to the chronostratigraphic correlation of dysoxic–anoxic events, a detailed isotope study was carried out in a shelf carbonate sequence of Sergipe–Alagoas Basin (Fig. 8.2), at a time when no detailed biostratigraphic studies were regionally available, and the sequence was demonstrated to be coeval with the black shale deposition episode (Takaki and Rodrigues, 1993). Whole rock carbon isotope results point out to an abrupt increase of the $\delta^{13}C$ values from burial depths of 700 m to 600 m, followed by a decrease

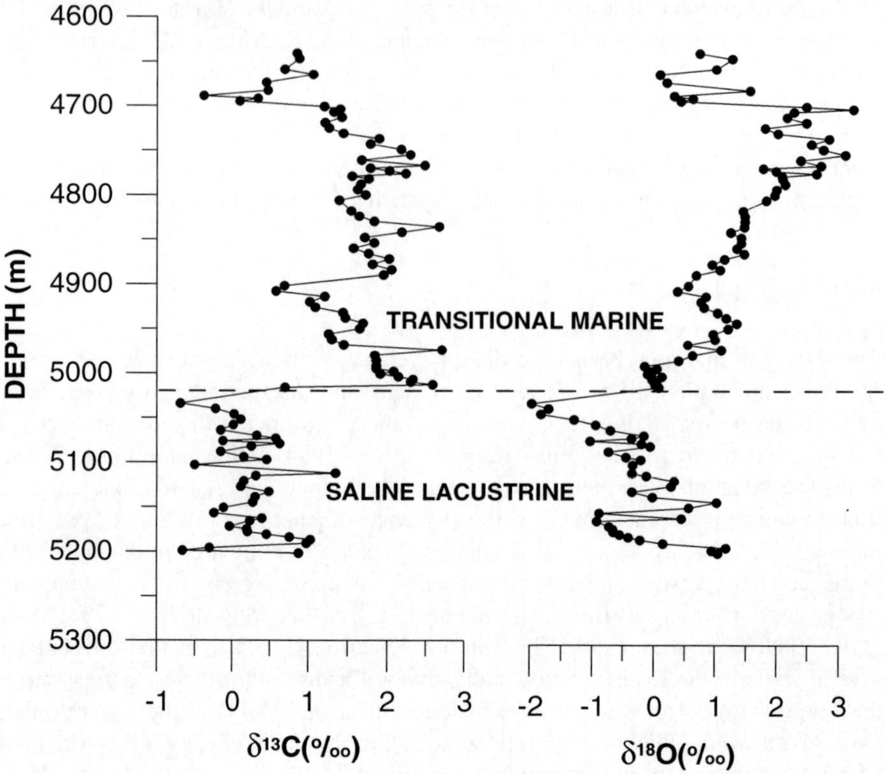


*Figure 8.5 Carbon and oxygen isotope signatures of Lower Aptian carbonates, Campos Basin, southeast Brazil.*

towards 500 m (Fig. 8.7). Carbon isotopic variations of 2.5‰ in inorganic fractions of Upper Cretaceous sequences have only been recorded for the dysoxic–anoxic event at the Cenomanian–Turonian boundary (e.g. Scholle and Arthur, 1980; Pomerol, 1983). Based on these assumptions, the Cenomanian-Turonian boundary was tentatively placed at about 700 m (Takaki and Rodrigues, 1993). Subsequent biostratigraphic dating of the same event by Cunha (2000) confirms the isotope-based age estimates, as shown in Figure 8.7. The more negative $\delta^{18}O$ values suggest that the dysoxic–anoxic event took place under mild climatic conditions, associated with higher relative temperatures of sea-water.

Following the carbon isotope event, a sharp increase in the manganese content of carbonates appears to characterize the end of the anoxic period. Most of the manganese would have remained in solution during anoxia, and later precipitated in larger quantities during the shift from dysoxic-anoxic to oxygenated environmental conditions (Fig. 8.7).

The $\delta^{13}C$ fluctuations, turning gradually less positive in Turonian and younger sequences, might reflect a decrease in organic productivity of the ocean. This

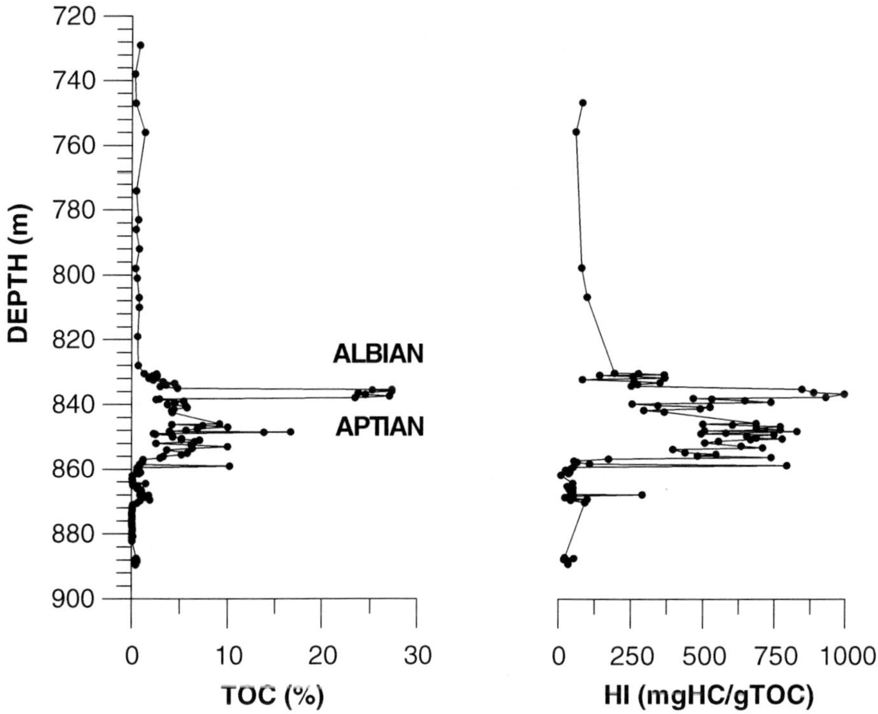


*Figure 8.6 Dysoxic–anoxic event as shown by the total organic carbon (TOC) and hydrogen index (HI) data, Aptian–Albian transition, Parnaíba Basin, northeast Brazil.*

phenomenon was probably related to the continuous cooling of oceanic waters in response to the progressive opening of South Atlantic Ocean, as suggested by the increasing $\delta^{18}O$ trend from Turonian to Santonian (Fig. 8.7), but that reached the Maastrichtian.

### 8.5.4 The Cretaceous–Tertiary Boundary

The Cretaceous–Tertiary (K–T) boundary event is well recorded in several places of the Brazilian continental margin. The sudden decrease in $\delta^{13}C$ values near K–T boundary appears to result from a corresponding fall in the rate of primary productivity in surface waters, due to mass mortality of oceanic plankton and the accumulation of nutrients, including dissolved $CO_2$, in the photic zone.

The Cretaceous–Tertiary boundary was sampled in several wells drilled in the Brazilian offshore area. This event in northeast and southeast offshore sedimentary basins is normally very well marked, and represented by a sharp increase in total organic carbon values and by $\delta^{13}C$ and $\delta^{18}O$ negative excursions. In the example shown in Figure 8.8, the organic carbon content of the Cretaceous section is around

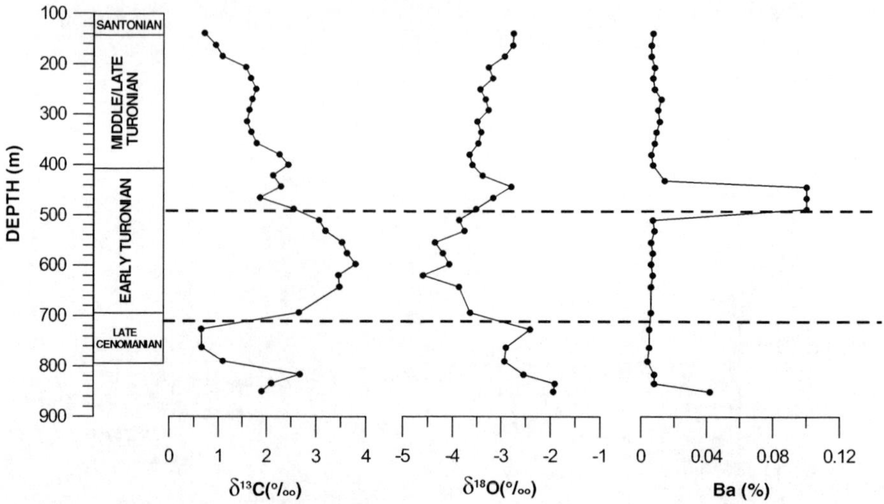


*Figure 8.7* *Characterization of Cenomanian–Turonian dysoxic–anoxic event based on carbon and oxygen isotope data, and Mn content, Sergipe-Alagoas Basin, northeast Brazil.*

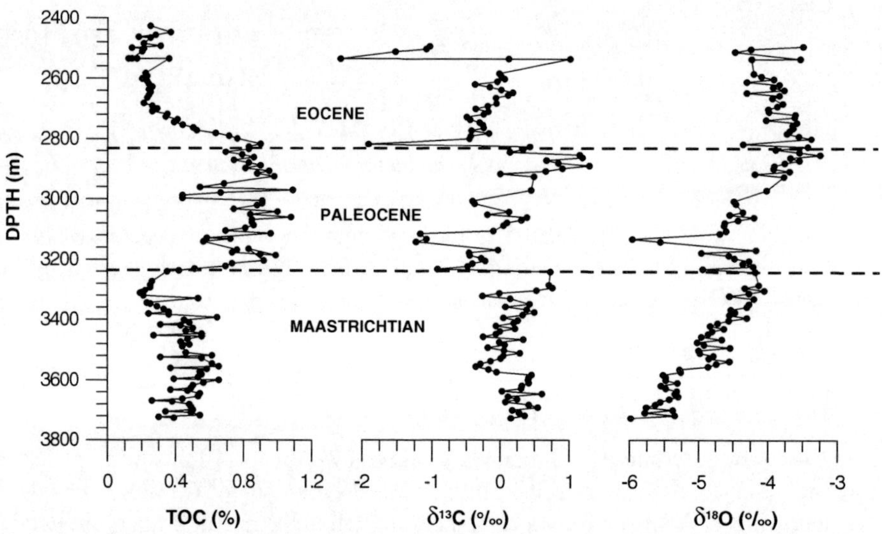


*Figure 8.8* *Total organic carbon (TOC) and isotopic data at the Cretaceous–Tertiary and Paleocene–Eocene boundaries, Cumuruxatiba Basin, northeast Brazil.*

0.4%, being followed by a prominent increase up to 0.8% and 1.2% values. These organic carbon contents in the K–T boundary are not similar everywhere in Brazilian offshore basins, but as a rule they are two to three times higher in the Paleocene. The very low hydrogen index of the same samples indicate a predominance of oxidized organic matter, probably derived from the continental area. The negative oxygen and carbon isotopic excursions (1.0 to 2.0‰) are believed to be related to changes in ocean thermal conditions (greenhouse effect) and the result of rapid reduction in the rates of primary productivity and nutrient accumulation (mass mortality of oceanic plankton) due to the effects of a possible meteorite impact (e.g. Boersma *et al.*, 1979; Hsü and Mckenzie, 1985; Schmitz *et al.*, 1992).

### 8.5.5 The Cenozoic

The increased $\delta^{13}C$ and $\delta^{18}O$ values documented for the late Paleocene are interpreted as a recovery of surface water productivity and to changes in the climatic conditions to the levels that existed before the K/T boundary event.

As shown by the $\delta^{18}O$ isotope data, the rapid climatic change of the latest Paleocene was one of the most dramatic warming events in the geological record (Fig. 8.8). Oxygen isotope values decrease during late Paleocene to reach a minimum in the earliest Eocene, thus suggesting maximum sea surface temperatures at that time, as pointed out by Shackleton and Kennett (1975). Synchronous with the decrease in $\delta^{18}O$ values was a abrupt negative excursion in the $\delta^{13}C$ values. The Paleocene–Eocene boundary event is isotopically similar to that of K–T boundary, probably reflecting a similar origin. In general, the Eocene are characterized by continuous cooling of sea-waters, as suggested by the continuous increasing of $\delta^{18}O$ values (Figs. 8.8, 8.9).

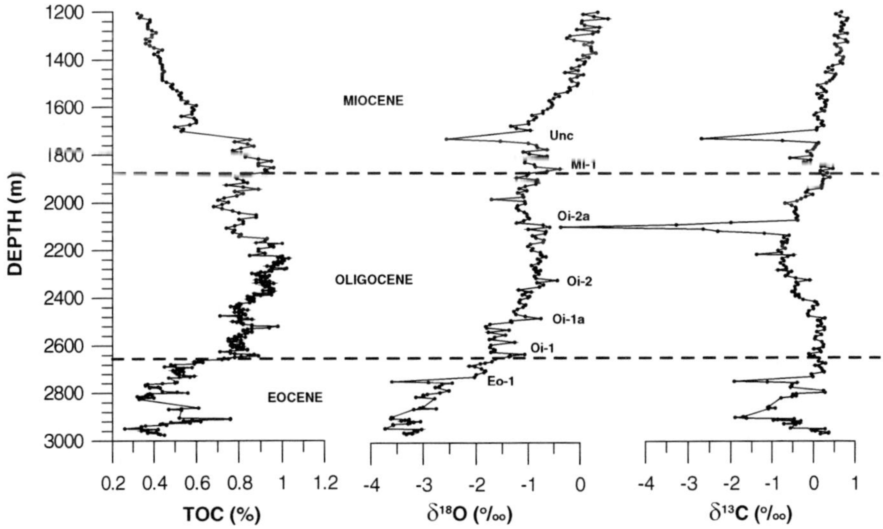


*Figure 8.9 Total organic carbon (TOC) and carbon and oxygen isotope data, Tertiary of Campos Basin, southeast Brazil.*

In the Campos Basin, the Tertiary oxygen isotope record shows the first Cenozoic $\delta^{18}O$ maxima at ~40 Ma, probably linked to Antarctica's ice growth event. As a consequence of this glacio-eustatic sea-level fall, an unconformity at the middle-upper Eocene boundary is suggested by a sharp decrease of $\delta^{18}O$ and $\delta^{13}C$ values, related to meteoric diagenetic effect (Eo-1 in Fig. 8.9). The Oi-1 oxygen isotope event of Miller *et al.* (1991) in the lowermost Oligocene, also observed in the Campos Basin (Oi-1 in Fig 8.9), is considered to represent the earliest robust evidence of the glaciation-deglaciation event in Antarctica (Miller *et al.*, 1991). The lower Oligocene is characterized by a set of higher frequency glaciation–deglaciation events. These are probably related to Milankovicht cycles of 400 Kyr. with two $\delta^{18}O$ maxima at 35 Ma and 30 Ma, the later corresponding to the lower-upper Oligocene boundary (Oi-1a and Oi-2, respectively in Fig. 8.9). The decrease of $\delta^{18}O$ values in the uppermost Oligocene points out to a warmer event, that precedes the $\delta^{18}O$ maxima of the lowermost Miocene (Miller's Mi-1 in Fig. 8.9). Following the Mi-1 event, an unconformity is suggested by the abrupt, negative $\delta^{18}O$ and $\delta^{13}C$ excursion at approximately 20 Ma in thc lower Miocene (Unc in Fig. 8.9).

The total organic carbon data are an another interesting point to be considered. The coeval variation between the oxygen isotope data and the total organic carbon content, associated with organic matter predominantly derived from higher plants, suggest an increase of terrestrial organic matter transported to the sea during the sea-level fall (Fig. 8.9).

## 8.6 Acknowledgments

I thank José Henrique Gonçalves de Melo and Eugenio Vaz dos Santos Neto (Petrobras-Cenpes, Rio de Janeiro) for their careful review and helpful comments, which much improved the manuscript.

Chapter 9

# Paleobotany and Paleoclimatology

*Part I: Growth Rings in Fossil Woods and Paleoclimates*

LAUREEN SALLY DA ROSA ALVES[1] and MARGOT GUERRA-SOMMER[2]

[1] Universidade do Estado do Rio de Janeiro (UERJ), Faculdade de Geologia Departamento de Estratigrafia e Paleontologia, Rua São Francisco Xavier, 524/2032-A, Maracanã, 20559-900 Rio de Janeiro, RJ, Brazil.

[2] Universidade Federal do Rio Grande do Sul (UFRGS), Instituto de Geociências, Caixa Postal 15001, 91501-570 Porto Alegre, RS, Brazil.

*Part II: Leaf Assemblages (Taphonomy, Paleoclimatology and Paleogeography)*

TÂNIA LINDNER DUTRA

Universidade do Vale do Rio dos Sinos (UNISINOS), Av. UNISINOS, 950, 93022-000 São Leopoldo, RS, Brazil.





*E.A.M. Koutsoukos (ed.), Applied Stratigraphy*, 179-202.

## 9.1 Introduction

Paleoclimate interpretation from fossil plants is based on two aspects: their adaptive morphology and the relationship between vegetal communities distribution and present day climate zones. The effects of latitude, continent position and relief, together with other climatic factors have been used to characterize the present day climatic zones of the Earth. These zones are also defined by ecosystem or ecoregion distribution. Nowadays, the main criteria used to delineate ecoregions are the floral composition and distribution (Tricart and Cailleux, 1972; Bailey and Cushwa, 1981). By this criteria there are several domains, which comprise divisions that are further subdivided into provinces. The province concept is basically founded on Köppen's (1931), later modified by Trewartha (1969), large climatic zones (the polar, wet temperate, dry and wet tropical domains). Paleofloristic interpretations based on the Actualism Principle consider distinct paleoclimates to be important elements in depicting paleofloristic provincialism, as they should restrict vegetation from spreading between regions. Within present day floras it can be observed that, whereas latitude imprints a global pattern on the distribution of specific climate zones, topography plays a role in the delineation of distinct environments. In any particular area, characterized by its own type of soil, topography and microclimate, distinctive species live together with evidely tolerant species, the latter ones occupying broader regions. These factors combine to give an ecosystem diversity and biographic distribution complexity that are difficult to characterize from fossil floral studies alone.

Therefore, not all plants are useful as paleoclimate indicators because some of them are very tolerant to differing environmental conditions. Comparative analysis between paleoflora and present day flora is also limited by some morphostructural features that are present in fossil plants but absent in modern forms due to evolutionary processes. Taking into account that many species are able to migrate throughout as well as between continents, as a result of tectonism and consequent climatic changes, studies based on homology may be misleading. Floras from middle latitude highlands are similar to those from high altitude lowlands. About this subject, Raup and Stanley (1978) reported that within sedimentary successions deposited in low latitudes an upland local fossil flora could erroneously be interpreted as an indication of a regional temperate climate.

Wherever the essential conditions for development of the taphonomic processes are ensured, lowland associations dominantly represent the paleofloristic register due to their higher preservation potential. Plants from hydrophilous, hygrophilous and mesophilous environments have the greatest potential to autochthon or hypoautochthon preservation. On the other hand, plants from xerophilous environments, normally represented by hypoautochthonous or, less commonly, allochthonous elements of the paleofloristic associations, must be carefully considered, as suggested by the taphonomic studies performed by Scheihing and Pfefferkorn (1984) on terrestrial plants of the Orinoco delta, and by Rich (1989) on Holocene lakes. These authors observed that the aerial parts of the plants tend to be buried relatively close to their growth place hence endorsing the already well-established concept of the dominant hypoautochthonous nature of the vegetal remains of paleofloristic associations. Several factors play a role in the selection of the fragments to be preserved, including texture, transport agent and energy, nature of the trapping sediment, season and water temperature. In view of these peculiarities, it is possible to conclude that it may not be completely safe to interpret paleoclimate based entirely on an actualistic conception.

# Part I: Growth Rings in Fossil Woods and Paleoclimates

## 9.2 Fossil Woods and Paleoclimatology

Terrestrial plants are highly dependent on environmental conditions due to their sedentary way of life, which makes them directly influenced by climate changes. On the other hand, higher plants are composed of several organs, which are quite often separately incorporated into sediments. The distinct organs are detached from each other through processes related to the plants' life cycle, biological agents or natural catastrophes. They also have different potential of fossilization, which is more closely related to their chemical composition rather than their physical structure. Thus, diverse kinds of fossilization may occur as a consequence of specific taphonomic processes closely related to specific depositional environment characteristics.

In terms of wood petrification, fossilization of organic structures can result in inorganic mineral replicas that preserve microscopic anatomical features. This sort of fossilization, widespread around the world within rocks of different ages, has been studied to obtain paleoecologic and paleoclimatic information. Scurfield *et al.* (1974), Scurfield (1979), and Scurfield and Segnit (1984) developed comprehensive studies concerning the mechanism of silicification. They compared the obtained results with the processes occurring during chemical pulping of wood, and indicated that silica penetration initiates on the walls, through its micropores, providing a mineral framework able to keep the three-dimensional stability of the fossil wood.

Either the presence or absence of growth rings in fossil woods has been related to remarkable climate differences among distinct biogeographic realms (Elkins and Wieland, 1914; Arnold, 1947; Beck, 1953; Matten and Banks, 1967). Different authors carried out paleoenvironmental interpretations supported by the characteristics of growth rings in fossil woods.

Chapman (1994) compared fossil woods (Jefferson, 1981, 1982; Kelly and Moncrieff, 1992) from different areas of a fossil forest from the upper Albian of the Alexander Islands, Antarctica (paleolatitude 69°S), which grew close to the sea-level on a large delta plain. Some fossil woods with wide tracheids were considered to have thrived in places where the water supply was abundant during most of the growing season; in another area, the fossil woods present much smaller tracheids, and consequently the mean ring widths of these specimens are smaller than in the large-celled specimens. According to Chapman (1994), this may indicate that they come from a drier part of the delta plain.

Yao *et al.* (1994), in their study of Permian sequences of the Western Henan, Northern China, demonstrated the possibility of revealing regional paleoclimate evolution through the study of growth rings of fossil wood, in combination with miofloristic analysis and lithological data. According to these authors, fossil woods collected at the sequence base (*Araucarioxylon yichuanense*), in the Taiyuan Formation (Asselian/Sakmarian), as well as those taken from the overlying strata (*Protophylocladoxylon henanense*), in the Sanshi Formation (Artinskian), do not show growth rings. These evidences associated with other data such as the presence of *Lepidodendron* stems, indicates a warm humid, equable climate, most probably a tropical rainforest type. Secondary woods (*Araucarioxylon yamaense*) from the middle member of the overlying Schichienfeng Formation (Artinskian–Changhangian) show growth rings.The presence of growth rings was interpreted as the result of a pronounced climate change that took place by the end of the Late Permian in the Western Henan, Northern China, as also attested by the abrupt decline of the *Gigantopteris* flora. Previous studies of Yao (1983) reported the absence of the growth rings in woods of the lower Upper Permian, South China, in fossil floras dominated by *Gigantopteris*. According to Yao *et al.* (1994) based in wood analysis supported by results from lithological as well as chemical analysis, in the realm of the Permian Cathaysia Floristic Province, the North China Subprovince, entered into the arid zone of the Northern Hemisphere during the Late Permian while the South China Subprovince still remained in a tropical zone.

Two opposite interpretations about the position of the British Isles within the Early Carboniferous (Late Tournasian–Late Visean) paleoclimatic zones were reconciled by Falcon-Lang (2000) using anatomical wood ring evidence. Results obtained from the geological evidence (calcretes, vertisoils, vertically laminated lacustrine units, lagoonal facies and playa-like evaporites) supported paleoclimatic interpretations positioning the British Isles within the tropical zone, invigorating a tropical seasonal monsoonal climate (Golonka *et al.*, 1994). On the other hand, Chaloner and Creber (1973), based on growth ring analysis in fossil woods, indicate a humid or weakly seasonal tropical climate for this interval. Subsequently, Falcon-Lang (1999a, b) examined wood fragments from these assemblages, and observed that only about 50% exhibited growth rings. Falcon-Lang (2000) demonstrated that the studied gymnosperms had a long leaf longevity. In consequence, no growth rings were produced, or growth rings had very subtle ring boundaries, not reflecting annual growth periods. These irregular and subtle growth rings are formed in modern araucarian conifers in the monsoonal tropics of Northern Australia (Ash, 1983, 1985).

The aforementioned studies indicate that growth rings can be used as an important tool for interpreting ancient climate and environments, but its significance must be interpreted considering other available geological data.

## 9.3 Gymnosperm Growth Rings

Gymnosperm secondary xylem is composed of a tissue system spreading both horizontally (radial extent) and vertically (axial extent) in the stems, branches and roots.

Its function includes water and minerals transport, storage of substances and structural support. The axial system of present day gymnosperms is represented by the tracheids and, less often, by the axial parenchyma.

Within tropical regions, plants with active cambium during their entire development have continuously subdividing cambium cells, the resultant cells gradually differentiating to form the xylem and phloem. However, not all tropical plants show a continuous cambial activity (Fahn, 1974). Kramer and Kozlowski (1972) indicated that, even in the more favorable climates, trees do not grow continuously all year round. Nonetheless, although growth rings can be formed in tropical regions with dry and wet seasons (Worbes, 1989), most species do not form distinct growth rings and, often, these rings do not correspond to xylem increments produced during actual growth seasons (Jacoby, 1989).

Among present day tropical floras of some regions, such as Northern Australia, India, Argentina, Central America, Western Africa, Kenya and South Africa, woods displaying growth rings occur and allow a more detailed analysis. This characteristic is ascribed to both the seasonal nature of some areas, with environmental conditions favoring dormancy, and the plant's genetic propensity to cyclically develop growth rings (Jacoby, 1989). These rings are formed during the longer dry season and clearly express the cambial activity rhythim.

Détienne (1989) recorded a sometimes macroscopically imperceptible growth rings of tropical trees. The author dismissed the possibility of determining the plant's age from growth ring counting since many false growth rings can be formed.

In temperate climates, cambial activity regulation is controlled by the photoperiod, which cyclically triggers the activity of the hormone auxin. Cambial dormancy always occurs during periods of environmental stress, and plant growth reinitiates whenever this phase ends. Thin young leaves lose much water due to evapotranspiration, requiring a high conduction capacity from the xylem, which is supported by the wide tracheids of the early wood. During the subsequent growth season evapotranspiration decreases as a consequence of cuticle thickening. Therefore, a smaller sap volume per time is required, its transport being efficiently done by the thicker-walled tracheids, typical of the late wood. The latest cells of the growth ring can be identified by their reduced lumen and very thick secondary walls.

The foundation for wood growth ring analysis is related to density differences between the wood produced in the early stages relative to that produced by the end of the growth season. The early wood is less dense, has wider cells and thinner walls. Within a growth ring, the transition from the early to the late wood may be gradual and in perceptible, whereas the transition from the late wood of one ring to the early wood of the next ring is always abrupt and clearly delineated (Raven *et al.*, 1996). Studies of Creber (1977), Keller and Hendrix (1997) and Parrish and Spicer (1988) comproved that the markedness of the ring boundary (in particular, percentage latewood) has been used to indicate the intensity of climatic seasonality or as a measure of the favourability of conditions towards the end of the growing season.

The width of each growth ring changes each year has been considered as a result of luminosity, temperature, rainfall and soil moisture variations. Under favorable growth conditions, such as during periods of adequate rainfall, wider growth rings are formed whereas under less favorable circumstances narrower growth rings are created (Larcher, 1986).

Abrupt changes in water availability, or in any other environmental parameter, can cause the formation of more than one ring during a single cycle. For this reason, the indiscriminate use of the term "annual rings" is inadequate to name any feature related to the periodical growth of woody plants (Creber and Chaloner, 1984; Raven *et al*., 1996).

Different studies in woods growing under similar conditions in actual floras, as well as analysis in fossil assemblage from the same geographic demonstrate that the origin of growth ring in woods is not influenced only by the climatic seasonality, but also by genetic control (Tomlinson, 1980; Kumagai *et al*., 1995). Taking in to account this control, classification schemes of modern conifer woods use growth rings patterns as a specific character (Barefoot and Hankins, 1982; Greguss, 1972). La Marche (1982) showed that growth rings pattern changed within different conifer families, and observed that this character has some correlation to regional climate. These results correlate nature of growth rings with partially genetic control. The relative importance of the intensity of climate seasonality and of genetic control, in the generation of grow rings patterns has not yet been clarified. In a detailed study, Falcon-Lang (2000) presented numeric data to support the relationship between growth ring markedness and leaf longevity. Data presented suggest that ring markedness may be strongly influenced by leaf longevity in addition to reflecting intensity of climate seasonally. This hypothesis provides a possible endogenous basis for the relationship between ring markedness and leaf longevity. A strong inverse linear relationship between median leaf longevity and growth ring markedness (RMI) is demonstrated for different genus of conifers in South England. So, leaf longevity may be an important factor that must be taken into account in the inference of paleoclimate from growth rings.

On the other hand, Francis and Poole (2002) using analyses on growth rings of fossil woods (Cretaceous and Paleogene of Northern Antarctica Peninsula) supported by sedimentary and geochemical evidence concluded that, despite possible taxonomic and internal control on growth ring formation, a strong external climate signal can be a detect which matches global climate.

Irregular exchange activity related to changes in environmental conditions can cause the formation of false growth rings. Unusually, due to extremely unfavourable conditions, an entire growth ring related to one development period can be completely missing. Barefoot and Hankins (1982) emphasized that indistinct growth rings, which are apparently definite when observed macroscopically, can occur in the southern hemisphere temperate zones (Figs. 9.1A, B, 9.2).

Puri *et al*. (1983) concluded that dense forests protect trees from extreme temperatures. Even under dry conditions, the temperature within a forest may be very low but the relative humidity is kept high. Therefore, trees are protected and the possible damage caused by freezing is less dramatic (Geiger, 1965). Evidence of cambial alterations caused by freezing is shown by the presence of a dark zone of compressed or crushed cells around the branch replacing a growth ring that should have been formed (Chapman, 1994).

A

*Figure 9.1A Thin section of morphogenus B: (A) false growth ring; (B) true growth ring. Irati Formation, State of São Paulo, Brazil. Scale bar equal to 0.86 cm – see colour version of this figure in Appendix..*

B

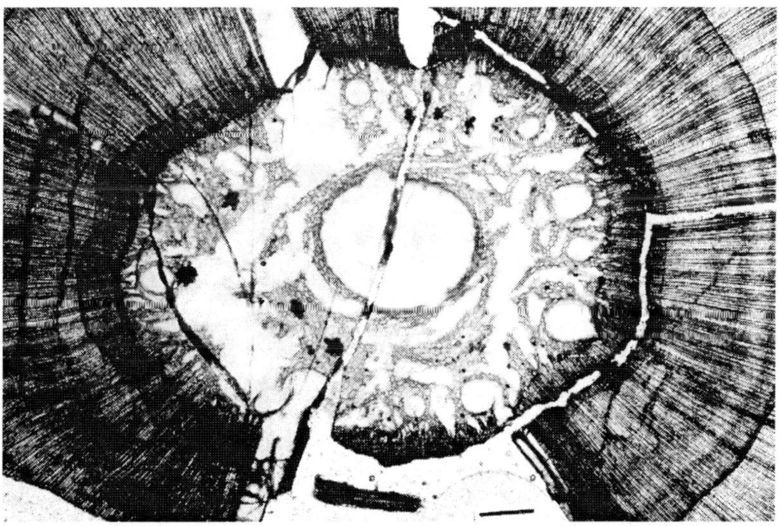

*Figure 9.1B Morphogenus B: Detail of pith and growth rings of Figure 9.1. Scale bar equal to 0.26 cm – see colour version of this figure in Appendix.*

The use of the mean sensitivity analysis in growth rings of modern woods aims to establish the global growth trend for a specific association. This measurement takes into account changes in the growth ring width within a succession and may range from zero, when every growth ring has the same width, to more than one. Trees having a mean sensitivity less than 0.3 are classified as complacent, whereas figures greater

*Figure 9.2* Scleromedulloxylon batoviense *Guerra-Sommer 1978. Apparent growth ring. Serra alta Formation, State of Rio Grande do Sul, Brazil. Scale bar equal to 330 μm – see colour version of this figure in Appendix.*

than 0.3 characterize sensitive trees. The definition of the mean sensitive trends of fossil woods can be a useful a indicator of the regularity of the length of the growing season and a criteria to interpret the environmental conditions under which an association has grown, specially the degree of environmental stress experienced at the growing site (Creber, 1977; Francis, 1994). However, of the efficiency method relies on the presence of abundant and correlative fossil material (Creber, 1977; Jefferson, 1981, 1983).

Chapman (1994) highlighted that fossil woods present some benefits in terms of their use as paleoenvironmental and paleoclimatic indicators, such as: preservation: due to its lower density than water even large pieces of wood can float for an undetermined period of time before being incorporated into sediments. Fossilized wood debris associated with carbonate matrix or marine fossils therefore must be carefully analysed as they could be very far from their place origin. Nevertheless, the paleolatitude information that can be inferred from a trunk association preserved within a sedimentary succession still remains very important (Jefferson, 1981; Basinger, 1991). Complexity: in contrast to leaves, pollen or spores, wood does not represent a single organ. Roots, as well as different parts of the stems and branches, have particular features that require distinct appraised whenever paleoenvironmental or paleoclimatic information is searched. The function of wood: the main energy source of the plants is obtained by the leaves through photosynthesis. The energy used for growth ring development (wood production) varies according to the plant species, amount of energy required by the other parts of the plant, amount of saved energy during the previous year and energy production during the current year. These variations generate a very

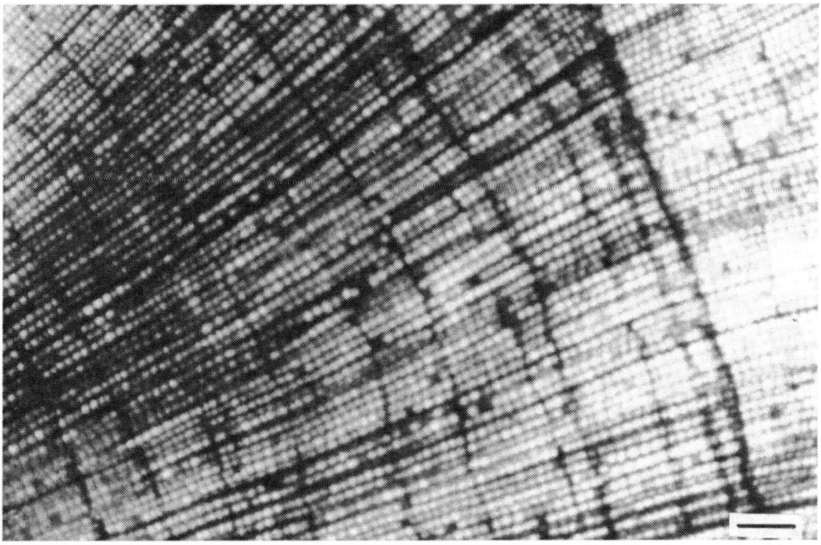

*Figure 9.3* Myelontordoxylon glandulifera *Pessôa 1985. Detail of transverse section of growth ring of probably a root. Serra Alta Formation, State of Rio Grande do Sul, Brazil. Scale bar equal to 98 μm – see colour version of this figure in Appendix.*

complex pattern that must be considered where growth rings are used for climate interpretation purposes.

According to Chapman (1994), based in Bannan (1954a, b), Carlquist (1975), Creber (1977), Creber and Chaloner (1984), the main characteristics of different wood organs, important for identification of isolated fossil material can be summarized as follows: twigs and branches: narrow rings, narrow tracheids, multiple rings traumatic parenchyma, reaction wood, upper trunk: rings regular, tracheids wide, reaction wood unusual; lower trunk: rings regular, tracheids wide, one ring per year near base, frost damage near base.; stump: rings irregular, tracheids wide, one ring per year; root: rings narrow and faint in outer parts; tracheids very wide, late wood cell walls thin, very few late wood cells (Fig. 9.3).

## 9.4 Case Study: *Barakaroxylon* (Surange and Maithy) Kulkarni, Maithy and Surange 1970 in Upper Permian Sequences of the Paraná Basin, Brazil

Analyses performed on permineralized paleoxylogical associations that occur within Late Permian sequences of the Paraná Basin show a significant compositional variation as well as excellent anatomic preservation (Guerra-Sommer, 1977, 1978; Costa, 1981).

The Paraná Basin, situated on the western border of South America, is an intracratonic basin filled with Paleozoic strata, Mesozoic sedimentary rocks and basalt lava and Cenozoic deposits. It overlies an area larger than 1.400.000 $km^2$ that includes part of the territory of Brazil, Eastern Paraguay, Argentinean Mesopotamia and Northern Uruguay.

The material studied here was collected from the southern portion of the Paraná Basin included in the Carboniferous–Lower Triassic Megasequence defined by Milani *et al.* (1998) for the Paraná Basin, based on the paradigm of sequence stratigraphy. According to lithostratigraphic criteria, the material comes from the Serra Alta Formation and Assistência Member of the Irati Formation (Schneider *et al.*, 1974).

The main aims of this study are:

- to define relative differences in growth rings of the fossil wood collected in both the Irati and the overlying Serra Alta Formations, in the same geographic area; and
- to infer the climatic conditions during the plant's life cycle and to correlate different growth rings patterns with climate changes.

Considering that processes related to growth ring formation are a result of both genetic and environmental parameters, the material studied here is strictly related to the morphogenus *Barakaroxylon* (Surange and Maithy) Kulkarni, Maithy and Surange, 1970.

**Level 1:** The samples related to *Barakaroxylon resiniferum* (Fig. 9.4), collected in the Passo São Borja (RS) outcrop come from a succession characterized by an alternation of carbonate and terrigenous mudstone and related to the Assistência Member of the Irati Formation (Schneider *et al.*, 1974).

A characteristic feature of *Barakaroxylon resiniferum* is the presence of a narrow band of late wood, generally composed of one to four cells, which differs from the wide early wood, which comprises 55 to 92 cells. Growth ring boundaries are well defined and within some growth rings the transition of the early to the late wood is gradual, although the boundary between successive layers is always abrupt. The presence of growth rings in the secondary wood suggests cyclical climatic variations. The analysis of the narrow, true growth rings indicates a climate with well-defined seasons and moisture deficiency during the growth season (Fig. 9.5).

The cambium activity initiation marks the beginning of the wet season hence early wood production; at the end of this probably short-term season, and under decreasing of humidity, late wood was produced. The subsequent dry season was accompanied by vegetative stagnation and consequent cambium inactivity. The return of the humid season was coeval with the re-initiation of the cambium activity. The regular alternation of dry and wet seasons is attested by the uniform thickness of the growth rings. The lack of false growth rings is a response to the well-defined, uniform and clearly marked seasons.

Based on the aspects presented here, the appraisal of the *Barakaroxylon resiniferum* growth rings the inference that this plant grew in a winter wet biome,

*Figure 9.4* Barakaroxylon resiniferum *Guerra 1976 (Alves, 1994). Detail of the pith and growth rings. Irati Formation, State of Rio Grande do Sul, Brazil. Scale bar equal to 1.5 cm – see colour version of this figure in Appendix.*

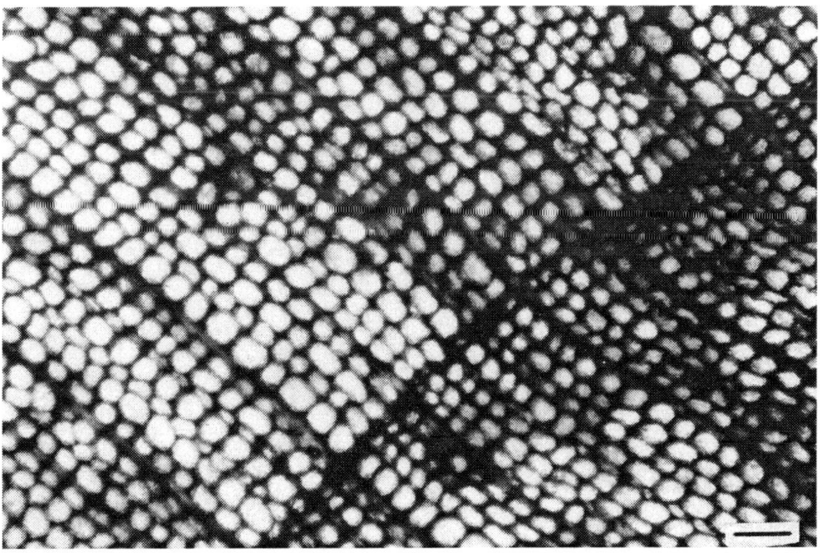

*Figure 9.5* Barakaroxylon resiniferum *Guerra 1976 (Alves, 1994). Limit of the growth ring. Irati Formation State of Rio Grande do Sul, Brazil. Scale bar equal to 98 μm – see colour version of this figure in Appendix.*

a floral association identified by Walter and Box (1976) for present day floras, during the development of a Mediterranean-like climate. It is possible to assert that the Irati Formation, at least in the São Gabriel region, was deposited under a climate characterized by well defined and uniform seasons throughout the years. The growth ring structure suggests a warm and relatively dry climate, similar to the present day Mediterranean region, as narrow growth rings indicate climatic conditions not favorable in terms of water supply.

Strahler and Strahler (1989) said that the present day Mediterranean climate, in both hemispheres, is characteristic of latitudes ranging from 30° to 45°, such as Central and Southern California, Mediterranean coast, Western and Southern coast of Australia, Chilean coastline, Cape Town and southern regions of Africa. According to these authors, the Mediterranean climate is characterized by a wide range in water supply (semi-arid, sub-humid and humid conditions), rainy winter and dry summer. For them, this climate is also portrayed by temperature oscillations of moderate magnitude, with hot summers and moderate winters. Demangeot (1987) stressed that in the wintertime the Mediterranean regions are affected by polar-related cyclones whereas in the summertime these regions are influenced by subtropical high pressure cells that cause arid conditions similar to those of deserts. According to them, these extreme seasonal alternations constitute the world's most contrasting climates.

**Level 2:** The *Barakaroxylon guerrae* Monteiro, 1979 and *Barakaroxylon* sp. (Fig. 9.6) specimens were collected in an outcrop located in the São Gabriel-Batovi region.

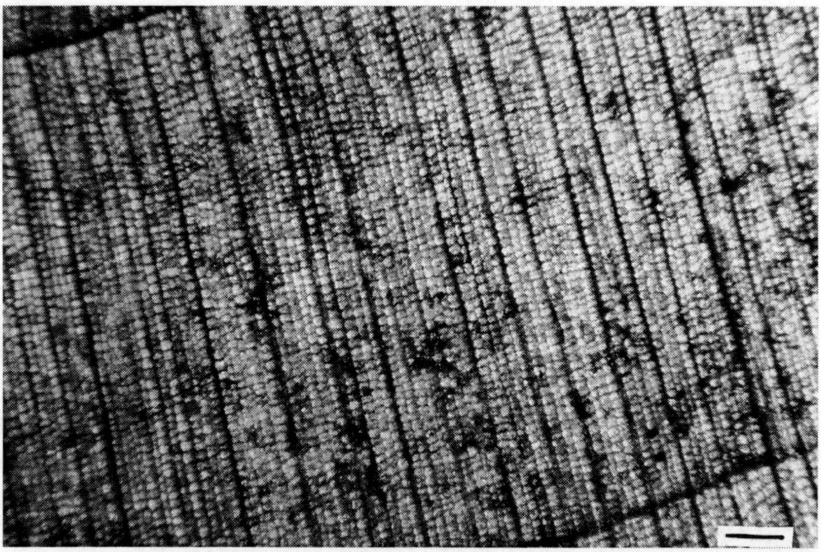

*Figure 9.6* Barakaroxylon *sp. Transversal section. Limit of growth ring. Serra Alta Formation, State of Rio Grande do Sul, Brazil. Scale bar equal to 330 μm – see colour version of this figure in Appendix.*

This exposure is related to the Serra Alta Formation (Schneider *et al.*, 1974; Lavina *et al.*, 1991) and includes a succession composed of siltite, gray shale and carbonate concretions. Their growth rings are wider than those previously described, displaying more than 150 cells per ring and a narrow late wood (not more than 8 cells).

This species presents well-defined growth ring limits. In the studied specimens, the transition from the early to the late wood can be either gradual or diffuse, although the boundary between the layers is always abrupt. Late wood tracheids are rectangular and flattened, displaying a diameter decrease and wall thickening. These features infer that cyclical climatic conditions were favourable, and growth seasons were long (Fig. 9.7).

The simple presence of growth rings displaying similar features to those mentioned above can not be conclusively correlated to present day climates. These attributes can be either produced under pluvial temperate climate, such as in the Californian Sequoia and Southern Chile Larch forests, both located in the Humid Temperate Domain, such as in the Atlantic Forest in Torres (RS, Brazil), also located in the Humid Tropical Domain. However, a humid pluvial climate, with long-term growth seasons, can be safely concluded from the growth rings analysis.

The studied material confirms the 50°S paleolatitude of the Brazilian Gondwana during the Kazanian, as formerly proposed by previous authors (Scotese *et al.* 1979; Ziegler *et al.*, 1996) in their paleogeographic reconstruction studies.

The paleoclimatic conditions inferred from both of these woods improve the simulated climate of late Permian for Pangea B. configuration of Fluteau *et al.* (2001). Based on atmospheric general circulation model, using Walter and Box (1976)

*Figure 9.7* Barakaroxylon *sp. Detail of transversal section of Figure 9.6. Limit of growth ring. Serra Alta Formation, State of Rio Grande do Sul, Brazil. Scale bar equal to 98 μm – see colour version of this figure in Appendix.*

*Table 9.1 Upper Permian fossil woods from southern Brazilian Gondwana: implications for paleoclimatic interpretation. (1), (2) Alves (1994); (3) Alves (1999); (4) Milani et al. (1998)*

| | | ***Barakaroxylon*** **– GROWTH RINGS PATTERNS (1)** | **OTHER PALEODATA** | **LITHOSTR.** | **CHRONOSTR.** | **PALAEOENVIRONMENT / PALAEOCLIMATE (4)** |
|---|---|---|---|---|---|---|
| **BATOVI REGION RS** | ***B. guerrae*** | General pattern: large growth rings (more than cells) thick early wood narrow late wood (1–5 cells)<br>Hidric disponibility: cyclical climatic conditions and favorable long casting development terms – good hidric disponibility | | **SERRA ALTA FORM.** | **TATARIAN** | **PLUVIAL TEMPERATE**<br>**NERITIC** |
| | | Climate: not defined. This kind of growth rings occur in both, humid temperate dominium of tropical humid dominium | | | | |
| | ***B.* sp.** | | | | | |
| **PASSO DE SÃO BORJA RS** | ***B. resiniferum*** | General pattern: narrow growth rings (96 cells) – narrow late wood (1–4 cells)<br>Hidric disponibility: short lasting development terms restrictions in hidric disponibility | ***Mesosaurus*** Mortality (2)<br><br>Highly Fragmented ***Botryococcus*** (2)<br><br>**Striated pollen grains with wide taenias (3)** | **IRATI FORMATION** | **KAZANIAN** | **IMPROVMENT OF WATER DISPONIBILITY**<br>**RESTRICT**<br>**SEA LEVEL RISE**<br>**MEDITERRANEAN LIKE** |
| | | Climate: Mediterranean-like. Rainy winter and dry summer; hot summer, moderate winters; winters storms and cyclones | | | | |

classification, Fluteau *et al.* (2001) included the studied area in a warm temperate, dry summer, to humid warm temperate climate.

Previously achieved climate inferences, based upon other paleontological and sedimentological data, corroborate the ideas extracted from the growth rings patterns. The widespread *Mesosaurus* mortality, registered in tempestites of the Passo São Borja (RS) outcrop, was interpreted as a result of storm-induced physical–chemical environmental changes (Lavina *et al.*, 1991). This interpretation was supported by microflora studies performed by Alves (1994) that indicated anoxic conditions related to phytoplankton collapse (Table 9.1). Morphological studies on striated pollen collected in the Passo São Borja outcrop (Irati Formation) has shown the presence of wide taenias that extend beyond the central body. This feature indicates a protective mechanism against desiccation due to severe climatic conditions related to seasonal aridity (Alves, 1994).

Most of the former authors agree that the environment in which the deposition of the Irati Formation took place was associated with an intracontinental water body (Hachiro, 1991; Santos-Neto, 1993). According to Santos-Neto (1993) a very shallow water depositional setting for the deposition of the Irati Formation can be deduced from the occurrence of some sedimentary structures, such as wave-generated cross-stratification, and features related to subaerial exposure.

According to Milani *et al.* (1998) the shallow intracontinental basin related to the shale deposition of the Irati Formation was flooded during the shale deposition of the Serra Alta Formation, denoting the last marine flooding in the Paraná Basin. Consequently, the regional events responsible for the climate change at the Irati–Serra Alta Formations transition were related to a sea-level rise that modified local climate and improved water availability. Therefore, a relatively isolated depositional system, in which recurring climatic alternations have played a decisive role on facies association, was slowly replaced, due to a slow base-level rise, by another, less confined depositional system characterized by a more humid climate.

Facies changes at the Irati–Serra Alta Formations transition, such as the disappearance of carbonate beds and bituminous shales and the occurrence of slightly coarser-grained deposits, indicate a less restricted environment and a more humid climate relative to the arid and semi-arid conditions that had characterized the deposition of the Assistência Member. The Boro content decrease at the Irati–Serra Alta formation transition indicates salinity reduction, probably ascribed to climate improvement (Table 9.1).

## 9.5 Conclusions – Fossil Woods

Although climate interpretations may be inferred from the presence or absence of growth rings in fossil wood, several limitations still remain because the controlling factors for growth ring formation have a very complex nature. Bearing in mind these constraints, however, the efficiency of the method it is possible to assert that data obtained from wood analysis can produce important information on continental environment, biological productivity and seasonal nature.

The analyzed case study shows that, although absolute ring width has little significance for climate inferences, it permits to evaluate relative climate changes in comparisons of assemblages from different stratigraphic units.

# Part II: Leaf Assemblages (Taphonomy, Paleoclimatology and Paleogeography)

## 9.6 Leaf Assemblages

Like fossil wood, leaf assemblages normally attest hypoautochthonous or parautochthonous depositional conditions, being more useful for local paleoflora reconstitution than the pollen record (microfossils), which generally represent a large-scale vegetation distribution (Table 9.2). According to Burnham (1993) taphofloras compounded by remains of leaf denote quasi–instantaneous and minimally transported accumulations or time slices, making them more suitable for comparison between paleocommunities and for analogy with modern biomes.

In terms of depositional environments, fossil assemblages help to portray the diversity of lowland habitats. An example of these are fluvial systems, where the inhabitants of the crevasse splays, levees and ox-bow lakes, lakes, marshes and swamps can be distinguished by its paleofloristic association. Sometimes, through river transportation, also the extrabasinal uplands and some proximal high areas could be represented (Greenwood, 1991).

Fossil leaf assemblages are also very useful in illustrate the evolutionary changes in plants (Hickey, 1984; Crabtree, 1987; Takhtadjan, 1991; Hill *et al.*, 1999). So, when accompanied by detailed fieldwork, they could also constitute a good biostratigraphic tool (Upchurch and Wolf, 1987; Hill, 1994).

But probably it is in obtaining information on paleoclimate that fossil leaves could be best used. Foliar physiognomy is very sensitive to environmental variation, as it modifies its anatomy and morphology in response to distinct climatic parameters. The close relation between vegetational types and climate has been long used as a method of mapping variations in humidity, relief, temperature and soils (Bailey, 1998). Thus, it may be a key to paleoclimatic (Chaloner and Creber, 1990; Wolfe, 1993, 1995; Chaloner and McElwain, 1997; Uhl and Mosbrugger, 1999; Wilf, 2000) and paleogeographical inferences (Durden, 1974; Wing and Bown, 1985).

The climatic response of leaves observed for the first time by Raunkiaer (1934) and Webb (1959), has been tested in a large number of modern biomes, and this has led to the identification of a group of foliar characteristics, the most visible of which are presented in Figure 9.8.

The shape and size sensitiveness of living leaves to moisture conditions have a great potential when applied to the fossil leaves of flowering plants. The correlation between leaf-area/mean annual precipitation and leaf-margin/mean annual temperature was used by Wilf *et al.* (1998) and Jacobs (1999), for example, in the evaluation

*Table 9.2 Taphonomic constraints of paleofloras and their meaning to the fossil record (after Kidwell and Behrensmeyer, 1993a; Burnham, 1993; Prentice, 1986 apud Barrón López, 1995; and Wing et al., 1995): (1) Barrón López (1995); (2) in subtropical and temperate forests, those with deciduous and mixed character have the greater potential to litter accumulation, nearly 4000kg/ha/year (Remmert, 1988); (3, some groups could have a distinct behaviour. The leaves of some modern beeches could survive 5 years or more in acid environments (Remmert, 1988)*

| | *Mode of Transport* | *Place of Life/Habit* | *Distance from the original biocenoesis* | *Time of Deposition* | *Taphonomy* | *Mean Time Involved* | *Problems* |
|---|---|---|---|---|---|---|---|
| MICRO FLORA | Mostly by wind (anemophilous pollen) and water (spores) | High areas / tall trees | Between 20 m and several kilometres | 5–50 years | Mainly alochthonous, mixing several biocenoeses | 100 to 1000 years | Lauraceae (1) and others that live in the understorey, are under-represented |
| MACRO FLORA | Natural dehiscence and incorporation in the soil, wind or water | Low and flat areas / low trees and shrubs | Parautochthonous assemblages: 0.1 to 0.5 ha (1–5 $km^2$) Litter: 1 to 3 $km^2$ (2) | Few months (rapid decay) (3) | Hypoautochthonous parautochthonous (more delicate, not surviving distant transport) | >1000 yrs to <1 Ma | Taphonomic processes can affect diversity. Sclerophyll leaves and deciduous plants are favoured |

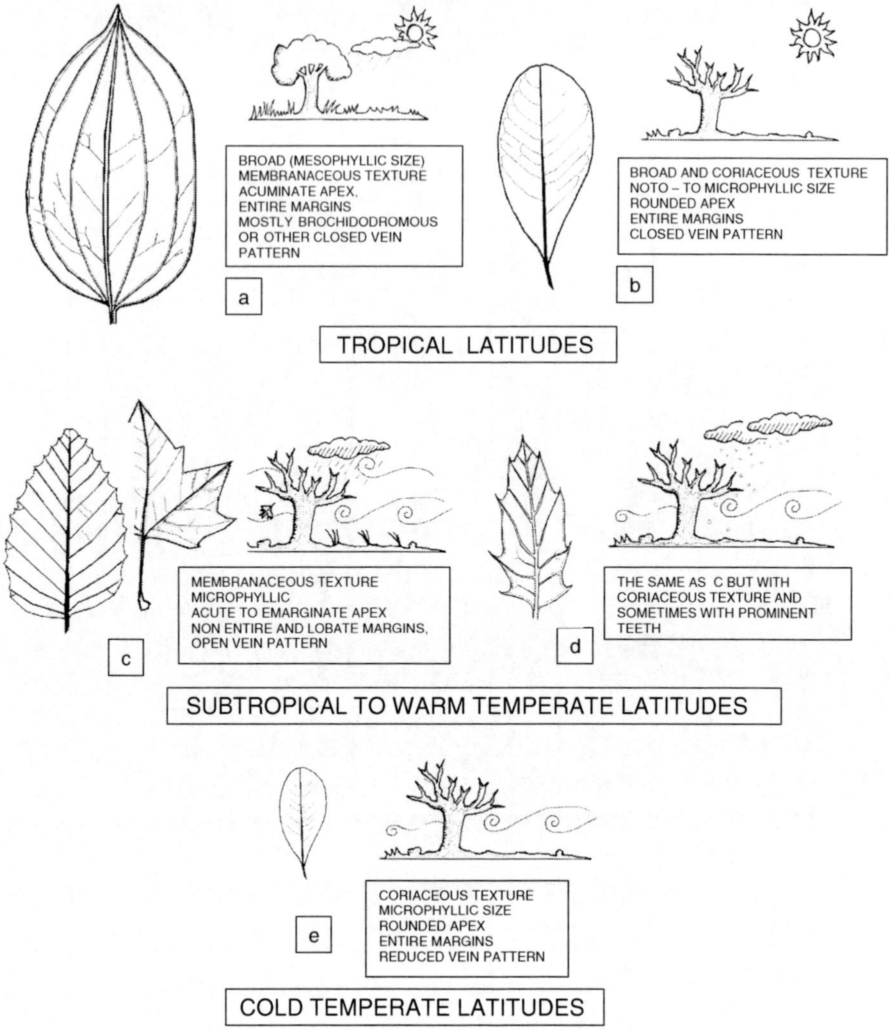


*Figure 9.8 Distinct leaf physiognomy and their relation with environmental parameters: (a) warm and wet; (b) warm and dry; (c) cold and wet; (d) seasonal (wet/cold) dry; (e) cold and dry.*

of climatic changes along the Paleocene–Eocene boundary in the United States and Africa, respectively. The presence of overarching papillae or reduced and deeply sunken stomata, like those showed by Hill (1994) in the Australian Araucariaceae, Proteaceae and Casuarinaceae during the Tertiary, are indicative of water stress or cold temperatures. The same could be inferred from thickened leaf margins (Esau, 1976) or reduced sizes of the leaf lamina (Wardle, 1967).

The use of CLAMP (Climate–leaf Analysis Multivariate Program by Wolfe, 1993) offers an accurate and precise method for obtaining climatic parameters from leaf assemblages. Its use in taphofloras has been extensively tested in Late Cretaceous and Tertiary paleofloras (Wolf and Upchurch, 1986, 1987; Greenwood, 1992, 1994; Wolfe, 1990, 1995; Wilf, 1997; Donohoo, 2001).

Moreover, the inverse relationship between atmospheric $CO_2$ and the stomatal index in leaves of $C_3$ plants can be used to infer paleo-$CO_2$ concentrations in the atmosphere (Beerling *et al.*, 1998; Royer, 2001; Kerp, 2002). In otherwise, high venation density normally have a positive correlation with wet climates (Hill, 1983).

Best resolutions can be obtained when the study includes the determination of common morphological characteristics between fossil leaves and those of comparative modern taxa ("nearest living relative method"). This approach is used for Tertiary assemblages, where many extant plant groups are represented (Mosbrugger and Utescher, 1997).

In view of those aspects, the plant macrofossils data from the Southern Hemisphere basins are of special interest. Although not as well known as those from the Northern Hemisphere, they are able to depict changes related to the Gondwana drift, especially between the late Mesozoic and Early Tertiary. This interval includes major changes in terms of paleogeography, such as the development of new interior seas, mountain buildings and onset of volcanic activity, which produced a deep impact on the world's life and climate (Francis, 1994).

The case study presented here uses one northern Antarctic Peninsula record to illustrate past environmental changes as reflected in a fossil flora.

## 9.7 Case Study: The Upper Cretaceous–Paleogene Paleoflora from King George Island, Northern Antarctic Peninsula: A Tool to Understand Autocyclic and Allocyclic Processes in High Latitudes Environments

### 9.7.1 The Context of Southern High Latitudes and the Evolution of Paleoflora

By the end of the Jurassic, latitudes higher than 50° in the northern and southern parts of the world were under the influence of wet and warm climate conditions (Hallam, 1985; Upchurch and Wolfe, 1987; Parrish, 1990; Greenwood, 1994). It provided ideal growth conditions for forests composed mainly of conifers, tree ferns, bennetittales and cycads. Although their diversity decreased as the timeframe moved into the Cretaceous, these floras showed a strong resistance to the mass extinction events that characterize the Cretaceous–Tertiary boundary (Crame, 1992).

The land masses, with extensive and continuous continental environments distributed from one pole to the other, produced a unique global climate that included wide tropical and subtropical belts and a narrow temperate one (Meyen, 1987; Vakhrameev, 1991). The polar ice cap was probably limited to the pole surroundings.

In the Southern Hemisphere, at latitudes between 60°S and 70°S, the Early Cretaceous was marked probably by an incipient system of westerly winds resulting for the onset of the drifting of continental masses. At the Pacific margin of the Antarctic Peninsula and southern South America (Elliot, 1988, 1991; Del Valle *et al.*, 1992; Hathway, 2000) their influence is magnified by the oceanic conditions of the climate. In addition, the continuous subduction movements create new high areas in the western border of the Peninsula that bar the winds and guarantee good precipitation rates – between 1000 and 2000 mm/year – favorable to wood growth (Jefferson, 1983; Francis, 1991, 1994) in those places, extensive to Australia and East Antarctica coast. Those processes give place to special ecotypes (*sensu* Margalef, 1983) or adaptive facies (Van Valen, 1986), which are useful in the paleogeographic and paleoclimatic reconstructions. Forests composed of Podocarpaceae, plants bearing *Classopollis* pollen (Francis, 1991), most probably Cheirolepidaceous conifers and pteridophytes were characteristic and cycads and broad-leafed ginkgos were common inhabitants of the coeval coastal lowlands (Archangelsky and Baldoni, 1972).

The first angiosperms arrived onto this scenery during the Albian. The first members of the fagalean lineage (Phillipe *et al.*, 1993; Cantrill and Nichols, 1996) were recorded on the South Shetland Islands (West Antarctic Peninsula). They were found between the Alexander Island assemblages, on the South Shetland Islands (70°S) showing two main foliar morphologies (palmate and pinnate craspedodromous), with mesophyllic to microphyllic leaf sizes, acute apices and entire margins that support the hypothesis of an adequate atmospheric humidity and temperature range. Growth rings of *in situ* fossil wood also attested high productivity although restricted to the summer months (Jefferson, 1982; Jefferson, 1983). The existence of a cold season is testified by the occurrence of the *Cyatheacidites* sp. and *Lycopodium clavatum* spores (Dettmann, 1986; Mohr, 1990).

By the end of the Campanian, leaf morphotypes (Fig. 9.9A) that could be related to modern *Nothofagus* (Nothofagaceae) made their first appearance in the Northern Antarctic Peninsula (Zastawniak, 1990; Francis, 1991; Dutra, 1997a; Dutra, 2000; Dutra, 2001; Dutra and Batten, 2000). The periodic and great environmental disturbance gave an opening to this opportunistic plant, which was able to colonize the new propitious volcanic soils and niches created. The genus went on to become the most characteristic of the Austral basin assemblages in the Tertiary (South America and Australasia) and today represents the most common element (with southern conifers) of the Southern Hemisphere temperate rainforests, where it maintains a preference for disturbed areas and thin volcanic soils (Veblen *et al.*, 1983; Hill, 1990; Hill, 1992; Hill, 1994; Veblen *et al.*, 1996).

### 9.7.2 Paleoenvironment, Assemblages and Physiognomy at King George Island Paleofloras

The testimony of those evolutionary and geological events are preserved in the South Shetland Islands archipelago, where the lithologies are better known on Alexander (Jefferson, 1982; Cantrill and Nichols, 1996), Livingston (Phillipe *et al.*, 1993) and King George Islands (Birkenmajer, 1980; Birkenmajer and Zastawniak, 1989;

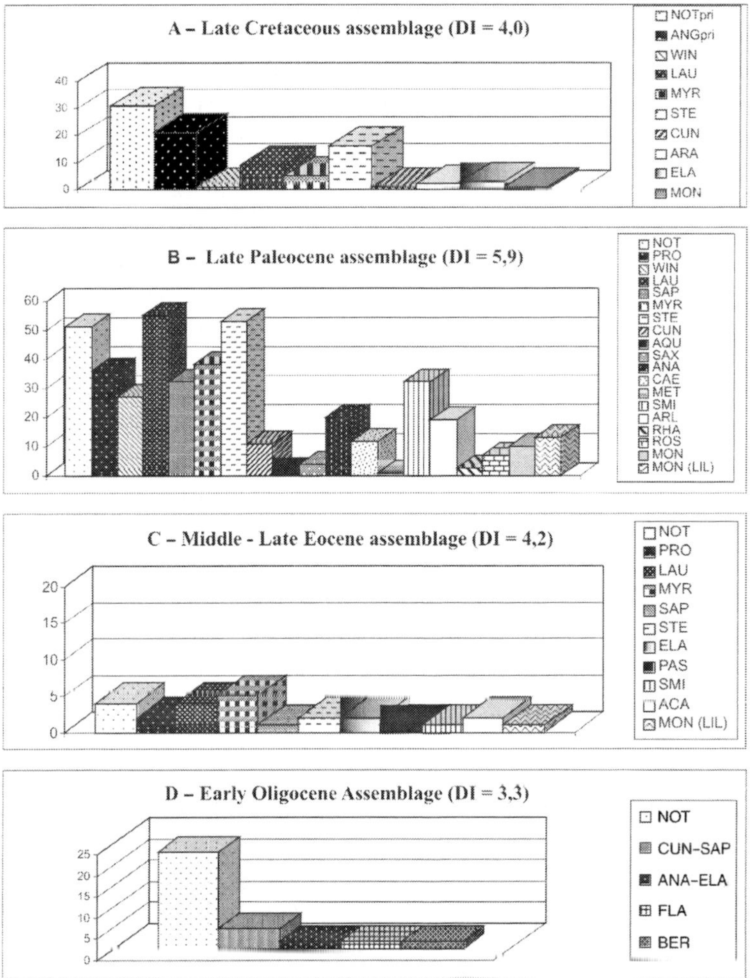


*Figure 9.9 Diversity index (DI) and proportional distribution of angiosperms (families) at King George Island paleoflora between the Upper Cretaceous and Lower Tertiary (ACA = Acanthaceae, ANA = Anacardiaceae, ANGpri = primitive angiosperms, AQU = Aquifoliaceae, ARA/ ARL = Araliaceae, BER = Berberidaceae, CAE = Caesalpinaceae, CUN = Cunoniaceae, ELA = Elaeocarpaceae, FLA = Flacourtiaceae, LAU = Lauraceae, MET = Melastomataceae, MON = Monimiaceae, MON (Lil) = Monocotyledons, MYR = Myrtaceae, PAS = Passifloraceae, RHA = Rhamnaceae, NOTpri = primitive* Nothofagus, *NOT = Nothofagaceae, PRO = Proteaceae, ROS = Rosidae, SAP = Sapindaceae, SAX = Saxifragaceae, SMI = Smilacaceae, STE = Sterculiaceae, WIN = Winteraceae).*

Shen Yanbin, 1994). The few periods of sedimentologically controlled regimes, specially in those levels dating to the end of Cretaceous, show that the fossil leaf preservation was commonly associated with braided river systems, floodplains and lakes (Cantrill and Nichols, 1996; Dutra *et al.*, 1996; Dutra, 1999).

The changes in paleofloristic content and from marine to continental environments during the late Jurassic and Lower Cretaceous were diachronous in the South Shetland Island area. An Albian age in the south (Alexander) and a Campanian age in the north (King George Island) is given and suggests either the gradual northwards uplifting of the magmatic arc (Elliot, 1988) or the presence of a hot spot under the Drake Sea, as proposed by Birkenmajer *et al.* (1986a).

These also allow the observation of a short cold interval that characterized the South Atlantic Ocean during the Cretaceous–Tertiary transition (Barrera *et al.*, 1987; Pirrie and Marshall, 1990; Klasz and Koutsoukos, 1991; Barrera, 1994). At this time, the western part of Antarctic Peninsula (South Shetland Islands) and the Pacific side of the Antarctic continent, showed micro-thermal conditions of climate during part of the year, and this was imprinted in the fossil wood by a short-term change of the growth ring pattern (Francis, 1991).

In the east (Larsen Basin *sensu* Del Valle *et al.*, 1992), in an area that was more protected by the maintenance of the proximity of the South America–Antarctic Peninsula, these changes were less remarkable. The first occurrence of *Nothofagus* pollen in the Maastrichtian of Southern South American (Menéndez and Caccavari de Filici, 1975; Romero, 1978) and Australian paleofloras (McLoughlin and Hill, 1996) confirms this physical continuity and the presence of a biogeographical province (Weddelian Province from Case, 1988). It also endorses the existence of a short cooling interval at the end of the Cretaceous that allowed this genus to migrate to the north. This event could be similar to that proposed by Villagrán (1990) for the cyclical climate changes occurred in South America during the Quaternary. She observed that during the glacial phases, the sub-Antarctic flora migrated northwards in such a pronounced way that taxa of the Magallanes tundra can be found in Central Chile associated with grass pollen.

At the end of Paleocene (Thanetian), the King George Island region was characterized by widely uniform environments resulting from a short interval of no intense volcanic activity. The simultaneous climate improvement, gave way to a more diversified paleoflora similar to what happened in several other parts of the world (Wolfe, 1990; Wise *et al.*, 1991; Thomson, 1992; Askin, 1990). The diversity indices found for the King George Island floras (Figs. 9.9B and 9.10) are also similar to those from the Late Paleocene of Northern Hemisphere (Wing and Bown, 1985; Wing *et al.*, 1995; Wilf, 2000) and other Austral floras (Hill, 1990; Greenwood, 1994). The same occurred in terms of the number of taxa present in the paleoassemblages, which reached 49 to 50 at this moment.

These changes were enhanced by an uplifting pulse of the magmatic arc, which caused a depositional break in the early Eocene record that can be followed along many sectors of the Northern Antarctic Peninsula (K–Ar ages between 8–52 Ma, according to Elliot and Trautmann, 1982; Birkenmajer *et al.*, 1986b; Soliani Jr. *et al.*,

1988; Shen Yanbin, 1994). From this time on, an altitudinal stratification of the forest types was established, accompanied by distinct taphonomic responses to the volcanic events (Askin, 1990; Dutra, 1999; Dutra, 2001; Poole *et al.*, 2001). Also a decrease in terms of diversity and equability followed each of the distinct volcanic phases (Fig. 9.9C).

The Middle to Late Eocene flora from King George Island will show a striking resemblance to those from the Eocene of Tasmania (Carpenter *et al.*, 1994) and New Zealand (Pole, 1994) and with the modern vegetation of Western Patagonia along the Andean Mountain Belt (Veblen *et al.*, 1983; Villagrán, 1990; Armesto *et al.*, 1995), exhibiting a specialization to altitudinal stratification and to cooler environments. In the flat areas of the eastern side of the Antarctic Peninsula (Larsen Basin) the appearance of palm tree pollen suggests the onset of the South Atlantic coastal environments (Askin, 1992; Baldoni and Barreda, 1986; Baldoni and Medina, 1989).

By the end of the Eocene the first microphilic and deciduous (with plicate vernation) *Nothofagus* leaves appeared in the assemblages of all northern Peninsula region, therefore pre-announcing the coldness that was going to characterize the Eocene–Oligocene transition (Case, 1988; Gazdzicki and Stolarski, 1992; Ditchfield *et al.*, 1993). This tendency is confirmed by the low diversity of Upper Eocene. Early Oligocene flora of King George Island (Fig. 9D), which is composed by taxa with leaves of coriaceous texture (Berberidaceae) and prominent teeth (Cunoniaceae and/or Rosidae). The foliar physiognomy and microphily of the *Nothofagus* leaves are very similar to those found in *N. antarctica* and *N. gunnii*, the modern southernmost elements of this genus.

Beds with shell layers and striated pebbles of the Late Early Oligocene as well as coarser-grained beds (in well logs) at the Weddell Sea support both this climatic inference and the assumption of a sea level fall (Ehrmann, 1991; Doktor *et al.*, 1988) at the

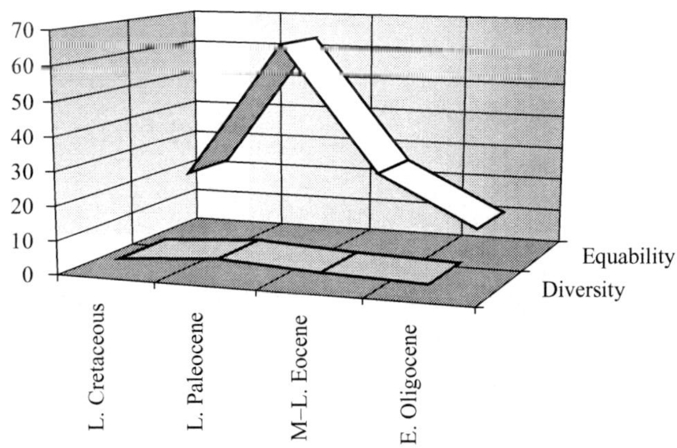


*Figure 9.10 Diversity and equability of King George paleoflora between Upper Cretaceous and Tertiary.*

end of the Rupelian. Birkenmajer (1988) considered it as the most expressive Cenozoic glacial event in the Antarctic Peninsula.

Although a climatic amelioration had taken place during the early Miocene (Ehrmann, 1991) there were no more vegetational elements capable of reconstructing the previous landscapes (Dutra, 1997b). The South Shetland Islands drift from the Antarctic Peninsula and Southern South America, caused by the opening of the Bransfield Straight and Drake Sea, probably hindered vegetation from re-colonizing the area.

## 9.8 Acknowledgements

We are grateful to Marion K. Bamford (University of the Witwatersrand, Johannesburg) and Jennifer C. McElwain (The Field Museum, Chicago) for their critical reviews and numerous suggestions for improvement of this manuscript, and to José Henrique G. Melo (Petrobras-Cenpes, Rio de Janeiro), for his helpful suggestions. Laureen Alves gratefully acknowledges support from a post-doctoral research grant of the Brazilian Conselho Nacional de Pesquisas (CNPq).

Chapter 10

# Palynofacies Analysis and its Stratigraphic Application

DAVID J. BATTEN[1] and DARRIN T. STEAD[2]

[1] Institute of Geography and Earth Sciences, University of Wales, Aberystwyth SY23 3DB, and Department of Earth Sciences, University of Manchester, Manchester M13 9PL, UK.

[2] Wellstrat Services Ltd, 1 Castle Grange, Wrexham Road, Caergwrle, Wrexham LL12 9HL, UK.



## 10.1 Introduction

During the past 25 years there has been much interest in the way in which the composition of acid-resistant organic matter, or palynological matter (PM), in sedimentary rocks can be used to aid the interpretation of depositional environments. Linked to this is its proven value for indicating the potential of a particular stratum as a generator of hydrocarbons. The former requires that the investigator is a palynologist by training. Depending on the degree of precision required, the latter can be determined by palynologists, coal petrographers or organic geochemists. More recently, the composition of PM recovered from rock samples, otherwise known as the palynofacies associated with a deposit, has also been used for stratigraphic determinations on a local scale for fine correlation of reservoir sections within oil-fields, and to some extent on a regional scale, particularly in areas and within successions in which more conventional biostratigraphic markers are scarce or lacking.

The environmental and source-potential aspects of palynofacies analyses have been discussed previously at some length by Tyson (1995), Batten (1996a, b, 1999a) and others.



*E.A.M. Koutsoukos (ed.), Applied Stratigraphy,* 203-226.

The purpose of this chapter is to concentrate on the stratigraphic applications. Consideration of palynofacies in their stratigraphic context can help in the interpretation of the processes controlling deposition and provide a basis for paleogeographic reconstructions. Although it is often possible to infer depositional environments from lithofacies characteristics, including their shelly faunal content, if any, palynofacies data may enable a more accurate appraisal, particularly in otherwise unfossiliferous deposits, which in turn may also have local or regional stratigraphic significance. Before this approach can be discussed satisfactorily, it is first necessary to put it into context by commenting, albeit briefly, on facies in general, palynofacies in particular, and the factors affecting their composition. In this connection, of fundamental importance are methods of sample preparation and the basis for distinguishing between, and naming, organic particles and masses in a microscope slide that is viewed in transmitted light.

## 10.2 Origin, Deposition and Diagenesis of Organic Matter

The origins of sedimentary organic matter are extremely diverse. Much is derived from land plants, some comes from associated life forms, such as fungi and beetle carapaces, and some will have been eaten and excreted by herbivores, both large and small. The remains of aquatic plants and algae, especially in an amorphous condition, are also abundant, and some of this too is preserved as faecal pellets, but there is also a wide variety of resistant bodies, such as the cysts of phytoplankton, that are readily preserved as fossils.

The geographic and stratigraphic distribution, and facies relationships of sedimentary organic matter vary widely. Only a very small fraction of what is potentially available for deposition becomes incorporated in sediment and preserved, and this is commonly at least partly decayed. Most organic material is rapidly degraded by a variety of means (as discussed in some detail by Batten, 1996a, pp. 1012–1016). It may still be destroyed by microbial activity or diagenetic processes even when it has been buried in sediment (Batten, 1996a, pp. 1016, 1018; 1996b).

The physical and chemical changes that take place both before and after deposition lead to four main kinds of organic matter being preserved in all but the most thermally altered of sedimentary rocks (see Batten 1996a, b and references therein). These are: (1) the protective walls of spores, pollen grains, prasinophyte algal bodies and other microfossils of distinctive morphology; (2) more or less structured fragments such as cuticles and woody tissues; (3) unstructured materials having an amorphous appearance; and (4) soluble compounds that can be extracted using organic solvents and which often contain what geochemists call ‘biological markers’ or ‘biomarkers’. The last of these is beyond the scope of this chapter.

## 10.3 Facies and Palynofacies

The term ‘facies’ has been much discussed since its introduction in the late 1830s. Nowadays it is generally accepted that it is a body of rock of a specified character. On its own it covers numerous features so prefixes are commonly added to limit and/or clarify the sense in which it is being used. Hence, 'lithofacies' refers to all the characters of a rock record of any sedimentary environment; 'biofacies' to the composition and preservation of the organic components of a rock or unconsolidated sediment; 'freshwater facies' to the environment that the sediment reflects or in which the sediments that make up a particular type of rock or suite of mixed rocks is thought to have accumulated.

'Palynofacies' (or 'palynological facies') is most widely applied in a general way to mean organic matter that is recovered from a rock or unconsolidated sediment by the standard palynological processing technique of digesting a sample in HCl and/or HF (see sample preparation below), i.e. in the sense of Combaz (1964, 1980). All sedimentary deposits containing organic matter have, therefore, an associated palynofacies, be it a few charcoal particles or a wealth of miospores (small spores and pollen grains) and phytoclasts (fragments of plants; see section 5). Ideally, it is a distinctive association of acid-resistant organic components of a deposit that accumulated under certain conditions and reflects a particular process or environment. In practice, however, many palynofacies are not especially distinctive and, as a result, not particularly useful from either biostratigraphic or paleoenvironmental viewpoints. Hence, they inevitably vary in their interpretative value.

As discussed by Batten (1996a), some authors have provided other definitions of palynofacies, among which are those that include mention of (inferred) depositional environment. In our opinion it is better to separate palynological facts (the composition of palynofacies) and the labelling of recognizable types from their (paleo)environmental interpretation. The latter should not be considered without reference to other characteristics of a deposit unless for some reason this information is unavailable, because the conclusions drawn will inevitably be more equivocal than when these data are combined. An understanding of the sedimentary context of a particular palynofacies is, therefore, essential if an environmental interpretation is to be more than broadly based. In a paleopalynological context, this means the difference between, for example, backswamp and merely non-marine.

The recognition of a particular type of palynofacies, and hence the pointers it provides to depositional conditions, must depend upon its recovery on at least several, and preferably many, occasions from different sedimentary successions of varying age. Repetitive character combinations are thus of fundamental importance and can only be obtained if sufficient numbers of samples have been examined. Detailed interpretations require that a large amount of time be spent identifying palynomorphs and determining relative abundances of both these and the associated organic detritus, and the more samples considered the better.

A vertical succession of conformable sedimentary deposits of varying character must indicate accumulation in changing conditions. At any two levels within it, the different environments represented will probably have been present elsewhere within the area or basin of deposition. If erosive or other breaks are present, these may represent the passage of one or more different environments, the evidence for which, if any, was removed subsequently. Even if the junction between two beds is merely sharp the environments they reflect may not have been geographically neighboring.

The fact that palynofacies, in common with lithofacies, do occur repeatedly enables some order to be brought to the otherwise continuous variation in composition of sedimentary deposits and the organic matter they contain. In a rhythmically or cyclically deposited succession several kinds of palynofacies may be encountered together although not necessarily in the same order. Such facies associations are fundamental to environmental interpretation, whether usually sequential or apparently occurring at random.

The first step towards establishing palynofacies associations in outcrop or borehole/well sections is to examine the color, bedding, composition, texture, fossil content and sedimentary structures; in other words, the lithofacies. Determinations of palynofacies types may lead to refinements of, and provide evidence for, changing environmental conditions inferred from such 'conventional' facies analysis, but only if based on reliable criteria generated from the accumulation of large quantities of data.

Computer-based numerical methods and statistical techniques can be useful for handling palynofacies (and lithofacies) data when the amount of information available becomes difficult to manage and assess. In our opinion, they should not, however, generally be used as the main means of identifying palynofacies, as Batten (1996a, 1999a) has noted previously. It is more satisfactory from a practical viewpoint if they are categorized simply on the basis of relative abundances and other data on the various components of an organic preparation. If the information accumulated is not characteristic of any particular environment or group of lithofacies then it is unlikely that a numerical analysis will produce much that is useful.

Commonly there is a need to modify the data in various ways so that they can be handled effectively, and more abundant or more obvious components tend to be stressed at the expense of data that are close to being statistically insignificant but nevertheless geologically important. Although such modifications can be taken into account by re-examining the raw data after an analysis has been completed, it is at this stage that numerical methods and statistical tests are better applied, as a means of checking or reinforcing rather than determining palynofacies identifications.

## 10.4 Sample Preparation for Palynofacies Analysis

Methods of sample preparation for palynofacies, source-potential and organic-maturation studies have been considered previously by Batten and Morrison (1983) at some length, and more recently by Tyson (1995), Batten (1996a, 1999a, b) and others. Treatment of this topic here will, therefore, be restricted to the salient points.

Palynofacies studies rely on comparisons between occurrences and relative abundances of the organic components of a palynological preparation. In view of the number of variables involved it is most important to maintain as routine a preparation procedure as possible; hence consistency is of the essence. Standard weights of sample should be used: 5 g is normally sufficient but if the rocks or sediments to be dealt with are unlikely to yield much organic matter, such as some limestones and chalks, and sandy facies, multiples of this amount should be processed. It is then possible to measure the quantity of organic matter yielded by each sample, at least in a general way.

Digestion of the mineral matrix is commonly achieved using concentrated HCL and HF, the amounts of acid required and the duration of the treatment varying according to the lithology of the sample. A limestone will normally need to be immersed in HCL considerably longer than, for example, a shale containing a few ostracod fragments. A quartz-rich sandy siltstone must be subjected to HF for a lengthier period than an argillaceous chalk. A coal comprising mainly gelified woody tissues and an oil-shale containing a large quantity of amorphous organic matter resulting from bacterial degradation of algae and aquatic plants may not respond to treatment in either of these acids, disaggregation being achieved only by physical crushing or subjection to oxidizing reagents. The latter should not, however, be carried out before the first 'kerogen' slide is made. This is prepared whether or not there has been any reaction with HCL and HF. Exceptionally a slide may, in addition, be made prior to HF treatment if the sample is mainly carbonate. We prefer glycerine jelly as a mounting medium but other substances can be used if preferred or necessary (e.g. in a hot climate).

One of the problems with the residues remaining after the onslaught of HCL and HF is that they often contain large amounts of finely disseminated detritus less than 10 μm in diameter. The resulting slide preparation may, therefore, not only look unattractive under a microscope but also be difficult to analyse because at least some of the larger phytoclasts and palynomorphs are smothered and rendered indeterminate, their characters being obscured. Since categorization of the coarser fraction can be vital to the interpretation of both depositional environment and source potential for hydrocarbons, it is important to remove at least some of the offending debris. This may be done by filtering using a sintered glass funnel (porosity 2) or by sieving through a fine (e.g. 10 or 15 μm) screen. The former allows the preparator to remove as much or as little finely particulate matter as necessary. The latter tends to result in 'cleaner' preparations because most of the particles smaller than the diameter of the mesh are removed. Since the presence of finely divided organic matter is critical to analyses of both paleoenvironments and source potential, its loss can lead to misinterpretations. Despite this risk, most palynologists find it more convenient to use sieves.

The preparation procedure may be taken a stage further if required (and as advocated by Batten and Morrison, 1983 and Batten, 1996a), such as by subjecting the remainder of the residue to brief ultrasonic and/or oxidative treatment. Depending on the composition and maturity of the organic assemblage, the ultrasonic vibration will break up coagulated fine-grained detritus, dislodge it from the surface of palynomorphs, cuticles and other phytoclasts, and cause fragmentation of brittle components. The end product, after further sieving and filtering, is usually a preparation that is at least slightly cleaner. Oxidation may have similar effects and, in addition, may brighten or lighten the color of the palynomorphs, making it easier to study their morphology. A second 'standard' slide is made following this treatment.

Unlike HCL and HF, use of oxidising reagents and alkalis can effect considerable changes in the general aspect of a preparation. Care must, therefore, be taken in deciding how much of each is required. This will depend on both the composition and the maturity of the organic matter that is preserved in the first slide. It is because of the risk of altering the proportions of organic components of a preparation that most published palynofacies studies have been based on unsieved and/or sieved organic matter isolated from rock or sediment samples solely by HCl and/or HF treatment.

## 10.5 Classification of Sedimentary Organic Matter

The basis for classifying and naming components of sedimentary organic matter has been the subject of numerous papers, and was discussed in detail by Tyson (1995) and Batten (1996a). Many of these are concerned with characters that are observable in reflected light, the terminology of which is now widely agreed. Unfortunately no such consensus exists for the non-palynomorph components of palynofacies as seen under a transmitted light microscope. The number of terms potentially available is greater than required for effective categorization. The following are among those we consider are acceptable, with accompanying abbreviations; all may be directly or partially correlated with maceral groups and subdivisions thereof that are generally accepted by coal petrologists:

1. Structured organic matter (STOM). This consists of phytoclasts and zooclasts. Phytoclasts include wood (black and brown), charcoal and other black particles, cuticles, bark and cork, other (non-cuticular) tissues, tubes, filaments and hairs, and fungal hyphae. Zooclasts include egg cases and fragments of beetle carapaces.
2. Unstructured (structureless) organic matter (USTOM). This comprises amorphous organic matter (AOM), which may have either terrestrial (AOMT) or aquatic (AOMA) origins, gelified matter, resin and amber, and solid bitumen. AOM commonly has bacterial and algal components, and may be dominated by them. It is typically fluffy and/or homogeneous in appearance, often uniformly granular, and commonly very finely laminated, but depending on the structure of the original constituents it may also appear rather membranous or fibrous, have palynomorphs embedded within it, and/or fluoresce strongly when examined in blue or ultraviolet light (see Tyson, 1995; Batten, 1996a, b).

An aspect of classification not dwelt on previously by Batten is the 'palynomaceral' grouping of Whitaker (1984; also Whitaker *et al.*, 1992), which has been used (sometimes in a modified form) successfully by others (e.g. van der Zwan *et al.*, 1993; Fig. 10.1), including one of us (Stead), in a number of investigations. This concerns the transmitted-light characteristics of phytoclasts especially as they relate to their hydrodynamic properties (buoyancy) and degradation, but with support from ultraviolet, reflected-light analyses. Although they may have a variety of origins, palynomacerals have been considered to equate in a general way to biological and coal-petrographic types. By applying basic sedimentological principles to the terrestrial input and paleontological principles to the aquatic (marine or non-marine) content, depositional environments have been interpreted by this means.

In Whitaker *et al.* (1992) four palynomacerals are recognized. These are summarized diagrammatically, and in a non-marine paleoenvironmental context (after van der Zwan *et al.*, 1993), in Figure 10.1. Palynomaceral 1 was considered to comprise orange-brown or dark brown structured or structureless dense material of variable shape and preservation. In palynological preparations Whitaker *et al.* (1992, p. 172) included "structured plant debris (mainly resinous cortex material), humic gel-like substances, resinous substances, and algal detritus (mainly degraded *Botryococcus* spp.)", which were considered to be of low buoyancy because of their "frequently large size and higher specific gravity", with the "spongy cortex nature" rendering it susceptible to waterlogging. They noted that the maceral is not resistant to physical abrasion and would have been readily destroyed in high-energy environments.

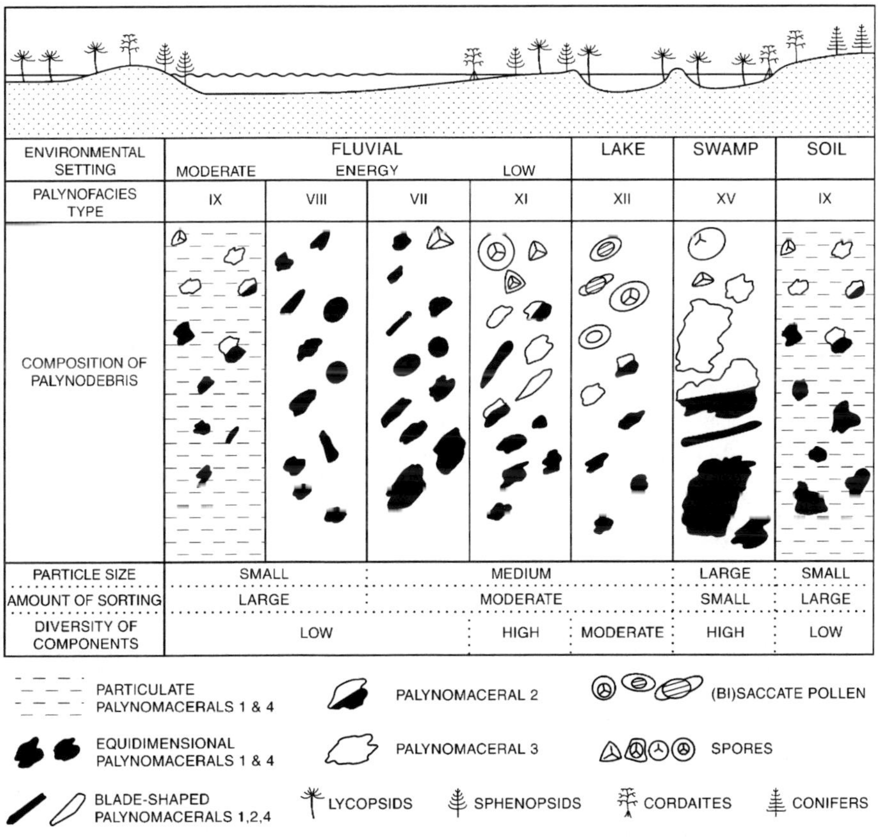


*Figure 10.1 Carboniferous non-marine palynofacies, based on figure 6 in van der Zwan* et al. *(1993), modified, showing categories of palynomacerals and their relative abundance in a succession comprising terrestrial and fluvio-lacustrine deposits.*

Of this heterogeneous mixture of organic particles the most surprising inclusion is *Botryococcus*. This is because dead specimens of extant species are known to form a floating scum on lake surfaces (Batten and Grenfell, 1996). They are also difficult to destroy by chemical means in the laboratory.

Palynomaceral 2 was described (p. 173) as "Brown-orange structured or structureless material of irregular shape. It may include structured plant material (some leaf, stem or small rootlet debris), algal detritus and, to a lesser extent, humic gel and resinous substances. Buoyancy is considered to be higher relative to palynomaceral 1 because of its thinner and often lath shaped character."

Palynomaceral 3 was categorized (ibid.) as "Pale, relatively thin, irregularly shaped, usually structured material, occasionally bearing stomata. It is considered the most buoyant of palynomacerals 1–3. It may include structured plant material (mainly of leaf origin, which may or may not bear a waxy surface coating), and degraded aqueous plant material."

Finally, the characteristics of palynomaceral 4 were regarded (ibid.) as "Black, or almost black, equidimensional, blade- or needle-shaped material, which is usually uniformly opaque and structureless, but might occasionally show cellular structure." The origins of this palynomaceral are many and varied, and include "compressed humic gels, charcoal (resulting from forest fires), reworked charcoal and geothermally fusinized (occasionally semi-opaque) material." In addition, Whitaker *et al.* also noted: "Opaqueness, or darkening of organic material can also be the result of secondary effects such as staining by migrating fluids within the sediments."

The blade-shaped forms were considered to reflect the preferential breakup of larger pieces of "oxidised (mainly charcoal) woody debris parallel to the long axis of elongate cellular structure typical in stem material." As they noted, such material is particularly resistant to degradation and very buoyant. It can, therefore, be transported over long distances, and is "especially concentrated in [deposits that reflect] high energy palaeoenvironments." Needle-shaped material was also regarded as buoyant, but less likely to be preserved in high-energy settings. Equidimensional forms of palynomaceral 4 were described (p. 174) as "frequently intermediate in character to palynomacerals 1 or 2 and may thus have a relatively lower buoyancy."

The forms of palynomaceral 4 have been recorded by Batten (unpublished data) on many occasions but the other three palynomacerals are regarded by him as being too heterogeneous for satisfactory application, although as noted above, Stead has made use of them with success. We both recommend that whenever possible a project using the palynomaceral approach to palynofacies should be carried out by only one person so that recording of the various categories noted above is consistent. This is because what one person might regard as relatively thick palynomaceral 2 another might consider to be palynomaceral 1. As a result, when two people are involved in investigating a series of well sections, for example, such recording differences might mean that some palynofacies correlations between the sections would not be recognized.

Despite Batten's reservations about palynomacerals 1–3, a point of agreement with Whitaker *et al.* is that it is useful to take into account the buoyancy of palynomorphs in paleoenvironmental interpretations. Saccate, especially bisaccate, pollen grains are known to be particularly buoyant. Large numbers of bisaccates have been encountered in marine deposits far removed from any source vegetation, reflecting long-distance transport by both wind and ocean currents (e.g. Traverse, 1988, p. 379). Flimsy, thin-walled spores may also be transported and concentrated in fine-grained sediments hundreds of kilometers offshore. On the other hand, thick-walled spores and larger plant products (e.g. megaspores, seeds) tend to behave more like silt or very fine sand particles in water and, hence, become incorporated in coarser deposits closer to their terrestrial source.

It is also useful to take into account the diversity of palynomorph assemblages, which can aid significantly the interpretation of environments of deposition. Commonly fossil spores and pollen grains are abundant and varied in sediments that accumulated in near-shore marine, lagoonal and non-marine settings whereas impoverished assemblages, limited in both numbers and diversity, tend to be associated with deposits far from land. On the other hand, palynofacies containing common representatives of only a few taxa may indicate close proximity to the source vegetation. In such circumstances their distribution reflects the composition of the local vegetation and the quantity of spores and pollen grains produced by the parent plants concerned more than their size and buoyancy.

## 10.6 Palynofacies Types

Detailed sedimentological analyses of subsurface successions and surface exposures coupled with an equally thorough documentation of the preservation and distribution of palynological constituents can often lead to the determination of a variety of environments and subenvironments that may be characterized palynologically. The basis for recognizing and labelling palynofacies types has been published in papers by a number of authors. An early attempt was made by one of us (Batten, 1973) to classify palynofacies recorded from the largely non-marine Lower Cretaceous, Wealden succession of southern England. Other approaches to identifying and classifying have also tended to be limited in their application to the particular successions being investigated.

Whitaker *et al.* (1992, p. 172) used the relative abundances of the main palynomorph groups they recorded ("non-saccate sporomorphs, freshwater algae, fungal spores, saccate sporomorphs, dinocysts, acritarchs, marine algae, microforam[iniferal] text linings"), "structureless (sapropelic) organic matter" and their four palynomacerals as a basis for defining five palynofacies associations. These range from association 1, which consists of a very small amount of PM, mostly or entirely small particles of palynomaceral 4, and devoid of, or containing very few palynomorphs, to association 5, which is dominated by 'structureless organic matter'. In between these two extremes are preparations that consist of: large, mainly equidimensional and irregularly shaped palynomaceral 4, again with only a few or no palynomorphs (association 2); a mixture of large and small palynomacerals 1–4 together with palynomorphs (association 3); and assemblages dominated by palynomorphs (miospores and/or algae) with subordinate small palynomacerals 1–4 (association 4).

To these associations they added three environmental prefixes (M, marine; B, brackish; T, terrestrial) and a variety of morphological suffixes (e.g. e, notably equidimensional; s, exceptionally small; 1s, one type of spore type dominant; 2d, two dinoflagellate cyst species dominant) as qualifiers. Hence, for each major environmental type, associations 1–5 were considered (p. 175) to reflect "a gradual decrease in energy, oxygenation and proximity to a terrestrial source".

However, since palynofacies are infinitely variable in composition, their characteristics grade into one another, which means that intermediate stages and atypical associations also need to be identified. According to the system of Whitaker *et al.*, intermediate stages can be represented by noting the next closest type in brackets, e.g. M3(M4). Other variations can be expressed in a similar way. An association that is typical of marine palynofacies 4 but which contains an abundance of large palynomaceral 2 would, for example, be logged as M4(M2). In both cases the general environmental setting is indicated by the M for marine. Hence, the system of Whitaker *et al.* is flexible enough to accommodate such variations, although the more complex the combination of characters the more complicated the label becomes.

Whether or not one chooses to use such a labelling system, as noted previously it is important not to link a palynofacies type too closely with an environment of deposition. Leaving aside the classification of Whitaker *et al.* from here on, we now consider some of the associations of palynomorphs, phytoclasts and AOM in deposits that reflect marine, brackish and non-marine/freshwater environments of deposition.

It makes sense to begin with a few of the most morphologically and numerically limited types of palynofacies, namely those consisting of only small assemblages of black or nearly black blade- and/or needle-shaped fragments of charcoal and no or very few palynomorphs or any other organic particles in association. They may be typical of either coarse or fine-grained lithologies; both reflect oxidizing conditions, the former in high-energy settings from which most of the organic content has been winnowed out leaving only a few trapped oxidized fragments behind, and the latter in low-energy settings. In freshwater and brackish environments, oxidizing and/or high-energy conditions may be associated with periodically exposed areas of floodplains and river beds, and lake or lagoon margins (Fig. 10.1). In the marine realm, equidimensional blade-shaped particles may occur in beach and near-shore sands and sandstones, and also far offshore in muds and mudstones that reflect very low rates of deposition.

Since there are commonly no, or only a very few, palynomorphs in these impoverished palynofacies, the environment represented by a particular bed within a sedimentary succession cannot be determined as marine, brackish or freshwater with any certainty unless there other indicators (e.g. invertebrate fossils) present in the deposit. If these are also missing, the relationship of the bed concerned with adjacent, palynomorph- or other fossil-bearing facies may help in this respect, though less positively, especially in marine basin-margin successions in which depositional environments may cover an entire spectrum from marine through variable salinity to freshwater and terrestrial (soils).

A variation on the organically impoverished theme is a phytoclast assemblage of similar composition but which consists mostly of comparatively large fragments of black woody detritus. This tends to indicate closer proximity to the source of the detritus, hence often implying less comminution and more rapid entrapment of phytoclasts. It can be common in fluvial systems where, for example, it may be associated with point-bar successions above channel-lag gravels.

The next stage in the progression towards greater diversity is an assemblage in which palynomorphs are present in larger numbers. It now becomes more difficult to generalize because of the increased number of variables involved. One feature common to most palynofacies containing large palynomorph assemblages is that varying amounts of tissues and less altered (non-charcoalified) woody detritus (vitrinite rather than inertinite according to the coal maceral terminology) are also usually present. Another is that they reflect generally lower energy environments than many impoverished woody assemblages.

Depending on the age of the deposit sampled, in the marine realm the palynofacies may contain palynomorphs derived from both land plants (spores and pollen grains) and aquatic organisms (dinoflagellate cysts and/or acritarchs). The logical expectation that the further offshore the deposit the fewer and smaller the land plant products (spores, pollen grains and phytoclasts) is commonly substantiated, although such a trend can be much modified by the freshwater input of major rivers and dispersal by ocean currents. Open marine associations are typically dominated by palynomorphs of marine origin together with subordinate organic matter of largely aquatic (algal) origin, amounts per gram of sediment varying from large in areas of oceanic upwelling to very small in regions where most of the organic matter that is potentially available for deposition is eaten or otherwise removed from the water column.

Nearer shore the composition of the marine palynomorph assemblages recovered from deposits reflecting similar low-energy conditions commonly changes, with fewer taxa and different proportions. Cysts of marine phytoplankton and remains of other marine organisms become even less numerous in brackish-water deposits although the number of individuals of the species concerned may be very great. Marine organisms will, of course, normally be absent from sediments that have accumulated inland, although freak storm conditions leading to temporary marine incursions may leave their mark in a lowland sedimentary succession that reflects mostly freshwater deposition.

Low-lying marginal-marine environments are very susceptible to minor changes in climate, sea level, and other local and regional conditions, all of which can have a major effect on the composition of the sediments that accumulate in the area. The very large number of sedimentary facies variations that are encountered in marginal-marine to freshwater and terrestrial successions is reflected in a great diversity of associated palynofacies as a result of numerous local influences. A few examples of these are: minor changes in the salinity of coastal lagoons and estuarine marshes; differences in the composition of the local vegetation; the susceptibility of components of the vegetation to desiccation and wildfire during dry periods; and the tendency for some water bodies to dry up during droughts and for others to become stagnant. On a lamination-by-lamination scale, the differences in palynofacies associations recovered may also reflect seasonal changes.

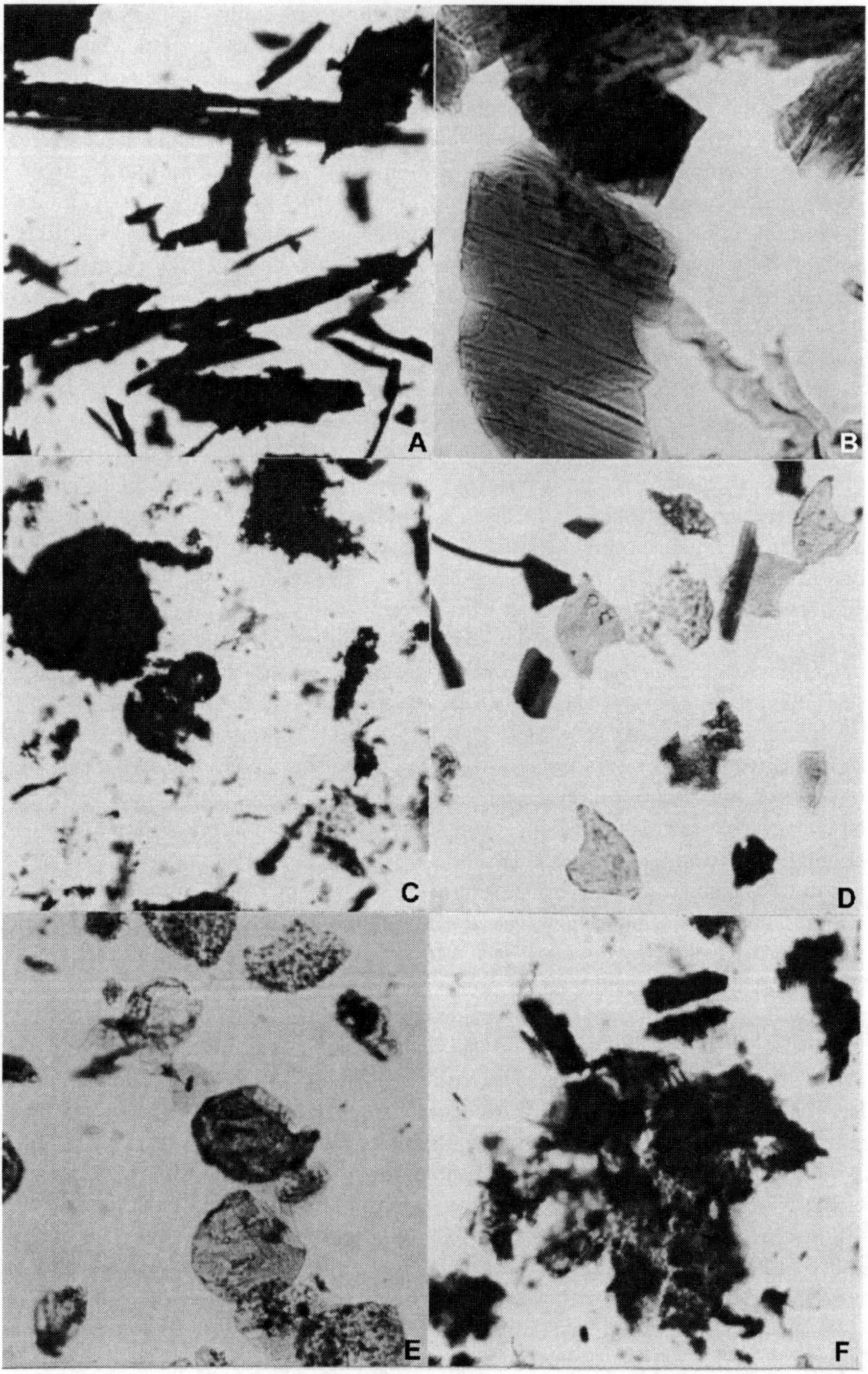
A
B
C
D
E
F

Stagnation is the final stage in the reduction of the energy of environments of deposition. There is so little disturbance that oxygen levels are lowered and the organic matter that accumulates is only partially consumed by bacteria or other organisms. Such low-energy settings in which bottom-water conditions and some (or much) of the overlying water column are dysoxic or anoxic are today characteristic of deep-water deposition in partially enclosed basins (e.g. the Black Sea) and regions of oceanic upwelling, and any area of stagnant water, such as a small pond within a swamp. In the geological record there is plenty of evidence of long periods of widespread stagnation in the marine realm, resulting from sluggish water circulation.

All of these observations are intended to give a general impression of the considerable variety of palynofacies that may be encountered and, hence, their potential in analyses of environments of deposition, which in turn can be used to varying degrees in a stratigraphic context. They are supplemented by illustrations of fields of view of palynofacies, and a few of their components, representing and reflecting different ages and depositional conditions respectively (Figs. 10.2, 3, 4; some of the photographs in these figures are cited individually in section 10.7). On a more detailed level, what an analyst records from a suite of samples from a particular sedimentary succession will depend to some extent on the general or dominant environment of deposition represented. For example, in Jurassic successions both on- and offshore in north-west Europe, parts are dominated by non-marine deposits, other parts by marine formations. The non-marine deposits pose a particular problem for biostratigraphers because very few of the palynomorphs encountered are stratigraphically useful on a fine scale. This is where a detailed assessment of variations in palynofacies can be useful (see below).

## 10.7 Stratigraphic Application

The application of palynofacies analysis to stratigraphy involves the recognition of the 'same' environment within an age-controlled sedimentary succession. This is particularly useful, and may be essential, for correlating approximately coeval sections within rocks or sediments that do not contain biostratigraphic markers or do not represent a sufficient amount of time for any significant changes in the composition of

*Figure 10.2 Examples of palynofacies and their components from successions of different ages reflecting a variety of depositional environments. A, palynofacies dominated by splintered, lath-shaped black wood; late Cretaceous, Bass Strait, Tasmania. B, brown wood; early Cretaceous, southern England. C, brownish black to black palynomorphs (acritarchs), phytoclasts and AOM form a deposit that is over-mature from the petroleum potential viewpoint; early Paleozoic, southern Sweden. D, dinoflagellate-rich* (Nannoceratopsis) *palynofacies, indicating marine deposition but perhaps in conditions of lower than normal marine salinity; mid Jurassic (Bajocian), North Sea. E, dinoflagellate-rich marine palynofacies; early Cretaceous, Australia. F, mixed black and brown wood, and AOM; late Jurassic (Oxfordian), England. All ×120 – see colour version of this figure in Appendix.*

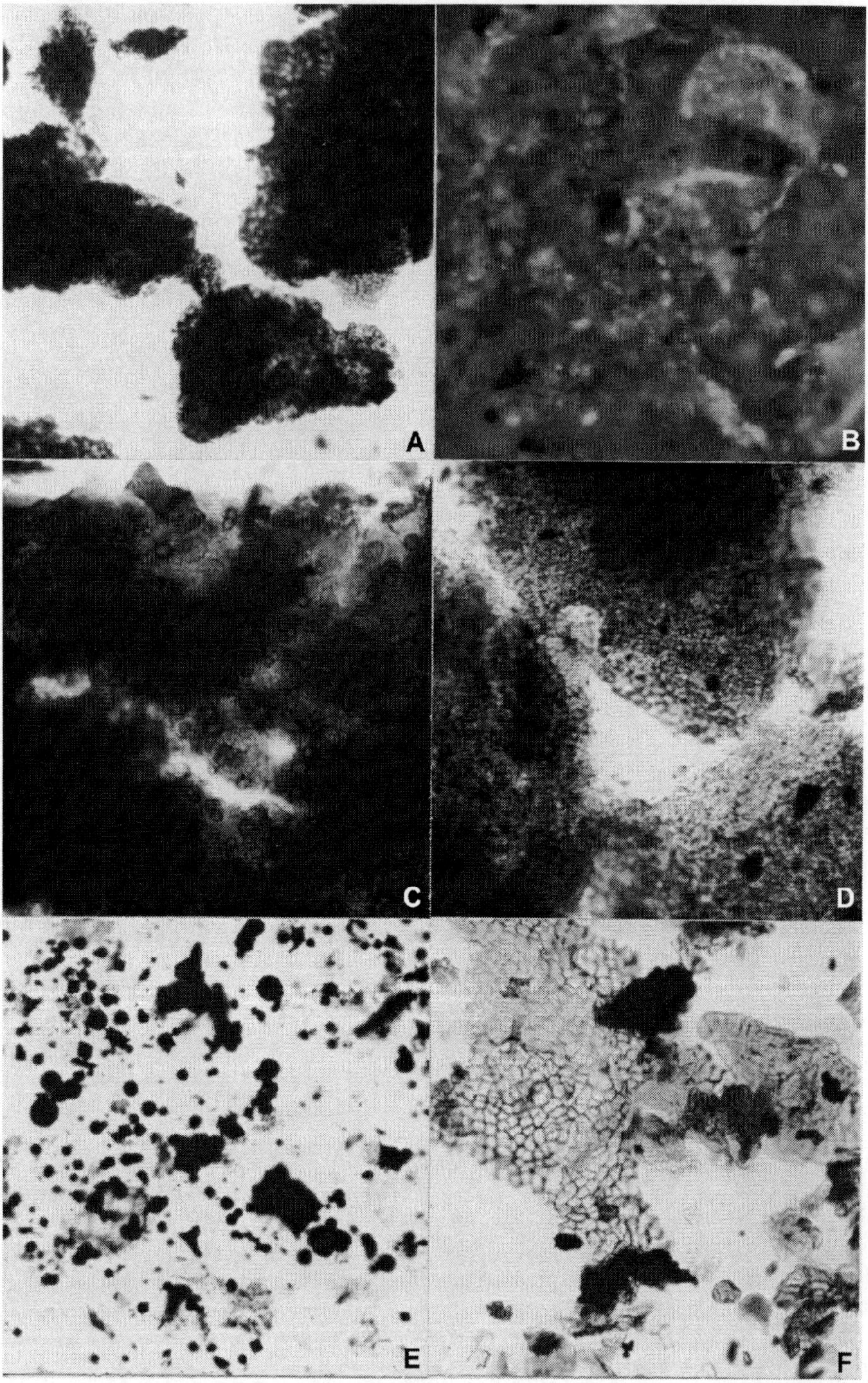
A
B
C
D
E
F

assemblages of palynomorphs, ostracods, foraminifers, ammonites or other fossils to be recognised. In a succession that is, for example, entirely early Bajocian in age, standard (bio)stratigraphic methods may not lead to any basis for subdivision, therefore hindering precise correlation with sections of similar age elsewhere. A high-resolution palynofacies analysis may enable subdivision of the succession into several units.

This type of analysis is particularly useful in horizontal drilling projects, which are driven by the need for reservoir-scale stratigraphic control. Production biostratigraphy aims to subdivide a reservoir into correlatable units. The target (e.g. reservoir) unit to be drilled may be very thin and beyond the dating resolution achievable through the standard palynostratigraphic or other biostratigraphic techniques that are traditionally used on a regional exploration scale. One or more distinctive palynofacies may, however, be associated with it, enabling continuous lateral recognition. For example, if deposits that accumulated in a lower shoreface environment are being targeted then the palynofacies for this environment need to be established. Above may be upper shoreface accumulations, and below beds that reflect deposition in near-shore marine conditions. By determining the palynofacies characteristics for each unit, the direction of drilling can be controlled. In this case, if the palynological matter changes to a more marine aspect it may be inferred that the drill bit has penetrated too deeply and that the direction of drilling needs to be altered. If a more proximal facies is recorded then it is possible that the drill bit has shifted upwards into strata overlying the critical unit. This approach to drilling along specific horizons is known as biosteering, and is now of major importance in analysing reservoirs prior to and during their commercial development (Fig. 10.5).

Some formations can generally be recognised from their associated palynofacies without the need to find particular biostratigraphic markers. For example, although the total organic recovery from the Kimmeridge Clay Formation in the North Sea Basin is often very large, only a small proportion usually consists of readily identifiable palynomorphs, dark AOM being the dominant component (Fig. 10.3A, B). Such palynofacies are also typical of recovery from onshore sections of this formation in north-east Scotland, and in eastern and southern England, thus indicating extensive lateral continuity of one type of palynofacies.

*Figure 10.3 Examples of palynofacies and their components from successions of different ages reflecting a variety of depositional environments. A, AOMA-rich assemblage from an oil shale; late Jurassic (Kimmeridgian), southern England. B, AOMA and a bisaccate pollen grain buried within it, fluorescing under UV illumination; late Jurassic (Oxfordian), England, C, degraded brown 'wood' full of resinous globules; Carboniferous, England. D, AOMA preserved in faecal pellets; late Jurassic (Oxfordian), England. E, palynofacies with AOM and abundant pyrite, much of which is preserved in framboids; early Cretaceous, Celtic Sea, offshore UK. F, palynofacies dominated by brown wood, cuticles and other membranous tissues; mid Jurassic, Denmark. All × 120 except B, which is × 300 – see colour version of this figure in Appendix.*

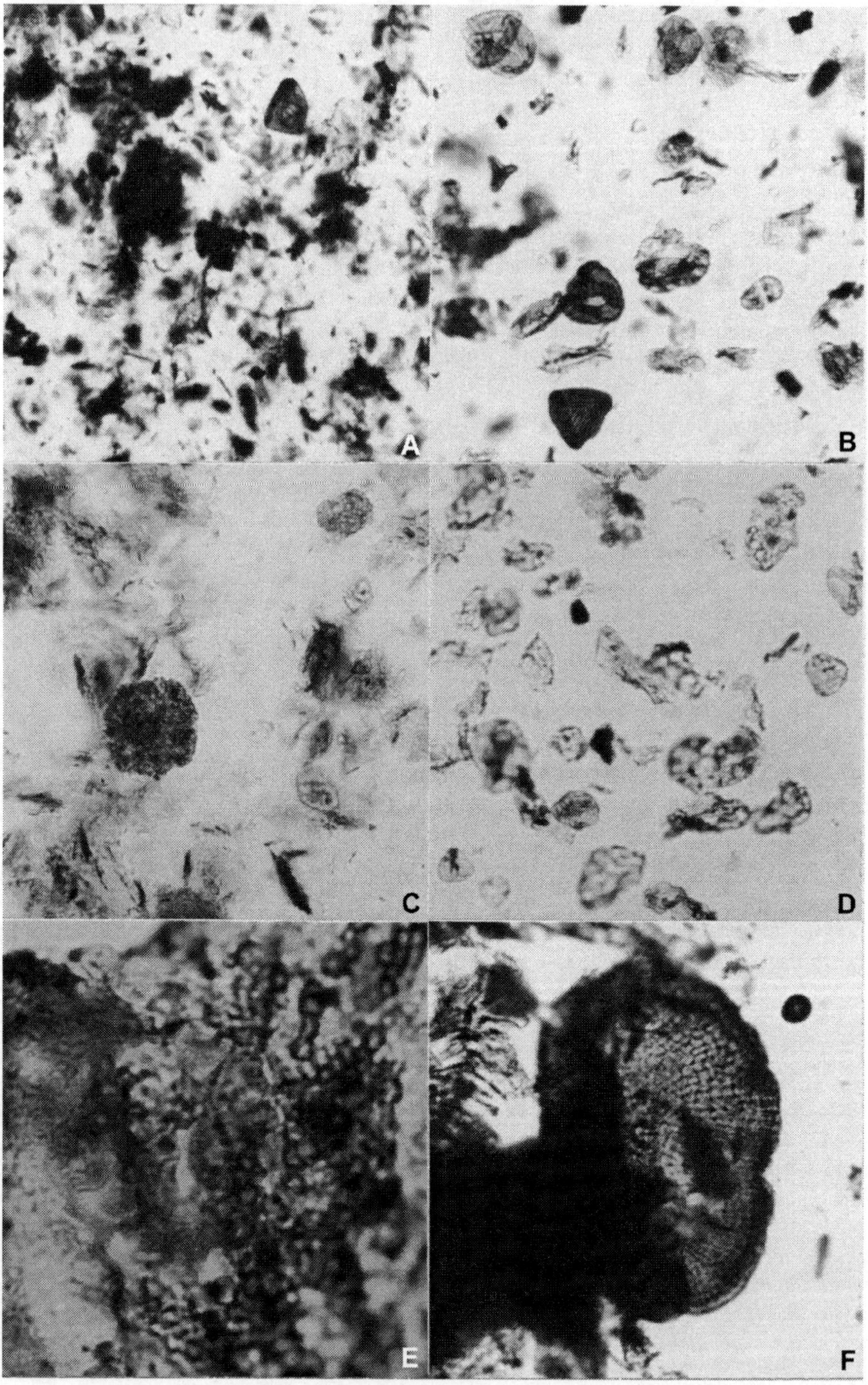
A
B
C
D
E
F

Another example from the North Sea concerns the Middle Jurassic Brent Group, the palynofacies characteristics of which have been documented by Whitaker *et al.* (1992, p. 179, fig. 6), Williams (1992) and Sawyer and Keegan (1996). Occurrences of individual palynomorph taxa are of little biostratigraphic value in the succession; as a result the palynostratigraphic assessment is based on abundance 'events' and palynofacies interpretation. Depending on which field within the Brent Formation is the subject of analysis, a large number of consistent and distinct palynofacies types can be recognised. The paleoenvironments represented are well understood. Typically these are offshore marine at the top (Heather Formation), passing downwards through marine/brackish (Tarbert Formation), lacustrine, mature marsh, lagoonal and swamp-lagoonal (Ness Formation, summarised as delta/prograding delta and lagoon/swamp), upper shoreface (Etive Formation), lower to upper shoreface – brackish/marine (Rannoch Formation), marginal to offshore marine (Broom Formation), and offshore marine (Drake Formation). These formations are generally recognizable on the basis of the palynofacies characteristics of the samples recovered from them (cf. Fig. 10.6).

As with all palynofacies work, the key to the stratigraphic application of palynological matter is careful observation and description of every aspect of the slide preparations. In addition to quantitative and/or semi-quantitative logging of the organic components, other points also need to be noted. These include the shape, size, color and state of preservation of phytoclasts, and the quantity of organic matter recovered. For example, in the Brent Group succession, the occurrence of blade/lath-shaped black wood (Fig. 10.2A) has been used to correlate events over a wide area.

In a number of well sections and surface exposures in basins that we have investigated, a general darkening of organic matter within certain stratigraphic units is a consistent feature. The significance of other assemblage characters that are commonly ignored in routine biostratigraphic analyses can also be taken into account, such as abundances of very thin, crumpled palynomorphs (Fig. 10.4D) and the occurrence of reworked material. Everything that can be seen in a preparation is potentially useful and should be documented. After a number of samples from different locations within a particular succession or basin have been analysed, characters that may have some stratigraphic application start to become apparent.

---

*Figure 10.4 Examples of palynofacies and their components from successions of different ages reflecting a variety of depositional environments. A, palynofacies associated with a fluvial channel-fill mudstone containing abundant AOMT and common miospores; early Cretaceous, southern England. B, as A but after brief oxidation and ultrasonic treatment, which has removed much of the amorphous matter. C, typical association of* Botryococcus *with non-marine AOMA; earliest Cretaceous, English Channel, offshore UK. D, oxidized, poorly preserved assemblage of miospores with some dinoflagellate cysts in a low salinity palynofacies; early Cretaceous, southern England. E, tissue showing evidence of fungal attack; late Cretaceous, Peru. F, epiphyllous fungus in a non-marine palynofacies; Cretaceous, Egypt. All ×120 except E, which is ×300 – see colour version of this figure in Appendix*

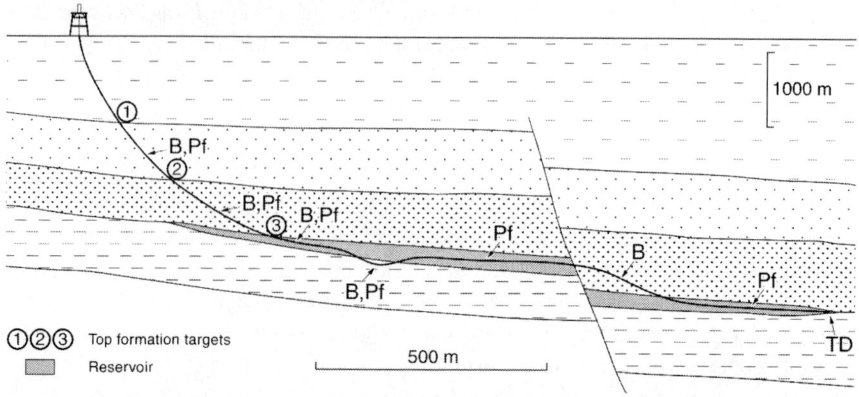


*Figure 10.5* *Structural cross-section along a well path to show the importance of biostratigraphy and recognition of palynofacies types during biosteering. Conventional biostratigraphy determines the angle of build of the well path; biostratigraphy and palynofacies, or palynofacies analyses on their own, keep drilling on target. B, biostratigraphy; Pf, palynofacies; TD, terminal depth.*

It is important to note that samples devoid of palynomorphs may still be rich in organic matter. These would be ignored in 'standard' biostratigraphic analyses, but may be usefully included in palynofacies investigations.

### 10.7.1 Limitations

The main limitations to applying palynofacies analyses to biostratigraphic problems are that palynological matter of very similar aspect may be encountered in deposits that are widely separated in time, whereas its composition may alter laterally within a single sedimentary unit because of changing environmental conditions.

By way of illustration of the first point: Jurassic deposits that reflect a lacustrine/lagoonal environment in which water circulation was rather restricted may well yield abundant AOM with *Botryococcus* spp. in association. An assemblage of identical aspect with the same environmental implications may also be recorded from a Cretaceous (Fig. 10.4C) or Cenozoic succession. Hence, a basic stratigraphic framework arrived at by 'traditional' means is still usually required if an analysis of palynofacies is to be applied successfully to stratigraphic problems, although the importance of this point may vary from basin to basin. Within a particular depositional area, for example, the occurrence of light-colored AOM with *Botryococcus* spp. in association may be immediately recognisable as Miocene because such a facies is not seen again at any other level. On the other hand, age control via standard biostratigraphic analysis is necessary in connection with the Brent Group in the North Sea, but once within this succession, palynofacies analysis becomes important. Within the lowest (Drake) formation of the group and below

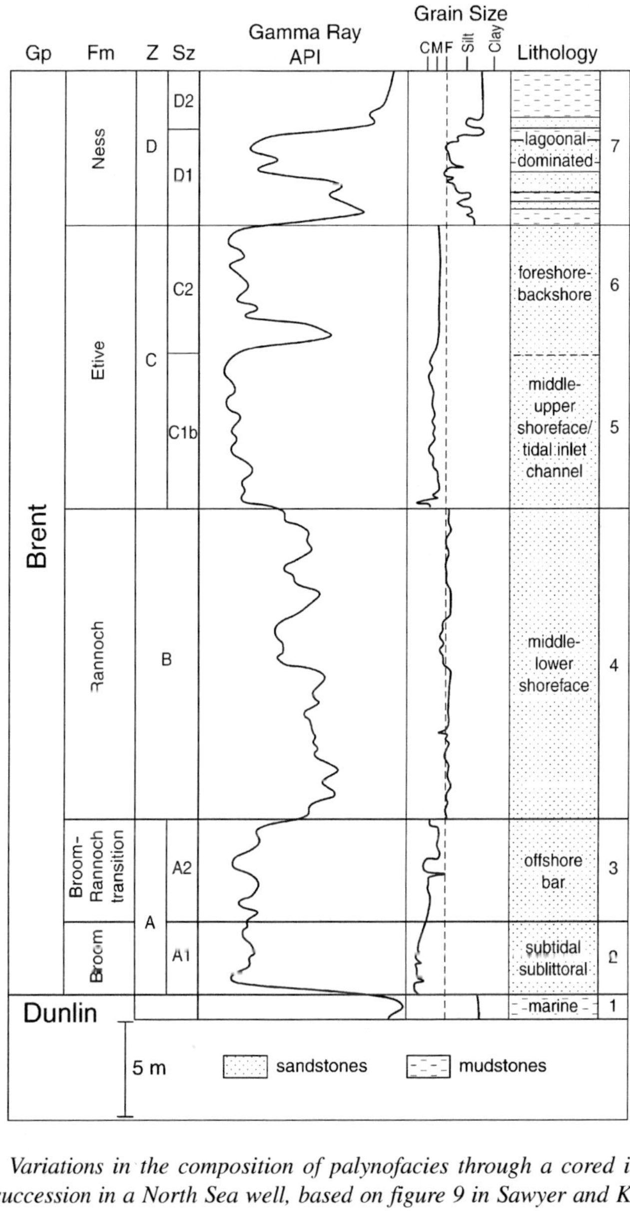


*Figure 10.6 Variations in the composition of palynofacies through a cored interval in the lower Brent succession in a North Sea well, based on figure 9 in Sawyer and Keegan (1996). Gp, Group; Fm, Formation; Z, Zone; Sz, Subzone; CMF, coarse, medium, fine. 1, marine palynofacies containing a diverse assemblage of dinoflagellate cysts; 2, major influx of bisaccate pollen, lath-shaped palynomaceral 4 common, influx of species of the dinoflagellate cyst* Nannoceratopsis; *3, increase in palynomaceral and palynofloral diversity; 4, large palynomaceral 4, minor influxes of* Nannoceratopsis; *5-6, limited recovery of organic matter, mainly small palynomaceral 4, often rounded; 7, common blade-shaped palynomacerals 2–4, rich miospore assemblage,* Nannoceratopsis *common.*

it, the deposits reflect more open marine conditions and traditional biostratigraphic methods are again important.

The laterally extensive marine Kimmeridge Clay Formation is potentially well suited to correlation based on palynofacies types. A modern example of a widespread depositional environment that will yield non-marine palynological matter of broadly similar composition is the extensive swamps of southern Sudan. Where deposits of a particular character are less extensive, the role of palynofacies as a stratigraphic tool is more limited, although even in successions in which facies changes are rapid both laterally and vertically, as, for example, in the non-marine Wealden succession of southern England, consistent associations of palynomorphs, phytoclasts and amorphous organic matter together with differences in preservation not only have paleoenvironmental significance but also can be used to identify different formations (Batten, 1973, 1975). In the marine realm, rapid facies changes are commonly associated with the building of deltas. Since blade-shaped black wood is very buoyant it can be widespread in sediments that accumulated in near-shore marine-deltaic environments, but it is inevitable that the palynofacies associated with samples taken from sediments of the same age from a delta outwards into a basin will reflect this change to more open marine conditions. In these circumstances it is necessary to look for overlapping features to effect correlations.

### 10.7.2 Case Study: Brent Group, Ninian Field, UK Sector, North Sea

Here we add a little more to the references that have already been made above to the Brent Group because its characteristics are such that they provide a good example of the sort of succession in which palynofacies analyses yield valuable stratigraphic and paleoenvironmental information where other methods fail or are less useful. The group is also of economic importance in that it forms a reservoir unit in the Ninian Oil Field.

The strata of which it is composed represent a progradational shoreface–deltaic wedge. The biostratigraphy is well established, a number of zones giving good age control having been recognized. Within the succession, the Broom, Rannoch, and Etive formations (noted above) are virtually barren of palynomorphs. These are all within the lower part of the group (Zone A) that Sawyer and Keegan (1996) targeted in their paper on the use of palynofacies in characterizing sand-dominated sequences. Their method involved analysis and photography of palynodebris in addition to conventional biostratigraphy of both mudstone- and sandstone-dominated units.

Individual sandstones typically reveal a recognizable palynofacies, the characters of which were a result, not surprisingly, of a combination of factors: the type of material available for inclusion in the sediment, the mode of transport, and subsequent environment of deposition. Sawyer and Keegan analysed the succession using a modified version of the approach of Whitaker (1984) combined with representative photographs. Use of the latter enabled the identification of subtle, minor changes in palynological content that often characterize sand bodies.

The approach adopted was used subsequently at the well site as a geosteering tool in order to keep the well on target. It has also been employed to improve the stratigraphic understanding of wells within an oil-field succession. Well-site biostratigraphy is crucial in projects of this sort. Palynofacies and palynological 'events' are identified and tied in with the gross zonal breakdown and sedimentary description of the succession.

### 10.7.3 Palynofacies and Sequence Stratigraphy

Much of the above discussion also applies to the use of palynomorph occurrences and palynofacies analyses in sequence-stratigraphic interpretations (e.g. Prauss, 1993; Steffen and Gorin, 1993; Helenes and Somoza, 1999). The characteristics of the PM recovered combined with other data on a sedimentary facies have been shown to be of considerable value in this respect. Lowstand systems tracts (LSTs) are typically indicated by palynofacies that contain abundant, often comparatively large phytoclasts and numerous spores and pollen grains that are not very well preserved, suggesting that they have been partly oxidised. Black, charcoalified, woody detritus is commonly an important component of such palynofacies. By contrast, transgressive systems tracts (TSTs) are frequently associated with palynofacies in which dinoflagellate cysts are more important components whereas the assemblage of miospores and phytoclasts is smaller and less varied, and highstand systems tracts (HSTs) may contain a significant component of AOM, with dinoflagellate cysts being the most numerous of the palynomorphs recovered.

Among recently published sequence-stratigraphic studies is one on a Cretaceous succession in Ecuador by Vallejo *et al.* (2003) that is based on a lithofacies, biofacies and biostratigraphic analysis. These authors recognised a series of LSTs, TSTs, HSTs and sequence boundaries using the ratio of marine to terrestrially derived palynomorphs, the main components of the palynofacies, a number of calcareous nannofossil and palynomorph 'events', occurrences of oysters and planktonic foraminifera, and other fossil and sedimentological evidence, as shown in Figure 10.7. The paper demonstrates the value of taking a multidisciplinary approach to sequence stratigraphic studies and the contribution that palynofacies analyses can make to them.

## 10.8 Conclusion

Careful observation and recognition of repeated occurrences of types of palynological matter is the key to successful palynofacies studies, no matter what the goal of a particular project. What might appear at first to be insignificant detail may prove in the longer run to be vital; everything should be noted: palynomorph taxa, the size, shape and type of phytoclasts, the presence of amorphous detritus, even persistent mineral content (e.g. Fig. 10.3E) and evidence of fungal attack (e.g. Fig. 10.4E, F). Variations in the type and abundance of organic matter in fine-grained deposits both laterally and vertically are a result of sedimentological sorting and preservational differences, and therefore important for determining (paleo)environments and source rock potential

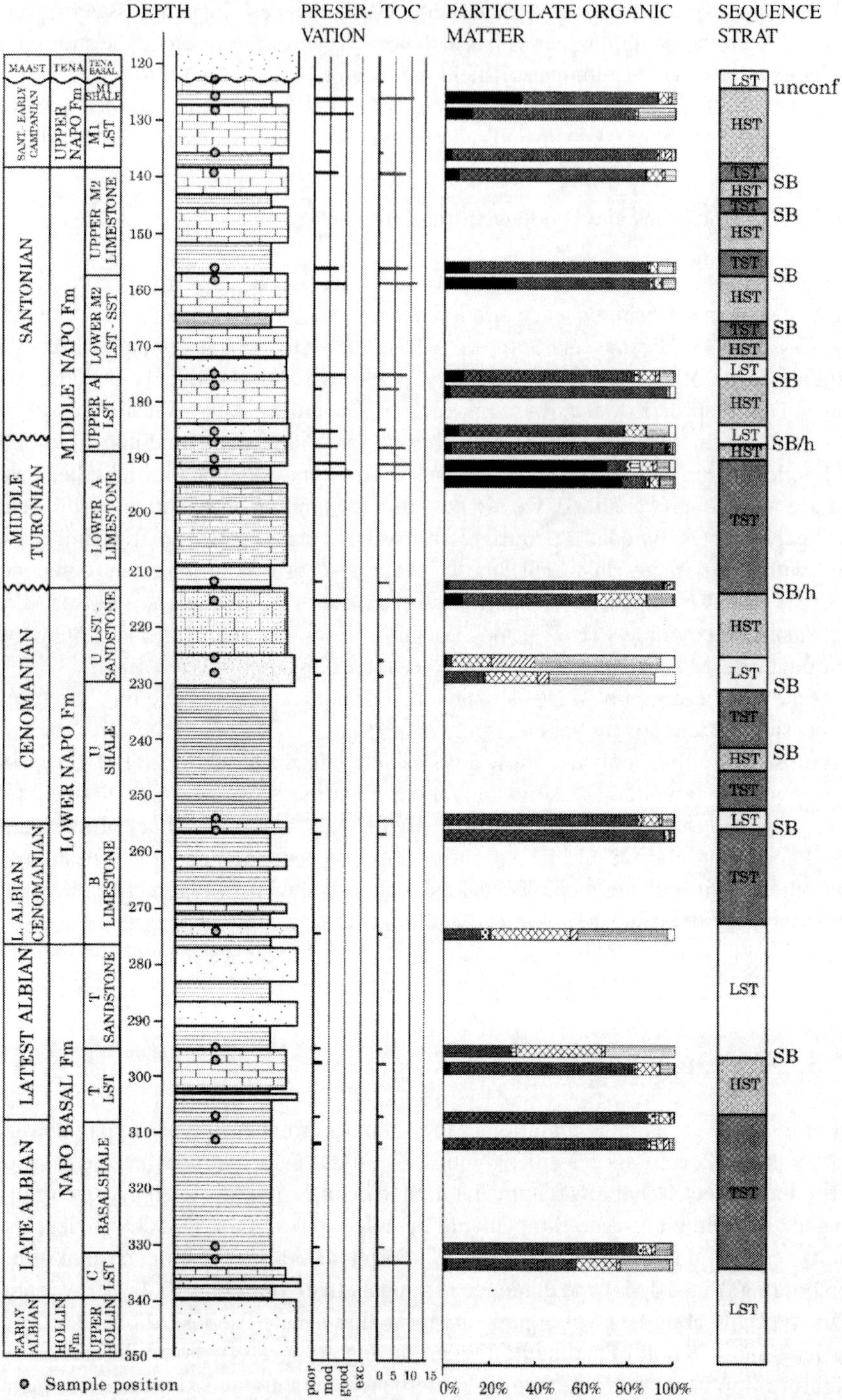


*Figure 10.7 Part of a stratigraphic succession encountered by a well drilled in Ecuador showing lithologies, biostratigraphically important 'events' in the calcareous nannofossil and palynomorph records, the total organic carbon (TOC) content, the preservation and occurrence of*

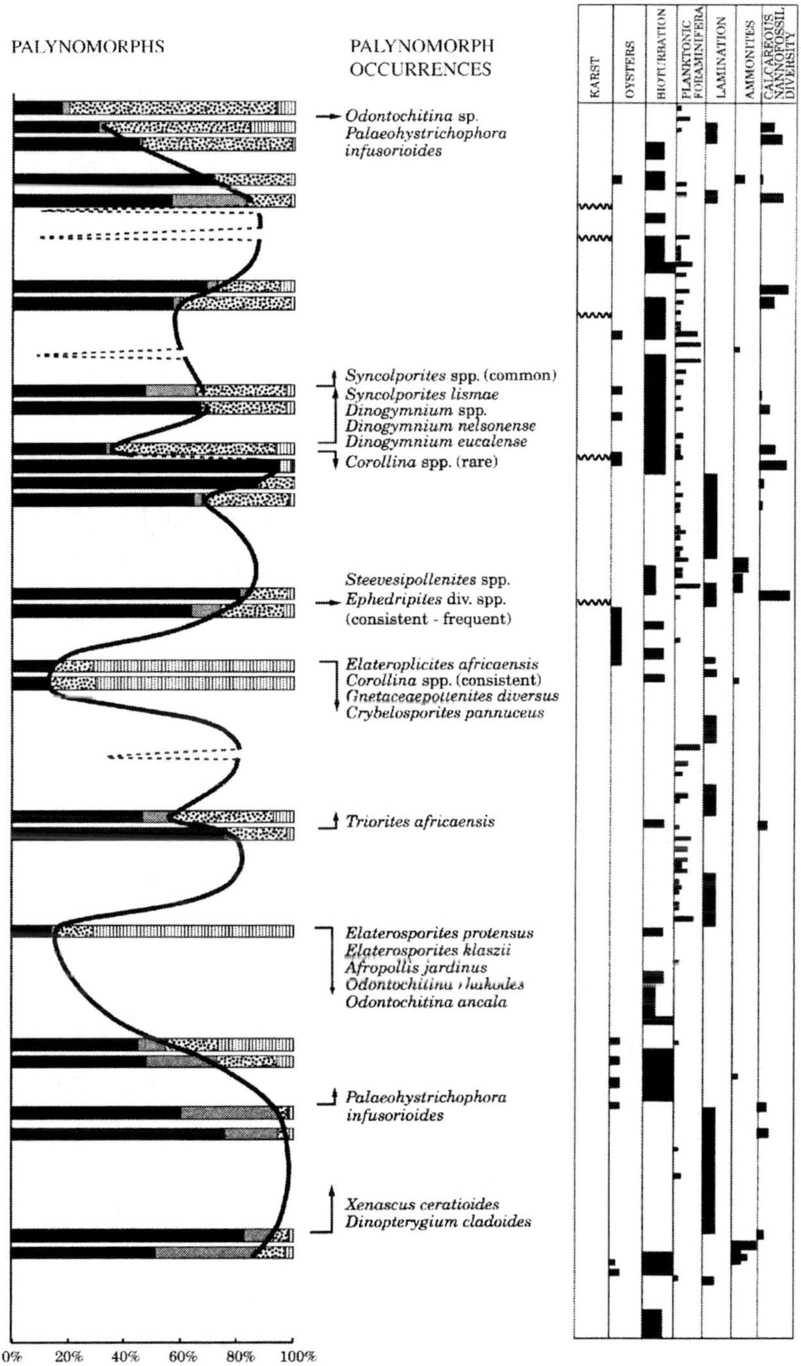


*Figure 10.7 Continued*

*palynological matter (particulate organic matter), sedimentary structures and other sedimentological and paleontological data, and their sequence stratigraphic interpretation (based on figs. 2 and 3 in Vallejo* et al. *2003, modified).*

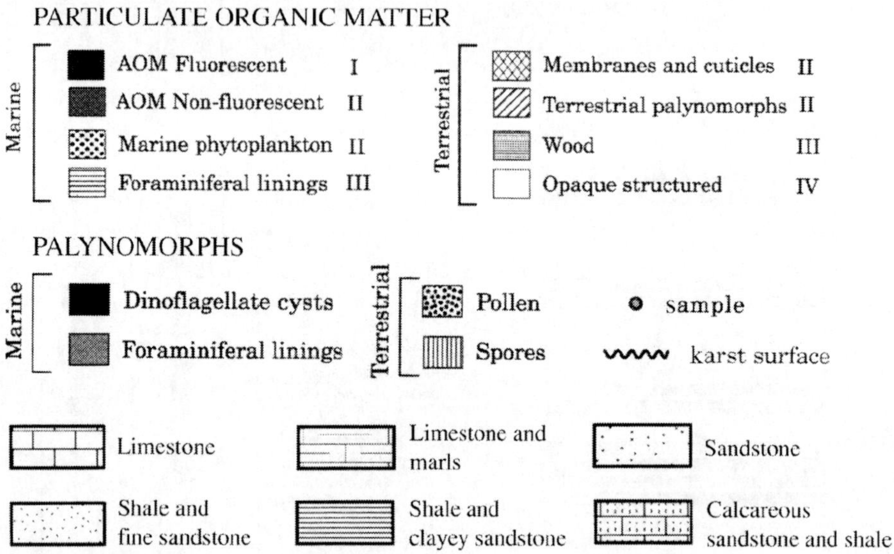


*Figure 10.7 Continued*

for petroleum. Identification of palynofacies in conjunction with conventional biostratigraphy can lead to an improved and more complete understanding of the stratigraphy of sand-dominated sequences. It can be employed in sequence-stratigraphic studies to recognize boundaries in successions within which there is apparently no sedimentological evidence for them. It can also be used to guide the well trajectory during drilling by discriminating between non-pay zones above, below and within reservoirs.

## 10.9 Acknowledgements

We thank Antony Smith (University of Wales, Aberystwyth) for drafting Figs. 10.1, 5 and 6, and for his work on Fig. 10.7.

Chapter 11

# Sequence Biostratigraphy with Examples from the Plio-Pleistocene and Quaternary

HILARY CLEMENT OLSON[1] and PETER R. THOMPSON[2]

[1] Institute for Geophysics, The University of Texas at Austin, 4412 Spicewood Springs, Bldg. 600, Austin, Texas 78759, USA.

[2] Computational Biochronology, P.O. Box 261039, Plano, Texas 75026, USA.

---





*E.A.M. Koutsoukos (ed.), Applied Stratigraphy,* 227-247.

## 11.1 Introduction

Sequence biostratigraphy is the use of (micro)fossil samples to describe the age and paleoecological nature of key bounding surfaces and depositional systems tracts and cycles within a sequence stratigraphic framework. Where possible, it can be used to delineate the temporal and spatial extent of these surfaces, systems tracts and cycles in concert with sedimentological, petrophysical, and geophysical data. Although sequence stratigraphy was originally perceived as a rival to biostratigraphy, both disciplines are now well integrated and iterative in nature (Emery and Myers, 1996; Simmons, 1998). Sequence biostratigraphy requires a deliberate effort between biostratigraphers, sedimentologists, petrophysicists, and geophysicists to develop the most parsimonious explanation of geological events, and yet one that honors all observational data.

Microfossil samples contain a wealth of biological and mineralogical information concerning biostratigraphy, paleoecology, and taphonomy. The biostratigrapher has the responsibility of using this myriad of data to sort out cogent patterns contained within the samples, and to recognize their stratigraphic significance. Depending upon the suite of available data and the age of the stratigraphic section, the tasks in sequence biostratigraphy consist of: (1) taxonomic identifications of microfossils; (2) quantitative estimation of species abundances standardized to known sample volume; (3) mineral-content evaluation; (4) age determinations/zonal assessments; (5) paleoenvironmental analysis; (6) glacial/interglacial interpretations; and (7) graphical display of the patterns obtained from these assessments against log character, core analysis, lithology, and possibly seismic data (van Gorsel, 1988; Powell, 1992). The goals of this paper are to discuss the important tasks of a sequence biostratigraphic study, to stress the need for their integration in sequence stratigraphic interpretation, and, finally, to demonstrate the application of sequence biostratigraphy to two end-member (well-cuttings *vs*. cores) examples from the Plio-Pleistocene and Quaternary records of the Gulf of Mexico.

## 11.2 Sampling/Processing Techniques and Data Collection

### 11.2.1 Collection Methods, Sample Spacing, and Sampling Interval

The spectrum of collection methods for microfossil samples ranges from outcrop hand samples to subsurface well-cuttings. In the subsurface, drilling fluids may obscure and dilute true sample volume of well-cuttings, and drilling-fluid additives may introduce contaminating microfossils or adversely damage the fossils by chemical reaction.

Both the sample spacing (distance between two samples) and sampling interval (length of stratigraphic section covered by one sample) may vary greatly. In very detailed piston core or outcrop studies (e.g. section 11.6.2) sample spacing may be on the order of 10's of cm and sampling intervals on the order of 1–2 cms. This detailed stratigraphic control is in stark comparison with typical studies of subsurface wells (e.g. section 11.6.1). In the subsurface, sample spacing may be as much as 20 m and sampled intervals may be amalgamated because fluid-circulated well-cuttings essentially produce a composite sample. To minimize uncertainty regarding the exact location of stratigraphic horizons, biostratigraphic data should be interpreted against a more continuous signal such as core or outcrop description, or wireline log.

### 11.2.2 Sample Volume: Standardization of Samples

Both sample volume and standardization are critical when comparing stratigraphic sections from outcrops, cores, and/or wells. Too small a sample volume may hinder recognition of rare, diagnostic taxa; too large a sample volume may prohibit thorough sample examination. Micropaleontological, quantitative data presented without regard to sample weight or volume can be very misleading because sample sizes are not standardized (Thompson, 1981). Frequently, foraminiferal samples are normalized for comparison by weight (dry weights measured both before and after washing/sieving; Shepard and Moore, 1954). Results are then reported with reference to "per gram," etc. (e.g. section 11.6.2). In some microfossil disciplines, standardization may be achieved through alternative means. For example, palynomorph concentration can be calculated by spiking samples of measured volume with a known number of exotic marker flora, such as the spore *Lycopodium clavatum* (Stockmarr, 1971; McCarthy and Gostlin, 2000), before chemical treatment.

### 11.2.3 Coarse Fraction

Coarse fraction is that portion of the washed sample sieved at 63 or 74 microns (Shepard and Moore, 1954). Normally, it is reported as a weight percentage of the initial sample weight. The relative proportions of minerals and biogenic components within the coarse fraction are a direct function of the depositional environment. For example, hemipelagic sediments have virtually no mineral content but are rich in microfossils; clastic sediments have a high dilution of fossil assemblages by terrigenous grains. Although coarse fraction data are often used to calibrate microfossil data with lithologic interpretations from logs, these data can also be used as proxies for sea-level fluctuations during glacial and interglacial cycles (see sections 11.4.2.3 and 11.6.2).

*11.2.3.1 Analyzing the Coarse Fraction Content: Microfossil Content*

Microfossil abundances are obtained either by absolute counts or quantitative estimates. Microfossils studied may include a diverse association ranging from calcareous to siliceous to phosphatic to organic remains (e.g. Jones, 1996; Davydov *et al.*, 1997; Ross and Ross, 1997; Buck *et al.*, 1999; Graefe, 1999; McCarthy and Gostlin, 2000; papers in Olson and Leckie, 2003). Each microfossil type dictates the specific counting techniques required for collecting and the quantitative analytical methods available for analyzing the data. Many approaches are discussed in this book as well as elsewhere in the literature (e.g. Lipps, 1993; Neal *et al.*, 1995; Klapper, 1997; Lagoe *et al.*, 1997; Buzas and Hayek, 1998; Bown, 1999; Jones and Rowe, 1999; Olson *et al.*, 2003).

*11.2.3.2 Analyzing the Coarse Fraction Content: Mineral Grains*

Microfossil sample processing extracts and concentrates not only one or more fossil types, but also mineral grains, from their matrix. For the sake of efficiency, visual estimates of mineral abundances are normally recorded. The abundance of minerals vs. fossils gives an indication of dilution, which, in turn, is a function of proximity to a clastic source, sediment accumulation rate, and accommodation space. Estimation of the coarse fraction constituents (commonly quartz, glauconite, mica, pyrite, coal, and lignite) provides an excellent means of differentiating depositional environments. In subsurface studies, the relative abundances of microfossils-to-minerals should be integrated with lithologic and depositional interpretations based on coarsening/fining trends on wireline logs.

## 11.3 Correlation Patterns in Sequence Biostratigraphy

### 11.3.1 Fossil Ranges

The value of the fossil record is in its non-repetitive continuity: every index fossil has a determinant beginning in time and, for all but extant forms, a time of extinction. For organisms that become fossilized, we can study their distribution both in time and space. The temporal distribution of a species, or Biochronozone, is the interval between that particular species' evolutionary appearance (First Appearance Datum or FAD) and evolutionary extinction (Last Appearance Datum or LAD). Latitudinal and other geographic obstacles prevent most organisms from a pole-to-pole occurrence so that shorter, partial ranges exist towards the edges of their spatial distribution (Fig. 11.1). These partial ranges are bracketed by local first occurrences (FO; compatible with FAD) and last occurrences (LO; compatible with LAD), which create the Biozone and are often used as correlation markers in local studies.

In core or outcrop, it is a straightforward matter to sample a measured interval and delineate the depth of first and last occurrences. In the rotary well-cuttings acquired by industry (Fig. 11.2), however, biozonal events are observed in rock chips whose precise depth cannot always be determined. When dealing with outcrops or borehole

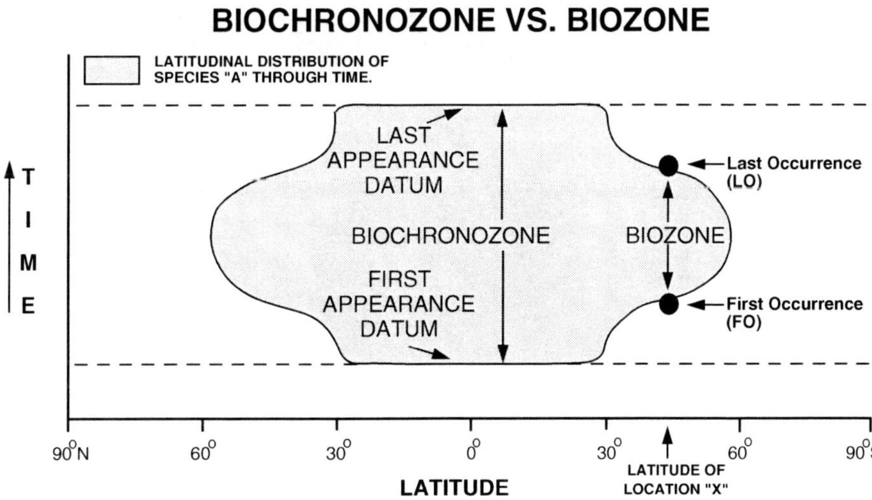


*Figure 11.1 The time range of the biozone (local range) for a particular species can vary significantly from its biochronozone (range of existence from inception to extinction). In this diagram, the biozone of species "A" at location "X" spans a significantly shorter period of time than the biochronozone.* Last Occurrence *(LO) is often used to signify the highest occurrence of a fossil (top of biozone) in a stratigraphic section, as distinct from the* Last Appearance Datum *(LAD), a term applying to the top of the biochronozone. Similarly,* First Occurrence *(FO) is often used to signify the lowest occurrence of a fossil (base of biozone) in a stratigraphic section, as distinct from the* First Appearance Datum *(FAD), a term applying to the base of the biochronozone.*

cores, both the FO and the LO are generally distinguishable. However, when examining well-cuttings, cavings from uphole can obscure the FO. Thus, the customary procedure is to identify the LO of a particular taxon (i.e. Fig. 11.2). If borehole mud pressures are well monitored, cavings are minimized and the FO may also be distinguishable.

### 11.3.2 Chronostratigraphic Correlation vs. Depressed Biostratigraphic Events

Sedimentary sequences on continental margins commonly bear a mixture of LADs, LOs, depressed LOs, reworked LOs, and elevated LOs, and it is the task of the biostratigrapher to derive a chronostratigraphy from these datums. Because the LO and FO of biostratigraphic markers do not always represent the same chronostratigraphic age, the chronostratigraphic framework is established through the transfer of *select* (i.e. high confidence) biostratigraphic events onto a lithologic section, wireline log profile, or seismic line. Although, many geoscientists are aware of the discrepancies between biostratigraphic markers, the daunting question remains: "How do I know which tops or events are most reliable and which are in fact suspect?"

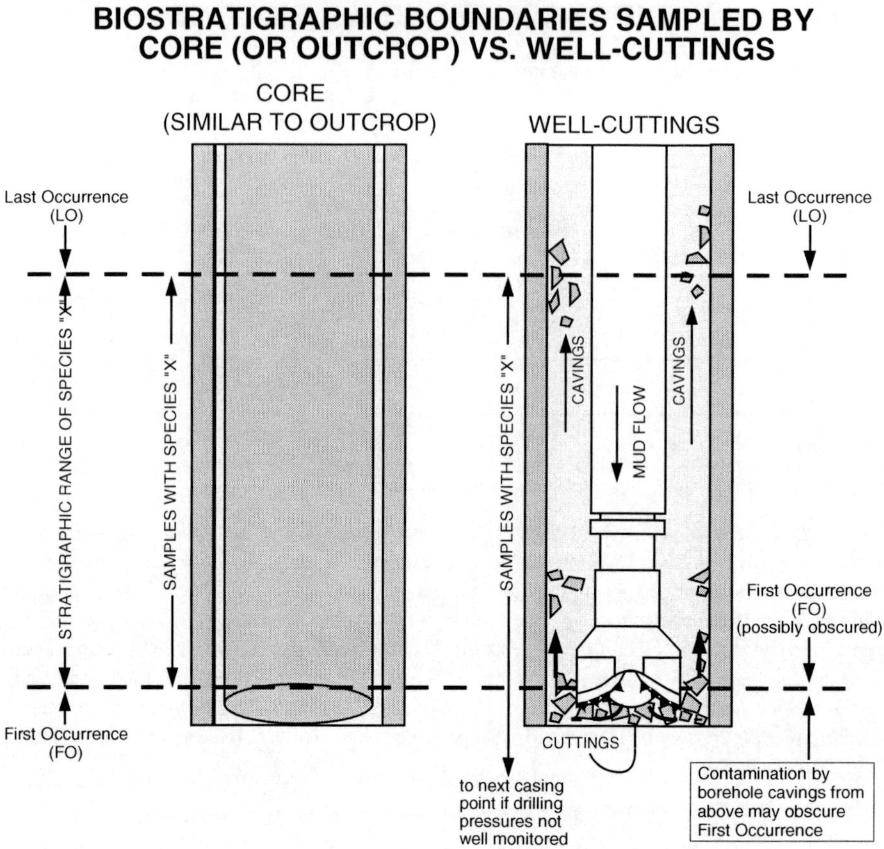


*Figure 11.2* *Fundamental differences exist between biostratigraphic data obtained from core or outcrop samples (left) vs. well-cuttings (right). Note that samples from core or outcrop data allow recognition of both the First Occurrence (FO) as well as the Last Occurrence (LO) of species "X." In boreholes with poorly monitored mud pressures, the LO may be the only discernable boundary for species "X" due to contamination from borehole cavings.*

### *11.3.2.1 What is a Depressed Event?*

Changing paleoenvironment through time impacts the occurrence of age-diagnostic fossils. Specifically, the term *depressed* event is often used to describe an *apparent* extinction event that is in fact only a local emmigration of a species due to a change in environmental parameters. Figure 11.3 shows the distribution of a marine taxon, Species A, with evolutionary range from just before Time 1 to slightly after Time 3. During this time interval, a sandy delta lobe migrates into the area. The species cannot tolerate the physiochemical properties of the clastic plume, such as low salinity, high turbidity, restricted food supply, etc., and is suppressed progressively from locations 2 through 6. As the lobe wanes higher in the section, the species immigrates in a landward direction

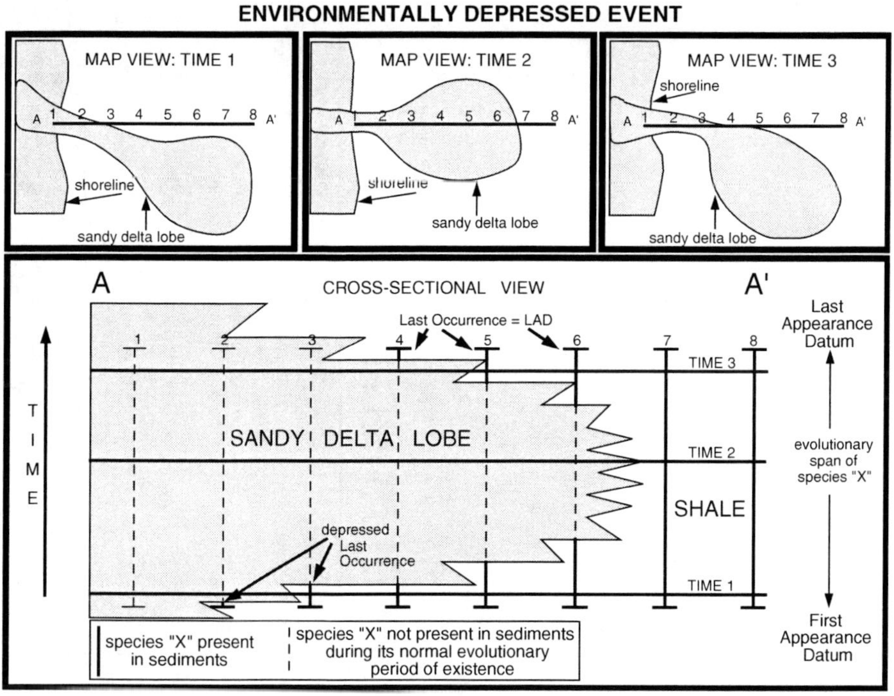


*Figure 11.3 A last occurrence is depressed updip because of the incursion of sediments from a sandy delta lobe. For this example, the local biozone is equivalent to the biochronozone. Map views at the top of this diagram display a sandy delta lobe that migrates into the line of the section (cores from wells 1–8) during Time 2, and is essentially out of the line of the section at Times 1 and 3. This migration of the sandy delta lobe causes depressed Last Occurrences at wells 2 and 3 and the total absence of species "X" at well 1. Notice that a depressed Last Occurrence is followed upsection by a Last Occurrence equivalent to the Last Appearance Datum in wells 4, 5 and 6. The singular Last Occurrence represents the Last Appearance Datum in wells 7 and 8.*

over time. A complete sampling of this depositional sequence underscores the discrepancy between the stratigraphic positions of the LO in locations 2–3 and the highest occurrence of the species in locations 4–8. A break in continuous occurrences is also apparent in locations 4–6. Depressed events 2–3 should not be correlated to the stratigraphically younger LOs in locations 4–8.

### *11.3.2.2 Distinguishing Between a Chronostratigraphic and a Depressed Event*

In the case where two correlation lines cross, both of the correlation lines cannot represent true chronostratigraphic horizons. Either one of the correlation lines represents a true time horizon and the other line does not, or both of the correlation lines are invalid with respect to chronostratigraphy. Figure 11.4 shows two wells along a dip

# RECONCILING CROSS-CORRELATING LINES

## Problem: Contradicting Biostratigraphic Correlation Patterns

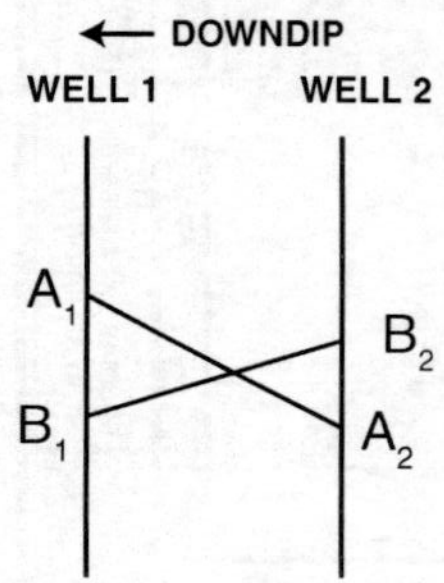


Two Last Occurrences (LOs), species A and B, are annotated in each of two wells in a cross section. Notice that when the datums are connected between the two wells, the correlation lines cross. Both of the correlation lines cannot represent true time lines. It is possible that both of the lines are invalid. For this discussion, we will assume that one of the correlation lines represents a valid time line, and the other line is invalid for a particular reason. In the examples which follow, we will assume that one of the correlation lines is valid and explain the resultant logic for why the other line could be invalid. We will then select the scenario which results in the most geologically and paleontologically plausible explanation.

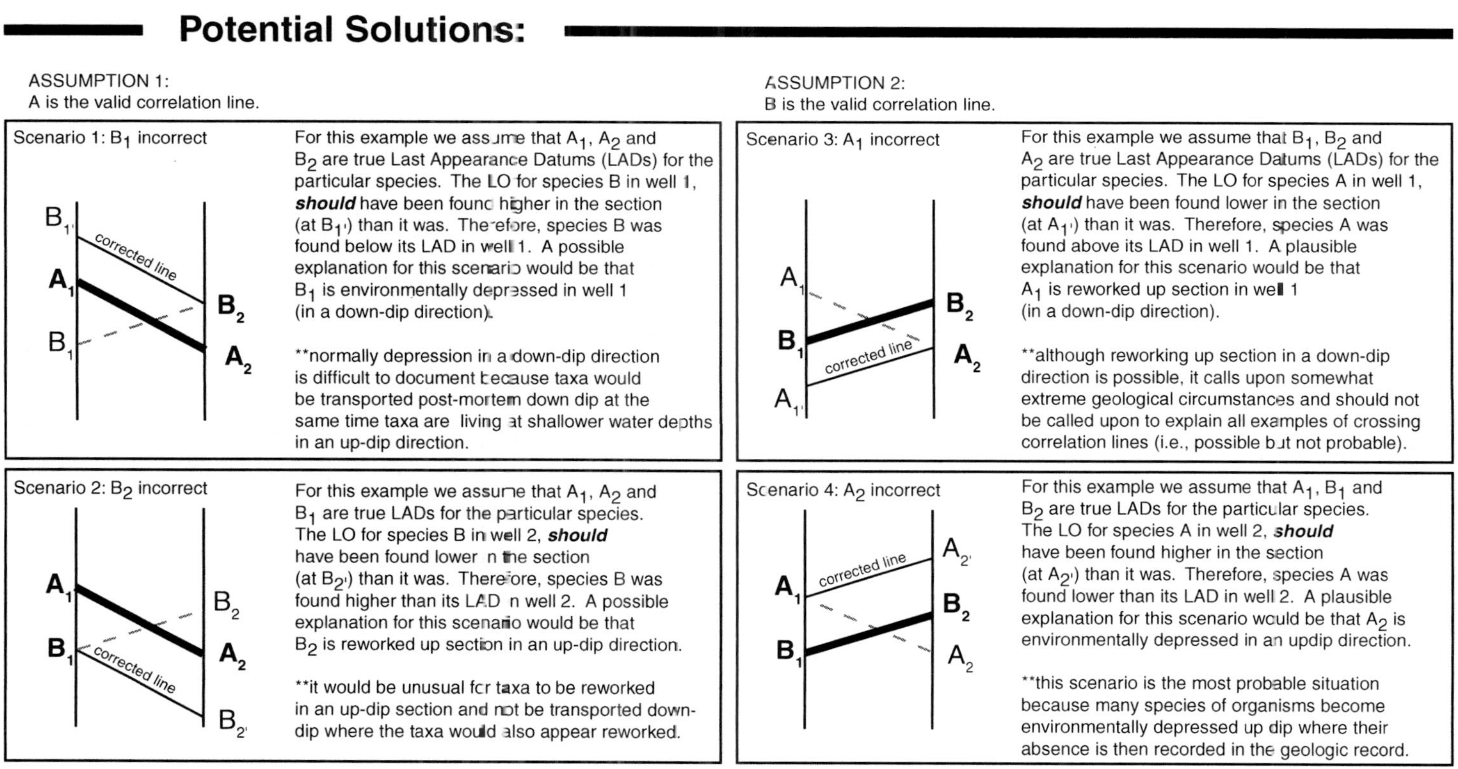


*Figure 11.4* *Reconciling cross-correlating biostratigraphic lines is a major task for most sequence biostratigraphic studies. A careful, logical approach enables the distinction between depressed biostratigraphic events and those events that are more likely to represent true chronostratigraphic markers. In this figure, biostratigraphic datums shown represent Last Occurrences in the stratigraphic section. For discussion purposes, Last Occurrences are analyzed assuming that they may represent Last Appearance Datums, until disregarded as chronostratigraphically insignificant.*

transect with two biostratigraphic datums recorded in each well. Correlation of datums A and B between the wells results in correlation lines that cross. Clearly the lines cannot both represent valid chronostratigraphic datums. When analyzing the various logical scenarios that could have resulted in these two crossing correlation lines, it is first important to note the direction of dip in the section. In the examples shown, well 1 is downdip from well 2.

*Assumption 1: A is the Valid Correlation Line.* Because we have now assumed that correlation line $A_1$–$A_2$ is chronostratigraphically significant, we correct our time line for datum B by reconstructing it parallel to datum A. We can achieve this in one of two ways: moving $B_1$ or moving $B_2$.
*Scenario 1: $B_1$ incorrect:* In this scenario, we correct datum B by moving $B_1$ to a position higher in the section in well 1, renamed $B_{1'}$ (Fig. 11.4). A possible explanation for the difference between the real position of the correlation lines and the "corrected" position would be that $B_1$ is environmentally depressed downdip in well 1. We consider this scenario improbable because, although depression downdip may occur, it is difficult *to document* in the stratigraphic record: taxa are easily transported post-mortem down dip.
*Scenario 2: $B_2$ incorrect:* Once again, assume that correlation line $A_1$–$A_2$ is chronostratigraphically significant. However, in order to correct our time line for datum B in this example, we move $B_2$ to a lower position in the section in well 2, renamed $B_{2'}$ (Fig. 11.4). This scenario is not very probable because it would call on a species to be reworked in an up-dip section (well 2), and somehow not be transported down-dip where it would also appear reworked at the same chronostratigraphic horizon in well 1.

*Assumption 2: B is the Valid Correlation Line.* For this scenario, we assume that correlation line $B_1$–$B_2$ is the valid chronostratigraphic horizon. We correct our time line for datum A by reconstructing it parallel to datum B. Similar to the two scenarios above, we can achieve this in one of two ways: moving $A_1$ or moving $A_2$.
*Scenario 3: $A_1$ incorrect:* In this scenario, we correct datum A by moving $A_1$ to a position lower in the section in well 1, renamed $A_{1'}$ (Fig. 11.4). A possible explanation for this scenario is that $A_1$ is reworked up section in well 1, i.e. in a down-dip direction. Reworking is a possible but not highly probable explanation because it calls upon somewhat extreme geological circumstances. Higher confidence would be given to a "reworking down dip" explanation if several species were found reworked.
*Scenario 4: $A_2$ incorrect:* Again, we assume that correlation line $B_1$–$B_2$ is the valid chronostratigraphic line. However, in this scenario we correct our correlation line for datum A by moving $A_2$ to a position higher in the section in well 2, represented by $A_{2'}$ (Fig. 11.4). A plausible explanation for this scenario would be that $A_2$ is environmentally depressed updip. This scenario is the most probable of the four presented because there are many organisms (both benthic and planktonic) that become environmentally depressed updip (i.e. towards a shallower environment) where their absence can then be recorded in the stratigraphic record.

## 11.4 Biofacies Patterns in Sequence Biostratigraphy

### 11.4.1 Biogeography: Living Microorganisms are Proxies for their Environment

Benthic organisms are sensitive proxies for the environmental properties of water masses and substrates in a particular location; and the complex environmental gradient associated with depth in the ocean leads to benthic communities that are depth-stratified. In contrast, planktonic organisms occupy marine surface waters (upper 100–200 m), and their environmental distribution is largely determined by temperature and salinity factors. They are largely absent or rare in coastal waters and abundant in open marine waters. Thus, the ratio of planktonic to benthic foraminifera (P/B ratio) increases across most continental shelves onto the slope (Grimsdale and van Morkhoven, 1955).

### 11.4.2 Paleoecology: Fossil Organisms are Proxies for Paleoenvironment

#### *11.4.2.1 Paleo-Water Depth*

Paleoenvironmental analysis, by appealing to a direct analogue with the distributions of living organisms, is the basis for interpretation of depositional environments. Paleoecology has a dual relationship with sequence stratigraphy: fossils provide tools for reconstructing patterns of environmental change that relate to sequence-forming processes (especially paleo-water depth estimations), and sequence stratigraphy provides a predictive framework that facilitates recognition and interpretation of patterns of biotic change in the geologic record (Brett, 1998).

Taphonomic processes can introduce fossils (i.e. by downslope transport) where they did not live, remove fossils that did exist at a particular site, or concentrate fossils at a site in "lags." Whereas, allocthanous and exotic species complicate paleoenvironmental interpretation (e.g. Denne and Sen Gupta, 1989), the displaced elements in the faunal assemblage provide information about the provenance of the sediments and, thus, information about depositional processes (e.g. Thompson, 1992; Lagoe *et al.*, 1992; case study in section 11.6.2).

#### *11.4.2.2 Oxygenation: Biofacies Record of Transgression and Regression*

An excellent example of how biofacies record the influence of transgressive and regressive episodes on oxygenation is the study by Nagy *et al.* (2001) on the Callovian Brora Argillaceous Formation of the Inner Moray Firth Basin. They interpreted oxygenation trends using five foraminiferal biofacies: (1) Transgressive-anoxic: diversity reduction, agglutinant-frequency increase, dominance increase; (2) Transgressive (MFS)-oxygen depletion: diversity reduction, high dominance, low frequency/absence of calcareous taxa; (3) Transgressive-improved environmental setting: diversity increase, dominance decrease, frequency increase in calcareous taxa; (4) Regressive-hypoxic improving to normal oxygenation: diversity increase, frequency/diversity increase in calcareous taxa, dominance decrease; and (5) Regressive-normal marine conditions: high diversity, high frequency of calcareous taxa, low dominance (Nagy *et al.*, 2001).

*11.4.2.3 Climatic Fluctuations: Insights from Sequence Biostratigraphy*
Climate change is a mechanism for effecting global environmental changes that are reflected in the biotic population. Concomitant changes in the fossil record illuminate the history of glacial/interglacial fluctuations and associated geologic processes.

*Response to Climate Fluctuations.* Cyclic patterns of cool- and warm-water taxa within a stratigraphic context shed light on climatic fluctuations, and their impact on such geologic factors as sea level, depositional processes, and sediment source/supply (e.g. section 11.6.2). Interpretations of palynological assemblages, characterizing dry and humid climatic conditions, such as those by Van der Zwaan and Spaak (1992) in the Triassic of Northwest Europe, link sedimentary cycles with sequence stratigraphy and Milankovitch orbitally forced cyclicity. Biostratigraphic patterns can be integrated with other climate signals, such as isotopic signatures (e.g. Baum *et al.*, 1994).

*Grain-Size and Carbonate Responses to Climate Fluctuations.* Numerous studies during the CLIMAP Project in the 1970's (e.g. CLIMAP, 1976; Cline and Hays, 1976) demonstrated that in the Atlantic, Caribbean and Gulf of Mexico, total carbonate ($CO_3$) curves in late Quaternary sediment cores could be used as proxies for oxygen-isotope curves (e.g. Damuth, 1975; Prell and Hayes, 1976). In deep water, increases in coarse fraction and $CO_3$ during interglacials reflect an increase in the concentration of foraminifera due to hemipelagic sedimentation processes. Decreases in coarse fraction and $CO_3$ during glacials are commonly the result of dilution of foraminiferal content by increased terrigenous supply to deep water. In section 11.6.2, we demonstrate the use of these patterns in determining glacial and interglacial deposits, and in determining the relationship between sea-level change and depositional processes.

## 11.5 Anticipations from Sequence Biostratigraphy

### 11.5.1 The Framework of Sequence Biostratigraphic Patterns

Historically, the most important contributions from micropaleontology were age control and paleo-water depth. In sequence biostratigraphy, we go beyond these two traditional roles and attempt to define the key bounding surfaces and depositional systems tracts within a sequence stratigraphic framework, and to describe the age and paleoecological nature of these surfaces and systems tracts.

In this section we discuss biostratigraphic patterns and their integration into a sequence stratigraphic framework. The resultant interpretations we discuss using the language of sequence stratigraphy, a discipline initially defined within the context of seismic scale observations. We employ the familiar seismic stratigraphic terminology of a clastic passive margin model (Vail *et al.*, 1977b; papers in Wilgus *et al.*, 1988), although it is equally applicable to the framework presented by Galloway (1989).

Since the revolution that has occurred in stratigraphy since AAPG Memoir 26 in 1977, stratigraphy has stressed the incompleteness of the rock record and sought to estimate this incompleteness (Holland, 1999). Biostratigraphy is key to this endeavor. Holland (1999) calculated that the completeness of the stratigraphic record could be as low as 10–40% at the 100-k.y. timescale and as low as 1–5% at the 10-k.y. timescale.

#### *11.5.1.1 Sequence Boundaries and Lowstand Systems Tracts in Passive Margin Basins*

The sequence boundary (SB) surface represents a major regression and is commonly accompanied by subareal exposure, erosion, and landward downcutting. As the hiatus associated with the SB is traced into the basin, however, less section is often absent until the surface becomes totally conformable in the marine basin. These surfaces may be identified and dated using marine microfossils (e.g. Poag and Commeau, 1995). However, where the duration of a hiatus is minimal, the missing section may fall within a single biozone and be difficult, if not impossible, to detect given the limitations of biostratigraphic resolution and the sample interval (Powell, 1992).

Biostratigraphic patterns associated with the SB include: (1) abrupt truncation or diminution of marine microfossil (e.g. foraminifera, nannofossils, dinoflagellates) abundance at the surface, (2) overlying increase in terrestrial pollen and spores and their diversity (e.g. Helenes and Somozoa, 1999; Rull, 2000), (3) overlying microfossils indicating cooler climate, (4) overlying decrease in P/B ratio, (5) overlying increase in reworked (older) microfossils derived either from the hinterland (rivers) or the slope (slumping), (6) overlying increase in coarse fraction, as well as its mineral component (more sand prone), and (7) well-expressed substrate-controlled ichnocoenoses resulting from firmground erosional surfaces (e.g. *Glossifungites* ichnofacies or *Thalassinoides*-dominated fabrics; Savrda, 1995; Savrda *et al.*, 2001a, b).

The lowstand systems tract (LST) is an accumulation of sediment that represents deposition along slope and base of slope subsequent to the marine regression represented by the SB. LST deposits range from turbidites to mass-transport deposits (e.g. slumps, slides, debris flows) and are often characterized by many of the same biostratigraphic patterns delineated in the above paragraph, specifically, a minimum in paleo-water depth and an increase in the sand content of the coarse fraction.

In cases where a SB is followed by a lack of increase in sand content, the pattern suggests lowstand sediments are not present. In such a case, the SB may be placed at or near the point at which paleo-water depth shifts to a minimum. In deep water, this horizon may be characterized as a faunal discontinuity where the transgressive systems tract (TST) lies directly on the highstand systems tract (HST) of the previous sequence.

#### *11.5.1.2 Transgressive Systems Tract*

During the rise in sea level following the regression, sedimentation eventually decreases along the slope and basin as marine waters flood across the shelf-break and lock the primary distribution of sediment onto the shelf. TST sediments are deposited during this phase of the marine transgression and are characterized by the following biostratigraphic patterns: (1) upward increase in marine microfossil abundance and paleo-water depth (e.g. Tye *et al.*, 1994), (2) FOs as species occupy new niches, (3) upward decrease in terrestrial spores and pollen and their diversity (e.g. Rull, 2000), (4) microfossils indicating a warmer climate, (5) upward increase in the P/B ratio, (6) upward decrease in reworked microfossils, (7) upward decrease in coarse fraction, as well as its mineral component (less sand prone), (8) possible upward increase in percentage of glauconite within the mineral component, (9) coquinas of skeletal debris often typical at the bases (Brett, 1995), (10) paleontological event beds during late transgressive stage as a result of episodic burial in sediment starved, and commonly low oxygen, settings favoring extraordinary preservation (Brett, 1995), and (11) *Thalassinoides*-dominated fabrics resulting from transgressive ravinement (Savrda, 1995). Limestones may be present resulting in washed residues with poor fossil assemblages due to cementation and nannofossil recrystallization, or rich assemblages due to the lack of detritus dilution.

#### *11.5.1.3 Maximum Flooding Surface*

The maximum flooding surface (MFS) above the TST represents the time of maximum marine incursion and is biostratigraphically characterized by: (1) profound pulse of deep water benthic microfossils and maximum paleo-water depth, (2) maximum incursion into shelf regions of planktonic foraminifera, nannofossils and dinoflagellates, (3) tops of various rare taxa because subsequent shallow-water ecology excludes taxa from the area or because increased clastic detritus dilutes the fossil abundance too much to find specimens, (4) minimum in terrestrial pollen and spores (e.g. Helenes and Somozoa, 1999; Rull, 2000), (5) minimum in reworked microfossils, (6) maximum in P/B ratio, (7) maximum shale content evidenced by coarse fraction analysis, (8) abrupt to subtle vertical ichnofabric successions that signal sea-level-mediated environmental change and possible "oxygen-deficient" ichnocoenoses (Savrda, 1995), and (9) *Thalassinoides* firmgrounds formed in response to transgression-induced sediment starvation (Savrda *et al.*, 2001a and b).

#### *11.5.1.4 Highstand Systems Tract*

After the maximum marine incursion, the HST is deposited as sea level begins to fall toward the level of the shelf-slope break. The sedimentary locus may shift as sediments prograde in the direction of the shelf edge and are biostratigraphically characterized by: (1) common marine microfossils and upwards-shallowing paleo-water depth, (2) irregular LOs due to problems of ecology and clastic dilution, (3) decrease in the P/B ratio, (4) increased mineral component (mostly fine-grained minerals and often lignitic) of coarse fraction content, and (5) limestones on low sediment-input shelves.

## 11.6 Sequence Biostratigraphic Case Studies

### 11.6.1 Gulf of Mexico Plio-Pleistocene Sediments and 3rd-Order Cycles: Evaluating Biostratigraphic Datums and Correlating Faunal Discontinuities Between Wells

Previous studies of the Plio-Pleistocene of the Gulf of Mexico illustrate the importance of sequence biostratigraphy. Here, we re-examine two wells from the Armentrout and Clement (1990) and Armentrout (1996) studies with emphasis on (1) the role of faunal discontinuities in interpreting SB and MFS candidates, (2) crossing biostratigraphic correlation lines, (3) biostratigraphic correlation lines as a guideline for correlating faunal discontinuities, and (4) the initial integration of log and seismic data.

Interpretive panels from two wells in the western Gulf of Mexico are illustrated in Figure 11.5: Mobil A-158 #1 and Mobil A-70 #1. Wavy lines denote faunal discontinuities on each well panel. In order to determine the geographic extent of these discontinuity surfaces, the faunal discontinuities must be correlated within a stratigraphic framework. Armentrout and Clement (1990) presented a correlation study of these wells, and we use their faunal discontinuity numbering system: FDA-1, FDA-2, etc. The methodology used for correlating biostratigraphic datums in this study (Armentrout and Clement, 1990), has not been previously documented, and is discussed below.

Biostratigraphic tops from well-cuttings are recorded using codes (key at bottom of Fig. 11.5) on each well. Twenty tops are posted for the Mobil A-70 #1 well and are correlated with their equivalents in the Mobil A-158 #1 well. These lines must be evaluated for their chronostratigraphic significance. Each of the three categories of microfossils (nannofossils, benthic foraminifera and planktonic foraminifera) has a different line pattern; lines interpreted to have higher chronostratigraphic confidence are shown using bolder and thicker lines. These higher-confidence correlation lines are then used as guidelines for correlating faunal discontinuities between wells.

Faunal discontinuities near the top of the section are constrained using correlation lines for *Trimosina denticulata* (2B) and *Angulogerina illingi* (1B). Notice that 2B and L (Fig. 11.5) are crossing correlation lines (see Fig. 11.5 for rationale). We interpret *Pseudoemiliana lacunosa* (L) to be depressed updip and choose *T. denticulata* as the preferred correlation line. Somewhat lower in the section LG (large *Gephyrocapsa*) and H (*Helicosphaera sellii*) are crossing correlation lines with 4B (*Trifarina rutila*). *T. rutila*, and just below, *Calcidiscus macintyrei* (M) are considered more valid correlation lines; LG and H are rejected due to updip environmental depression. These four correlation lines (2B, 1B, 4B and M) are used to constrain the correlation of faunal discontinuities FDA-3, -4 and -6 (Armentrout and Clement, 1990) between the two wells. Not every faunal discontinuity should be assumed present in every well: some may merge and some may be merely local in nature. In this particular example, FDA-3 and -6 are higher confident correlations than FDA-4 because of the proximity of biostratigraphic tops to the two former faunal discontinuities.

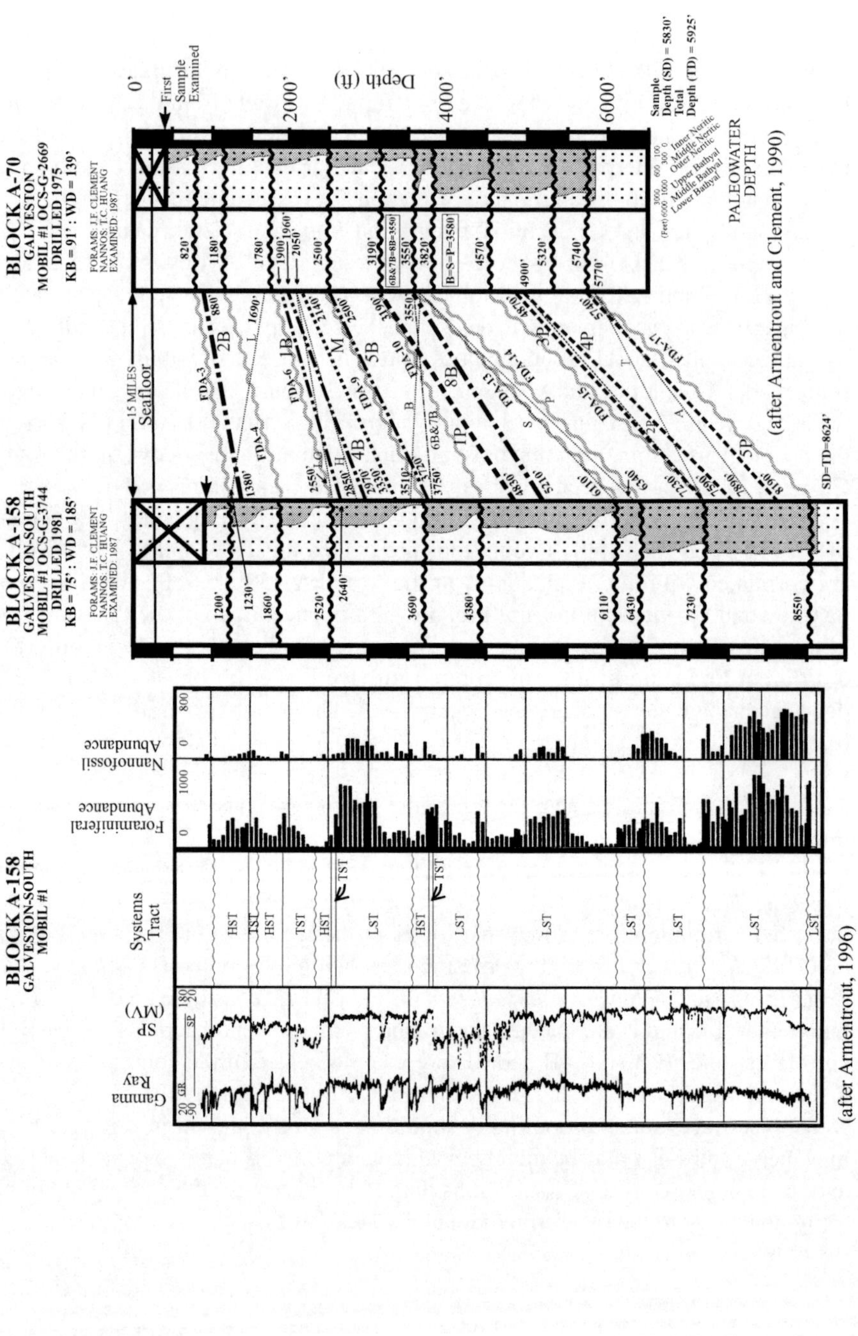
BLOCK A-70
GALVESTON
MOBIL #1 OCS-G-2669
DRILLED 1975
KB = 91' ; WD = 139'
FORAMS: J.F. CLEMENT
NANNOS: T.C. HUANG
EXAMINED: 1987
BLOCK A-158
GALVESTON-SOUTH
MOBIL #1 OCS-G-3744
DRILLED 1981
KB = 75' ; WD = 185'
FORAMS: J.F. CLEMENT
NANNOS: T.C. HUANG
EXAMINED: 1987
15 MILES
Seafloor
First Sample Examined
Depth (ft)
0'
2000'
4000'
6000'
Sample Depth (SD) = 5830'
Total Depth (TD) = 5925'
SD=TD=8624'
Inner Neritic
Middle Neritic
Outer Neritic
Upper Bathyal
Middle Bathyal
Lower Bathyal
PALEOWATER DEPTH
(after Armentrout and Clement, 1990)
BLOCK A-158
GALVESTON-SOUTH
MOBIL #1
Gamma Ray
SP (MV)
Systems Tract
Foraminiferal Abundance
Nannofossil Abundance
HST
TST
LST
(after Armentrout, 1996)

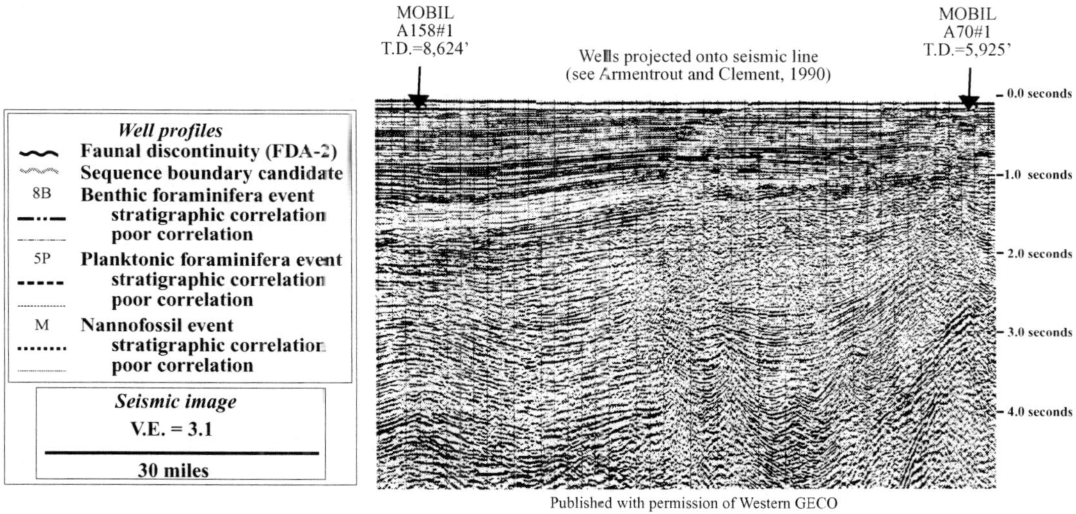


**Calcareous Nannoplankton**

| | |
|---|---|
| L | *Pseudoemiliana lacunosa* |
| LG | large *Gephyrocapsa* |
| H | *Helicosphaera sellii* |
| M | *Calcinodiscus macintyrei* |
| B | *Discoaster brouweri* |
| P | *Discoaster pentaradiatus* |
| S | *Discoaster surculus* |
| T | *Discoaster tamalis* |
| A | *Sphenolithus abies* |

**Planktonic Foraminifera**

| | |
|---|---|
| 1P | *Globorotalia miocenica* |
| 2P | *Globorotalia multicamerata* |
| 3P | *Globoquadrina altispira* |
| 4P | *Globorotalia margaritae* |
| 5P | *Globigerina nepenthes* |

**Benthic Foraminifera**

| | |
|---|---|
| 1B | *Angulogerina illingi* |
| 2B | *Trimosina denticulata* |
| 4B | *Trifarina rutila* 1 |
| 5B | *Trifarina rutila* 2 (=2nd occurrence) |
| 6B | *Alvarezina sinuata* |
| 7B | *Cristelleria subaculeata* (and *Marginulinopsis lacrimata*) |
| 8B | *Bolivina imporcata* |

*Figure 11.5* *Plio-Pleistocene wells, offshore Texas illustrate an application of the techniques for reconciling cross-correlating biostratigraphic datums (see text for discussion). Faunal discontinuities (FDA-1, etc.) are from Armentrout and Clement (1990). Biostratigraphic codes are used in the cross-section for efficiency of space and the legend is given at the bottom of the figure. Well log data and additional fossil abundance data displayed at the same scale as the wells are shown at left (after Armentrout, 1996). Multichannel seismic data displayed across the lower right show the principal geometric patterns observed in the section (above 2.0 s). Once biostratigraphic datums are reconciled, selected datums should then be posted directly onto the seismic data to form a chronostratigraphic framework for seismic stratigraphy.*

Proceeding lower in the section, *Discoaster brouweri* (B), *Alvarezina sinuata* (6B) and *Cristellaria subaculeata* (7B) are interpreted as depressed updip because they cross the second occurrence of *Trifarina rutila* (5B) and *Globorotalia miocenica* (1P). Notice that the depth of tops 6B and 7B is also correlative with the faunal discontinuity in the Mobil A-70#1 well, a relationship indicating potential environmental depression. Armentrout and Clement (1990) also interpreted 8B (*Bolivina imporcata*) to be depressed in the updip well. Although the 8B correlation line does not cross other lines, it is coincident with the faunal discontinuity at 3550 ft and a drop in paleo-water depth to middle neritic water depths. These characteristics would support a possible depression of 8B updip and could be confirmed with a more extensive evaluation including additional wells. Using the planktonic markers 1P and 3P (*Globoquadrina altispira*), as well as the benthic marker 5B, the following faunal discontinuities are correlated: FDA-9, -10, -13, -14 and -15. We may interpret that 3P is slightly depressed updip based on geometry and the coincidence of 3P with FDA-15 in the Mobil A-158#1 well.

At the base of the well, 2P (*Globorotalia multicamerata*) is depressed below 3P; whereas, T (*Discoaster tamalis*) and A (*Sphenolithus abies*) are depressed below 5P (*Globigerina nepenthes*), and T is depressed below 4P (*Globorotalia margaritae*). The lowest correlative faunal discontinuities form FDA-17. The faunal discontinuity present in the updip well at 5320 ft appears to be missing in the downdip well.

These correlations and candidate sequence boundaries must be integrated with logs and seismic data. The log for the downdip well, Mobil A-158#1, is shown in the upper left of Figure 11.5. A preliminary interpretation of systems tracts can be made using log patterns (gamma ray and spontaneous potential), as well as foraminiferal and nannofossil abundances. The faunal discontinuities in deeper paleoenvironments (below 3000 ft in the well) correspond quite well to SB picked out on the log display. However, the faunal discontinuities in shallower paleoenvironments (above 3000 ft in the well) correspond more closely to the transition from TST to HST, i.e. the MFS.

Patterns of large-scale evolutionary radiation and mass extinction may be related to large-scale, sea-level fluctuations (Brett, 1998). Transgressions provide increased habitat space and climates that foster evolutionary radiations. However, the spread of bottom-water anoxia during transgressions may result in the loss of seafloor habitat and produce large numbers of extinctions. Bioevents are commonly associated with either a SB or MFS, with major extinctions associated with habitat reduction during major regressions or with anoxic events during major transgressions. Generally, rising sea level correlates with evolutionary radiations. Thus, some ecological-evolutionary unit boundaries may correlate either with a SB or MFS (Brett, 1998). These patterns are recognized in the case study discussed above from the Plio-Pleistocene of the Gulf of Mexico.

More detailed interpretations can be made when the wireline logs are viewed at a larger scale than presented here (refer to Armentrout, 1996). Importantly, these depositional systems tracts interpreted from wireline logs, can now be correlated between wells based on the correlation lines defined using biostratigraphy. Finally, the larger scale systems can be incorporated into interpretations of seismic geometries. Notice that the seismic data do not allow for the same detail as shown in the logs. However, the seismic data are the best display of continuous geometries in the study area and illustrate an expanding stratigraphic section from right to left.

### 11.6.2 Gulf of Mexico Quaternary Sediments and 5th-Order Cycles: Climate History, Sea-Level Response, and Timing and Sourcing of Downslope Transport

Latest Quaternary sediments in the Gulf of Mexico intraslope basin province are situated in a region of complex structure that results from gravity tectonics and salt diapirism. One of the study objectives was to determine the relationship of depositional processes to sea-level fluctuations. Interpretation of over 71,000 km of high-resolution seismic data and 239 piston cores has been augmented through a sequence biostratigraphic approach. The most important tools applied were foraminiferal analysis, climate zonations (Z, Y, X, etc. of Ericson and Wollin, 1968), coarse fraction and percent $CO_3$, reworked faunal analysis, depositional processes analysis from cores, and integration with seismic facies analysis (Damuth and Olson, 2001; Olson *et al.*, 2001).

Foraminiferal biostratigraphy (especially within the *G. menardii* complex), and fluctuations of coarse fraction and $CO_3$ contents of the piston cores were used to determine ages, climatic zones and late Quaternary sea-level fluctuations. Piston Core IG41-17 serves as a stratigraphic reference for the project (Fig. 11.6). Fluctuations in the abundance of *Globorotalia menardii* complex demonstrate that some cores penetrate the Z zone (most recent interglacial [highstand]-high *G. menardii* abundance), Y zone (most recent glacial [lowstand]-low *G. menardii* abundance), and part of the X zone (previous interglacial [highstand]-high *G. menardii* abundance). Marker taxa include an interval containing *Globorotalia ungulata* near the top of the Z zone, disappearance of *Globorotalia inflata* at the top of the Y zone, and the LAD of abundant *Globorotalia flexuosa* at the top of the X zone. The *G. flexuosa* marker is consistent with the occurrence of a white volcanic ash found in several of the cores in the intraslope basins.

In addition to the *G. menardii* index, the percent coarse fraction and the percent $CaCO_3$ show the following general trends through the X, Y and Z zones: both high coarse fraction and $CaCO_3$ content during interglacials (highstands), and low coarse fraction and $CaCO_3$ content during glacials (lowstands). (These trends differ from scenarios situated closer to direct clastic sources, such as the Mississippi River, and detailed above in section 11.2.3.2. In the above case, the coarse fraction increases due to the increase in foraminiferal content, rather than an increase in clastic material.) A crossover between the coarse fraction curve and the $CO_3$ curve occurs commonly during glacials when a sandy interval is encountered in the core. For example, thin sand beds show up as an increase in coarse fraction, but a decrease in $CO_3$ in Core IG38-15 (Fig. 11.6). In this case, the increased coarse fraction is not a result of increased foraminiferal concentration, but rather, an increase in the sand content.

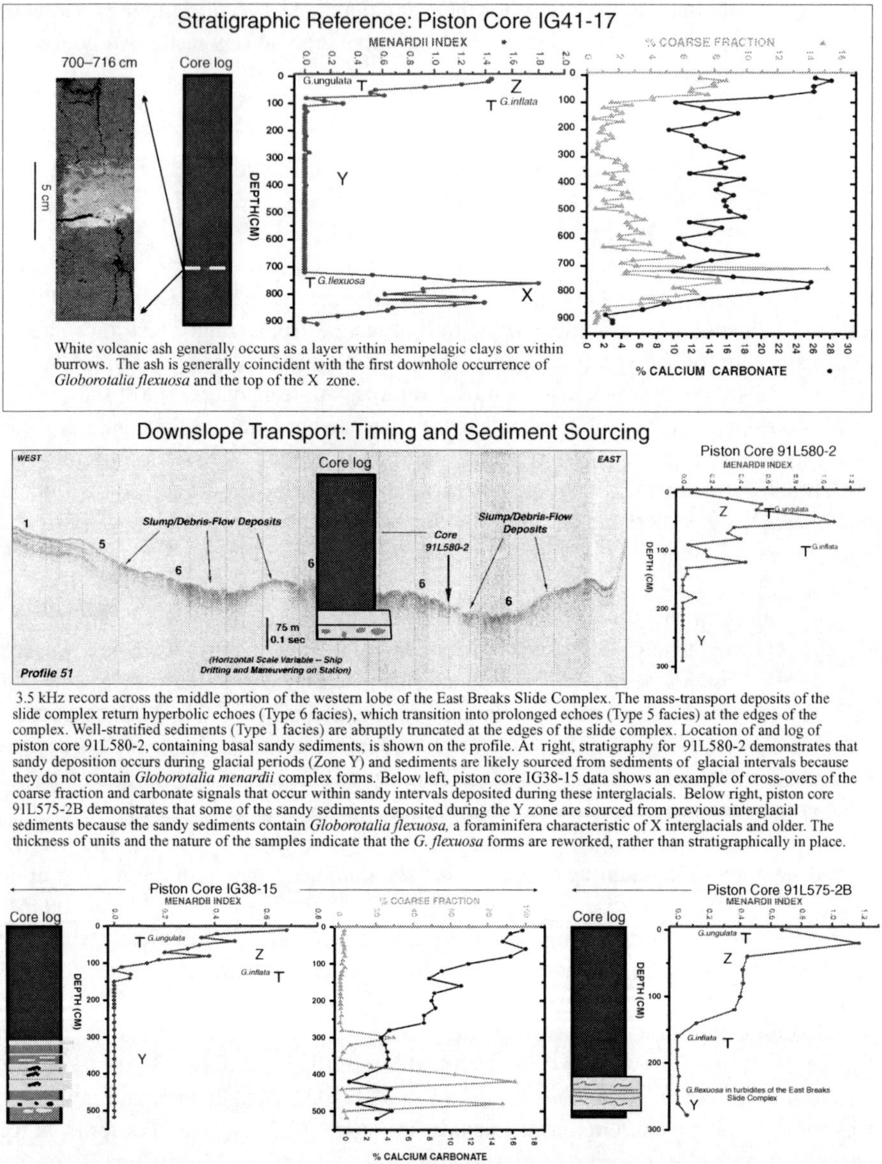


*Figure 11.6 Core IG41-17 at the top illustrates the various stratigraphic horizons utilized in the study of Latest Quaternary piston cores within the intraslope basins in the Gulf of Mexico. This core represents an excellent, stratigraphically coherent section illustrating the X, Y and Z zones of Ericson and Wollin (1968), as well as the* Globorotalia ungulata, G.inflata, *and* G. flexuosa *markers, and the white volcanic ash at the top of the X zone.* G. ungulata *is present at the top of the Z zone and* G. inflata *marks the top of the Y zone. The* G. flexuosa *marker is consistent*

Analysis of depositional processes and facies within the stratigraphic framework developed in this study sheds light on the timing of various depositional events within the intraslope basins. The majority of cores show that very little downslope transport occurs during interglacials. Most cores record hemipelagic deposition during the Z and X zones. However, sediments through the glacial Y zone record downslope transport processes in the intraslope basins. Several cores show a pattern of slumps and debris flows followed immediately upsection by sandy turbidites and deposits derived from channelized flow. For example, Core IG38-15 (773 m water depth) in Figure 11.6 contains numerous silt and fine sand laminae and thin beds from 300–469 cm representing overbank deposits derived from flows down a small leveed channel. These sands overlie deformed mud and sand beds that are apparently a slump/debris flow deposit. Both the mass transport deposits and the overbank deposits represent LSTs that were deposited in the Y zone.

Integration of reworked microfossils into the framework highlights the occurrence of remobilization of sediments during the Y zone. Redeposition is most spectacularly displayed in the region of the East Breaks Slide Complex. For example in core 91L575-2B, *G. flexuosa* (240 cm) is present in LST turbidites (220–270 cm) of the East Breaks Slide Complex. The Z zone in this core is approximately 150 cm thick and the *G. inflata* marker is found at 160 cm. Based on stratigraphic markers and thicknesses, *G. flexuosa* appears to be reworked and represents Y zone remobilization (during sea-level lowstand) of sediments originally deposited during the X zone.

## 11.7 Acknowledgements

The authors appreciate fruitful discussions with our many colleagues, in particular, John Armentrout, Jamie Austin, J. Fred Clement, John Damuth, Anthony Gary, Martin Lagoe (d.), Mark Leckie and Graham Moss. AGIP, Anadarko, BPAmoco, Conoco, Marathon, Texaco, TotalFinaElf, Unocal and Vastar provided financial support for the Gulf Intraslope Basins (GIB) project. The authors appreciate reviews from Ron Waszczak and Garry Jones (d.), which helped significantly improve the manuscript. Many thanks to the editor, Eduardo Koutsoukos, for his work in compiling this volume. This is contribution #1553 from the University of Texas at Austin, Institute for Geophysics.

---

*Figure 11.6 Continued*

*with the occurrence of a white volcanic ash found in several of the cores in the intraslope basins. Note the parallel coarse fraction and carbonate plots that are used in conjunction with the Menardii Index to zone the cores (see text for discussion). Additional cores for this study were used to calibrate the seismic facies (e.g. 91L580-2; facies types after Damuth, 1980) – see colour version of this figure in Appendix.*

Chapter 12

# Taphonomy – Overview of Main Concepts and Applications to Sequence Stratigraphic Analysis

MICHAEL HOLZ[1] and MARCELLO G. SIMÕES[2]

[1] Instituto de Geociências, UFRGS, Porto Alegre, RS, Brazil. Universidade Federal do Rio Grande do Sul (UFRGS), Instituto de Geociências, Caixa Postal 15001, 91501-570 Porto Alegre, RS, Brazil.

[2] Instituto de Biociências, Universidade Estadual Paulista (UNESP) – Botucatu, Caixa Postal 510, 18618-000 Botucatu, SP, Brazil.



## 12.1 Introduction: Taphonomy – What is it About?

Paleontologists study the fossil record to learn about ancient lifeforms, their evolution and interaction in the geological past. However, their efforts are constrained by the uneven and incomplete nature of the rock and fossil record. As discussed in this chapter, Taphonomy and related disciplines, such as paleoecology and stratigraphy, are historical sciences (Martin, 1999) concerned with the history of Earth and the evolution of life recorded in the rocks.

When looking to the origin of a particular skeletal concentration in the geological record, taphonomists try to identify the historical pathways to explain their genesis, and to access their temporal resolution and spatial fidelity with obvious implications for paleoecological and evolutionary studies. However, taphonomic pathways could be very complex, because the origin of the fossil concentrations depends on an array of processes and conditions operating at different scales, including (Behrensmeyer *et al.*, 2000): (a) rate of input, total volume, and composition of biological remains; (b) selectivity and intensity of modification by physical, chemical and biological agents; (c) the rate and permanence of burial; (d) diagenetic conditions in the upper part of the



*E.A.M. Koutsoukos (ed.), Applied Stratigraphy,* 249-278.

sedimentary column, and (f) tectonic setting, determining rates of sediment aggradation and compaction, and structural deformation. Thus, in Taphonomy, laws that are applicable to the genesis of all fossil concentrations without restrictions are absent. However, some empirical generalizations, principles or rules, can be determined (Wilson, 1988; Martin, 1999), such as: (a) organisms with hard parts are more likely to be preserved; (b) rapid burial, particularly by fine-grained particles, in the absence of decay and scavenging, favor preservation; (c) taphonomic loss in shallow-marine environments is mediated by dissolution and bioerosion, whereas in terrestrial and fluvial settings it is largely enhanced by transport, disarticulation, sorting, and breakage; (d) small hard parts do not necessarily outnumber large ones; and (e) big hard parts do not necessarily last longer than small ones (see Martin, 1999, for a more complete list of principles).

In the context above, the term Taphonomy, in its *sensu stricto* (literally "burrowing laws"), is a misnomer (Cadée, 1991; Martin, 1999), and a more realistic definition is that Taphonomy is the study of how biological, chemical, and physical processes operating between each stage or pathway preserve or destroy organic remains and affect information in the fossil record (Behrensmeyer and Kidwell, 1985; Behrensmeyer *et al.*, 2000). Viewed from this perspective, the ultimate value of Taphonomy is that there are patterns and processes that can only be documented by the fossil record, supplying relevant information on sedimentary environments and processes which themselves have left no traces in the stratigraphic record (Fernández-López, 2000). Determine and understand these are the main challenge of taphonomists.

Although as old as Paleontology (Cadée, 1991), applied taphonomic studies began only in the twentieth century with the work of German school of *Aktuopaläontology* (mainly during the twenties, e.g. Wasmund, 1926; Weigelt, 1927a, b; Richter, 1928) and later on, by the work of the russian geoscientist Efremov (1940). Taphonomy spread from Russia and Europe to United States and has been intensively applied to Vertebrate Paleontology, initially by researchers, such as Olson (1958), Schaefer (1962), Lawrence (1968), Brain (1969) and Voorhiess (1969), and more recently by Boy (1977), Behrensmeyer (1978, 1988), Hill (1979), Bishop (1980), Hanson (1980), among others. Taphonomy found its entry to Invertebrate Paleontology by the work of Seilacher (1959, 1973), Kidwell *et al.* (1986) and Flessa (1987); and to Paleobotany by the work of Spicer (1980), Gastaldo (1986) and Pfefferkorn *et al.* (1988), among others. Nowadays, studies in taphonomy are applied to almost all paleontological branches. The most recent tendencies in taphonomy include a revival of some of the actuopaleontological concepts of the German school of the beginning of the century, which led to development of the branch of *Experimental Taphomomy* (e.g., Lask, 1993; Kidwell and Labarbera, 1993; Briggs, 1995). The tendency toward the study of temporal resolution of fossil concentrations, which began in the early eighties, were employed not only to marine faunas (e.g, Fürsisch and Aberhan, 1990; Flessa *et al.*, 1992; Kidwell, 1993; Kowalewski, 1996; Olszewski, 1999), but also to terrestrial fossil assemblages (e.g. Behrensmeyer, 1988; Kidwell and Behrensmeyer, 1993b; Aslan and Behrensmeyer, 1996; Holz and Souto-Ribeiro, 2000). A very complete overview on taphonomy and its role in modern Geosciences and Biosciences is given by Behrensmeyer *et al.* (2000).

Since the advent of Sequence Stratigraphy in the late 1980's (e.g. Wilgus *et al.*, 1988) and its increasing application in almost all types of marine and terrestrial depositional systems, taphonomy is experiencing a strong tendency towards integration with sequence stratigraphy studies. The integration between the dynamic concepts of stratigraphic analysis (e.g. identification of parasequences and depositional sequences) and the analysis of the fossil record (e.g. appearance and disappearance of fossils in the geological column, temporal mixture of fossils etc.) and the increasing integration between sequence stratigraphy and applied taphonomy are the key points of the present chapter. In other words, we will focus on the integration between sequence stratigraphy and taphonomy, in order to demonstrate that the stratigraphic control on fossil distribution and preservation is optimally analyzed and interpreted within a genetic stratigraphic framework, obtained by high-resolution stratigraphic analysis.

## 12.2 Concepts and Methods

In the last 20 years, the renewed interest in Taphonomy produced several studies where the basic concepts and some taphonomic methods were proposed and discussed (see, for example, Brett and Baird, 1986, 1997; Kidwell *et al.*, 1986; Speyer and Brett, 1986; Kidwell and Bosence, 1991; Kidwell and Behrensmeyer, 1993b; Kidwell and Flessa, 1995; Kowalewski, 1996, 1997; Martin, 1999; Fernández-López, 1999; Behrensmeyer *et al.*, 2000). Among them, the concepts of *Taphonomic Feedback*, *Spatial Fidelity*, and *Time-averaging* and its implications to paleoecology were intensely debated and critically reviewed by several authors (e.g. Kidwell and Bosence, 1991; Kidwell and Flessa, 1995; Kidwell, 1998; Kowalewski, 1996, 1997; Kowalewski *et al.*, 1998; Behrensmeyer *et al.*, 2000; Kidwell, 2002).

In sequence stratigraphy and applied taphonomy, genetic terms such as stratigraphic cycles, stratigraphic sequences, parasequences and taphofacies are employed to denote rock bodies of the stratigraphical record (see also Fernández-López, 2000, for the concepts of taphocycles, taphosequences and taphorecords). Here, one of the most important concepts is that fossil concentrations and their taphonomic traits change along environmental gradients. The differential preservation of bioclasts among sedimentary facies reflects the differential activity of taphonomic processes in different depositional settings. In this context, Fürsich and Oschmann (1993) proposed a *genetic classification* of the fossil concentrations based on the two main features: *biofabric* (three-dimensional arrangement of skeletal remains in the matrix, including orientation, packing and sorting of bioclasts) and *taphonomic signatures* that express the quality of preservation. Both features records biotic and abiotic parameters of the environment, along an onshore–offshore gradient. While the taphonomic signatures of skeletal elements often reflect the original habitats of the species, the biofabric is determined by the final concentration process (Davies *et al.*, 1989; Fürsich and Oschmann, 1993). Thus, the use of the biofabric and the taphonomic signatures resulted in a genetic classification, including nine genetic types (see the next section).

*Taphofacies or taphonomic facies* (Speyer and Brett, 1986) correspond to "suites of sedimentary rock characterized by particular combinations of preservational features of the contained fossil" (Brett and Speyer, 1990). Taphofacies are defined and interpreted by considering (a) the state of preservation of bioclastic particles, such as the degree of abrasion, dissolution, bioerosion and mineralization, and (b) the biostratinomic style of skeletal accumulations that varies predictably along environmental gradients (Speyer and Brett, 1986, 1988, 1991; Brandt, 1989; Fürsich and Oschmann, 1993; Kowalewski *et al.*, 1994). Thus, the taphofacies concept could be used as a predictive tool (see Speyer and Brett, 1988, for an example) not only in the marine sequences, but also to continental deposits (see Behrensmeyer, 1988). However, it should be noted that after fifteen years of the publication of the Speyer and Brett's (1998) paper a critical look into the taphonomic literature shows that no consensus has emerged on *taphofacies methodology* (see Kidwell *et al.*, 2001, for a critical review). Anyhow, taphonomic signatures (Davies *et al.*, 1989; Fürsich and Oschmann, 1993) are still the main tools to determine the state of preservation of organic particles. These were documented for a variety of organisms both in the modern and ancient sedimentary environments. However, consensus does not exist in the literature regarding the taphonomic variables scored (e.g. degree of articulation, bioerosion, breakage), how damage states are quantified (grades), and which methods of data analysis must be employed (Kowalewski *et al.*, 1994; Best and Kidwell, 2000a, b; Kidwell *et al.*, 2001), making comparisons among different works difficult. In fact, only very recently, some methodological strategies based on the sensitivity of taphonomic signatures to sampling, data-collection, and data analysis were tested using the skeletal accumulations from subtidal mud, sand, and shell gravel in the San Blas Archipelago of Caribean sea, in order to standardize the methods in taphofacies studies (see Kidwell *et al.*, 2001). However, as observed by the authors the protocol suggested deserves testing in the fossil record (Kidwell *et al.*, 2001).

Keeping the methodological problems above aside, it should be noted that several studies (Speyer and Brett, 1986, 1988, 1991; Fürsich and Oschmann, 1986; Brett and Baird, 1997; Fernández-López *et al.*, 2000) have consistently demonstrated that the genetic types of skeletal concentrations agree well with the stratigraphical, sedimentological, paleoecological, and ichnological data of a given sedimentary basin (see e.g. Fürsich and Oschmann, 1993; Fürsich and Pandey, 2003).

## 12.3 Taphonomy and Sequence Stratigraphy in Paralic Depositional Systems

The processes and events that are subject of applied taphonomic studies are essentially the same as those that control the genesis of a given sedimentary deposit. Thus, the taphonomic analysis has to be closely related to stratigraphic analysis. In other words, only by fully understanding the stratigraphic framework and the processes that acted during its genesis one can practice paleontology in an adequate and secure manner.

In fact, many of the current models in paleontology (e.g. biozone analysis, mass extinctions *versus* stepwise extinctions, punctuated equilibrium) require robust stratigraphic knowledge in order to be applied. Hence, *every significant paradigmatical change in stratigraphic science affects paleontology, and specially taphonomy*. With the advent of sequence stratigraphy as an analytic tool aiming a genetic interpretation of the sedimentary record, the pathway to taphonomic interpretation has changed – the dynamic manner of stratigraphic interpretation inspired taphonomists to look at the fossil record in a new manner. This includes the fossiliferous rock record at the level of facies and depositional systems as well as on the level of parasequences and systems tracts.

The recognition that the sedimentary record is highly punctuated, where the average day-by-day sedimentation is obliterated by sedimentary events of high magnitude, has led taphonomists to practice a new type of analytical approach on fossil occurrences – an approach that is clearly stratigraphy-orientated. For example, Brett and Seilacher (1991) depicted a model of storm taphonomy where most of the classical *fossillagerstätten* of the world are interpreted in terms of tempestites. Onshore zones preserve concentration-*lagerstätten* (e.g. the Triassic *Muschelkalk* coquina), while under deep water conditions conservation-*lagerstätten* are generated (e.g. the Solnhofen or the Hunsrück fossils). O'Brien *et al.* (1994) applied this concept to crinoid-bearing mudstones from the Silurian of New York and Soares (2000) used it to explain the unique *Mesosaurus* taphocoenosis of the Permian Paraná Basin, Brazil.

The popularity of sequence stratigraphy analysis increased during the nineties and with its almost universal application - from marine to continental settings – taphonomists now are aware that the record of fossils is clearly conditioned by the depositional history as depicted by sequence stratigraphic analysis. The insight came from invertebrate taphonomists due to the proximity of their subject of study – the marine fossils – and the pioneer sequence stratigraphic model. Skeletal concentrations, as shell-beds and bone-beds in shallow marine facies, were thought to be the product of storms and other reworking events, as pointed out, for instances, by Mongin (1959), Reif (1971, 1982), Aigner (1985) and Kidwell (1989).

### 12.3.1 Shell-Beds, Taphofacies and Sequence Stratigraphy

Sequence stratigraphy has its root in the oil industry since it has originated from seismic stratigraphy (e.g. Payton, 1977), but with the increasing spread and popularity of the sequence stratigraphic concepts in the academia, taphonomy and sequence stratigraphy began to recognize their mutual importance. Key surfaces within a stratigraphic framework can be marked with the help of certain types of skeletal concentrations, and its origin can be adequately analyzed and understood by the application of concepts of sequence stratigraphy.

Banerjee and Kidwell (1991) were the first authors to state this clearly. Their study was based on the investigation of the systematic association between shell beds and parasequence-scale flooding surfaces in a Cretaceous third-order sequence, linked to a coal-bearing unit (Mannville Group of the Alberta Foreland Basin, Canada). Parasequences within the transgressive and highstand systems tracts of the studied interval have shell beds positioned in stratigraphically controlled levels (Fig. 12.1).

The authors recognized that shell-beds may develop at the base and at the top of parasequences. The fossil concentrations at the base of parasequences, named the "base-of-parasequence" or BOP shell-beds, are simple, single-event storm concentrations composed of comminuted, exotic shell debris and constitute valuable clue to flooding surfaces, specially within the more distal part of the transgressive systems tract, where the flooding surfaces may be lithologically hard to recognize. The BOP shell-beds lay upon flooding surfaces. Shelly accumulations occur also at the top of parasequences. These so-called TOP shell-beds are thicker, microstratigraphically complex units that denote local reworking or indigenous concentrations. The TOP shell-beds are capped by flooding surfaces and are specially useful in the shallower parts of the transgressive systems tract, where the BOP shell-beds may be absent. A third type of shell bed is recognized in deep water facies between parasequences and commonly is associated with firmgrounds and hardgrounds. These "Mid-sequence shell-beds" are composed of lime mudstones with disperse shells in various stages of dissolution and replacement, and represent a terrigenous-starved hiatal concentration generated during episode of maximum flooding (Fig. 12.2).

We consider the work of Banerjee and Kidwell (1991) as a classic example of integration of sequence stratigraphy and taphonomy, because it clearly demonstrates that

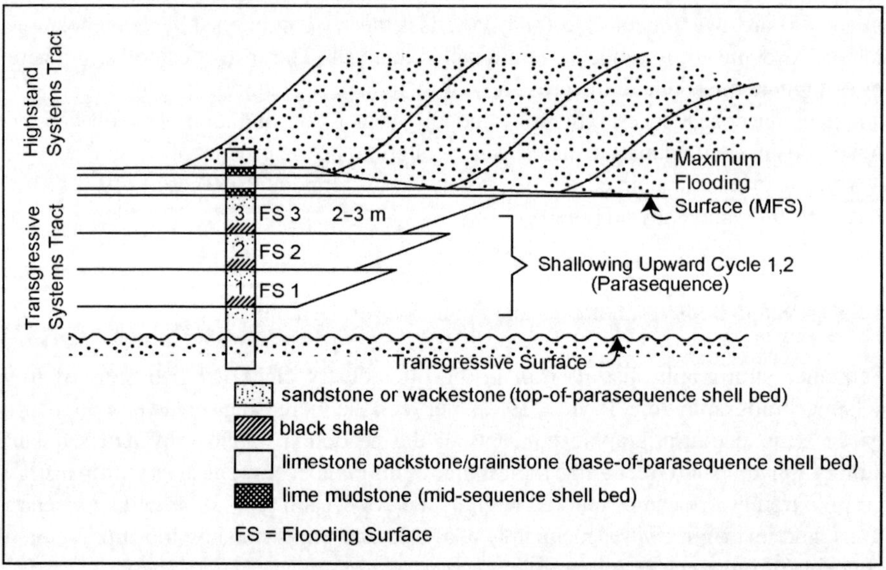


*Figure 12.1 Schematic diagram of parasequences of an interval within the Manville Group, Lower Cretaceous of Canada, showing the stratigraphically controlled occurrence of shell beds (after Banerjee and Kidwell, 1991, modified).*

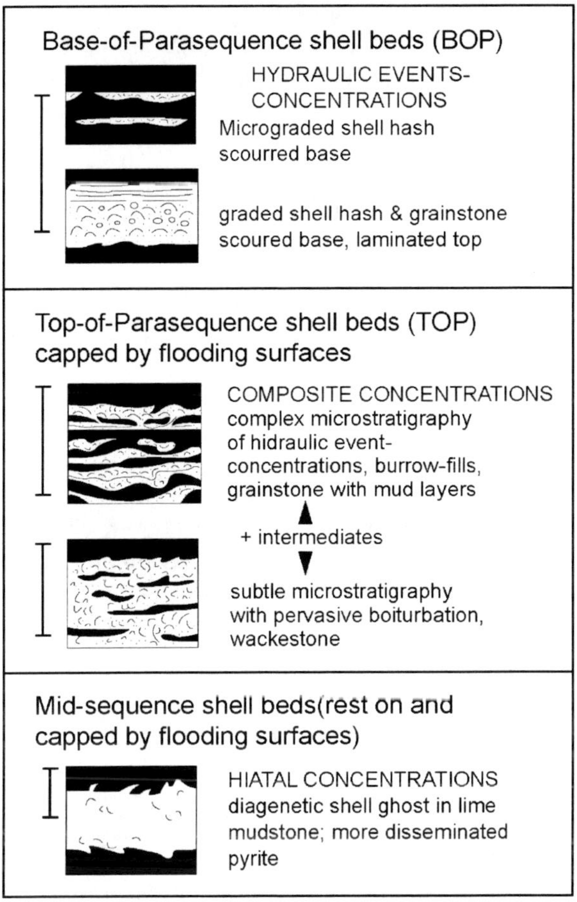


*Figure 12.2 Characterization of the different kind of shell beds associated with flooding surfaces (after Banerjee and Kidwell, 1991, modified).*

analyzing the shelly concentrations and interpreting correctly its genetic signature, the taphonomists may provide important clues for delimitation of flooding surfaces within the stratigraphic framework. On the other hand, the stratigraphers may aid the taphonomists interpreting some particular concentrations by providing stratigraphic information about the fossiliferous horizons.

Another example worth commenting is provided by Kidwell (1991), who defines three types of shell-beds, based on the dynamics of accumulation: (1) event concentrations, which are relatively thin and locally restricted coquinas originated during rapid events (e.g. a single mass mortality or a single event of storm reworking); (2) composite concentrations, which are thicker and wide spread coquinas resulting of multiple events, accumulated in shallow marine and coastal environments, such as tidal inlets, washover fans etc.; and (3) hiatal or condensed concentrations, formed in sites of very low to nil sedimentation rate. These three types of shell beds have a predictable distribution within a depositional sequence, as shown in Figure 12.3. Kidwell (1991) does not show the depositional sequence on a depositional systems tract level, but in the context of a transgressive–regressive cycle. The stratigraphic significant coquinas are as follows:

(1) a *base-of-cycle* coquina, mostly a composite or a hiatal concentration, which may coincide with a sequence boundary or, in cases where lowstand deposits are absent, with a transgressive surface;
(2) a *midcycle* coquina, generally a hiatal concentration associated with the condensed section of the depositional sequence;
(3) an *end-cycle* coquina, in general a composite concentration forming a regressive lag.

Fürsich and Oschmann (1993) combined the use of shell-bed taphonomy with a sequence stratigraphy framework, in a sedimentary basin analysis. For example, shell-beds within the Jurassic (Bathonian–Oxfordian) shallow-water pericratonic basins of Kachchh and Rajasthan, western India, were grouped in nine genetic types: (a) fair weather wave concentration, (b) storm wave concentration, (c) proximal tempestite, (d) distal tempestite, (e) current concentration, (f) primary biogenic concentration, (g) winnowed concentrations, (h) transgressive lag, (i) condensed concentration.

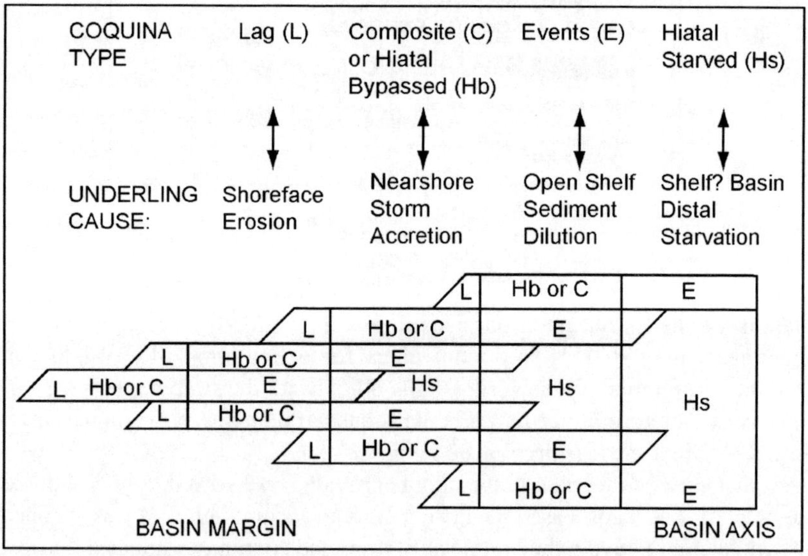


*Figure 12.3 Idealized distribution of shelly hiatal and lag concentrations within a marine sliliciclastic depositional sequence. (A) shows the proximal to distal trend of the different types of coquinas and (B) shows their distribution within retrogradational to progradational parasequences representing a regressive–transgressive cycle (after Kidwell, 1991, modified).*

These concentrations correspond to different bathymetric settings, and although some types occur in the same bathymetric range, their taphonomic signatures coincides well with the sedimentological data of geological sections where they are found.

By plotting the distribution of the nine genetic types of skeletal concentrations against the stratigraphical record of three sections (Kamaguna, Keera, Jumara) and supplementing it with sedimentological, paleoecological and ichnological data, Fürsich and Oschmann (1993) noted that the shell-beds records three orders of bathymetric trends. According to the authors, the first two orders are shallowing–deepening cycles, that may represents eustatic changes in sea level, and the third order corresponds to parasequences with a strong asymmetric sedimentary record.

The distribution pattern of skeletal concentrations in the Kamaguna section (Jura Dome), for example, encompassing rocks of Patcham and Chari Formations, allows the recognition of one large shallowing–deepening cycle, indicated by the replacement of storm-influenced concentrations by winnowed concentrations and by a decrease in the stratigraphic frequency of concentrations up-section (Fürsich and Oschmann, 1993). A second trend is noted by changes in grain size and sedimentary structures, and in the pattern of the nine types of skeletal concentrations. Finally, the third trend is manifested by the occurrence of transgressive lags, which follow abruptly skeletal concentrations that are indicative of shallow water conditions (Fürsich and Oschmann, 1993). At Jumara Dome, the rocks of the upper part of the Patcham Formation and Chari Formation, record cycles analogous to those of Jura Dome. However, the scarcity of skeletal concentrations generated under processes operating in shallow water conditions, and the low stratigraphic frequency of all identified types of skeletal concentrations are suggestive of a more offshore, deeper position of the area, showing that the overall trend is also modified according to their bathymetric position within the basin.

Gomez and Fernández-López (1994) focus on condensed sections and the presence of mixed taphocoenosis as a commonly employed criterion for the recognition of stratigraphic condensations. Within the basic sequence stratigraphy model, periods of maximum flooding lead to a temporal condensation of sediments in the offshore region, where the condensed section forms an important level for basin-wide correlation. These beds, recognizable by the presence of hardgrounds, high concentration of organic matter, offshore microfossils and autigenic minerals such as glauconite and siderite, are in general the first to be picked up during the initial stratigraphic analysis, and their relatively easy recognition gives the false idea that the condensed sections develop only during transgressive maxima. The authors drove attention to two issues: (1) other factors than major transgressions may condense sediments and (2) these factors should be recognized and taken into account in sequence stratigraphic analysis. Therefore, the authors recognize three categories of condensed sections: (1) stratigraphic condensation (due to decrease of the net sedimentation rate), (2) sedimentary condensation (due to decrease in the accommodation rate), and (3) taphonomic condensation (due to mixture of different age fossils).

Only the first category marks the transgressive maximum (near the R inflection point of the eustatic curve as depicted by the concept of Posamentier *et al.*, 1988, see p. 257).

The second is linked to low accommodation as during regressions and should be linked to lowstand systems tracts and late highstand systems tracts. The third category involves reworking of fossils in a high-energy environment and commonly occurs at the base of tempestites, i.e, sediments with high accumulation rate. Periods of taphonomic condensations are represented by concentrations formed by reworked fossils of different ages and different modes of preservation, and this type of condensation does not necessarily mark horizons of stratigraphic or sedimentary condensation. The authors concluded that the relative distribution of condensed sections cannot be used as a criterion for paleogeographic characterization or as indicators of base level changes, and that the presence of taphonomic condensed levels as an indication of stratigraphic condensation may not be a useful criterion.

Brett (1995, 1998) was one of the first to recognize and summarize the relationship of sequence stratigraphy to biostratigraphy, taphonomy and paleoecology. His main premise is that the predictive conceptual framework of sequence stratigraphy may be helpful to analyze and understand certain fossil concentrations, including fossil-lagerstätten. Brett (1995) shows the linkage between systems tracts and different taphonomic modes of preservation, using the classical Ordovician and Devonian records from New York State, as an example. He shows conclusively that taphofacies have a close relationship to sequence boundaries and flooding surfaces, because the sedimentation rate and environmental energy that characterize the periods of sequence boundary formation or flooding events determine the formation of different types of taphofacies (Fig.12.4). This concept has been applied by Simões *et al.* (2000) to the fossil concentrations of the Devonian sequence of the Paraná Basin, Brazil.

Brett (1995) states that corroded and fragmentary fossils are associated with sequence boundaries and with the development of the lowstand system tracts. Sediment bypass due to low accommodation, during lowstand systems tracts and initial transgressive systems tracts may lead to the development of coquinas with fragmented and abraded bioclasts. Thus, these shell-beds with high reworked elements are the most typical taphonomical expression for the base of parasequences, during lowstands and transgressive systems tracts, similar to what has been pointed out by Banerjee and Kidwell (1991). Reaching the period of maximum flooding and the development of the so-called condensed section, biohermas reef structures will overly major flooding surfaces in shallow water facies. In deeper water, fossils will not be significantly reworked because of the relatively low energy conditions of the offshore environment, but will be gradually destroyed by geochemical and biological processes (e.g. corrosion, dissolution, borings and incrustation), resulting in thin accumulation of geochemically stable skeletal fragments (e.g. conodont elements and vertebrate teeth). Highstand deposits, in turn, may display discrete thin skeletal hash beds. The author points out that the smooth transition between the early and the late highstand sediments, as predicted by the general model, does not occur in many sequences. Careful examination of several sections in the Paleozoic Appalachian basin revealed the occurrence of a distinct discontinuity, capped by a discrete condensed shell-bed separating the aggradational and the progradational phase of the highstand systems tracts. These so-called "precursor beds" would mark the initial phase of forced regression, bringing portions of the sea

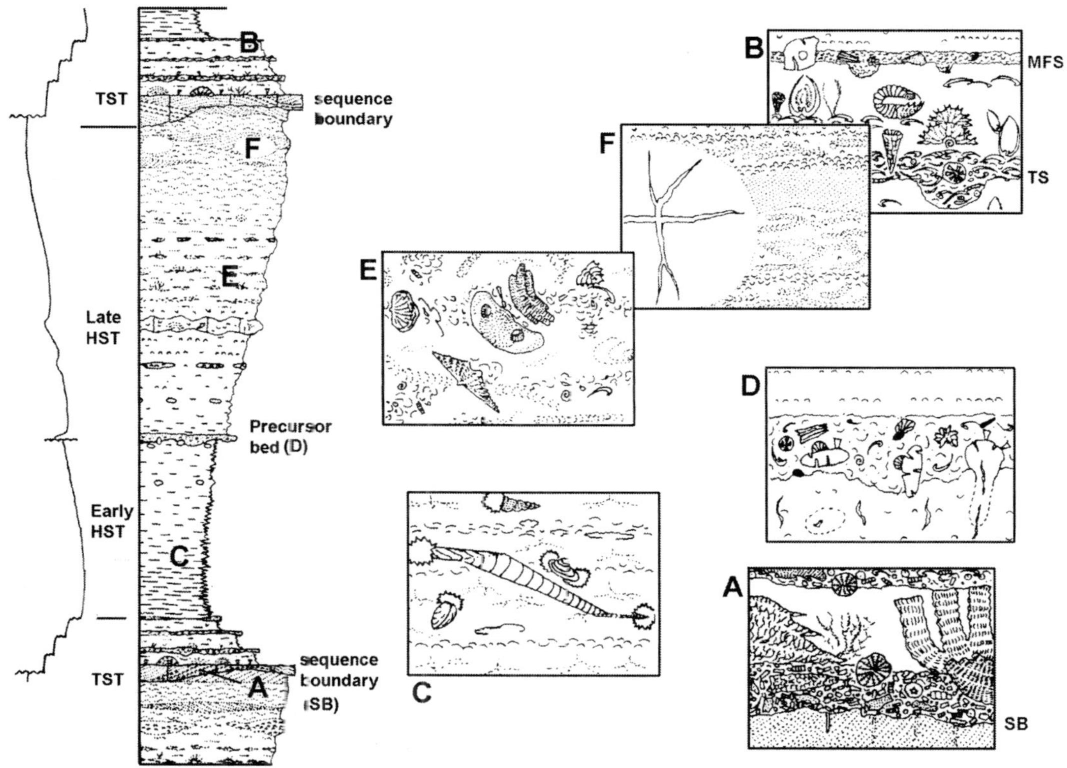


*Figure 12.4* *Taphofacies within a 10-meter thick section of the Middle Devonian of New York. (A) and (B) are transgressively reworked taphocoenosis overlying sequence boundaries, forming condensed accummulations here single burial event are almost absent. (C) is a dysoxic mud facies with a taphocoenosis formed within the early highstand, where* in situ burial *(e.g. bivalves) is characteristic; (D) is a reworked mid-highstand taphocoenosis interpreted as a "precursor bed" marking initiation of rapid sea level fall; (E) is a storm-winnowed, bioturbated taphocoenosis of the late highstand, and (F) is the "regressive cap" of the cycle, characterized by shelly sandstone with hummocky cross stratification (after Brett, 1995, modified).*

floor within the storm wave base and causing submarine erosion (Brett, 1995). It is worth to highlight the fact that the regressive phase of the late highstand and the formation of regressive systems tracts, as stated by several authors (e.g. Hunt and Tucker, 1992; Plint, 1996), is not predicted by the original model of sequence stratigraphy (e.g. Wilgus *et al.*, 1988) and, thus, the "precursor beds" of Brett (1995) represent an important contribution of taphonomy to sequence stratigraphic analysis.

### 12.3.2 Paleoecology and Sequence Stratigraphy

In his 1998 paper, Brett focuses on the relationship of sequence stratigraphy and paleoecology, since the two disciplines share partly their conceptual and methodological framework. One relationship is the usage of the analysis of paleoecological data as a sequence stratigraphic tool (e.g. Sagemann, 1992). Paleoecological information (e.g. salinity, water depths, bottom-water oxygenation) may indicate stratigraphic position within a depositional sequence. Concentration of echinoderms and cephalopods may be an indicator for open marine conditions, typical for late transgressive and early highstand deposits, and may even permit the recognition of sequence stratigraphic key surfaces (e.g. the appearance of marine ichnofossils as indicative of marine incursions associated with maximum flooding events; Fig. 12.5).

An inverse relationship does exist: the conceptual framework of sequence stratigraphy allows studies and predictions of the patterns of biological changes in the geologic record. The chronostratigraphic character of the framework of sequence stratigraphy can be used to make distinction between ecological successions (changes in ecosystems over a time-span of tens to few hundreds of years, i.e. short-term ecological changes) from community replacement (community change over tens to hundreds of thousand years, i.e. a long-term ecological change).

In summary, different kind of shell beds occur not randomly within a paralic succession, but have their distribution and taphonomic signature controlled by the sedimentation regime, as it can be analyzed and predicted by sequence stratigraphy. Figure 12.6 shows an overview on shell bed occurrence as discussed by above cited authors. It is a very schematic representation, which tentatively shows the distribution of coquinas in a ramp margin setting.

The sequence boundary is characterized by abraded and fragmented fossils due to forced regression, which reworks and transports bioclasts basinwards; the transgressive surface may be marked by the base-of-cycle shell beds of Kidwell (1991), the maximum flooding surface is characterized by condensed, hiatal shell beds due to slow or nil sedimentation rate (equivalent to the mid-sequence shell beds of Banerjee and Kidwell, 1991); and a occurrence of reworked shell beds, the precursor beds (Brett, 1995) mark the turnover from the initial to the late highstand systems tract.

Within each parasequence two kinds of shell beds may occur, the top-of-parasequence (TOP) and base-of-parasequence (BOP) shell beds as described by Bannered and Kidwell (1991). The occurrence and distribution of these shell beds is dictated by the normal weather and the storm weather wave base. In the offshore portion, well

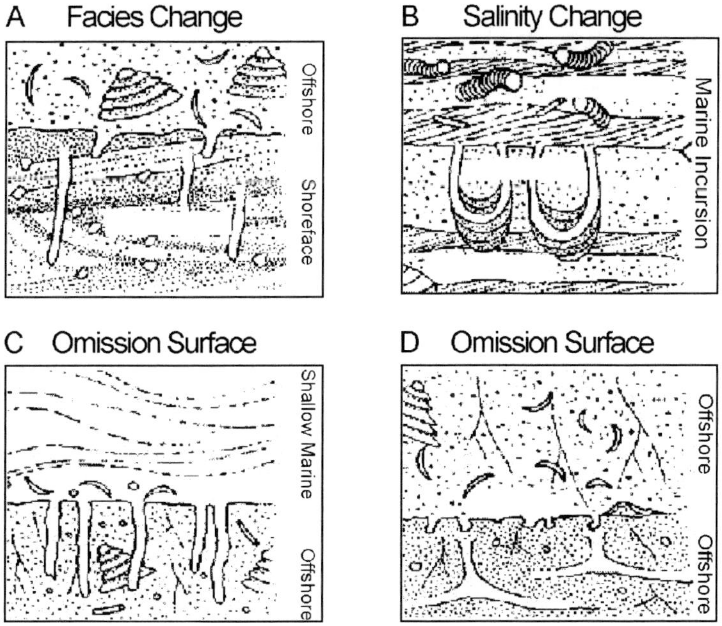


*Figure 12.5 Summary of body and trace fossils as indicators for significant stratal surfaces created by paleoenvironmental change. (A) the ichnological signature of a marine flooding surface, (B) the reflex of minor marine incursions into a non-marine environment, (C) an omission surface associate with shallowing and overlying by a "precursor bed", and (D) an omission surface recording an interval of sediment starvation and hardground development in offshore facies (after Brett, 1998, modified).*

below storm wave base, sedimentation of muddy, distal tempestites may preserve benthic fauna in life position, forming a zone of *in situ* burial (obrution deposits), as described by Brett (1995) and Simões *et al.* (2002).

Emery and Myers (1996) present an attempt to summarize the fossil signature of the different depositional systems tracts of the general sequence stratigraphic model, but the information is strongly biased towards the offshore facies. As pointed out by the authors, their review is limited by the small number of available studies where paleontology and sequence stratigraphy were fully integrated, and most of the examples come from biostratigraphic research of the Gulf of Mexico. Nevertheless, it is worth discussing and summing up (Table 12.1).

Another example that shows the fundamental importance of paralic sequence stratigraphy for the understanding of preservational aspect of taphocoenosis comes from the pioneer work of Holland (1995a, b). The author used computer added simulations of first and last appearances of fossils within hypothetical stratigraphic frameworks to show that the distribution of fossils are not random, but strongly controlled by the facies changes, ecological attributes of the organisms and episodic accumulation of sediments. In subsequent papers, Holland improved the model, predicting

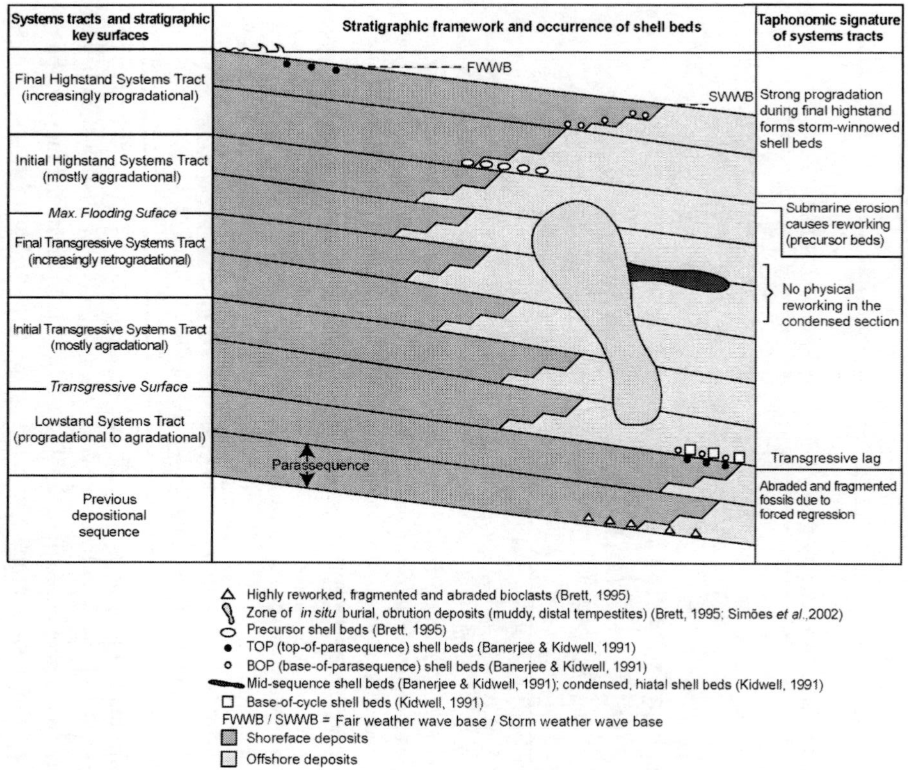


*Figure 12.6 General overview on shell bed distribution and their taphonomic signatures within a paralic depositional sequence formed in a basin with a ramp margin (discussion in the text).*

*Table 12.1 Paleontological signature of the different systems tracts within a depositional sequence (summarized after Emery and Myers, 1996)*

| Systems Tract | Fossil signature |
|---|---|
| Lowstand systems tract - lowstand fan - | Diverse plantonic and deep-water fossils, and reworked fossils reflecting sediment provenance, including slope fossil assemblages in rip-up clasts. Also benthic microfossils in hemipelagic drapes between turbidites. |
| Lowstand systems tract - lowstand wedge - | Reworked fossils and benthic fossils indicative of shallowing-upwards conditions, alternating with non-marine fossils in proximal settings. Biofacies boundaries will be diachronous. |
| Transgressive systems tract | Upward decrease in the amount of terrestrial fossils, while benthic fossils indicate deepening-upwards conditions. In proximal sites shoreface reworking will occur. Close to the trangressive surface, an abrupt change from non-marine to marine fossils will occur. Biofacies boundaries will be diachronous. |
| Highstand systems tract | Gradual upwards reduction in open-ocean planctonic fossils, while benthic fossils indicate shallowing-upwards conditions. Alternation and transition from shallow marine to brackish and terrestrial settings. Biofacies boundaries will be diachronous. |

stratigraphic distribution of first and last occurrences, changes in species abundance, changes in species morphology and distribution of gaps in the fossil record (Holland and Patzkowsky, 1999; Holland, 2000).

In order to understand the control on body fossil distribution in stratigraphic section, the author develops a series of models with increasing complexity and "geological reality". Preferred water depth of the taxa is used to model sedimentological control, by dividing 80 time steps into two identical shallowing upwards successions (=two parasequences) separated by a sharp flooding surface (Fig. 12.7). Assuming that all the 50 modeled taxa share the same facies control, the model shows that there is a clear control on the fossil distribution. Shallow water taxa cluster near the parasequence boundary, deep water taxa occur preferentially at the base of the parasequences.

As a depositional sequence is formed by parasequences arranged in sets with different stacking pattern, the author incorporates the previous model to the depositional sequence concept. Lowstand, transgressive and highstand systems tracts are modeled in terms of parasequence set, incorporating a hypothetical diversity of 1000 taxa. All parasequences have the same duration and rate of shoaling. Figure 12.8 shows the

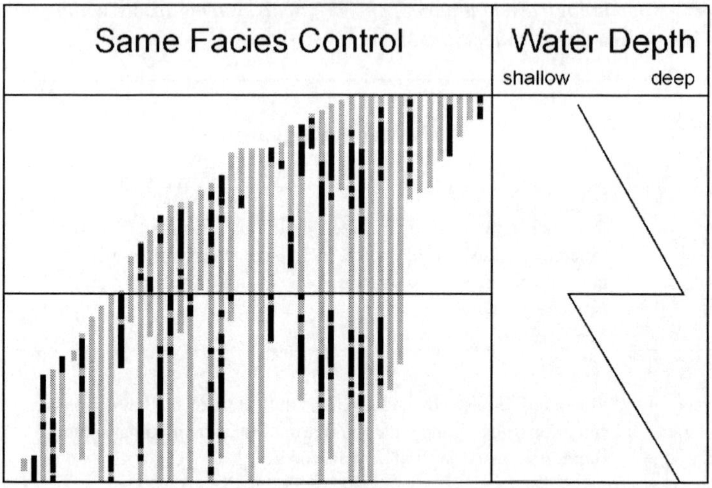


*Figure 12.7 Model of the stratigraphic distribution of fossils within a parasequence, showing that shallow-water taxa cluster near the parasequence boundary, while deep-water taxa occur preferentially at the base of the parasequences (after Holland, 1995a, modified).*

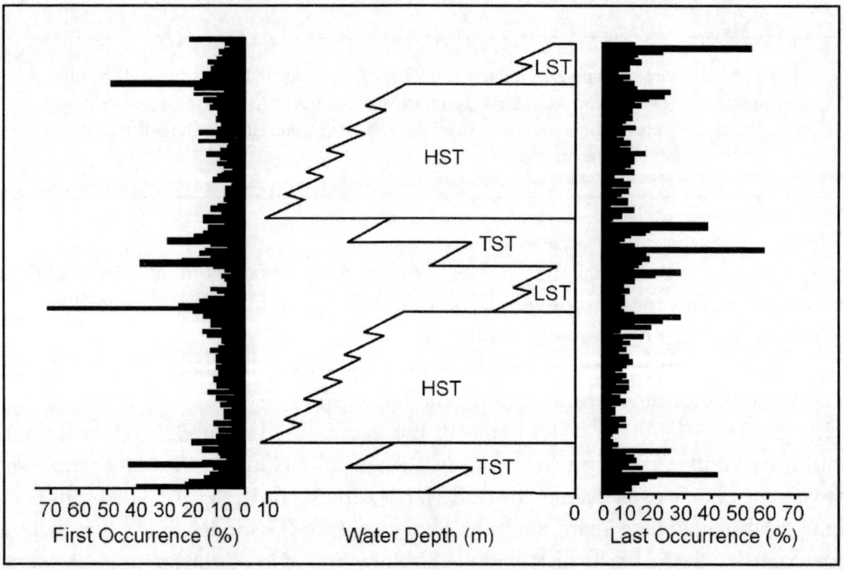


*Figure 12.8 Result of the simulation of a section with two identical depositional sequences. The first and last occurrences of taxa are clearly not random, but controlled by the stratigraphic key surfaces of the sequences: clusters of first occurrences occur above the sequence boundary and at all flooding surfaces of the transgressive systems tract, while clusters of last occurrences occur beneath the sequence boundary and the flooding surfaces at the base of the transgressive systems tract (after Holland, 1995b, modified).*

results of the simulation of a section with two identical depositional sequences. The first and last occurrences of taxa are clearly not random, but controlled by the stratigraphic key surfaces of the sequences: clusters of first occurrences appear above the sequence boundary and at all flooding surfaces of the transgressive system tracts, while clusters of last occurrences are present beneath the sequence boundary and the flooding surfaces at the base of the transgressive systems tract.

A preliminary test of the model is given by Holland (1995a, b), integrating sequence stratigraphy and available paleontological data from the Upper Ordovician section of southeastern Indiana, USA. The six sequences (Fig. 12.9) are composed by limestones and shales, yielding marine invertebrates, with many hundreds of species identified by previous workers. The author uses the paleontological data of Cummins and Galloway (1913, in Holland, 1995a, b), who present the stratigraphic range of over 230 species, most benthic forms, such as bryozoans, brachiopods, trilobites and mollusks. As shown in the above-mentioned figure, the distribution of the first and last occurrence spikes is clearly not random, but stratigraphically controlled. Most spikes are clustered near the sequence boundaries, others within the transgressive systems tract, corroborating the predictions of the model.

In summary, the history of a sedimentary basin influences strongly the distribution of fossils and introduces a bias towards preservation. First and last occurrences of

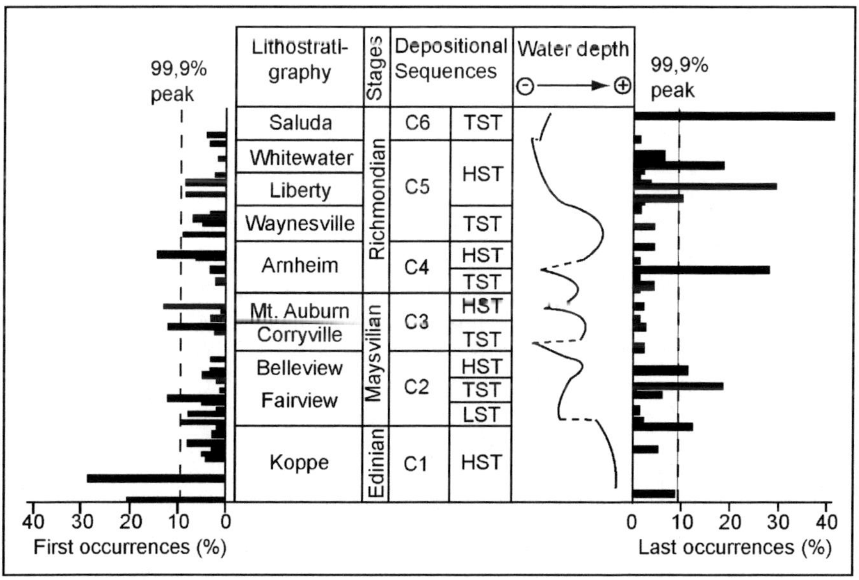


*Figure 12.9 First and last occurrences of marine taxa in the Upper Ordovician of Indiana/USA. The dotted 99,9% line indicates the maximum peak size to be expected most of the time, if the number of first or last occurrences were randomly distributed among the total number of horizons, indicating that the FADs and LADs are far more clustered than it would be expected from random distribution (after Holland, 1995b, modified).*

fossils within a geologic section are strongly controlled by base level changes, so that the their range and pattern of distribution can not be accepted at face value, at least for marine fossils. Terrestrial body fossils may have similar stratigraphic control, as we discuss further on in this chapter.

Taphonomists are aware that beside biasing factors such as selective transport, time-averaging, differential diagenetic influence and others, there is also a stratigraphic bias introduced by the above discussed factors that control the generation of a parasequence and its relative position within a depositional sequence. Holland (2000) recognizes unconformity bias, facies bias, and condensation bias as the main stratigraphic influence on fossil distribution (Fig. 12.10).

Unconformity bias occurs on the fossil distribution across sequence boundaries. First and last occurrences tend to cluster because of erosion and/or no deposition during the time span of life of the taxa. The bias increases with the duration of the hiatus represented by the sequence boundary (Fig. 12.10A). Facies bias causes first and last occurrences to cluster near flooding surfaces. Shallow water species tend to appear below and vanish towards the parasequence limit, while deep-water species tend to disappear toward the shallow facies of the parasequence and appear at or above the flooding surface (Fig. 12.10B). The condensation bias causes first and last occurrences to appear relatively clustered within the condensed section of a depositional sequence, where sedimentation rate is minimum (Fig. 12.10C). This bias alters the perception of the relative timing of the biotic events, making them more closely spaced than they actually are (Holland, 2000).

Another example concerning the stratigraphic control on fossil preservation in paralic settings is provided by Gregory and Hart (1992) studying palynomorph record. They tested the palynologic response to sea-level changes as recorded by the different systems tracts within a depositional sequence. The results from different tested areas were integrated into a model which can be summed up as follows.

The lowstand systems tract is characterized by a relatively high amount of terrigenous palynomorph, reflecting the domination of terrigenous forms even in outer shelf regions. The amount of reworked palynomorphs will also be high because of the remobilization of previously deposited sediments during the base-level drop and adjustment of the river profiles. The age of the recycled material will depend upon the severity of valley incision during the epoch of early lowstand deposition (i.e. the basin floor fan).

Preservation of terrestrial palynomorphs in the transgressive systems tract shows a decreasing trend because of the increasing distance between depositional site and the shoreline and the trapping of the terrigenous sediment in estuaries and bays of the coastal zone. The record of recycled material will also decrease, and will probably indicate erosion associated with the formation of a ravinement surface, or represent suspended load that escaped from the estuarine zone during times of extreme high river discharge.

In the highstand systems tract the inverse pattern occurs, because sedimentation is progressively more progradational, and the quality and quantity of the palynological record will depend upon the proximity of the incoming deltas. This concept has been applied by Holz and Dias (1998) to an Early Permian succession of the Brazilian Paraná Basin.

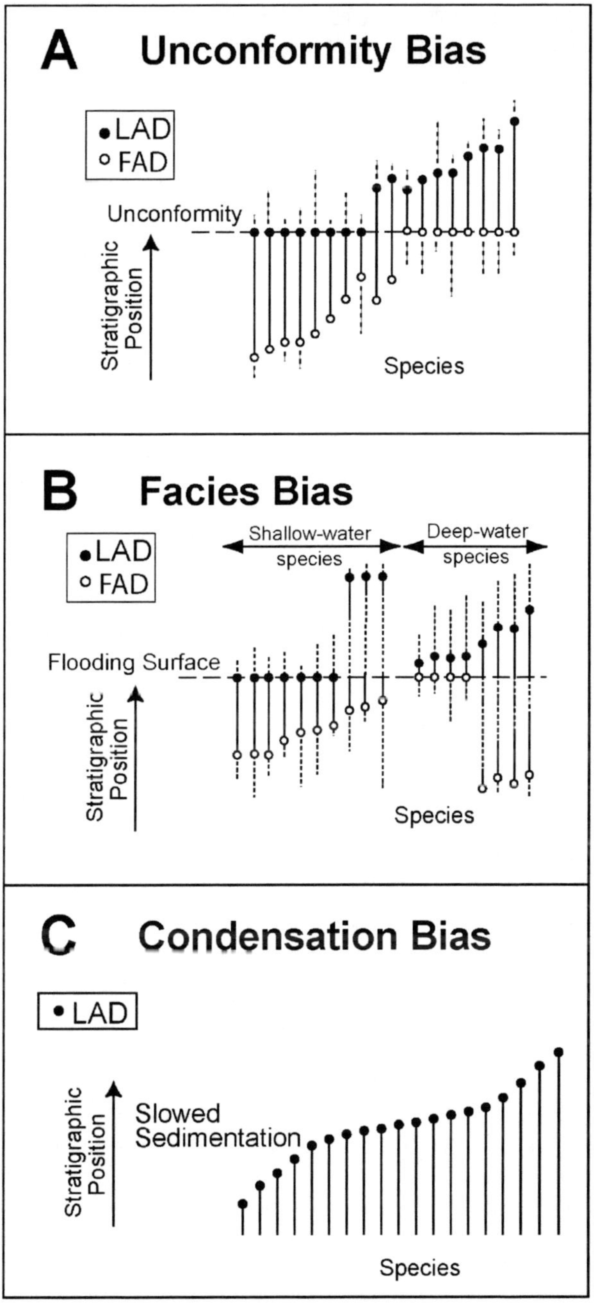


*Figure 12.10 Types of stratigraphic biases, as discussed in the text (after Holland, 2000, modified).*

## 12.4 Taphonomy and Sequence Stratigraphy in Non-Marine Depositional Systems

It is noteworthy that almost all above discussed researches towards integration of sequence stratigraphy and taphonomy are based upon fossil concentrations of nearshore to offshore systems. This is due to three main issues:

1. the basic model of sequence stratigraphy, as depicted by the so-called "Exxon-school" in the late eighties, is based on paralic systems, and most of the key surfaces of the model (mainly the parasequence limits and the maximum flooding surfaces) are readily recognized in those settings, but are hard to identify in fluvial settings;
2. the stratigraphic record of terrestrial systems is spatially more patchiness and vertically more discontinuous than the paralic record, hence increasing the difficulty of the stratigraphic analysis;
3. apparently, the intrinsic taphonomic complexity of a given taphocoenosis is better recorded by shells and other marine invertebrate remains than by bones in terrestrial settings.

However, sea-level is ultimate base-level, and, therefore, the application of the basic concept of accommodation space as the control factor of fluvial sedimentation has been investigated in the last few years, and stratigraphers are attempting to apply the basic concepts of sequence stratigraphy to fluvial settings.

In a seminal paper, Wright and Marriott (1993) discuss the role of accommodation and fluvial sedimentation. The authors stress the idea that a floodplain has a limited capability to store sediment. Since the accommodation space available for sediments to accumulate on a floodplain is controlled by the elevation of the channel and its bankfull depth, the base-level is also the main control to sedimentation in fluvial settings. When base level is stationary, floodplain deposits will rapidly aggradate and lateral accretion will replace aggradational sedimentation. Channels will, therefore, be more amalgamated. If the base level rises, the increasing accommodation favors storage on the floodplain and channels will be isolated. In addition, the development of hidromorphic soils may be favored due to the rising base level. Based upon these premises, Wright and Marriott (1993) propose a simple architectural model for a fluvial sequence (Fig. 12.11). In this model, the lowstand systems tract is characterized by coarse-grained channel in restricted areas and development of well-drained soils on terrace surfaces. During the initial deposition of transgressive systems tract, the accommodation, although increasing, is low enough to produce multistory sandbodies, while the deposition of the late transgressive systems tract, due to the high accommodation rate, leads to the development of isolated channel bodies. During the deposition of sediments under highstand system tract conditions, the rapidly decreasing accommodation space is responsible for the formation of amalgamated channels and higher rates of floodplain reworking.

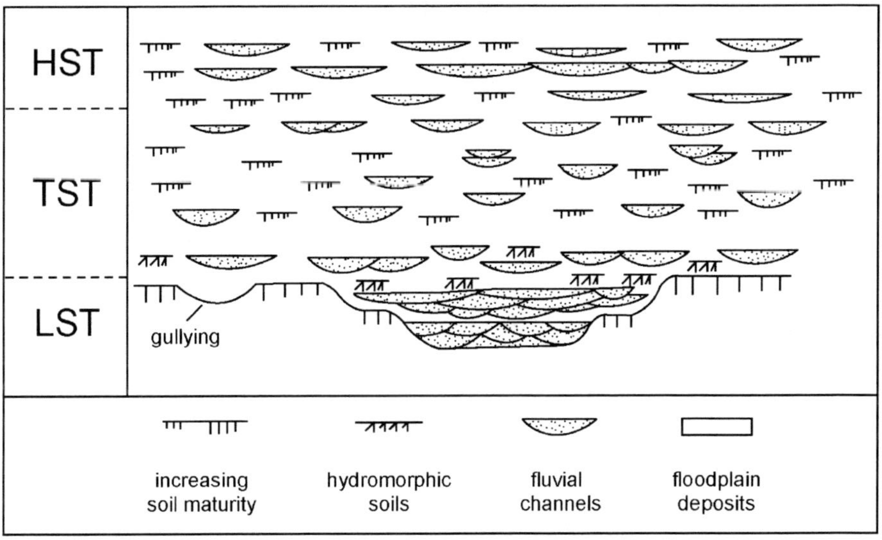


*Figure 12.11 A pioneer architectural model for fluvial sequence stratigraphy. The essential concept is that during phases of low accommodation rate, the channels will amalgamate (as during the LST and late HST), while during times of increasing accommodation rate (during the TST), channels will become isolated and floodplain deposits will be more widespread (after Wright and Marriot, 1993, modified).*

A similar model is depicted by Olsen *et al.* (1995) for the Upper Cretaceous Mesaverde Group of Utah, USA. Five unconformity-bounded fluvial sequences are recognized. The internal organization of a typical sequence (Fig. 12.12) shows amalgamated fluvial sandstone sheet at the base (corresponding to the lowstand systems tract of the previous model of Wright and Mariott), followed by a succession of increasingly scattered and isolated sandstone bodies which may culminate with marine or brackish transgressive deposits. This is topped by a slightly amalgamated sandbodies with a coarsening-upwards trend. While in the lower portion soil formation and large flood plain deposits are not easily preserved due to the reworking events, the middle part of the ideal sequence (the transgressive systems tract in Wright and Mariott' model) has a higher preservation potential, with well preserved bed forms and bars forms, and greater preservation of fine-grained deposits. In the topmost part of the idealized sequence, the decreasing accommodation leads to reworking and greater amalgamation of the sand bodies, and corresponding less preservation of fine-grained deposits.

The role of eustasy in controlling the creation of space for fluvial channels to develop and amalgamate was stressed by several authors (e.g. Wright and Mariott, 1993; Koss *et al.*, 1994; Olsen *et al.*, 1995; Emery and Myers, 1996), while others (e.g. Shanley and McCabe, 1994; Miall, 1996) advocated the role of tectonics, as the main factor that controls base level and, therefore, alluvial sedimentation. Anyway, the basic concept of sequence stratigraphy – that base level variations control accommodation – is

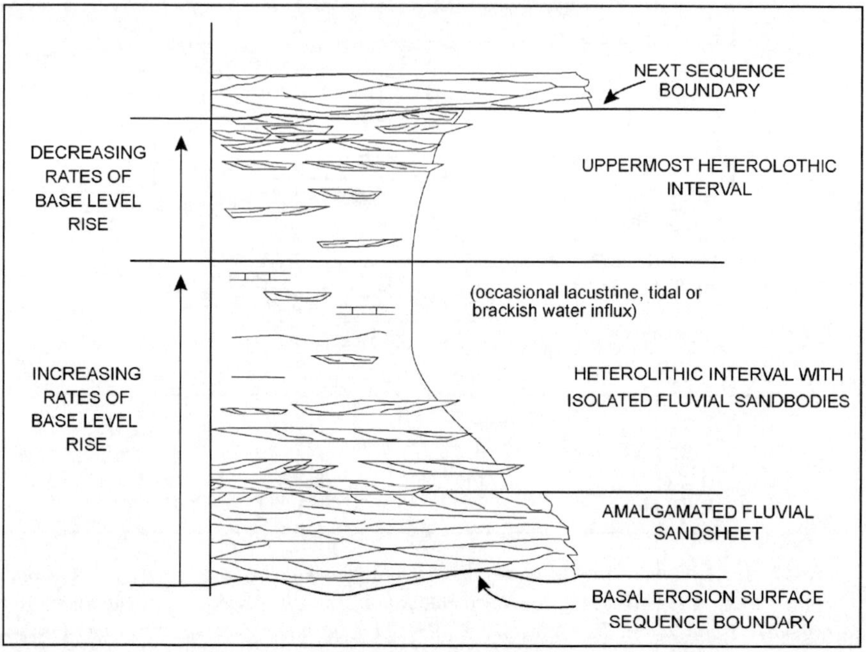


*Figure 12.12 Idealized alluvial sequence as depicted by Olsen et al. (1995). Compare with Figure 12.11.*

surely applicable to fluvial strata, as shown conclusively by the literature. We refer the reader specially to the collection of papers edited by Shanley and McCabe (1998).

This basic concept of a model for "fluvial sequence stratigraphy" has been further discussed and improved (e.g. Blum and Price, 1998; Legaretta and Ulina, 1998), but taphonomists working with terrestrial systems have never properly discussed the relationship between the taphonomic mode of fluvial preservation and the controls on sedimentation as depicted by sequence stratigraphy.

Vertebrate taphonomists are aware that terrestrial taphocoenosis preserved in fluvial facies have strong preservation bias that reflects vertical distribution and on time-averaging (e.g. Behrensmeyer and Hook, 1992). This bias is introduced by the depositional style of the channel and the overbank deposits, as shown by the following key studies.

Smith (1993) studied the fluvial facies of the Hoedemaker Member of the Late Permian Teekloof Formation of South Africa. Analyzing the proximal to distal floodplain facies, he found that each setting has vertebrate fossils with distinct taphonomic characteristics. The fossils of the channel bank environment are less weathered than those of the floodplain, where, in turn, fossils are more densely packed. Weathering increases and bone density decreases with the distance from the channel, so that in distal floodplain facies only scarce and weathered fossils do occur. Thus, the controlling factor of the taphonomic signature of the vertebrate fossils is the proximity of the burial site to the main channel combined with the frequency and intensity of floods capable to bury the skeletal remains in the floodplain. Similar observation has been made by Holz and Barberena (1994) studying the vertebrate fossil record of the Middle Triassic Santa Maria Formation of the Paraná Basin from Brazil.

From a combination of pedogenic and taphonomic data, concerning quantification of floodplain accretion rates and comparison between rates of accumulation and destruction of bones, Smith (1993) discusses the time resolution of the studied section. The findings of well-preserved *Diictodon* skulls in proximal floodplain facies are an indication of the deposition rate: since bones exposed on the floodplain do not withstand weathering for more than about 15 years (Behrensmeyer, 1978), the approximate time needed to bury a well preserved bone element can be calculated. The small (10 cm long) dicinodont skulls used by Smith (1993) indicate a floodplain accretion rate of 5.5 mm/y. The pedogenetic features of the paleosoils, in turn, indicate periods of about 10,000 years of very slow deposition, with an accretion rate of 0,035 mm/y.

Studying in detail the Miocene overbank deposits of the Chinji Formation of Pakistan, Willis and Behrensmeyer (1994) state that the preserved succession is not the result of a gradual aggradational sedimentation of the overbank fine-grained sediments, but rather a product of sedimentation followed by long periods of non-sedimentation and soil formation. Five hypotheses to explain the preservation of the floodplain deposits are presented (Fig. 12.13), starting with the rather simplistic model that the entire floodplain aggrades episodically. The second hypothesis postulates that rates of sediment aggradation is controlled by the proximity of the channel: in near-channel setting the sedimentation rate is greater than in the distal part of the floodplains. Well developed paleosoils covering the overbank successions would indicate depositional hiatuses generated during epochs when the main channel avulsed to distant places on the floodplain. This model matches the interpretation of the Hoedemaker Member by Smith (1993). The third hypothesis predicts episodes of degradation and incision, and posterior aggradation due to raise in base level. Channels and floodplain deposition is restricted to the valley until it is filled up, than the river system is free avulse to other locations, in a process similar to hypothesis 2. The fourth hypothesis predicts a continuous process of localized overbank deposition in several positions over the floodplain, filling up areas where accommodation is locally higher due to differential compaction or fault control, resulting in a patchwork of stratified floodplain sequences. The fifth hypothesis predicts that channel avulsion occurs during periods of rapid overbank deposition. The new channel belt creates its course across the floodplain, filling lower areas with stratified sediments and resulting in a depositional architecture similar to that of hypothesis four. Regarding vertebrate taphonomy, Willis and Behrensmeyer (1994) discuss the influence on distribution and pattern of fossil preservation by comparison of the depositional styles depicted from the five hypothesis and the possible taphonomic modes of preservation. The depositional mechanism of the first hypothesis would result in a large preservation, with the preserved fossils representing the floodplain fauna, as a whole. The preservation of

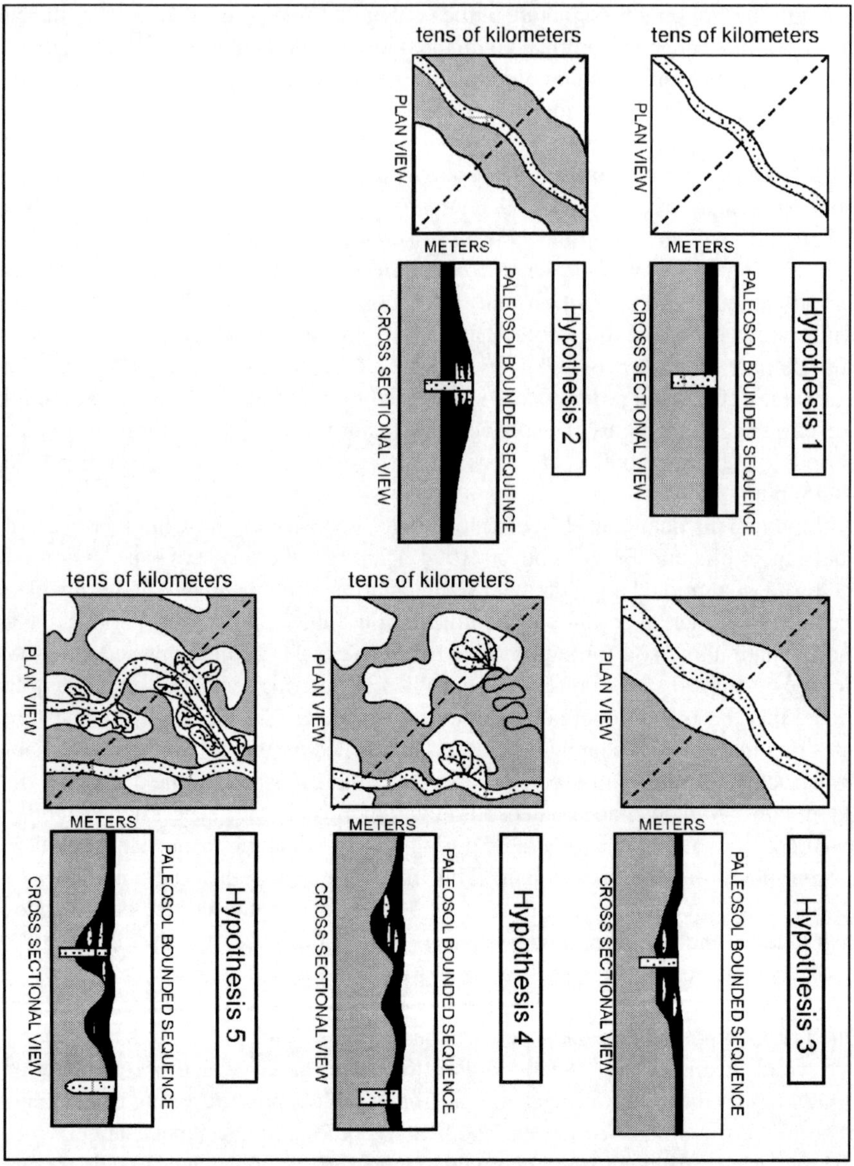


*Figure 12.13 Five hypothesis to explain preservation of floodplain deposits, as depicted by detailed study of the Miocene Chinji Formation of Pakistan, as presented by Willis and Behrensmeyer (1994). Discussion in the text.*

fossils in sediments deposited by the mechanisms of the second hypothesis would be controlled by the proximity of the channel. As the channel belt is relatively stable and floodplain sedimentation decreases with distance from the margins, a ecological gradient could be recorded.

For the third hypothesis, the fossil record would be biased by the alternation of habitats formed by the phases of fluvial incision, when habitats are to the incised valleys, and the phases of aggradation, when the environment would become more uniform. Lag deposits of reworked bones, representing time-averaged taphocoenosis, could be formed during the phases of fluvial incision. Hypothesis four and five predict similar geometry of the overbank deposits, with contemporaneous depositional and soil-forming areas. Vertebrate accumulations would be patchy and localized in floodplain channel deposits associated with crevasse-splay deposition, similar to the Permian taphocoenosis studied by Smith (1993).

In a study on bone beds in the Lower Cretaceous Wealden Group of England, Cook (1995) has shown that vertebrate accumulations in high-energy fluvial settings are characterized by fragmentary bones with well-developed abrasion features, resulting from several cycles of reworking by the river system and deposition as a winnowed lag deposit within the channel. In low-energy overbank settings the skeletal elements are less abraded and fragmented, but more weathered.

Thus, it is clear that the preservation of terrestrial vertebrate body fossils is controlled by the facies type of the fluvial system. But preservation of vertebrate trace fossils (tracks and footprints) is also facies-dependent, as shown by several studies (e.g. Cohen *et al.*, 1991; Smith, 1993; Nadon, 1993). In fluvial settings, the trace fossils preserve most likely in proximal floodplain settings.

Abridging the above discussion, one may say that the mode of the fossil record in fluvial systems is controlled by the development of channel and floodplain facies, as summarized in Table 12.2. The sub-environments of the fluvial settings have significant differences in sedimentation rate and channel features, leading to different taphonomic signatures.

The development of these settings, in turn, are controlled by variation of base level. Hence, well *versus* poorly drained floodplains and large amalgamated *versus* isolated sand bodies are a question of systems tract development. Thus, one can conclude that the taphonomic preservation, as well as the distribution of terrestrial body and trace fossils are a function of base level change. In this context, their preservation, as well as their first and last occurrences within a given stratigraphic unit may not be biologically, but stratigraphically controlled, in a manner analogous what Holland (1995a, b) demonstrated conclusively for the distribution of marine fossils.

Since the taphonomic signatures of terrestrial fossils are controlled by the type of fluvial settings, we may speculate that the overall characteristics of a taphocoenosis will show predictable variations within a fluvial depositional sequence. Figure 12.14 shows an attempt of integration between sequence stratigraphy and vertebrate taphonomy, derived form the previous discussion. Considering hypothetical taphocoenosis represented by bones collected in the three fluvial settings depicted in Table 12.2, the intrinsic taphonomic features of that concentrations would be controlled by its

Table 12.2 *Summary of taphonomic features of different fluvial settings*

| Fluvial setting | Fossil abundance | Kind of bone accummulation | Sedimentation rate | Disarticulation | Weathering stages * | Plant fossils and foot prints |
|---|---|---|---|---|---|---|
| ***Channel*** | low | hydraulically selected | low | very high | 1 to 3 | no preservation |
| ***Proximal floodplain*** | high to very high | hydraulically selected | high | low | 1 to 3 | rarely preserved |
| ***Distal floodplain*** | lowest | attritional | low | high | 4 to 6 | relatively common |

*Note:*
* numbers between parenthesis refer to the weathering stages of Behrensmeyer (1978).

position within the fluvial depositional sequence. Avulsion and crevassing of the fluvial channels is closely and directly related to the degree of disarticulation of the considered taphoncoenosis, and to the degree of time-averaging. Increasing channel amalgamation reduces skeletal articulation. This is the typical situation during low-stand and late highstand deposition. During times of transgressive systems tract development, the increasing accommodation will preclude channel amalgamation and enhance overbank deposition, and the fluvial style will change from braided to anastomosing. This will decrease disarticulation and mixture of faunal elements, hence decreasing time-averaging. The excision would be the "condensed section" of the fluvial sequence: during times of maximum accommodation rate (the "maximum flooding surface" in Fig. 12.14), most of the depositional site would be represented by distal floodplain deposits because of the scarcity and separation of the fluvial channels. The typical low sedimentation rate of this timing would lead to more prolonged exposure of the vertebrate remains, and therefore the bones of the taphocoenosis would be slightly more disarticulated and time-averaged than in the earlier part of the transgressive systems tract. Occasional tidal influence during the times of "maximum flooding", as predicted by some authors (e.g. Olsen *et al.*, 1995), would lead to a similar taphonomic signature because of the reworking caused by tidal currents.

We feel that some of the biostratigraphic problems, concerning both the temporal resolution and interbasinal correlation, may be due to the lack of integration between studies concerning fluvial sequence stratigraphy and taphonomy. Most of the biostratigraphers working with terrestrial faunas take first and last appearance of certain fossils as undoubted horizons for correlation. For instance, Lucas (1998) proposes that the first and last appearance datum (FAD and LAD) define a "land-vertebrate faunachron", an informal biochronological units used for biostratigraphical and biochronological correlation of the Triassic in a global scale. But sequence stratigraphy is completely absent as it is in almost all papers of this type, mostly because the lack of integration between paleontologists and stratigraphers. In this way, the appearance datum used by Lucas (1998) and others might be reflecting, at least in part, a stratigraphically controlled taphonomic history and therefore may not be such a accurate datum for correlation. We feel that this is a particularly important issue concerning correlation of the Triassic systems worldwide. If one studies the terrestrial fossil content of a transgressive systems tract of an alluvial setting of a certain basin and compares it with an alluvial succession developed during highstand systems tract situation in a different basin, the comparison between the appearance datum of the fossils may not match because of the different sedimentation regime, as discussed above. Floodplain accretion and soil preservation, as well as the reworking inside the channel setting will be completely different in the two situations, and may lead to the biostratigraphic problems concerning correlation of biozones of terrestrial vertebrate fossils, since almost all paleontologists use only lithostratigraphic columns as a geologic basis for their biostratigraphic study (mainly because in many instances these are the unique sources of data available, see Schultz, this volume, Chapter 6). Viewed from this perspective, we feel that taphonomy of fluvial fossil-bearing successions integrated to detailed sequence stratigraphic study of those succession might shed some light on the problems of correlation.

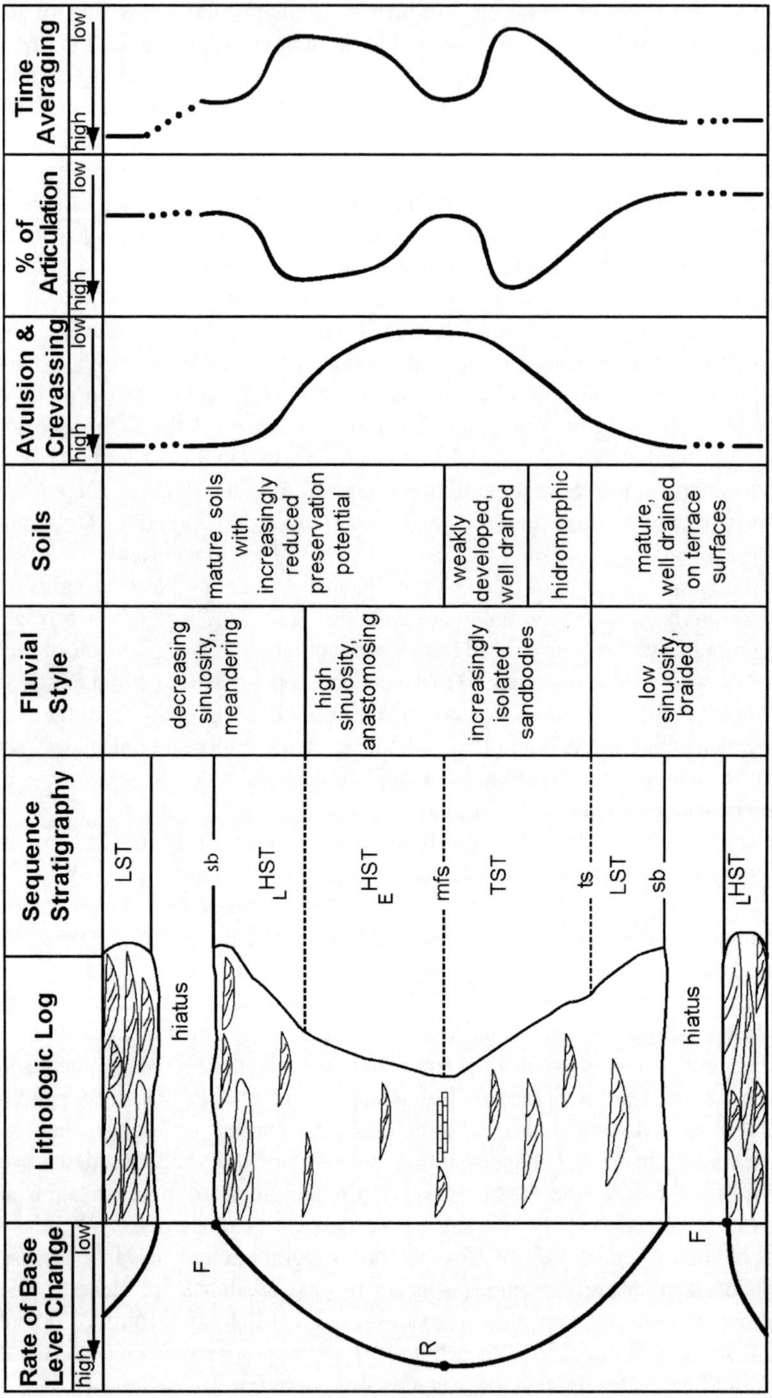


*Figure 12.14* *Schematic representation of a tentative integration of the basic fluvial sequence model and the taphonomic signature of terrestrial vertebrate's taphocoenosis.*

*Key: F/R = inflection points of the base-level curve; LST, TST, HST = lowstand transgressive and early to late (E/L) highstand systems tracts; sb = sequence boundary; ts = transgressive surface; mfs = maximum flooding surface.*

## 12.5 Conclusions and Afterthoughts

The advent of sequence stratigraphy – the geoscience that divides the infill of a sedimentary basin into *genetic* packages – revitalized and modified the traditional way in describing and interpreting the sedimentary record of a basin. Initially applied to paralic settings, nowadays sequence stratigraphic models arose also for non-marine systems, specially for fluvial settings. In a certain manner, stratigraphic science underwent a revolutionary process, where the traditional litostratigraphic approach, practiced by most stratigraphers until the 1980's, was replaced by process-orientated stratigraphic analysis.

As the processes and events that produce fossil accumulations are essentially the same as those that control the formation of the enclosing sedimentary deposit, the taphonomic analysis has to be closely related to stratigraphic analysis. Only by fully understanding the stratigraphic framework and the processes that acted during its genesis one can practice paleontology in an adequate and secure manner. Hence, *every significant paradigmatical change in stratigraphic science affects paleontology, and specially taphonomy*. With the advent of sequence stratigraphy as an analytic tool aiming a genetic interpretation of the sedimentary record, the pathway to taphonomic interpretation has changed, and now taphonomists look at the fossil record in a new manner.

Initially, most of the taphonomic studies under the sequence stratigraphy viewpoint concerned paralic setting, because the new stratigraphic paradigma was closely related to coastal and shallow marine settings. But the increasing sequence-stratigraphic research of terrestrial setting has led to models which clearly show that the basic concepts of sequence stratigraphy are also applicable to non-marine settings, specially to fluvial depositional systems. It is a fact that the exact relation between taphonomic signature and fluvial sequence stratigraphy is still unclear in comparison to taphonomy of paralic settings. However, taphocoenosis of these setting certainly are stratigraphically controlled as their paralic counterparts, and preliminary schemes as that one shown in Figure 12.14 are mainly suited to indicate the pathway of future studies in the taphonomy of the terrestrial realm.

We feel that our line of thinking leads to a afterthought concerning stratigraphical and paleontological correlation between basins with different geological character, what is specifically true for the Triassic period. Marine fossils are the basis for biostratigraphy and biochronology of this period, and can not be used for correlation and biostratigraphic studies of non-marine strata. But, if sequence stratigraphy would be used? Some of the German Triassic section have a very detailed sequence stratigraphic framework (e.g. Aigner and Bachmann, 1998). Attempts to correlate base-level variations and the consequent sedimentation pattern might be the only key towards a correlation between marine and non-marine strata of the Triassic period. This should be tested in an area were detailed stratigraphic and paleontological data are available for a marine succession laterally interfingered with an alluvial succession. Caution is necessary because of the non-contemporanity of the key surfaces along the depositional axis: sequence boundaries have a landwards increasing hiatus and the incised valley fill occurs only after the initial transgressive surface that separates the lowstand from the transgressive systems tract (e.g. Shanley and McCabe, 1994; Miall, 1996; Emery and Myers, 1996). Taking this into account, sequence boundaries and key flooding surfaces may be traceable laterally throughout the marine to non-marine succession and will provide a tool for correlation between terrestrial and marine guide fossils.

## 12.6 Acknowledgements

To editor Eduardo Koutsoukos for the invitation to contribute to this volume, and to the reviewers Gerhard Cadée and Michal Kowalewski for critics and suggestions that improved the text. M.H. and M.G.S. acknowledges the Brazilian National Research Council (CNPq) for research grants (352887/96) and (301023/94-8), respectively.

Chapter 13

# Significance of Ichnofossils to Applied Stratigraphy

S. GEORGE PEMBERTON[1] and JAMES A. MACEACHERN[2]
[1] Ichnology Research Group, Department of Earth & Atmospheric Sciences, University of Alberta, Edmonton, Alberta, Canada, T6G 2E3.
[2] Earth Sciences, Simon Fraser University, Burnaby, British Columbia, Canada V5A 1S6.



## 13.1 Introduction

Trace fossils and trace fossil suites can be employed effectively to aid in the recognition of various discontinuity types and to assist in their genetic interpretation. Ichnology may be employed to resolve surfaces of stratigraphic significance in two main ways: (1) through the identification of discontinuities using substrate-controlled ichnofacies,



*E.A.M. Koutsoukos (ed.), Applied Stratigraphy,* 279-300.

and (2) through careful analysis of vertical softground (penecontemporaneous) ichnologic successions (analogous to facies successions). Ichnological analysis is a valuable but highly under-utilized tool in genetic stratigraphic studies.

Stratigraphy, once considered to be a somewhat routine and mundane discipline, consisting mainly of the dry cataloguing of lithostratigraphic units, has undergone a dramatic renaissance. With the advent of genetic stratigraphic paradigms over the last two decades, stratigraphers have radically altered how we perceive and, therefore, interpret the rock record.

The stratigraphic utility of trace fossils can take on many guises and their significance varies depending on what stratigraphic paradigm is being employed. In the past, trace fossils were considered to be almost useless in stratigraphy because: most trace fossils have long temporal ranges; they are largely facies dependent; a particular structure may be produced by the work of two or more different organisms living together, or in succession, within the structure; the same individual or species of organism may produce different structures corresponding to different behaviour patterns; the same individual may produce different structures corresponding to identical behaviour but in different substrates, (e.g. in sand, in clay, or at sand–clay interfaces); and identical structures may be produced by the activity of systematically different tracemaking organisms, where behaviour is similar (Ekdale *et al.*, 1984).

These factors combine to make their biostratigraphic value negligible. Traditionally, it was thought that there were only three ways in which trace fossils could be utilized in chronostratigraphy: (1) tracing the evolution of behaviour; (2) as morphologically-defined entities (with no assumptions concerning their genesis); and (3) as substitutes for the trace-making organisms. In contrast, trace fossils are proving to be one of the most important groups of fossils in delineating stratigraphically important boundaries related to genetic stratigraphy (Savrda, 1991; Pemberton *et al.*, 1992; Pemberton and MacEachern, 1995) and event stratigraphy (Pemberton and MacEachern, 1997). Genetic stratigraphy lies at the core of three main stratigraphic paradigms: genetic stratigraphic sequences, allostratigraphy, and sequence stratigraphy. The recognition of stratigraphic breaks is essential in any genetic stratigraphic paradigm but is commonly a difficult task, particularly in subsurface analysis. Interpreting the origin of the discontinuity can be vital in resolving depositional environments of associated deposits, and in determining the stratigraphic framework of the system. To accomplish this requires the integration of ichnofacies relationships, physical sedimentology and sequence stratigraphic techniques.

## 13.2 The Conceptual Framework of Ichnology

Trace fossils are biologically produced sedimentary structures that include tracks, trails, burrows, borings, fecal pellets and other traces made by organisms. Excluded are markings that do not reflect a behavioral function. Owing to their nature, trace fossils can be considered as both palaeontological and sedimentological entities, thereby bridging the gap between two of the main subdivisions in sedimentary geology.

Summaries dealing with general ichnological principles can be found in Ekdale *et al.* (1984), Pemberton *et al.* (1992, 2002) and Bromley (1996).

The importance of ichnology to the fields of stratigraphy, palaeontology, and sedimentology stems from the following characteristics displayed by trace fossils: (1) long temporal ranges; (2) narrow facies range; (3) not commonly transported; (4) occurrence in other wise unfossiliferous rocks; and (5) creation by non-preservable soft bodied biota. These characteristics are exceedingly useful in facies analyses, including reconstruction of individual palaeoecological factors, sedimentary dynamics, and the documentation of local and regional temporal facies changes.

### 13.2.1 Behavioural Classification

Perhaps the single most important ingredient of ichnology is the functional interpretation of individual traces. Fundamental behavioral patterns can be dictated and modified by prevailing environmental parameters. Ekdale *et al.* (1984) recognized seven basic categories of behaviour; resting traces (*cubichnia*), locomotion traces (*repichnia*), dwelling traces (*domichnia*), grazing traces (*pascichnia*), feeding burrows (*fodinichnia*), farming systems (*agrichnia*), and escape traces (*fugichnia*). Ekdale (1985) added predation traces (*praedichnia*), and Frey *et al.* (1987) emphasized the importance of equilibria (*fugichnia*) to all other behavioral patterns. The fundamental behavioral patterns are genetically controlled but are not phyllogenetically restricted. The basic ethological categories have persisted throughout the Phanerozoic. Individual tracemakers may have evolved but basic benthic behavior has not. This ability to discern the behavioral trends of benthic organisms represented in the rock record greatly facilitates environmental interpretations.

### 13.2.2 The Ichnofacies Concept

Perhaps the essence of trace fossil research involves the grouping of characteristic ichnofossils into recurring ichnofacies. The concept, developed by Adolph Seilacher in the nineteen-fifties and nineteen-sixties, was based originally on the fact that many of the parameters that control the distribution of tracemakers change progressively with increased water depth. eight recurring ichnofacies have been recognized in the marine realm, each named for a representative ichnogenus (Fig. 13.1): *Trypanites, Teredolites, Glossifungites, Psilonichnus, Skolithos, Cruziana, Zoophycos*, and *Nereites*. Distinct non-marine associations are now being recognized, e.g. the *Mermia* ichnofacies, the *Termitichnus* ichnofacies, and the *Scoyenia* ichnofacies (Frey *et al.*, 1984; Smith *et al.*, 1993; Buatois and Mángano, 1995). Ichnofacies reflect adaptations of tracemaking organisms to environmental factors such as substrate-texture and consistency, food supply, hydrodynamic energy, salinity, water turbidity, sedimentation rates, temperature, and oxygen levels among others (Frey *et al.*, 1990; Bromley and Asgaard, 1991).

## *Seilacher's Concept of Recurring Ichnofacies*

**TRACE FOSSILS** ➨ **BEHAVIOUR** ➨ **ENVIRONMENT**

*Trace fossils are a manifestation of behaviour which can be modified by the environment.*

### *ECOLOGICAL CONTROLS*

The distribution and behaviour of benthic organisms is limited by a number of interrelated ecological controls, including:

**1. Sedimentation Rate**

**2. Substrate Coherence**

**3. Salinity**

**4. Oxygen Level**

**5. Turbidity**

**6. Light**

**7. Temperature**

**8. Water Energy**

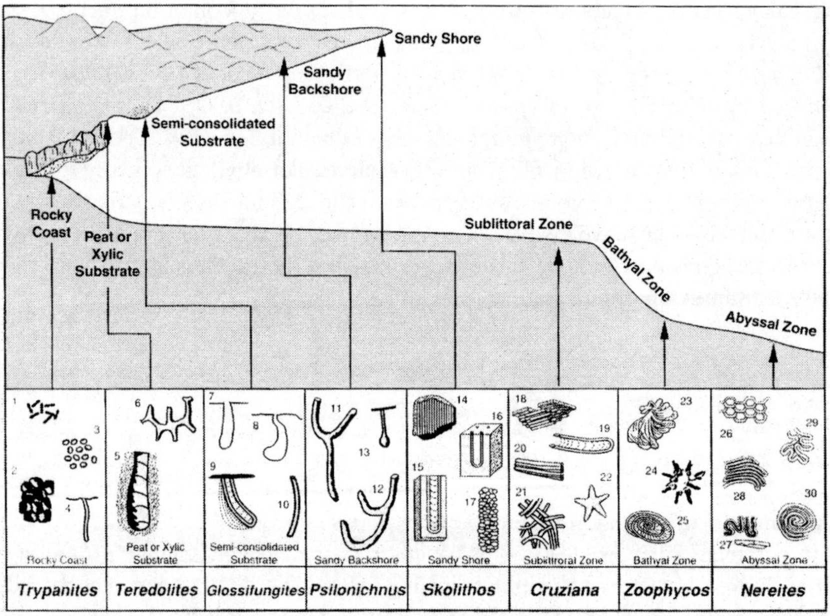


### *Distribution of Common Marine Ichnofacies*

Typical trace fossils include: *1) Caulostrepsis; 2) Entobia; 3) echinoid borings; 4) Trypanites; 5) Teredolites; 6) Thalassinoides; 7, 8) Gastrochaenolites or related genera; 9) Diplocraterion (Glossifungites); 10) Skolithos; 11,12) Psilonichnus; 13) Macanopsis; 14) Skolithos; 15) Diplocraterion; 16) Arenicolites; 17) Ophiomorpha; 18) Phycodes; 19) Rhizocorallium; 20) Teichichnus; 21) Planolites; 22) Asteriacites; 23) Zoophycos; 24) Lorenzinia; 25) Zoophycos; 26) Paleodictyon; 27) Taphrhelminthopsis; 28) Helminthoida; 29) Cosmorhaphe; 30) Spirorhaphe.*

*Figure 13.1 The concept of recurring ichnofacies (modified after Pemberton et al., 1992).*

Ichnofacies are part of the total aspect of the rock and therefore, like lithofacies, are subject to Walther's Law. Ichnofacies stand today as one of the most elegant but also most widely misunderstood concepts in ichnology. Marine ichnofacies are archetypal models based upon recurring trace fossil assemblages and are not intended to be paleobathymeters (Frey *et al.*, 1990). The non-marine assemblages (*Scoyenia, Mermia*, and *Termitichnus*) are general and in need of revision; brackish water assemblages have not as yet been named but they are distinct; the marine soft ground ichnofacies (*Psilonichnus, Skolithos, Cruziana, Zoophycos* and *Nereites*) are well-defined and distributed in response to numerous environmental parameters; and the traces in firmground (*Glossifungites*), woodground (*Teredolites*), and hardground (*Trypanites*) are distributed on the basis of substrate type and consistency.

## 13.3 Substrate-Controlled Ichnofacies and the Recognition of Stratigraphic Discontinuities

Three substrate-controlled ichnofacies have been established (Ekdale *et al.*, 1984): *Glossifungites* (firmground), *Trypanites* (hardground), and *Teredolites* (woodground). In clastic settings, most of these trace assemblages are associated with erosionally exhumed (dewatered and compacted or cemented substrates, and hence, correspond to erosional discontinuities. Depositional breaks, in particular condensed sections, may also be semilithified or lithified presumably at the upper contact (or downlap surface) and may be colonized without associated erosion. In general, however, the recognition of substrate-controlled ichnofacies may be regarded as equivalent to the recognition of discontinuities in the stratigraphic record.

Although certain insect and animal burrows in the terrestrial realm may be properly regarded as firmgound (e.g. Fürsich and Mayr, 1981) or more rarely, hardground suites, they have a low preservational potential and constitute a relatively minor component in the geologic record of such associations. The overwhelming majority of these assemblages originate in marine or marginal marine settings. A discontinuity may be generated in either subaerial or submarine settings; but colonization of the surface may be regarded to be under marine influence, particularly in pre-Tertiary intervals. This circumstance has important implications, pertaining to the interpretation of the origin of the discontinuity.

The substrate-controlled ichnocoenoses typically cross cuts a pre-existing softground suite and, hence, reflect conditions post dating both initial deposition of the unit and erosion of that unit. Such suites also correspond to a hiatus between the erosion event (which exhumes the substrate) and deposition of the overlying unit. During this time gap, organisms colonize those substrates. By observing (a) the softground ichnofacies (contemporaneous with deposition of the unit), (b) the ichnofacies of the exhumed substrate, and (c) the ichnofacies of the overlying unit, it is possible to make some interpretation regarding the origin of the surface and the allocyclic or autocyclic mechanisms responsible (Fig. 13.2).

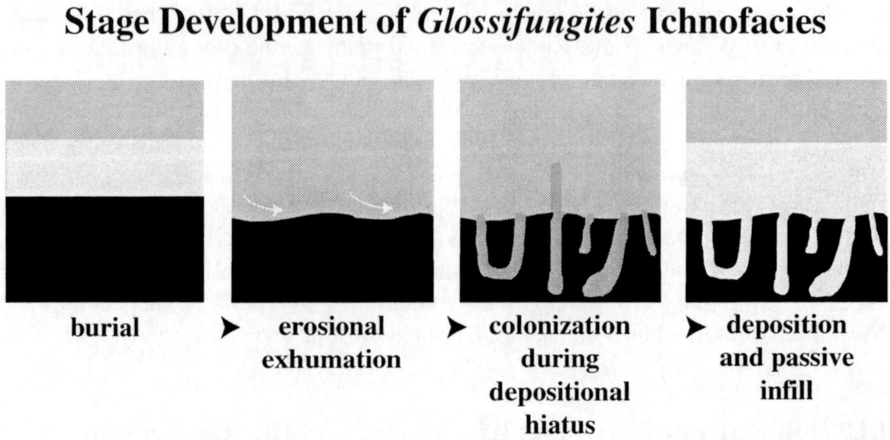


*Figure 13.2 Schematic development of a* Glossifungites *demarcated erosional discontinuity. A: The muddy substrate is buried and dewatered, resulting in a compacted stiff character. B: The shaly bed is erosional exhumed, resulting in development of a firm substrate. C: Colonization of the discontinuity surface by trace makers of the* Glossifungites *ichnofacies proceeds under marine conditions during a depositional hiatus. D: The structures are passively filled during a succeeding depositional episode (modified after Pemberton and MacEachern, 1995).*

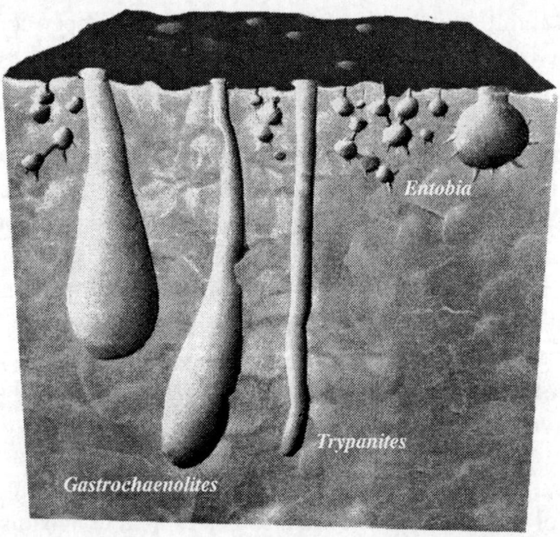


*Figure 13.3 Trace fossil association characteristic of the* Trypanites *ichnofacies.*

### 13.3.1 The *Trypanites* Ichnofacies

The *Trypanites* ichnofacies (Fig. 13.3) develops in fully lithified substrates such as hardgrounds, reefs, rocky coasts, beachrock and other omission surfaces. As such, development of this ichnofacies also corresponds to discontinuities that have major sequence stratigraphic significance. The traces are characterized by: (1) cylindrical to vase, tear or U shaped to irregular domiciles of suspension feeders or passive carnivores; (2) raspings and gnawings of algal grazers and similar organisms (mainly chitons, limpets, and echinoids); (3) moderately low diversities, although the borings and scrapings of individual ichnogenera may be abundant; and (4) borings oriented perpendicular to the substrate which may include large numbers of overhangs. In contrast to the *Glossifungites* ichnofacies, the walls of the borings cut through hard parts of the substrate rather than diverting around them.

### 13.3.2 The *Teredolites* Ichnofacies

The *Teredolites* ichnofacies (Fig. 13.4) consists of a characteristic assemblage of borings or burrows in woody or highly carbonaceous substrates. Woodgrounds differ from other substrates in three main ways: (1) they may be flexible instead of rigid; (2) they are composed of carbonaceous material instead of mineral matter; and (3) they are readily biodegradable (Bromley *et al.*, 1984). Such differences dictate that the means by which, as well as the reasons for which these two types of substrates are penetrated are different. Because currents can raft woody substrates, it is important to determine

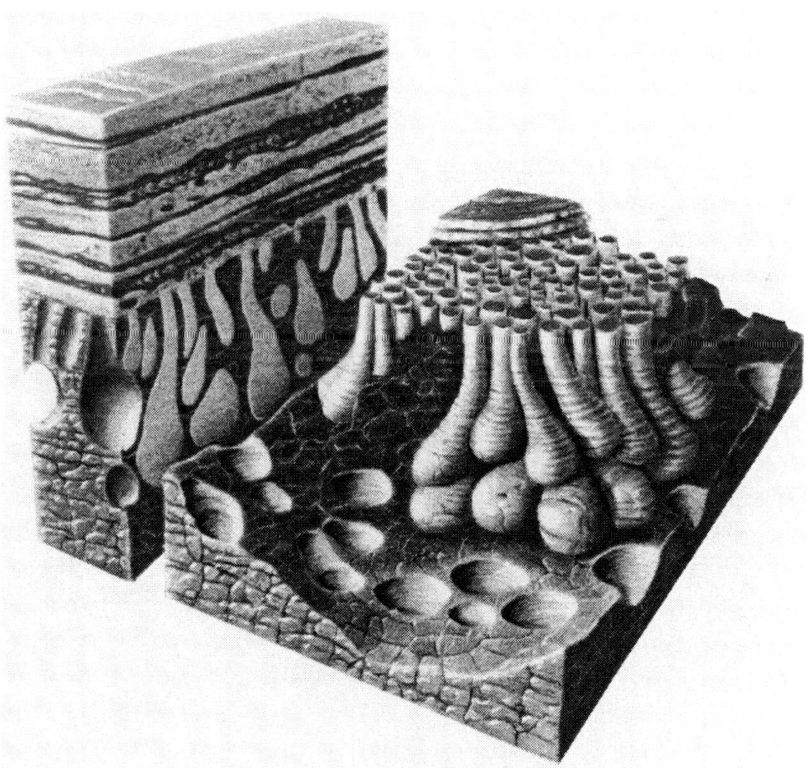

*Figure 13.4 Trace fossil association characteristic of the* Teredolites *ichnofacies.*

whether the borings are autochthonous or allochthonous. Only the autochthonous forms are true members of the *Teredolites* ichnofacies. These assemblages may also be important in defining sequence and parasequence boundaries (Savrda, 1991).

The *Teredolites* ichnofacies is characterized by: (1) sparse to profuse club shaped borings; (2) boring walls that are generally ornamented with the texture of the host substrate (i.e. tree ring impressions); (3) stumpy to elongate subcylindrical excavations in marine or marginal marine settings; and (4) shallower, sparse to profuse nonclavate etchings (isopod borings) in freshwater settings.

### 13.3.3 The *Glossifungites* Ichnofacies

The *Glossifungites* ichnofacies is environmentally wide ranging, but only develops in firm, unlithified substrates such as dewatered muds or compacted sands. Dewatering results from burial and the substrates are made available to tracemakers if exhumed by later erosion (Pemberton and Frey, 1985). Exhumation can occur in terrestrial environments, as a result of channel meandering or valley incision, in shallow-water environments as a result of meandering tidal channels, coastal erosion, erosive shoreface retreat, or as a result of submarine channels cutting through previously deposited sediments (Pemberton and Frey, 1985). Such horizons commonly form at bounding discontinuities and may be critical in the evolving concept of sequence stratigraphy (Pemberton *et al.*, 1992).

The *Glossifungites* ichnofacies (Fig. 13.5) is characterized by: (1) vertical, cylindrical U or tear shaped pseudo borings, sparsely to densely branching dwelling burrows, and/or mixtures of burrows and pseudo borings; (2) protrusive spreiten in some burrows that develop mostly through animal growth (funnel shaped *Rhizocorallium* and *Diplocraterion* [formerly *Glossifungites*]); (3) animals that leave the burrow to feed (e.g. crabs) as well as suspension feeders; and (4) low diversity, but commonly abundant individual structures.

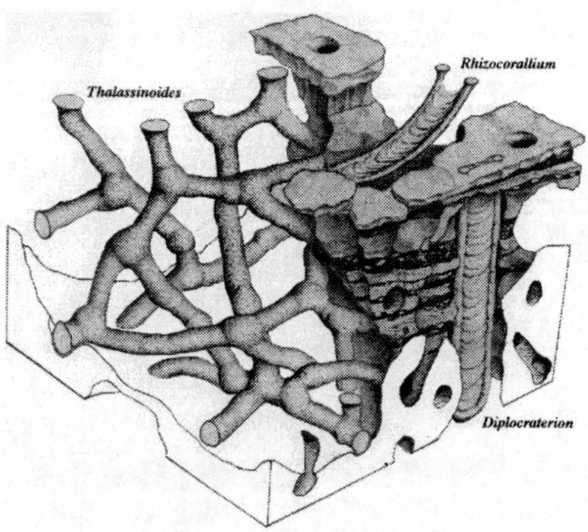


*Figure 13.5 Trace fossil association characteristic of the* Glossifungites *ichnofacies.*

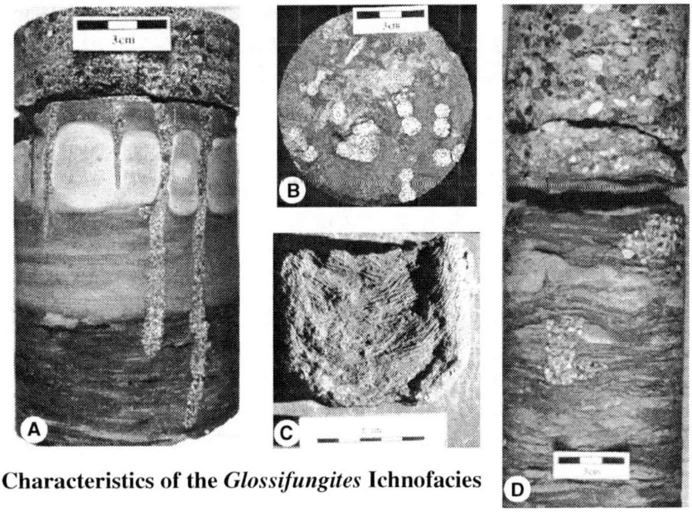


*Figure 13.6 Characteristics of trace fossils that are associated with the* Glossifungites *ichnofacies. A. The burrows tend to be robust, unlined domiciles, B. Are found in very high densities, C. Commonly display scratch marks; and D. Cross cut the original soft ground trace fossil assemblage – see colour version of this figure in Appendix.*

Firmground traces are dominated by vertical to subvertical dwelling structures of suspension feeding organisms. The presence of vertical shafts within shaly intervals is anomalous, as these structures are not capable of being maintained in soft muddy substrates. *Glossifungites* trace fossils are typically robust, commonly penetrating 20–100 cm below the bed junction, they tend to be large in diameter (e.g. 0.5–1.0 cm), and are sharp-walled and unlined (Fig. 13.6). Further evidence of substrate stability, atypical of soft muddy beds, is the passive nature of the burrow fill demonstrating that the structure remained open after the tracemaker vacated the domicile, thus allowing material from subsequent depositional events to passively accumulate in the open burrow. The post-depositional origin of the *Glossifungites* suite is clearly demonstrated by the ubiquitous crosscutting relationships with the previous softground assemblage. The final characteristic of the suite is the tendency to demonstrate colonization in large numbers. In numerous examples, seven to fifteen firmground traces, most commonly *Diplocraterion habichi*, have been observed on the bedding plane of a 9 cm (3.5 inch) diameter core, corresponding to a density of between 1100 to 2300 shafts per m$^2$.

## 13.4 Ichnological Applications to Genetic Stratigraphy

The applications of ichnology to genetic stratigraphy are mainly twofold. The most obvious use is in the demarcation of erosional discontinuities. The second use is subtler, and is concerned with the paleoenvironmental implications of the trace fossil suites, both with respect to the softground and the substrate-controlled assemblages.

### 13.4.1 Sequence Boundaries

Although subaerial exposure and/or erosion during relative sea level lowstand may produce widespread development of dewatered, firm or cemented substrates, most are unlikely to become colonized by substrate-controlled trace fossil suites unless the surface is subsequently exposed to marine or marginal marine conditions. In Cretaceous strata of the Western Canada Sedimentary Basin, deposition of significant thicknesses of nonmarine strata precludes development of such suites. These nonmarine deposits are, themselves, conspicuously unburrowed. For the most part, lowstand deposits rarely dominate incised valley complexes, since the system is largely a zone of sediment bypass (Van Wagoner *et al.*, 1990). Much of the sediment accumulation in these systems occurs during transgression, and typically rests on a marine flooding surface, typically, a ravinement surface amalgamated with the sequence boundary.

#### *13.4.1.1 Incised Submarine Canyons*

In the exceptional case of lowstand incised submarine canyons, the erosional discontinuity lies within a marine setting at the time of its excavation, and colonization of the walls and floor has a higher probability than in terrestrial valleys. Outcrops of the lower Miocene Nihotupu and Tirikohua formations in Northland, New Zealand, contain a noteworthy *Glossifungites* association related to submarine canyon incision (Hayward, 1976). The underlying Nihotupu Formation consists of volcanogenically derived siltstones, sandstones and subaqueous mass flow conglomerates, together with submarine andesite pillow-pile complexes. The softground assemblage is sparse, characterized by localized individual occurrences of *Thalassinoides, Planolites* and *Scalarituba.* These deposits are interpreted as turbidites emplaced at bathyal water depths (based on faunal content) within an inter-arc basin on the lower eastern flanks of the west Northland volcanic arc.

The contact with the overlying Tirikohua Formation is sharp and erosional, and exhibits visible relief. The exhumed substrate is demarcated by a *Glossifungites* assemblage, consisting of *Skolithos, Rhizocorallium*, and *Thalassinoides*. Hayward (1976) interpreted the erosional discontinuity as a submarine canyon wall, excavated into bathyal to neritic inter-arc sediment gravity flow deposits, due to tectonic uplift of the basin margin. Colonization of the canyon walls by the firmground tracemakers preceded eventual burial by canyon floor and neritic turbidite deposits of the Tirikohua Formation, probably corresponding to late stage relative sea level lowstand or early transgressive fill of the submarine canyon.

#### *13.4.1.2 Forced Regression and Lowstand Shorefaces*

Regressive surfaces of erosion generated below forced regression shorefaces or sequence boundaries formed beneath lowstand shorefaces (cf. Plint *et al.*, 1988; Posamentier *et al.*, 1992; MacEachern *et al.*, 1998) are cut within marine settings and therefore favour colonization of the discontinuity by substrate-controlled assemblages.

These settings have been proposed for sharp-based sandstones of the Viking Formation in the Joarcam Field (Posamentier *et al.*, 1992), the Garrington Field (Davies and Walker, 1993), as well as the Kaybob Field (Pemberton and MacEachern, 1995).

In the forced regression and lowstand scenarios, the regressive surface of erosion and the sequence boundary are cut by wave erosion, and extend only as far seaward as fairweather wave base (in storm-dominated systems, storm erosion may obscure this surface). Seaward of these positions, the erosional discontinuity passes into a non-erosional correlative conformity. As a result, the *Glossifungites* assemblage does not develop in positions where the overlying facies reflect deposition below fairweather wave base. In these basinal positions, coarse-grained lag deposits are absent as well.

Forced regression and lowstand shoreface deposits tend to be fairly thin, in response to the diminished accommodation space associated with relative lowstand of sea level. Lowstand shorefaces may be slightly thicker because they may be developed during late lowstand, where a slow rise in relative sea level may be initiated with associated increased accommodation space. Otherwise, these shoreface successions are virtually identical to the shoreface deposits of the highstand and the transgressive shorelines. Bioturbated silty shales and sandy shales comprise the base of the succession. These facies contain thin hummocky cross-stratified sandstone beds, and display distal to archetypal *Cruziana* assemblages. The silty and sandy shales constitute the lower and upper offshore deposits of the succession, reflecting deposition below fairweather wave base. In these positions, the basal contact is the non-erosional correlative conformity, typically bioturbated but lacking both a *Glossifungites* assemblage and a pebble lag. In proximal positions, the sequence boundary is erosional and directly overlain by bioturbated muddy sandstones and hummocky cross-stratified sandstones (Fig. 13.7). A fully marine, diverse proximal *Cruziana* suite, grading

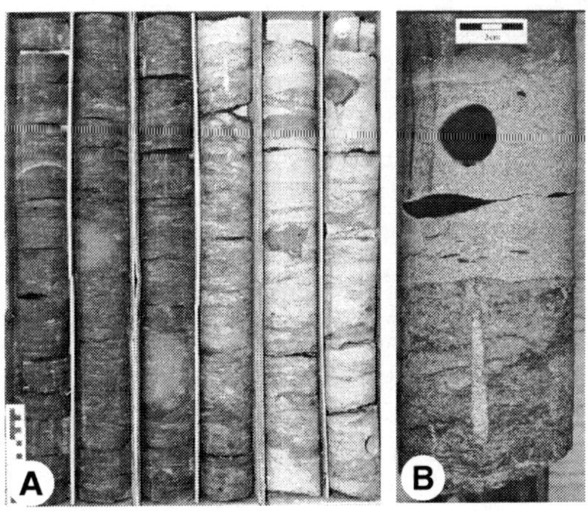


*Figure 13.7 Expression of the* Glossifungites*-demarcated sequence boundary at the base of the Kaybob forced regression shoreface. A. Kaybob interval in Well 11-35-61-20W5. B. Detail of firmground* Skolithos *at sequence boundary, 11-35-61-20W5, 1759 m – see colour version of this figure in Appendix.*

upward into a *Skolithos* assemblage, dominates thoroughly burrowed successions. Where a more complete succession occurs, trough cross-bedded sandstones, containing a sporadically distributed *Skolithos* suite, reflecting upper shoreface environments, may directly overlie the sequence boundary or grade upward from the middle shoreface sandstones. The lower shoreface, middle shoreface, and upper shoreface deposits overlie the erosional expression of the discontinuity, which may be demarcated by the *Glossifungites* ichnofacies.

### 13.4.2 Transgressive Surfaces

Transgressive surfaces are manifest by 1) mainly non-erosional marine flooding surfaces (MFS) and 2) low relief, erosional (ravinement) surfaces. The ravinement surfaces are referred to as transgressive surfaces of erosion (TSE).

#### *13.4.2.1 Marine Flooding Surfaces (MFS)*

Marine flooding surfaces (MFS) are typically abrupt contacts across which there is evidence of an increase in water depth. These surfaces are mantled with dispersed sand, granules or intraformationally-derived rip-up clasts, indicating some erosion. The preservation of underlying markers indicates that the degree of erosion is minimal. MFS are typically characterized by the abrupt juxtaposition of offshore, shelf or prodelta shales on shallow marine sandstones and are easily picked on geophysical well logs.

The Lower Cretaceous Viking Fm. in Western Canada contains numerous MFS separating coarsening-upward, regionally extensive parasequences. In contrast, the regional Viking parasequences are interpreted to reflect shoreface progradation under fully marine conditions. Three facies comprise a complete coarsening cycle, although the minor cycles rarely comprise a complete cycle. The basal facies consists of intensely bioturbated silty mudstone. Trace fossils are uniformly distributed and diverse, constituting a distal *Cruziana* assemblage. Bioturbated sandy mudstone facies grade upward from the silty mudstones and are intensely burrowed with a uniformly distributed and highly diverse *Cruziana* assemblage (18 ichnogenera). Muddy sandstone facies grade upward from the sandy shale facies and are intensely bioturbated with a diverse (18 ichnogenera) and uniformly distributed, proximal *Cruziana* suite. The cycles reflect both coarsening upward of facies, and an increase in diversity of ichnogenera, under fully marine conditions. Each major cycle is interpreted as progradational lower offshore to lower shoreface successions, developed during highstand conditions.

The marine flooding surfaces in the major cycles are commonly marked by the return to lower offshore or shelf shale deposition, and are typically abrupt. These flooding surfaces are unlikely to be disrupted by the diminutive trace makers that characterize the lower offshore settings. In other cases, minor cycles may show much biogenic modification of the MFS, particularly where lower shoreface deposits are overlain by upper offshore sandy mudstones. Such contacts may appear gradational, owing to the biogenic homogenization of the surface by the more robust and penetrative tracemakers.

Elsewhere, the upward transition from shallow to deeper water deposits may occur over intervals of several decimeters or more, reflecting, "slow" relative sea level rise.

### *13.4.2.2 Transgressive Surfaces of Erosion (TSE)*

Transgressive surfaces of erosion (i.e. ravinement surfaces) afford the most elegant manner of generating widespread substrate-controlled assemblages, because the exhumed surfaces are produced within a marine or marginal marine environment (Fig. 13.8B). This favour firmground colonization by organisms as the ravinement surface is excavated, and prior to accumulation of significant thicknesses of overlying sediment. The upper portion of the Albian Viking Formation in the subsurface of central Alberta contains numerous transgressive surfaces of erosion (TSE), recording a complex history of transgression, which culminated in maximum flooding of the Western Interior Seaway.

The recognition of discrete TSE is difficult on the basis of sedimentology alone, particularly when dealing with the upper Viking Formation, where there exist abundant, sharp-based pebble stringers and thin, trough cross-stratified, coarse-grained sandstones, intercalated with interbedded sandstones, siltstones and shales. A few of these coarse stringers could reflect the veneer on transgressive ravinement surfaces, but due to their abundance, picking which ones have regional stratigraphic significance is difficult. However, virtually every TSE incised into, or ravined across, shaly sediments and exhibits a *Glossifungites* suite. Many firmgrounds also appear to have been developed on siderite-cemented intervals within the shales. Whether the siderite is a function of the ravinement, a chemical response related to deep penetration by the tracemakers of the *Glossifungites* suite, or that pre-existing, siderite-cemented bands formed resistant layers through which the TSE could not incise, is uncertain. In the latter case, soft-bodied fauna would presumably find it difficult or impossible to penetrate a cemented layer.

These *Glossifungites* assemblages are manifest by firmground examples of *Diplocraterion*, *Rhizocorallium*, *Skolithos*, *Arenicolites*, and *Thalassinoides* (Fig. 13.8). The *Glossifungites* assemblages record suspension-feeding behaviour associated with the period of higher energy during active ravinement. Colonization of the exhumed surface post-dates erosive shoreface retreat but occurs prior to significant deepening. These transgressive surfaces of erosion are commonly overlain by conglomeratic lags, or erosionally-based, highly burrowed, marine pebbly and sandy shales and more rarely, by muddy sandstones.

### *13.4.2.3 Transgressively Incised Shorefaces*

Several Viking Formation oil and gas fields in central Alberta produce hydrocarbons from NW–SE trending, sharp-based sandstones interpreted as incised shoreface deposits. Many of these shoreface deposits are believed to rest upon a transgressive surface of erosion, including Chigwell (Raychaudhuri *et al.*, 1992), Joffre (Downing and Walker, 1988; MacEachern *et al.*, 1998), Gilby (Raddysh, 1988), and Giroux Lake (MacEachern and Pemberton, 1992).

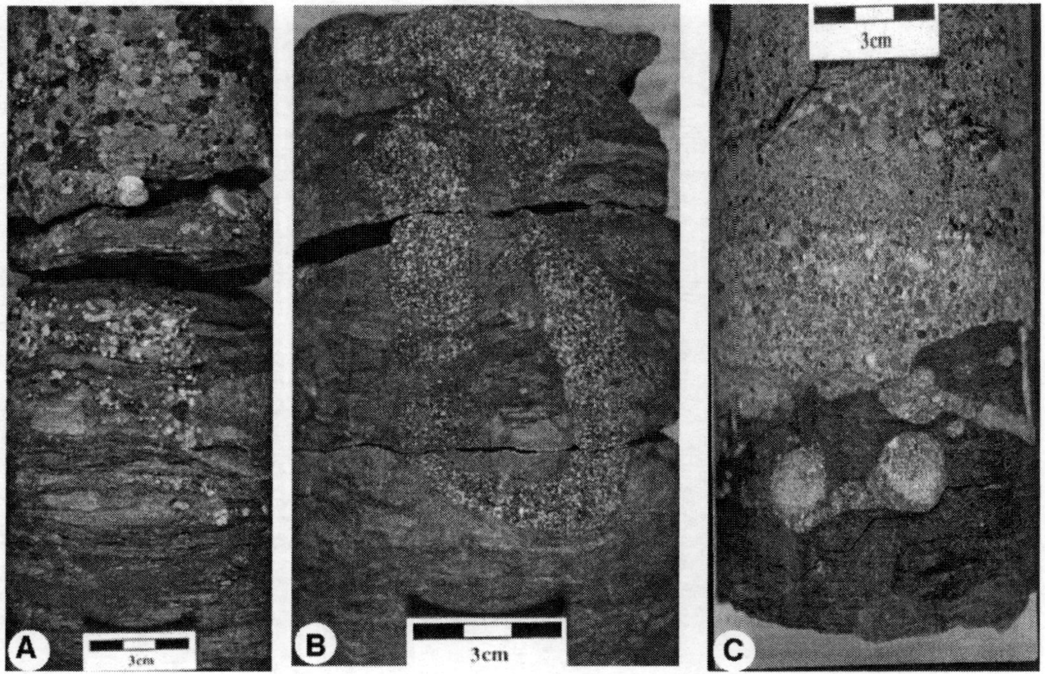


*Figure 13.8* Glossifungites *assemblages. A. Transgressively Incised Stillstand Shoreface. Distal, lower offshore shales truncated by conglomerates of a stillstand shoreface. The erosional discontinuity (E5 of Plint et al., 1988) is marked by a* Glossifungites *suite of conglomerate-filled* Thalassinoides *(arrow). Surface is interpreted as a TSE, associated with initial transgression, overlain by a stillstand progradational shoreface. Cardium Fm., Pembina Field, 12-9-51-10W5, depth 1596.2 m. B. Firmground* Arenicolites *marking a transgressive surface of erosion (TSE), Cretaceous Viking Formation (Core 07-19-62-19W5, 1652 m); and C.* Glossifungites *ichnofacies consisting of* Rhizocorallium *excavated into offshore shales and crosscutting a resident softground suite of* Helminthopsis, Planolites, Schaubcylindrichnus, Chondrites, *and* Zoophycos. *The firmground suite marks the base of an incised valley fill. Albian Viking Formation, Willesden Green Field, Alberta, 10-35-40-7W5, 2327 m.*

The facies that comprise these Viking transgressive shoreface successions are virtually identical to those of the forced regression complexes. The main differences lie in the thickness of the offshore to upper shoreface succession; transgressive systems tend to be thicker due to the increased accommodation space available.

The other principal difference between the forced regression complex and the transgressively incised shoreface complex lies in the erosional extent of the basal discontinuity. In the transgressive scenario, lower and upper offshore deposits, reflecting deposition below fairweather wave base, can overlie the erosional component of the basal discontinuity. Transgressive ravinement permits the generation of an erosional discontinuity which ultimately lies seaward of fairweather wave base during subsequent periodic progradation. This is because the modified surface was cut prior to shoreface progradation, while sea level lay at a stratigraphically lower position (Fig. 13.9).

A *Glossifungites* assemblage commonly demarcates the transgressive surface of erosion, even in positions where the overlying facies reflect deposition well below fairweather wave base. In these positions, a coarse-grained lag is also likely to be associated with the discontinuity. The transgressive surface of erosion passes seaward into a non-erosional marine flooding surface (MFS). In settings characterized by intense burrowing the contact may be obscured. Only the full integration of sedimentology, ichnology and stratigraphy permits the reliable recognition and interpretation of the discontinuity.

The Viking Joffre Shoreface Complex of the Gilby–Joffre trend (MacEachern *et al.*, 1998) contains a sharp-based shoreface, incised into stacked marine parasequences of the regional Viking. The incision surface (an FS/SB) slopes steeply along its landward edge and flattens out seaward, forming an asymmetric, one-sided erosional scarp. Granules and small pebbles of chert locally mantle the erosional discontinuity. More commonly, the surface is demarcated by firmground *Thalassinoides*, *Diplocraterion* and *Skolithos* of the *Glossifungites* ichnofacies, in both proximal and distal positions (Figs. 13.8A, C). The FS/SB is overlain by a coarsening-upward (shallowing-upward) succession of gritty sandy shales and muddy sandstones, containing a fully marine, diverse and uniformly distributed, archetypal to proximal *Cruziana* ichnofacies, reflecting an incised, weakly storm-influenced shoreface complex. The transgressive origin of the shoreface is supported by the presence of a *Glossifungites* suite, which demarcates an erosional expression of the discontinuity, overlain by deposits accumulated below fairweather wave base. This indicates that the surface was cut while sea level was lower, and was prograded across (cf. Fig. 13.9). The Joffre Shoreface Complex is interpreted to have prograded northward during a pause in transgression.

### 13.4.3 Amalgamated Sequence Boundaries and Marine Flooding Surfaces

Amalgamated sequence boundaries and transgressive surfaces are commonly colonized by substrate-controlled tracemakers. The lowstand erosion event typically produces widespread firmground, hardground, and woodground surfaces. The following

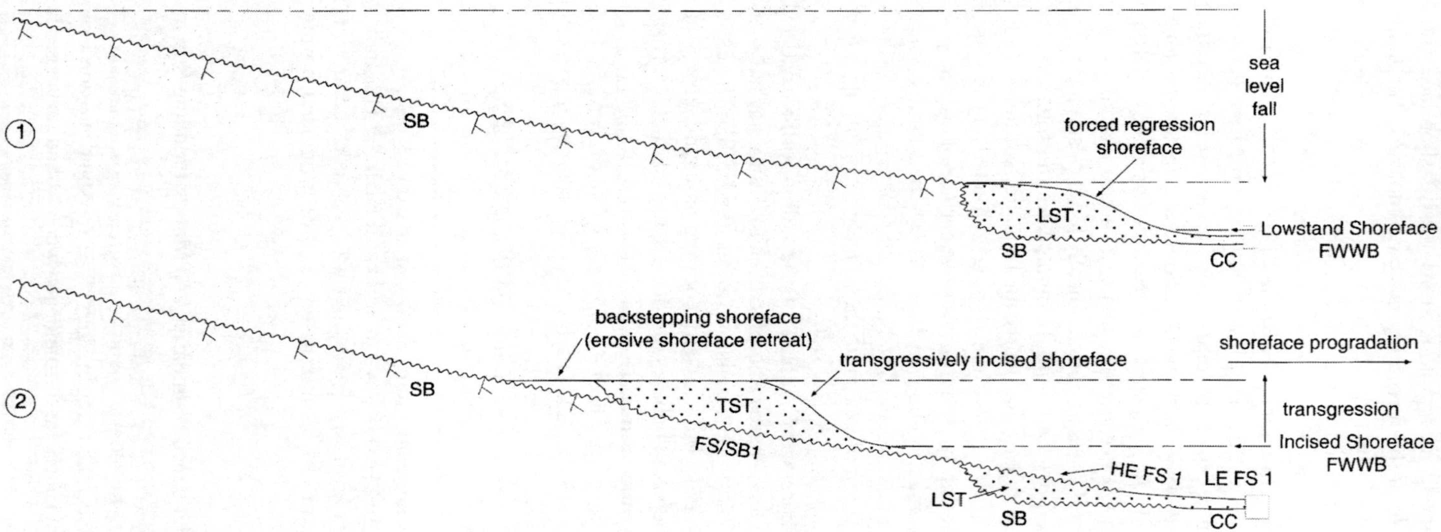


*Figure 13.9* *Schematic model of Forced Regression vs Transgressively incised shoreface successions. Note that seaward of fairweather wave base (FWWB), the sequence boundary (SB) is manifest by a correlative conformity (CC)(1). In the transgressively incised shoreface, the erosional nature of the transgressively modified sequence boundary (FS/SB1) can persist seaward of the fairweather wave base of the overlying prograding system (2) (modified from MacEachern et al., 1998).*

transgressive event, commonly accompanied by erosion, generates a TSE that tends to remove most or all of the lowstand deposits, and exposes the discontinuity to marine or marginal marine conditions and permits organisms to colonies the re-exhumed substrate. The discontinuity may correspond to subaerially exposed areas, such as delta plains, fluvial floodplains, interfluves, or incised valleys.

*13.4.3.1 TSE Across Subaerially Exposed Surfaces*

The Dunvegan Formation in the subsurface of the Jayar Field, central Alberta, contains a TSE, cut into rooted and subaerially exposed delta plain deposits (Bhattacharya and Walker, 1991). The erosional discontinuity is demarcated by a *Glossifungites* suite of *Thalassinoides*, passively filled with coarse-grained sands infiltrated from an overlying transgressive sand sheet. An excellent example of this type of surface also occurs at the Lower Albian Mannville Group–Joli Fou Formation contact in the Kaybob Field of central Alberta. There, rooted incipient paleosols are crosscut by robust, firmground *Thalassinoides*, passively filled with muddy sand and large siderite-cemented clasts. The overlying silty shales contain a distal *Cruziana* to *Zoophycos* trace fossil assemblages, recording deposition in proximal shelf to lower offshore environments. The amalgamated surface corresponds to an interfluve (Van Wagoner *et al.*, 1990) that was transgressively over run.

*13.4.3.2 Bay-Head Delta/Channel and Embayment Deposits*

In the Viking Formation of the Joffre Field area, an amalgamated sequence boundary and flooding surface with a scarp-like geometry truncates underlying regional Viking parasequences and the transgressively incised Joffre Shoreface Complex. A *Glossifungites* assemblage dominated by *Skolithos*, *Diplocraterion* and rare *Thalassinoides* locally demarcates this erosional discontinuity.

The deposits overlying the discontinuity constitute the Viking reservoir facies at Joffre. These deposits are oriented NW–SE, are at least 35 km long, and 8.5–9.0 km wide. The reservoir facies are dominated by trough cross-stratified and low angle planar stratified sandstones, pebbly sandstones and conglomerates, concentrated along the southern margin of the amalgamated sequence boundary and flooding surface. The coarse clastics progressively interfinger with, and ultimately pass into, interbedded mudstones and fine-grained sandstones in a northward and eastward direction. Near the base, the coarse clastics contain glauconite, siderite-cemented mudstone interbeds, mud interlaminae and resistant mudstone rip-up clasts, and display moderate to low degrees of burrowing, diminishing in intensity upward. The trace fossil suite corresponds to the *Skolithos* ichnofacies. Upward, facies are dominated by well-sorted, unidirectional trough cross-bedded and low angle planar stratified coarse clastics, locally in fining upward cycles with scoured bases. The clastics contain mudstone rip-up clasts and thin mudstone interbeds. Burrowing is of low abundance, and reduced diversity, with *Diplocraterion*, *Skolithos*, *Palaeophycus* and *Ophiomorpha* of the *Skolithos* ichnofacies. The interbedded mudstone and sandstone beds contain oscillation ripple and wavy parallel lamination and are weakly burrowed with a sporadically distributed, low diversity (salinity stressed?) trace fossil suite of the mixed *Skolithos–Cruziana* ichnofacies (MacEachern and Pemberton, 1994).

Detailed ichnological, sedimentological and stratigraphic analyses demonstrate that the coarse clastics overlying the discontinuity comprise at least 3 parasequences. These parasequences onlap on the discontinuity in a southwest direction and interfinger with mudstones to the northeast. Toward the north end of the field, erosional amalgamation of the coarse clastics is more pronounced and parasequence boundaries cannot be delineated easily. Near the southern end of the field, these parasequences partition the reservoir along the seaward (and structurally up-dip) edge of the deposit. Amalgamation of the reservoir facies at the north end limits the degree of partitioning.

The final parasequence of the embayment complex is truncated by a regionally extensive flooding surface, typically manifest by a wave ravinement surface. The wave ravinement surface is commonly demarcated by the *Glossifungites* ichnofacies. Facies overlying the marine flooding surface reflect fully marine conditions.

*13.4.3.3 Incised Valley Complexes: Demarcation of Valley Surfaces*

Five Viking Formation fields, namely Crystal, Willesden Green, Sundance, Edson, and Cyn-Pem, contain facies associations interpreted to reflect estuarine incised valley deposition. The facies successions and their distributions indicate that they accumulated in a barrier estuary or wave-dominated embayed estuary setting, in the sense of Dalrymple *et al.* (1992). In most of the incised valley systems of the Viking Formation, the valley base and walls are demarcated by a *Glossifungites* assemblage, indicating that the valley probably did not fill until the ensuing transgression (Fig. 13.8C). Either the valley served as a zone of sediment bypass and possessed no fluvial deposits, or any lowstand deposits were eroded and reworked during the subsequent transgression, producing an amalgamated (co-planar) sequence boundary and initial transgressive surface of erosion. The transgressive surface of erosion most likely reflects tidal-scour ravinement. The base of the estuarine valley fill serves both as the sequence boundary and as the base of the transgressive systems tract. In addition to the valley base, the incised estuarine complex also contains additional erosional discontinuities (Fig. 13.10).

## 13.5 The *Glossifungites* Ichnofacies and Porosity/Permeability Trends

The utility of trace fossils is not limited to palaeoenvironmental interpretation. As we have just noted, recent work has shown that ichnology is important in delineating surfaces that may have stratigraphic significance. In particular, the *Glossifungites* ichnofacies is used to demarcate burrowing into firm substrates (firmgrounds) that may be exposed where sediment accumulations are erosionally exhumed during changes in sea level.

Because of their discretely packaged permeable fill, *Glossifungites* surfaces can enhance the permeability of a relatively impermeable substrate. Permeability enhancement occurs where open burrows emplaced into a firmground are filled with sediment exhibiting textural characteristics distinctly different from the matrix. In such instances

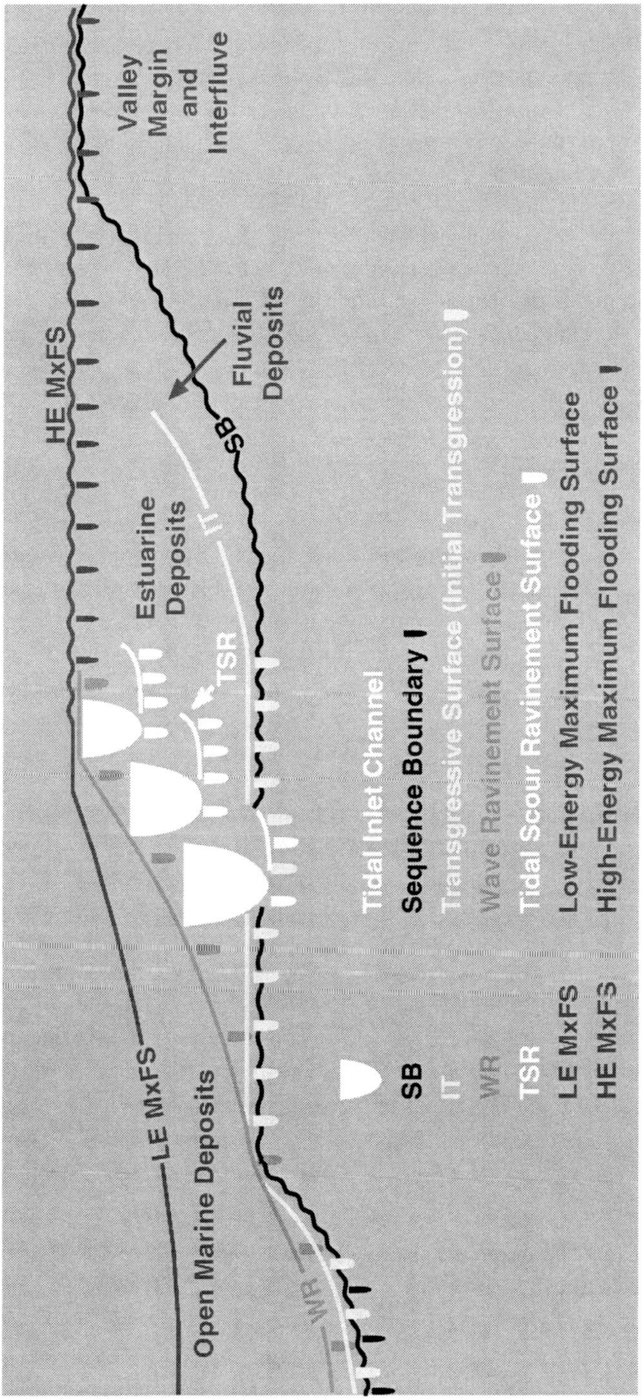


*Figure 13.10* *Schematic model of incised valley surface types commonly demarcated by the* Glossifungites *ichnofacies (modified after MacEachern and Pemberton, 1994).*

*Glossifungites* surfaces potentially alter the physical character of subsurface hydrocarbon reservoirs. The ability to characterize the resulting effective permeability provides a reservoir development tool that may be applied in a predictive or qualitative framework. However, the problem of determining effective flow parameters (such as permeability) in such highly heterogeneous media is difficult (Gingras *et al.*, 1999). This is an especially important consideration when heterogeneous elements, such as sand-filled burrows, are discrete and continuous. Such conduits essentially bypass the groundmass, controlling virtually all of the fluid flow parameters. Gingras *et al.* (1999) evaluated the impact that a *Glossifungites* surface might have on effective permeability. Key to this is the determination of the effective vertical permeability ($k_v$), and the effective horizontal permeability ($k_h$). Assessment of the effective permeability should be based on quantifiable characteristics of the bioturbated media, including the permeability of the matrix ($k_m$) and the burrows ($k_b$), trace fossil connectivity, and burrow density.

Both primary and diagenetic processes potentially alter or preserve higher porosity and permeability in and adjacent to trace fossils. This is contrary to the commonly held belief that bioturbation reduces overall permeability, and is the result of trace fossils locally introducing heterogeneity's to the rock matrix that improve the overall flow characteristics of that rock. *Glossifungites* surfaces provide a striking example of bioturbated substrates that potentially alter the flow characteristics of a reservoir.

Preliminary work indicates that such effects can be associated with both clastic and carbonate regimes. Research has been done on carbonate systems in the Cretaceous Shuaibia Formation in Qatar, and the Jurassic Ghawar Field in Saudi Arabia. In the example from Saudi Arabia the differences result in what is referred to as "super-perm".

In clastic systems, these concepts have been successfully applied in the Triassic Sag River Formation in Alaska. The top of the Sag River is a series of transgressive surfaces of erosion culminating in the deposition of the Kingak Shale (Fig. 13.11). These surfaces are characterized by distinct *Glossifungites* surfaces that are regionally correlatable. Detailed mini-permeameter work indicates that the matrix permeability is approximately 50 md as measured by core plugs. The burrows, however, display permeability 5 to 6 times (250–300 md) that of the matrix. This results in a dual permeability system that can have effects on deliverability, reserve calculations and secondary recovery considerations. In the case of the Sag River Formation the *Glossifungites* systems acted as conduits for enhanced fluid flows (Fig. 13.12). Measurements indicated that in the case of the *Glossifungites* surfaces at the top of the Sag River Formation the $K_v$ was greater than the $K_h$ and it was recommended that the zone be drilled and drained horizontally.

## 13.6 Conclusions

The main applications of ichnology to applied stratigraphy are twofold. The most obvious use lies in the demarcation of erosional discontinuities. To date, substrate-controlled

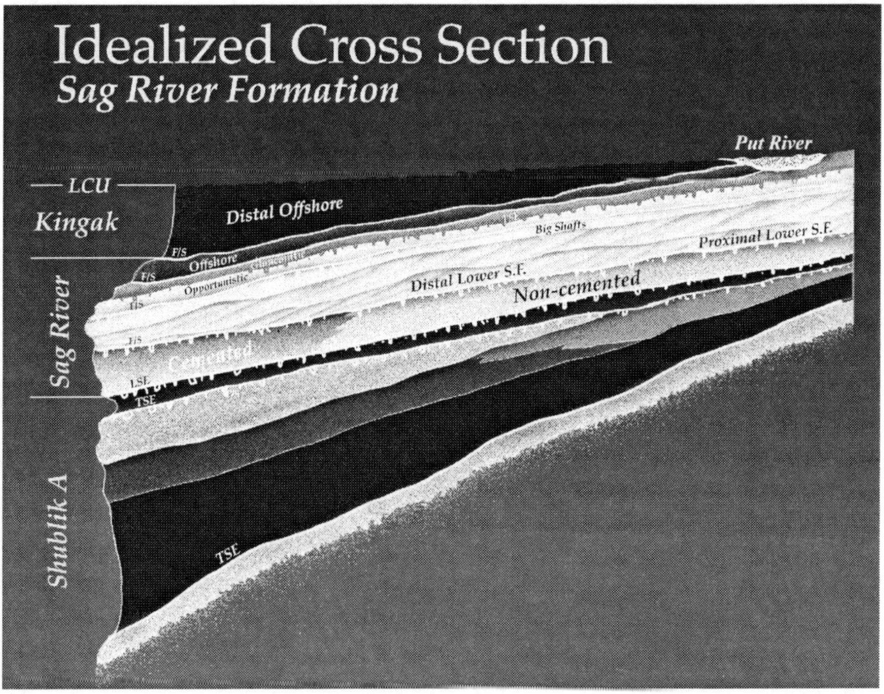


*Figure 13.11 Schematic representation of the relationship of the Sag River Formation to the Shublik and Kingak Formations. The cross-section shows the distribution of the internal surfaces and the interpreted depositional environments.*

ichnofacies have been under-utilized as a means of recognizing and mapping these stratigraphically important surfaces, both in outcrop and the subsurface. Locally, many surfaces are obvious on the basis of sedimentology alone; however, their appearance can change markedly across the study area, making correlation difficult. The *Glossifungites* ichnofacies is important in both the recognition and genetic interpretation of erosional discontinuities in marine-influenced siliciclastic intervals, as the many Cretaceous examples cited from the Western Canada Sedimentary Basin demonstrat. In many cases, the genetic interpretation of the discontinuity has come principally from the ichnofossil assemblage associated with the discontinuity and the overlying units. The continued integration of substrate-controlled ichnofacies with detailed stratigraphic and sedimentologic analysis will undoubtedly enhance and refine developing genetic stratigraphic paradigms.

The second use is subtler, and is concerned with trace fossil behaviour and their paleoenvironmental implications. Trace fossils, when used in conjunction with primary sedimentary structures, are useful in the delineation and interpretation of facies and facies associations. When these behavioral and substrate-controlled aspects of ichnology are integrated fully with other sedimentologic and stratigraphic analyses,

*Figure 13.12* Glossifungites *assemblage trace fossils (mainly the shrimp burrow* Thalassinoides) *characterize a number of surfaces in the Sag River Formation. Detailed mini-permeameter measurements indicate that in some instances these burrows have much greater permeability than the rock matrix and they may represent permeability conduits. Such systems would not generally show up on most permeability logs models. The* Glossifungites *assemblage burrows therefore could act as the conduits for enhanced fluid flow in this reservoir – see colour version of this figure in Appendix.*

the result is a powerful approach to the recognition and genetic interpretation of discontinuities in rock record.

## 13.7 Acknowledgements

We would like to thank Tom Saunders for his dynamic artwork depicted in Figs. 13.3, 13.4, 13.5, 13.11, and 13.12. The authors would also like to thank the Natural Science and Engineering Research Council of Canada (NSERC) for research funding. The senior author would like to thank the Canada Research Chairs program for their support of his research. The paper was improved through the efforts of reviewers Murray Gingras and John-Paul Zonneveld and we are grateful for their insight.

Chapter 14

# Cyclostratigraphy

MARTIN A. PERLMUTTER[1] and NILO C. DE AZAMBUJA FILHO[2]
[1] ChevronTexaco Energy Technology Company, 4800 Fournace Place, Bellaire Texas 77401, USA.
[2] PETROBRAS-CENPES, Cidade Universitária, Quadra 7, Ilha do Fundão, 21949-900 Rio de Janeiro, RJ, Brazil.





*E.A.M. Koutsoukos (ed.), Applied Stratigraphy,* 301-338.

## 14.1 Introduction

### 14.1.1 Definition

Cyclostratigraphy is the subdivision of geologic study that investigates cyclic depositional patterns. At present, there are two main areas of inquiry that have developed in cyclostratigraphy. One area focuses on understanding how climatic and sedimentary processes interact with tectonic processes to produce the observed stratigraphic cycles. The second focuses on the techniques of time measurement of cycles and the development of a geochronologic framework. In order to understand the causes and frequency of stratigraphic cycles both areas need to be integrated.

### 14.1.2 History

Although the stratigraphic record has been interpreted as having cyclic patterns for well over one hundred years and the terms cycle or cyclic stratigraphy have been used to describe these patterns for decades, the first use of the term cyclostratigraphy was relatively recent. Kalmykova *et al.* (1979) used the term in a paper that described its application to the study of Permian marine strata on the Russian Platform. Fischer (1986) later used the term to describe a method for the time correlation of stratigraphy using orbital cycles. Perlmutter and Matthews (1989) also used the term to describe a technique developed in the mid 1980's at Texaco's Research Center that used an understanding of the effects of orbital cycles on depositional systems and lake and sea level to predict lithostratigraphy.

### 14.1.3 The Geologic Processes that Cause Cyclicity

Three factors have been associated with the cause of stratigraphic cycles; sea level (or lake level) change, tectonism (uplift and subsidence), and climate change. Sea level change was probably the first factor to be recognized to have the potential to create repetitive sedimentary sequences. In 1874, Newberry suggested that the rise and fall of sea level produced fining upward sequences overlain by coarsening upward sequences. In 1936, Wanless and Shepherd proposed that glacioeustasy and associated climate cycles caused the cyclothems in the Paleozoic. In the recent past, the analytical tool known as sequence stratigraphy was developed by relying on sea level change as the dominant control of marine sedimentary processes and depositional patterns (Vail *et al.*, 1977a; Goldhammer *et al.*, 1987; Posamentier and Vail, 1988). Causes of global sea level change are associated with mid-oceanic ridge volume, a long-term tectonic effect, and ice mass at the poles, an effect that has both long-term and short-term impact, and can have tectonic and climatic components.

Variations in the uplift and subsidence of provenance areas and sedimentary basins have also been identified as fundamental causes of stratigraphic cycles (Williams, 1891; Twenhofel, 1932; Krynine, 1942; Weller, 1956; Bott and Johnson, 1967; Veizer and Jansen, 1985; Blair, 1986). These tectonically induced changes can alter rates of erosion and sediment transport and shift the distribution of depocenters. Cycles produced in this manner can range in frequency from tens of thousands of years to tens of millions of years, but are episodic, not regular, tending to occur randomly in time.

Climate change was first suggested, as a primary cause of sedimentation cycles, by Gilbert in 1895. This work was followed by many others who recognized that regularly occurring depositional cycles could be produced by recurrent climatic changes associated with periodic variations in the Earth's orbit around the Sun (Penck, 1914; Bradley, 1929; Twenhofel, 1932; Krynine, 1942; Garner, 1959; Van Houton, 1964; Crook, 1967; Young *et al.*, 1975; Fairbridge, 1976; Olsen, 1980; LeTourneau, 1985; Herbert and Fischer, 1986; Dean and Gardner, 1986; Perlmutter and Matthews, 1989). Climate cycles cause stratigraphic cycles by: (1) changing sea level by varying the volume of polar and alpine glaciers; (2) changing lake level by altering the local balance of precipitation and evaporation; (3) changing sediment production rates, grain size and mineralogy by varying the weathering conditions in the drainage basin; (4) changing sediment transport rates by altering the hydrodynamic conditions in the river system; and (5) changing and shifting the environments of deposition by altering the climate in the receiving basin.

## 14.2 Quantitative Analysis of Cycles

The periodic variations in the Earth's orbit around the Sun were first quantified by Milankovitch in 1941. These cycles, now called Milankovitch cycles, have three main components related to aspects of the Earth's orbit and tilt that affect the receipt and distribution of heat from the Sun; eccentricity, obliquity and precession. Eccentricity, the variation of the ellipticity of the Earth's orbit around the Sun, has two primary frequencies, about 413 kyr and about 100 kyr. Obliquity, the variation in the amount of tilt of the Earth's axis, has one primary frequency, about 41 kyr. Precession of the Equinoxes, the progressive change in when seasons occur relative to the position of the Earth's orbit around the Sun, has two primary frequencies, 19 kyr and 23 kyr. The interference patterns of all these cycles can produce a long-term climate cycle that varies but can be up to 2.4 myr long (Matthews and Frohlich, 1991). Milankovitch cycles are discussed in more detail in the section on climate.

The objectives of quantitative cyclostratigraphic analyses are to identify and reconstruct the astronomical periodicities responsible for observed stratigraphic patterns. Commonly, this is done by evaluating changes in bed thickness and composition of the components of a cycle within a calibrated timeframe. Statistical tools have been developed to facilitate these correlations and to determine the chronology of the cycles under investigation. Schwarzacher (1993) presents an excellent review of the statistical and mathematical techniques that are commonly used to evaluate cycles, including spectral analysis. He also discusses the data requirements for analysis and problems inherently related to the conversion of bed thickness into time. Lacunarity, a relatively new technique, is a multi-scaled measure of translational invariance or the extent to which a pattern varies by location (Plotnick *et al.*, 1993, 1996). Lacunarity has been successfully used to evaluate component frequencies of cyclic sequences and the relative contribution of each frequency to the variability of the cycles (Perlmutter and Plotnick, 1997, 2002).

The most critical aspects affecting quantitative cycle analysis are the completeness of the stratigraphic record under investigation and density of the sampling regime. A record in which cycle components have not been related properly to the processes that are forcing the changes in thickness or lithology over the span of a cycle, or by a record that contains many difficult-to-distinguish erosional or hiatal surfaces, can cause erroneous evaluations. In addition, flawed answers can result from improper resolution of the stratigraphic system caused by a sampling interval that is greater than the variability of the cycles themselves. Great care must be taken to ensure the viability of the dataset.

To extract time and periodicity properly from sedimentary cycles, it is important to understand the causes of their origin. This is accomplished by identifying the significant

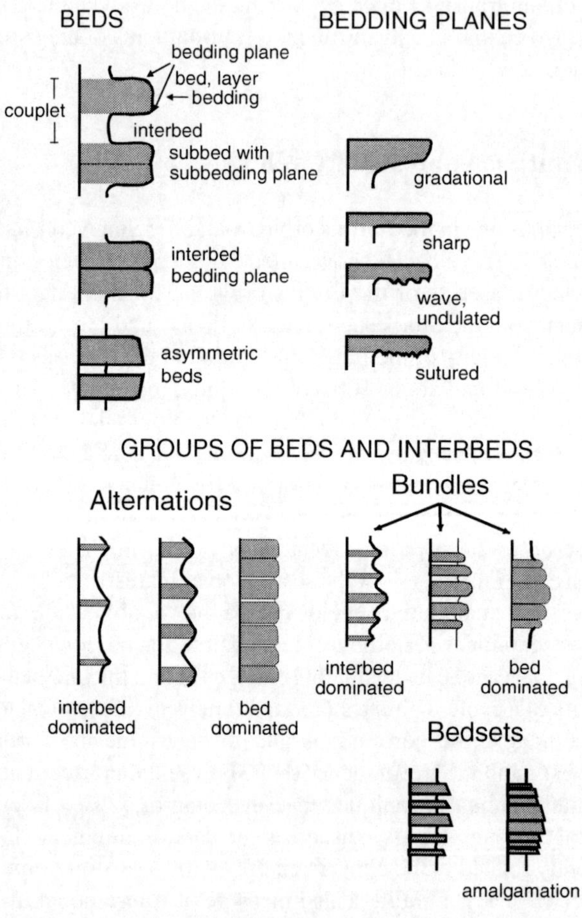


*Figure 14.1 Descriptive terms for bedding phenomena from Einsele et al. (1991).*

sedimentary characteristics of the cycles and associating these characteristics with the sedimentary processes that created them. For example, to assist this analysis, Einsele *et al.* (1991) identified sedimentary bedding types and proposed that these bed types be classified into groups based primarily on bed thickness and cyclicity (Figs. 14.1, 14.2, and 14.3).

Another important issue that needs to be taken into account is the restricted lateral extent of the data commonly available for the analysis of cyclicity, such as wells, cores

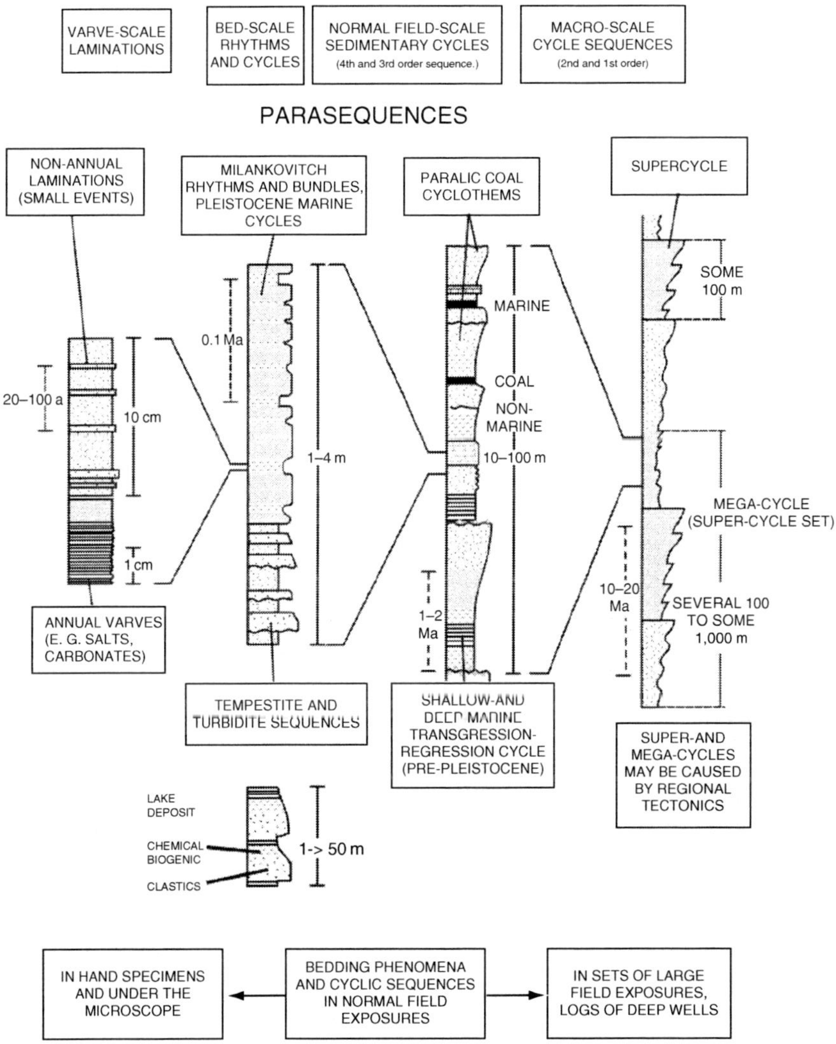


*Figure 14.2 Nomenclature used in Einsele et al. (1991) to describe rhythmic and cyclic sedimentary sequences of different origins, thicknesses and time periods in the field. Orders of cycles in parentheses are after Vail et al. (1977a) and Haq et al. (1987). The term megacycle is used for "supercycle set" in Haq et al. (1987).*

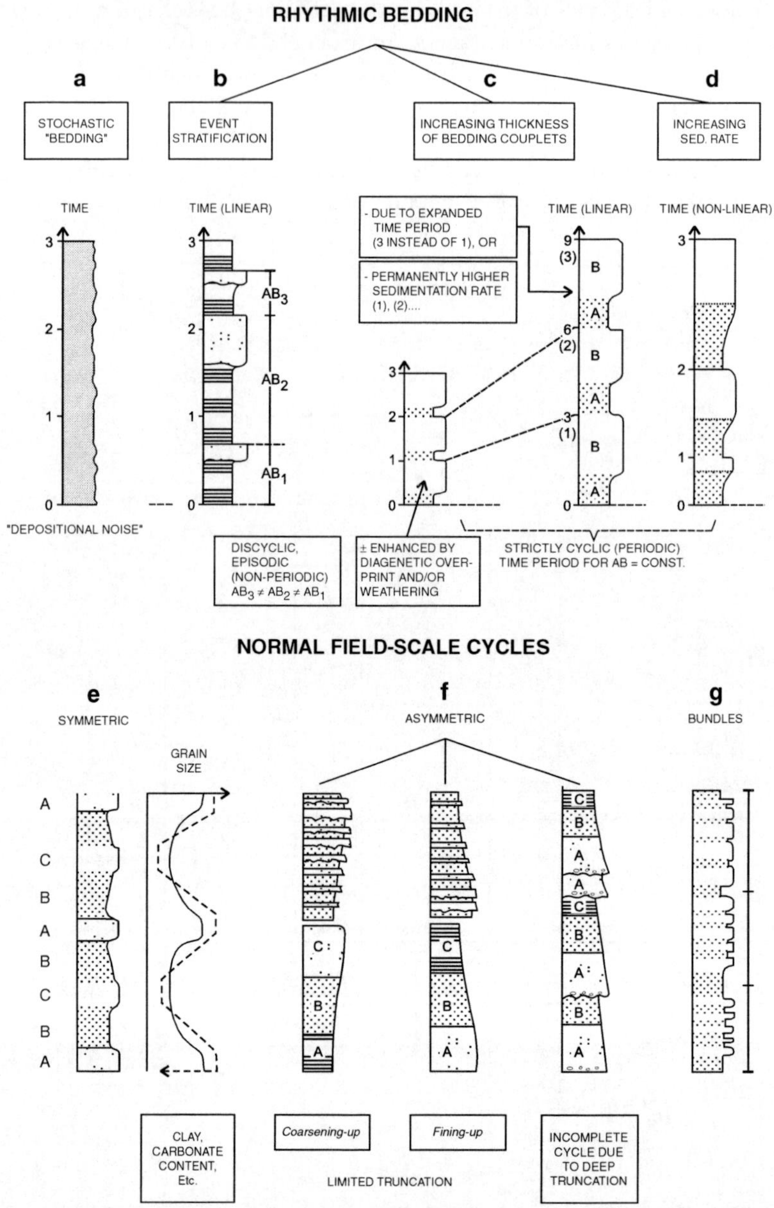


*Figure 14.3 Nomenclature used in Einsele et al. (1991) to describe stochastic bedding (a), and different types of rhythmic (b–d) and cyclic (e–g) sedimentary successions. All sequences may be either strictly cyclic or periodic (c and d) or non-periodic (b). The thicknesses of corresponding bed types can change from being cyclic due to either variations in the time period (c) of succeeding cycles or in the rates of sedimentation (c and d). Rhythmic bedding caused by depositional events (b) may show similar alternations as cyclic sequences but they are always non-cyclic or non-periodic.*

or outcrops. It is crucial to integrate additional data, such as seismic sections, to complement the understanding of lateral variations.

## 14.3 The Impact of the Resolution of Data and Tools on Interpretation

We are going to focus our discussion on sequences controlled by allocyclic forces (those generated from outside the sedimentary system such as orbital cyclicity and climate change) as opposed to autocyclic forces (those operating within a sedimentary system such as river avulsion, turbidite flows, or storms). The time span recorded in a sedimentary cycle caused by orbital cycles can vary due to regional or local factors such as climate succession. It is, therefore, important to identify these variations in order to make an accurate evaluation of the time span and frequencies of a cycle. If this is done properly, cyclostratigraphy can then be used for geochronology.

However, determining depositional variations within a cycle is not always straightforward. For example, changes in accumulation rates can cause mistakes in the interpretation, especially when rates are interpreted from wire-line logs. The inherent variation in bed thickness, combined with the limits of the vertical resolution of the logging tool itself, can cause the interpreter to inadvertently combine cycles of different periodicities. The lack of resolution of the tool thus produces errors by undersampling the sequence, generating cycle frequencies that are not linked with the processes that caused them (Davis, 1986; Fig. 14.4). To eliminate this problem, higher resolution tools need to be used. Alternatively, a technique can be applied that consists of building a matrix in which possible periodicities are plotted against possible bed or cycle thickness (vertical resolution) at varying accumulation rates. This matrix, compared with average accumulation rates of the total analyzed succession, can indicate intervals that could produce interpretation errors related to the vertical resolution (Table 14.1).

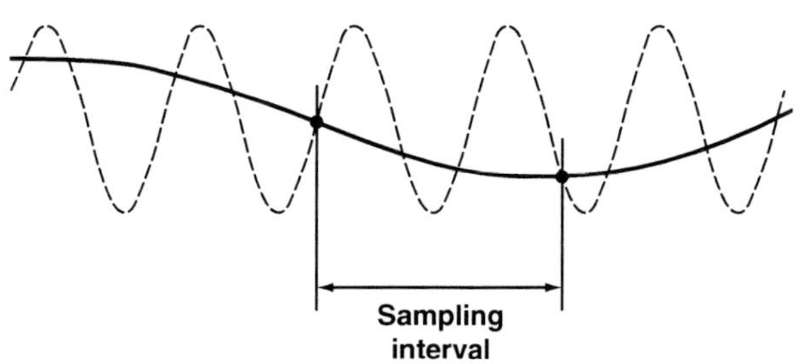


*Figure 14.4 Sinusoidal wave of high frequency (dashed line) sampled at discrete intervals, creating a new wave with an apparent lower frequency (solid line). The frequency of the original line is above the Nyquist frequency for that sampling interval.*

*Table 14.1 Matrix comparing accumulation rates, periodicities and cycle thickness*

| Periods | PERIODICITIES kyr | | | | | | | | | | |
|---|---|---|---|---|---|---|---|---|---|---|---|
| | Milankovitch Band kyr | | | | | Solar cycles | | | | | El Niño |
| | Long Eccentricity | Short Eccentricity | Obliquity | Precession | Precession | King-Hale | Gleiss-berg | Double Hale | Hale | Shwabe | |
| | 410 | 100 | 41 | 23 | 19 | 180 yr | 80 yr | 44 yr | 22 yr | 11 yr | 5 yr |
| Rates cm/kyr | Cycle Thickness in m | | | | | | | | | | |
| 1 | 4.1 | 1 | 0.4 | 0.2 | 0.2 | 0.0018 | 0.0008 | 0.00044 | 0.00022 | 0.00011 | 0.00005 |
| 2 | 8.2 | 2 | 0.8 | 0.5 | 0.4 | 0.0036 | 0.0016 | 0.00088 | 0.00044 | 0.00022 | 0.0001 |
| 5 | 20.5 | 5 | 2.1 | 1.2 | 1 | 0.009 | 0.004 | 0.0022 | 0.0011 | 0.00055 | 0.00025 |
| 10 | 41 | 10 | 4.1 | 2.3 | 1.9 | 0.018 | 0.008 | 0.0044 | 0.0022 | 0.0011 | 0.0005 |
| 15 | 61.5 | 15 | 6.2 | 3.5 | 2.9 | 0.027 | 0.012 | 0.0066 | 0.0033 | 0.00165 | 0.00075 |
| 20 | 82 | 20 | 8.2 | 4.6 | 3.8 | 0.036 | 0.016 | 0.0088 | 0.0044 | 0.0022 | 0.001 |
| 30 | 123 | 30 | 12 | 6.9 | 5.7 | 0.054 | 0.024 | 0.0132 | 0.0066 | 0.0033 | 0.0015 |
| 50 | 205 | 50 | 21 | 12 | 9.5 | 0.09 | 0.04 | 0.022 | 0.011 | 0.0055 | 0.0025 |
| 100 | 410 | 100 | 41 | 23 | 19 | 0.18 | 0.08 | 0.044 | 0.022 | 0.011 | 0.005 |
| 1000 | 4100 | 1000 | 410 | 230 | 190 | 1.8 | 0.8 | 0.44 | 0.22 | 0.11 | 0.05 |
| 2000 | 8200 | 2000 | 820 | 460 | 380 | 3.6 | 1.6 | 0.88 | 0.44 | 0.22 | 0.1 |

traditional log resolution (GR, resistivity, Sonic, Density)

## 14.4 Sedimentation Rates *vs*. Accumulation Rates

The direct use of modern sedimentation rates to estimate accumulation rates in stratigraphic sequences is not a straightforward exercise because the intervals of accumulation include times of erosion or non-deposition. In general, the longer the time-span of a stratigraphic interval, the greater the number of periods of erosion and non-deposition that is included in that interval. As a consequence, modern deposition rates may be significantly higher than ancient ones (Kukal, 1990; Sadler and Strauss, 1990). A method to evaluate the completeness of a stratigraphic section was proposed by Schwarzacher (1993). He defines completeness as $C(n) = 1 - U/n$, where $C(n)$ is the completeness of the stratigraphic record related to the total time, U is the number of stratal units without deposition and n is the time interval. In order to simplify the analysis, deposition rates are assumed to be non-varying when this equation is applied. This assumption may not be accurate. A way to estimate possible errors associated with these analyses is to use the random walk model proposed by Schwarzacher (1993).

## 14.5 Tools and Techniques Used to Evaluate Cyclicity

### 14.5.1 Spectral Analysis

Once the completeness of the stratigraphic section is understood, either time series analysis or power-spectral density analysis may be used to quantify the periodicity of

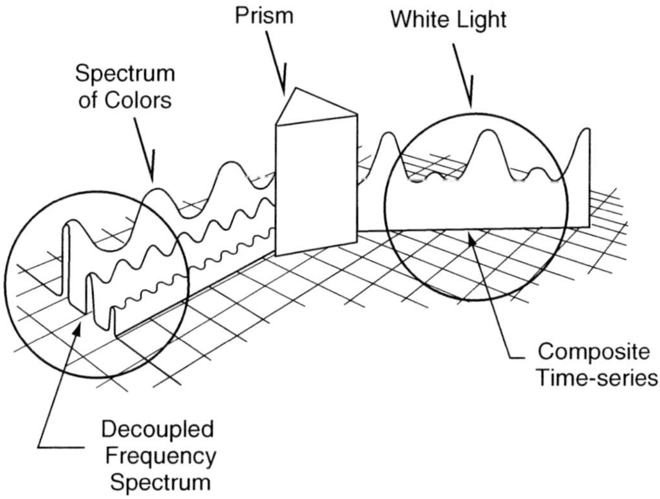


*Figure 14.5 A wave composed of many frequencies can be deconvoluted into its component waves. This is similar in principle to the way prisms break down white light into a spectrum of colors (Davis, 1986).*

a sequence. In time series, cycles in the sedimentary record are analyzed for their harmonics or component cycles (Chatfield, 1989). Spectral analysis is essentially a modification of Fourier analysis to make it suitable for stochastic values which can be only partially predicted by past values, rather than deterministic functions of time in which a time series can be predicted (Schwarzacher, 1975). The recognition of hidden cycles by spectral analysis has been compared to the way a prism reveals the component colors of white light (Davis, 1986; Fig. 14.5).

### 14.5.2 Application of Spectral Analysis in Sedimentary Sequences

There have been many studies that link stratigraphic cycles to climate oscillations and orbital cycles, as well as to higher frequency changes such as solar and tidal cycles. Spectral analysis of Pleistocene sediments has indeed shown that climatic oscillations are linked to the Earth's orbit, and can range from thousands to millions of years.

As indicated previously, however, care must be taken when applying spectral analysis in sedimentary successions because thickness is commonly substituted for time. Most sedimentary processes do not produce continuous accumulations, causing a distortion in the conversion of cycle thickness into time. In addition, changes in weathering and depositional conditions over a climate cycle can produce variations in lithology. These lithologic changes may be very significant depending on the location and may cause the number of beds in a cycle to be misinterpreted (Perlmutter and Matthews, 1989). The selection of sections with rhythmic bedding is very important to avoid misinterpretations, sections not showing this property will be very prone to lead the interpreter to the wrong results.

One method to convert between thickness and time is to reconstruct the astronomical periodicities responsible for observed stratigraphic patterns. This can be done as follows: (1) Select a stratigraphic interval in a well that has sedimentary cycles of relatively uniform thickness (Fig. 14.6). If there are many variations in cycle thickness in the well, divide the section into shorter intervals; (2) Estimate the time interval of the section under investigation using biostratigraphy; (3) Determine the average accumulation rate of the interval based on biostratigraphic data; (4) Determine the vertical resolution of the tool used to measure the data and compare with the cycles expected for the interval based on the average accumulation rate calculated in item 3 (Table 14.1); (5) Take samples at uniform intervals; (6) Calculate sedimentary cycle thickness with spectral analysis by dividing the most significant frequency values by the sampling interval (Fig. 14.7); (7) Build a matrix of thickness and compare it with the matrix composed of periodicities of the Milankovitch bands (Tables 14.2 and 14.3). Select only the values of the matrix that have similarities greater than 90% to 95% to be more selective; (8) Plot the selected ratios and corresponding thickness–Milankovitch periods against the corresponding average accumulation rates of the analyzed interval; (9) Choose those with the best statistical fit. This implies having cycles with similar ranges of sedimentation rates (Table 14.4); (10) Count the number of cycles in the interval and multiply by the periodicity to establish the preserved time span of the analyzed section. An unidentified, but significant hiatus in the interval will create a discrepancy in the accumulation rates obtained from the Milankovitch band compared with the one calculated only based only on biostratigraphy.

Another type of discrepancy can occur caused by the difference between the thickness of the calculated long eccentricity cycle (18.66 m) relative to the thickness of this cycle in the log (Figure 14.6) (21 m), and the corresponding accumulation rate in (Table 14.4). This inconsistency is explained by the fact that the total analyzed interval is not long enough to get a sufficient number of data points for the spectral analyses to properly identify the 21 m cycle, which would correspond to an accumulation rate of 5.08 instead of 4.52, closer to those of the other cycles. Selecting a larger interval to be analyzed would solve this discrepancy.

### 14.5.3 Frequency and Amplitude in Sedimentary and Stratigraphic Models

A sequence stratigraphic model proposed by Mitchum and Van Wagoner (1991; Fig. 14.8) uses the interaction of eustasy and subsidence in order to form parasequences and high frequency sequences. This model erroneously implies that the amplitudes of the higher frequency cycles are always smaller than the lower frequency cycles. However, it is well documented from the Pleistocene record of oxygen isotopes and the sea level change that higher frequency cycles can also have very high amplitudes (Imbrie and Imbrie, 1979). Data indicate that: (1) over the last 2 myr,

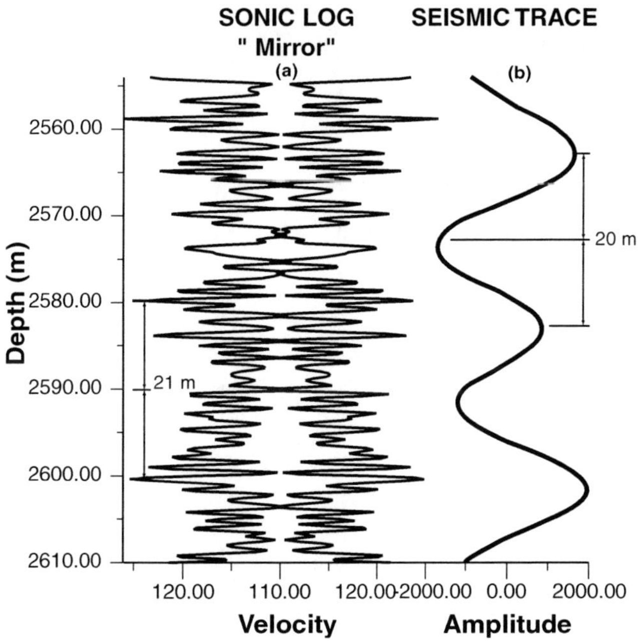


*Figure 14.6 Selected interval from a well showing "mirror" sonic log (a) and an amplitude seismic trace (b). Note the well-defined cyclicity in both records (modified from Castro, 1999).*

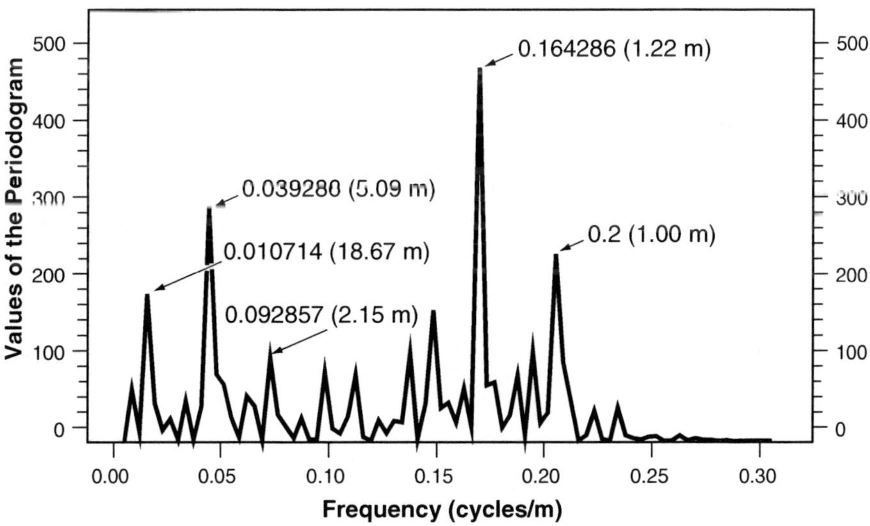


*Figure 14.7 Periodogram showing the main frequencies of the spectral analysis of the sonic log of the interval from Figure 14.6 and corresponding cycle thickness of a segment of the Petrobras nannofossil biozone N-560a (lower Miocene).*

*Table 14.2 Matrix showing the ratios of cycle thickness in meters. Adjacent to the ratios the corresponding higher similarities (>95%) of the Milankovitch ratios are shown in brackets*

**MILANKOVITCH MATRIX**

| ka | 410 | 100 | 41 | 23 | 19 |
|---|---|---|---|---|---|
| 410 | 1.00 | 4.10 | 10.00 | 17.83 | 21.58 |
| 100 | | 1.00 | 2.44 | 4.35 | 5.26 |
| 41 | | | 1.00 | 1.78 | 1.78 |
| 23 | | | | 1.00 | 1.21 |
| 19 | | | | | 1.00 |

*Table 14.3 Matrix showing the ratios of the Milankovitch periodicities*

| m | 18.66 | 5.09 | 2.15 | 1.21 | 1 |
|---|---|---|---|---|---|
| 18.66 | 1.00 | 3.66 (4.10) | 8.68 (10.00) | 15.42 (17.83) | 18.66 (21.58) |
| 5.09 | | 1.00 | 2.36 (2.44) | 4.20 (4.35) | 5.09 (5.26) |
| 2.15 | | | 1.00 | 1.77 (1.78) | 2.15 (2.16) |
| 1.21 | | | | 1.00 | 1.21 (1.21) |
| 1 | | | | | 1.00 |

sea level repeatedly rose and fell tens of meters at a cycle frequency of about 20 kyr (precession cycle); and (2) the largest rises in sea level, over a hundred meters, were actually the most rapid, occurring during the warming phase of a precession cycle associated with the shift from glacial to interglacial. What the sedimentary record actually shows is that 3rd order cycles are composed of bundles of these high

*Table 14.4 Correlation of cycle thickness with Milankovitch periodicities and corresponding accumulation rates.*

| Sedimentary cycle thickness (m) | Corresponding Milankovitch period (ka) | Accumulation rate (cm/ka) |
|---|---|---|
| 18.66 | 410 | 4.55 |
| 5.09 | 100 | 5.09 |
| 2.15 | 41 | 5.25 |
| 1.21 | 23 | 5.29 |
| 1.00 | 19 | 5.26 |

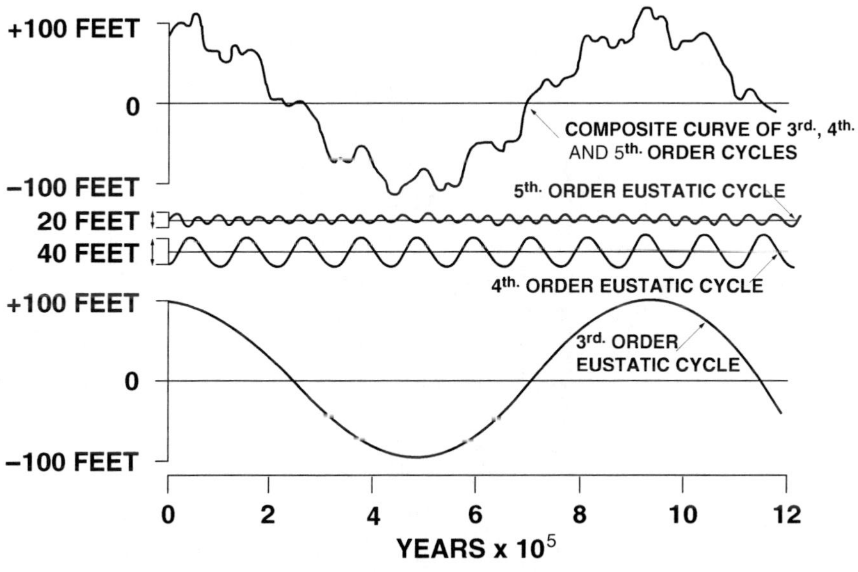


*Figure 14.8 Model of a composed curve of relative sea level to produce parasequences using different eustatic frequencies and constant subsidence in a basin (modified from Michum and Van Wagoner, 1991). This model is discussed in the text.*

frequency cycles, rather than single, lower frequency phenomenon with high amplitude. However, due to the low vertical resolution of the seismic data available for the studies, the shifts in the sedimentary record are interpreted as 3rd order cycles or sequences.

## 14.6 Applications

### 14.6.1 Campos Basin: The Cyclicity of the Oligocene–Miocene Section

The Oligocene–Miocene section of the Campos Basin, offshore Brazil, consists mostly of marlstones and mudstones with some chalk and turbidite layers, deposited in a bathyal environment (Fig. 14.9 – Oligocene). Part of this sedimentary section (lower Miocene) was used to illustrate the methodology discussed above in Tables 14.2–4 and Figures 14.6 and 14.7.

An interval of the Oligocene, composed of rhythmically intercalated marlstones and mudstones was cored (Fig. 14.9). The interval shows a well-defined variability in the percentage of the $CaCO_3$ content based on samples collected at 5 cm spacing (Fig. 14.10). The spectral analyses performed on these variations showed $CaCO_3$ cycle thicknesses of 20–25 cm, 42–50 cm, 120 cm and 420 cm. Assuming an average accumulation rate of 1–1.2 cm/kyr for the Oligocene section in this area, the conversion of cycle thickness into time displays good agreement with Milankovitch periodicities (Fig. 14.11). These carbonate cycles are primarily related to the

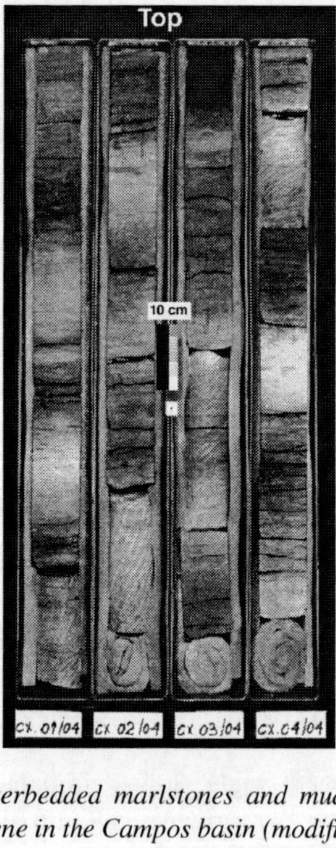


*Figure 14.9 Rhythmically interbedded marlstones and mudstones deposited in a bathyal environment during the Oligocene in the Campos basin (modified from Azambuja Filho, 1990). Note that the carbonate–mudstone cycle is around 20–25 cm thick. The average regional accumulation rate for this interval is 1 to 1.2 cm/kyr. The vertical scale is 10 cm long.*

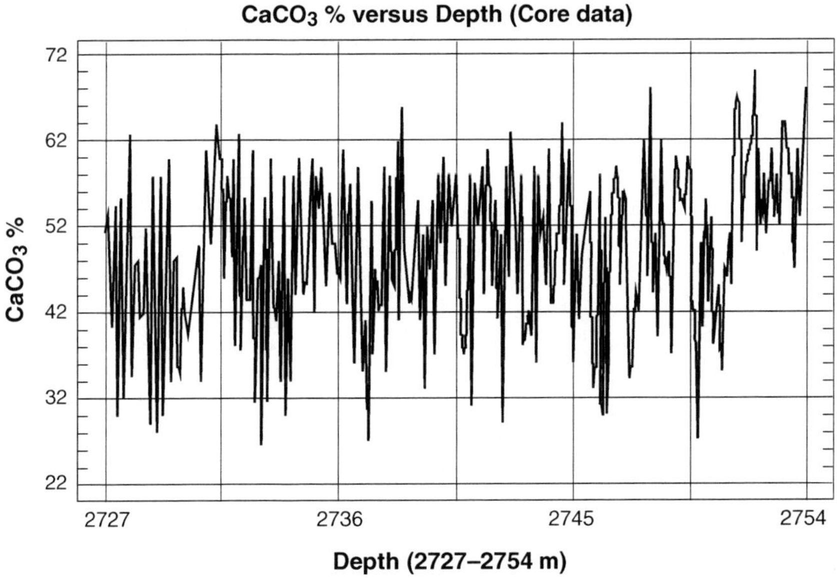


*Figure 14.10* *Percent carbonate of the cored interval, shown in Figure 14.9, plotted versus depth. Note two well defined frequencies or cycles. The highest frequency corresponds to centimeter-scale alternations or couplets of approximately 25 cm thick, defined by color variations from lighter to darker in the core. The carbonate content is related to the relative abundance of nannofossils and foraminifer related to mudstone (modified from Azambuja Filho, 1990).*

eccentricity and precession, while obliquity was only locally important (Azambuja Filho, 1990; Azambuja Filho and Azevedo, 1995).

Similar analyses were also carried out on the same Oligocene–Miocene sedimentary section making use of a seismic amplitude trace processed at a sampling interval of 2 milliseconds (Castro, 1999). These data were compared to a sonic log from a nearby well (Figs. 14.6, 14.12, and 14.13). There is a general increase in the accumulation rates vertically from the lower Oligocene up to the present. This is indicated by the progradational feature clearly shown in the seismic section and is further established by comparing accumulation rates from cored intervals and logs from the Oligocene (Fig. 14.9 – accumulation rate of 1 cm/kyr) and the Miocene (Fig. 14.14 and Table 14.4 – accumulation rate of 5.2 cm/kyr). The marlstone–mudstone couplets in the Miocene are much thicker than those of the Oligocene as can be seen in the Figure 14.14. High accumulation rates in the Miocene, combined with relatively low interval velocities, allowed good vertical seismic resolution, making it possible to identify periodic oscillations in the $CaCO_3$ content and their link to the orbital cycles (Table 14.4). Both the 100 and 413 kyr eccentricity cycles were identified by combining seismic and sonic logs (Figs. 14.6 and 14.13). The sonic log additionally allowed the identification of the precession and obliquity cycles due to its higher vertical

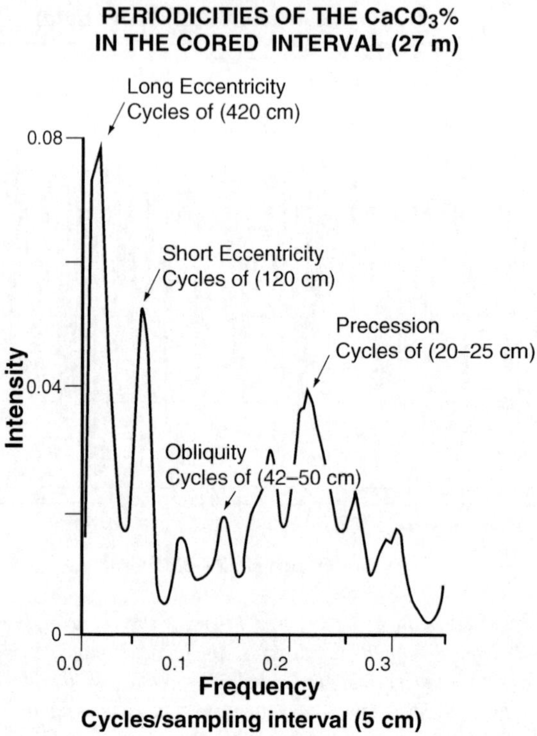


*Figure 14.11 Periodogram showing the dominant cycle frequencies and corresponding Milankovitch periodicities of the cored interval. The conversion of cycle thickness into time was performed by dividing the value of the main cycle frequencies by the sampling interval (5 cm).*

resolution. These results indicate that seismic is a potential tool for cyclostratigraphic analyses when sedimentary successions have very high accumulation rates and low interval velocities.

This type of cyclostratigraphic analysis was also applied to refine the geochronology of the Oligocene–Miocene turbidite systems existing in the Campos Basin. It indicated that these turbidite deposits were generated in well-defined depositional stages. Each stage is characterized by sudden deposition of sand with a sharp termination. This is expressed as a box-like wire-line log pattern (Fig. 14.15a). Subsequently, each system was overlain by the deposition of pelagic and hemipelagic sediments represented by cycles of marlstone and mudstones. The frequencies of variability of the turbidite systems were compatible with that of Milankovitch cycles. Within the pelagic and hemipelagic sediments interval minor turbidite events can be identified. They have occurred with a periodicity of about 400–500 kya (Fig. 14.15b), while the major turbidite events occurred within time intervals on the order of about 1.2–1.5 Ma

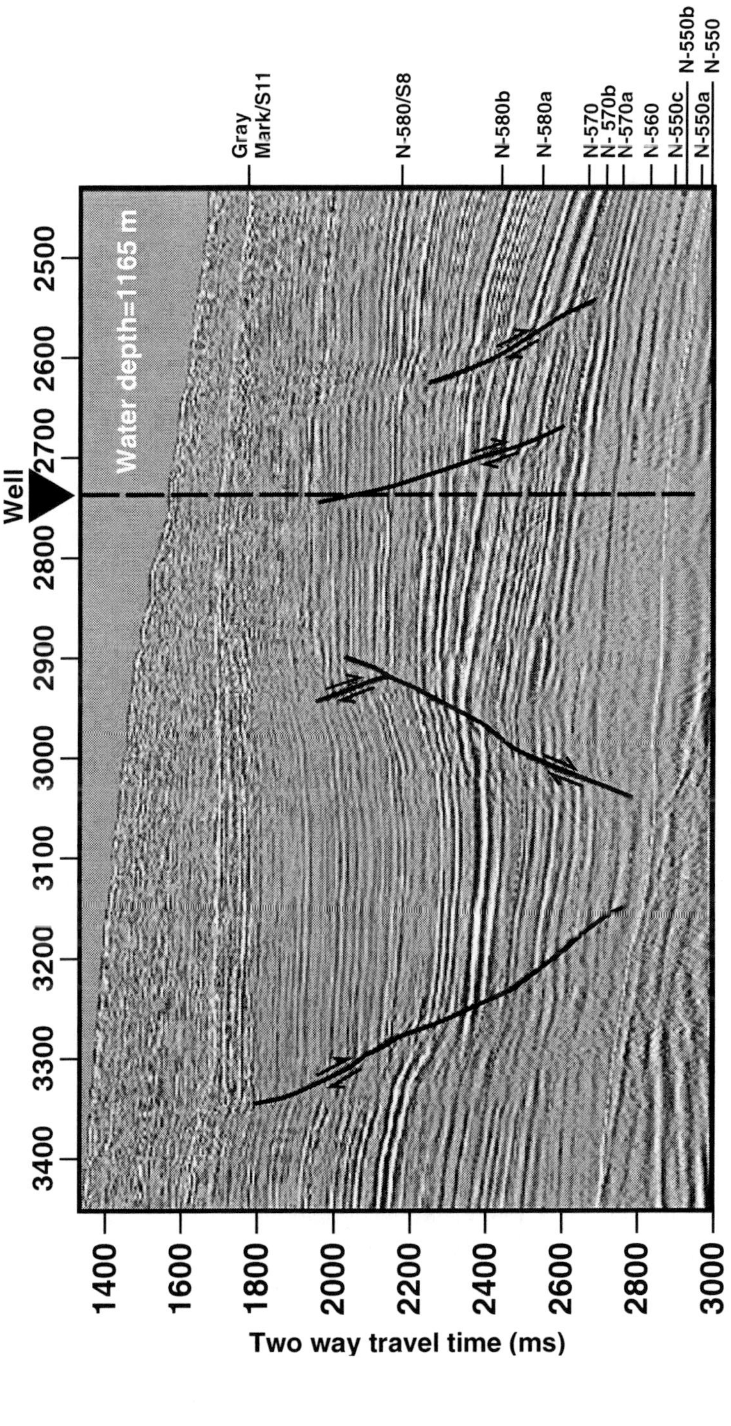


*Figure 14.12* Seismic section showing the well location and the dominant progradational feature of the Miocene section. N-560 – Petrobras nannofossil zonation (Castro, 1999).

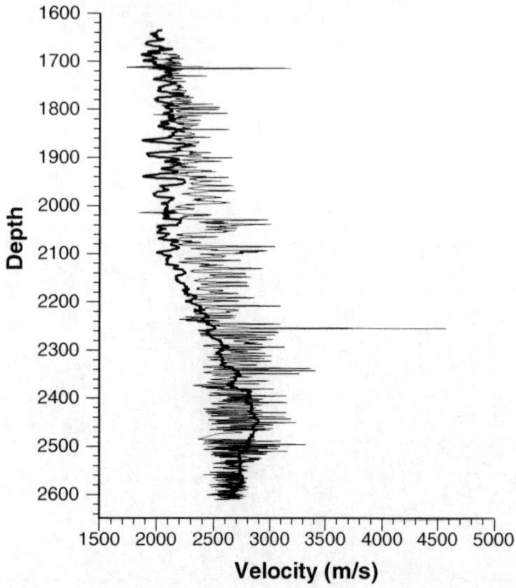


*Figure 14.13* *Comparison of sonic log frequencies with the nearby seismic amplitude trace. In the lower part of the well, below 2200 m, there is a substantial decrease in the seismic frequencies due to a velocity increase caused by an increase in the $CaCO_3$ content. This was due to a decrease in the accumulation rates (modified from Castro, 1999).*

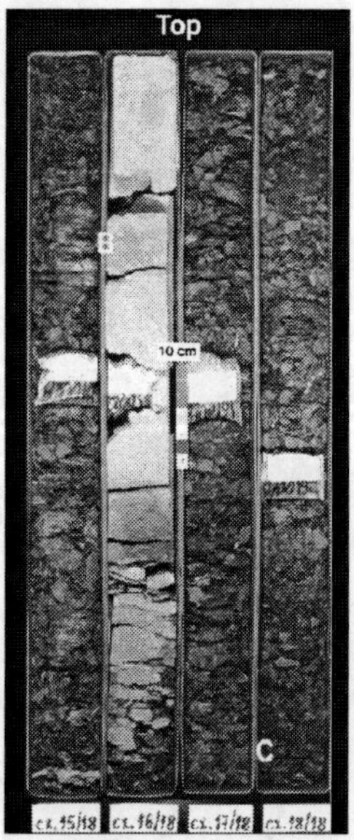


*Figure 14.14* *Lower Miocene upper bathyal mudstones and marlstones. Note the increase in the marlstone–mudstone cycle thickness compared to Oligocene (Fig. 14.9). This is due to an increase in the accumulation rates mainly in mudstones. The vertical scale is 10 cm.*

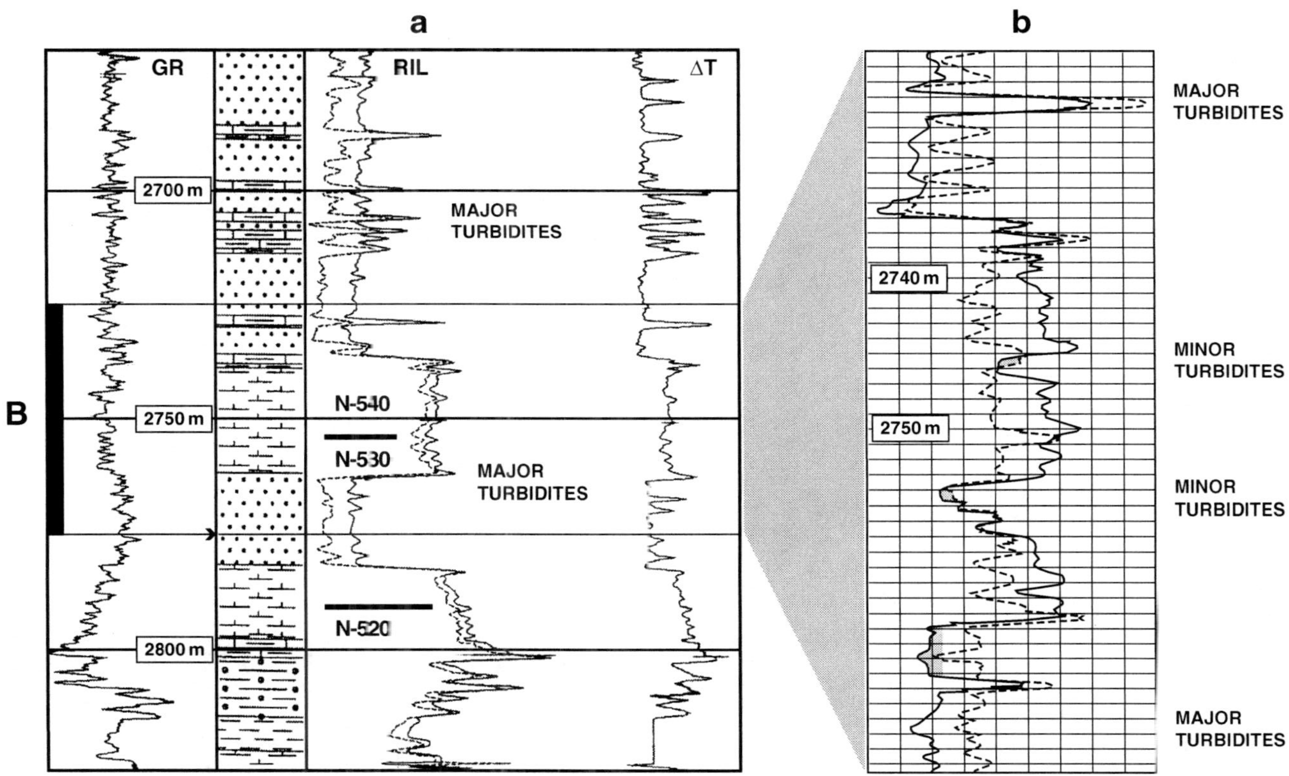


*Figure 14.15* *(a) Box-like wire-line log pattern, mainly in the resistivity and sonic, representing the major Oligocene turbidite events (modified from Azambuja Filho, 1990). (b) Detail of the density-neutron logs showing the existence of minor turbidite events within the hemipelagic sedimentation (modified from Azambuja Filho, 1990).*

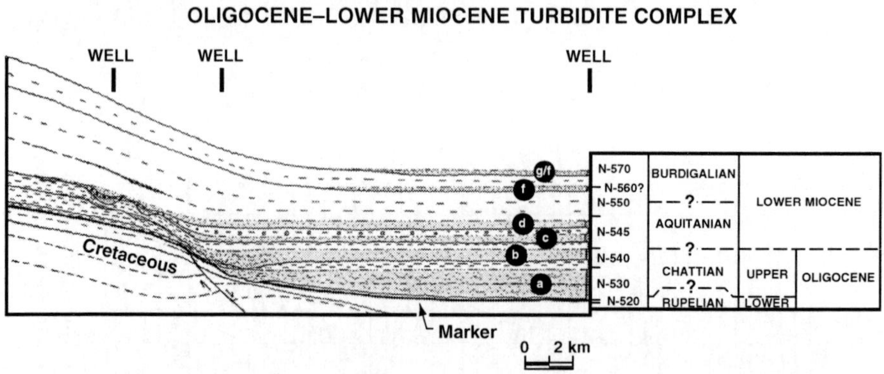


*Figure 14.16 Schematic representation of the main turbidite systems of the Oligocene and early Miocene age, in the Campos basin (modified from Azambuja Filho, 1990).*

(Azambuja Filho, 1990). During the Oligocene–Miocene of the Campos Basin the recurrence of this type of events were recorded as a series of sand-rich turbidites, which are schematically represented in Figure 14.16.

## 14.7 Climate

### 14.7.1 Milankovitch Orbital Oscillations: The Core of Climate Cycles

Climate cycles are produced by the interaction of the Earth's orbital parameters, eccentricity, precession of the equinoxes, and obliquity, which periodically alter the seasonal distribution of insolation (Milankovitch, 1941; Berger, 1980; Lockwood, 1980; Fischer, 1986; Kukla and Gavin, 1992; Perlmutter and Plotnick, 2002; Fig. 14.17). Eccentricity causes the orbit of the Earth around the Sun to periodically vary from elliptical to almost circular. When the orbit is elliptical, the distance from the Earth to the Sun varies throughout the year. Precession progressively shifts where in the Earth's orbit the closest (perihelion) and farthest (aphelion) distance to the Sun occurs. In other words, the time of the year when perihelion and aphelion occur, changes. Presently, the Earth is closest to the Sun in December and farthest in June. The impact of the precession decreases and disappears as eccentricity nears zero. The result is that eccentricity bundles precession cycles by causing minimal or missed precession-related changes when the Earth's orbit becomes circular (Fig. 14.18). Obliquity causes seasonality to vary by altering the tilt of the Earth's pole of rotation, the higher the tilt the greater the seasonality. Obliquity also causes the seasons of the Northern and Southern Hemispheres to be six months out of phase. When it is summer in one hemisphere, it is winter in the other. Combined, eccentricity and obliquity cause the effects of precession to be out of phase between the hemispheres as well (Fig. 14.19). For example, when perihelion occurs during the Northern Hemisphere summer (June), aphelion will occur six months later during the Southern Hemisphere summer (December). This is important to understand because this means that when the Earth's

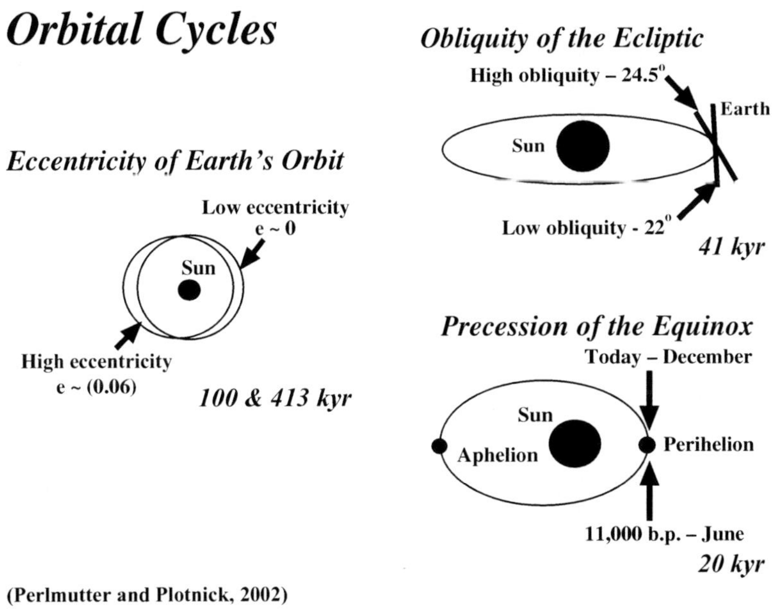


*Figure 14.17 Orbital cycles.*

orbit is eccentric, warm summers (perihelion in summer) occur with cold winters (aphelion in winter) in one hemisphere, while in the opposite hemisphere moderate summers (aphelion in summer) occur with moderate winters (perihelion in winter). This hemispheric pattern gradually reverses itself over the course of a precession cycle. The obliquity cycle may become dominant as eccentricity approaches zero.

The primary harmonics of the eccentricity cycle have periods of around 100 kyr and 413 kyr; precession, around 19 kyr and 23 kyr; and obliquity, around 41 kyr. Constructive and destructive interference of these primary cycles causes longer-term variations in insolation on the order of 1.6–2.4 myr in duration (Matthews and Frohlich, 1991). Equations to calculate orbital cycles for the last 10 million years were developed by Berger (1978) and Berger and Loutre (1988). When seasonal insolation cycles are plotted, it becomes apparent that the largest changes in insolation, up to 30%, occur at the time scale of precession (Fig 14.19). Although these equations are rigorous for last 10 million years only, cyclic variations in insolation have occurred throughout Earth history (Berger *et al.*, 1992). It is important to recognize, however, that the lengths of cycle frequencies and the amplitudes of Milankovitch cycles have gradually varied over geologic time (Berger and Loutre, 1988).

### 14.7.2 Climatic Response to Orbital Cycles

Changes in seasonal insolation can alter regional temperature patterns. The atmosphere accommodates changes in heating by adjusting the distribution of atmospheric pressure cells and circulation patterns. Combined, the changes in temperature, pressure and

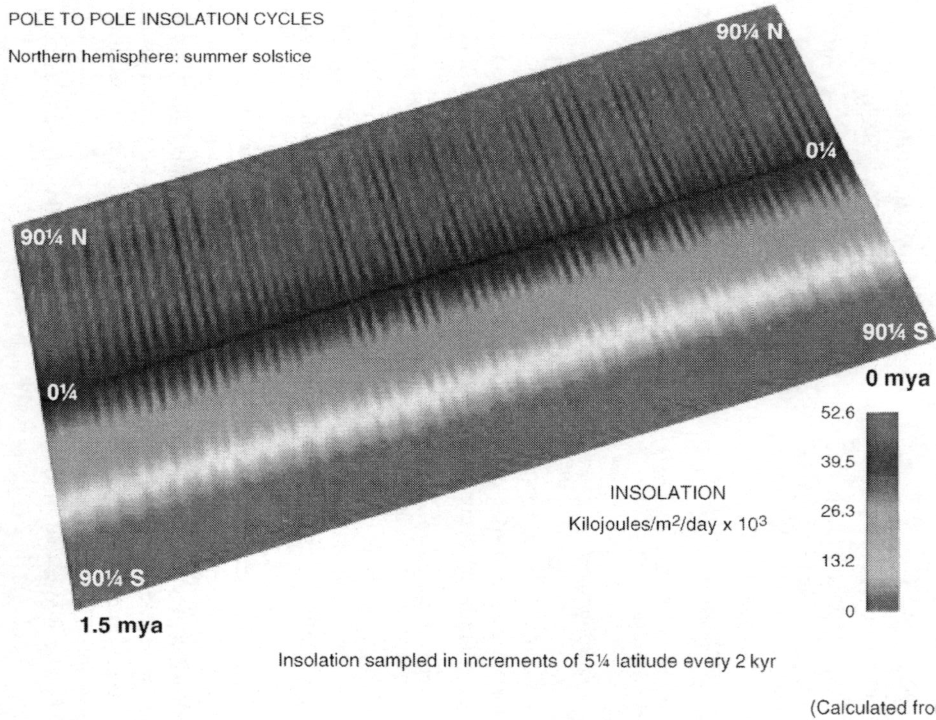


*Figure 14.18* *Pole to pole insolation. Plotted at the Northern Hemisphere summer solstice for a 1.5 myr interval. This figure shows the change in the summer heat received by the Northern Hemisphere as a result of the interaction of orbital cycles. Insolation is a function of eccentricity and the season of the year when the perihelion and aphelion occur. Red indicates warmer summers and blue cooler summers. A red and blue couplet is a precession cycle. Eccentricity bundles precession cycles into groups that contain a varying number of precession cycles. Precession-scale insolation cycles can change seasonal heating up to 30%. Insolation calculated from equations in Berger (1978) sampled at 5° increments of latitude every 2 kyr – see colour version of this figure in Appendix.*

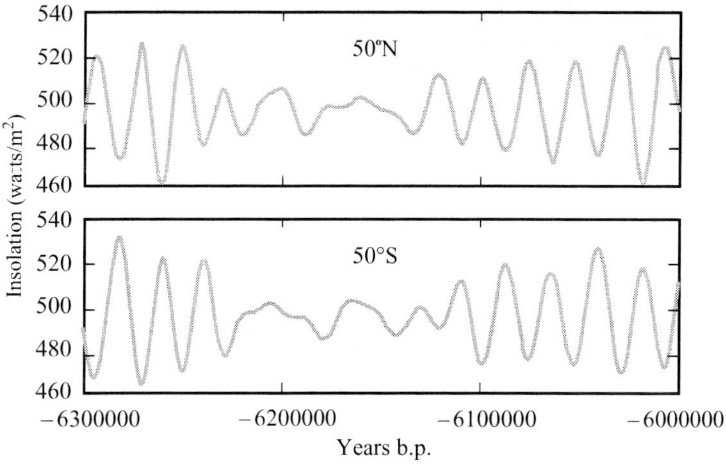


*Figure 14.19 Insolation cycles plotted at 50°N and S. The antipodal phase relationship at the precession scale disappears or inverts slightly when eccentricity approaches 0 between approximately 615,000 and 620,000 years bp. Insolation calculated from equations in Berger (1978).*

circulation alter patterns of precipitation and evaporation, shifting associated climate patterns and causing climate cycles (Glennie, 1984; Perlmutter and Matthews, 1989; Park and Oglesby, 1994; Perlmutter *et al.*, 1998). High frequency climate simulations of the Mid Cretaceous (Park and Oglesby, 1994) and the modern (Moore *et al.*, 2000) indicate that the largest climate shifts experienced by a region are indeed associated with changes that occur at the scale of precession. Eccentricity and obliquity cycles modulate the extent of the effect, as does the long-term interference of the main orbital harmonics. The result is a repetitive array of short-term, precession-scale climate cycles that vary between true end-member conditions determined by the longer-term interaction. The warmest end-member will be referred to as the climatic maximum and the coolest climatic end-member is referred to as the climatic minimum.

The response of the climate system of a particular region to a specific condition of insolation is not only a function of the phase of the insolation cycle itself, however. The response is inherently associated with location (latitude and longitude) and the (paleo)geography of the area (Perlmutter and Matthews, 1989; Park and Oglesby, 1994; Perlmutter *et al.*, 1995; Moore *et al.*, 2000). The result is that within a hemisphere not all areas become wetter or drier at the same point in an insolation cycle. For example, as insolation varies over an orbital cycle, equatorial areas may become drier while mid-latitude areas become wetter or one mid-latitude area may get wetter while another area gets drier. Climate variability is a function of the specific regional conditions (which can change over geologic time). Therefore, different zones of a single hemisphere can undergo different climatic successions in response to the same insolation cycle (Ibid.).

Regional paleoclimatic changes can be interpreted by mapping the global distribution of paleoclimate indicators and constructing seasonal climate maps for each phase of an insolation cycle (Perlmutter and Matthews, 1989, 1992; Matthews and Perlmutter, 1994; Perlmutter and Plotnick, 2002, 2003). To demonstrate this, the climates for the Wisconsin glacial period and the Holocene were mapped from paleoclimate indicators to compare the most recent shift of the global climate patterns (Fig. 14.20a,b; Perlmutter and Matthews, 1989, 1992). Figure 14.20 compares the differences interpreted from paleoclimate indicators for the last climatic minimum with present day conditions for the month of July. Note, that there is a well-defined poleward migration in atmospheric circulation, including the position of the intertropical convergence zone and the locations of pressure cells, oceanic surface currents and major climate belts.

To make the global climate pattern easier to explain and understand, seasonal climate maps have been simplified into a single map showing a series of climate belts (Perlmutter and Matthews, 1989). Each belt represents a zone of the Earth's surface that, to a first approximation, has similar climatic end-members over a climate cycle. The climatic succession of a belt is worked out by interpolating between the end-members. However, in order to use a climate belt map to evaluate the climatic range in an area, the regional to local conditions that may modify this first approximation of climate need to be incorporated. Conditions that need to be included are prevailing winds, proximity to an ocean or large body of water, ocean temperature and currents, orographic effects, and monsoonal effects.

The climate belt map for the last 3 million years is shown in Figure 14.21a. The climate succession of each belt by phase of the insolation cycle is shown in Figure 14.21b. Overall, equatorial and polar regions tend to become drier and mid-latitude regions tend to become wetter as conditions progress from the climatic maximum to minimum. However, even in this simplified view of climate change over an insolation cycle, a highly variable global pattern is apparent.

An additional level of complexity is added when one considers that the insolation cycles of the Northern and Southern Hemispheres are 180° out of phase on a precession scale (Perlmutter *et al.*, 1998; Perlmutter and Plotnick, 2002, 2003; Fig. 14.19). This causes the exact same climatic successions in opposite hemispheres to be out of phase with respect to time; about 10 kyr out of phase. To illustrate, the highest summer insolation for a hemisphere, the climatic maximum, occurs when eccentricity is high and perihelion occurs during the summer (Kukla and Gavin, 1992). Aphelion will be during winter, six months later. The condition of low summer insolation, the climatic minimum, occurs when eccentricity is high and aphelion occurs during the summer. Perihelion then occurs during the winter. Thus, hot summers and cold winters characterize the hemispheric climatic maximum, while cool summers and mild winters characterize a hemispheric climatic minimum. At the timescale of precession, the result is that, in general, when the Northern Hemisphere is at a climatic maximum, the Southern Hemisphere is at a climatic minimum. One consequence of such hemispheric differences could be the ability of a monsoon to produce summer rain. Monsoonal climates tend to respond directly to insolation and are strongest

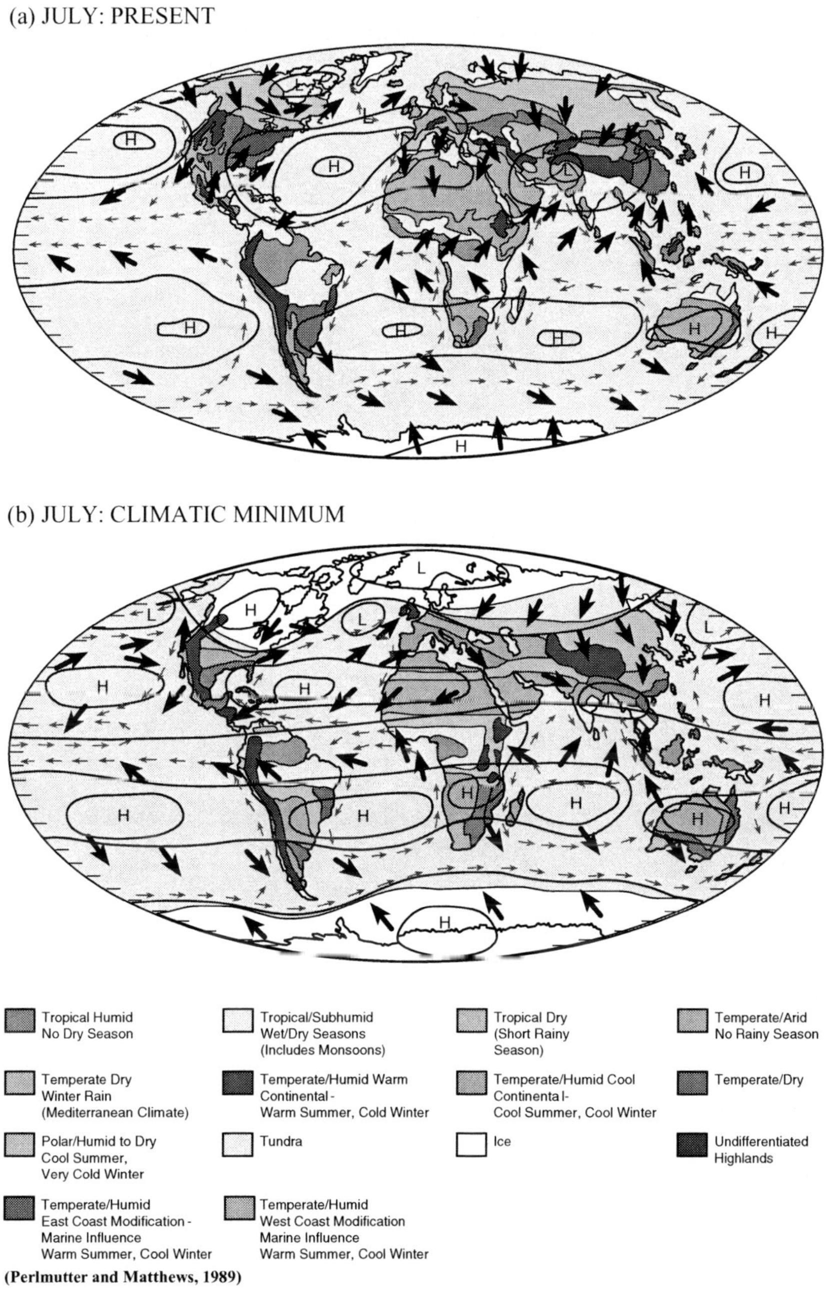


*Figure 14.20 Distribution of global climate patterns with the average position of July atmospheric pressure cells, wind patterns and ocean currents superimposed. (a) Present day. (b) Last climatic minimum estimated from paleoclimate indicators (after Perlmutter and Matthews, 1989) – see colour version of this figure in Appendix.*

(a) ***Zones of Similar Climate Ranges***

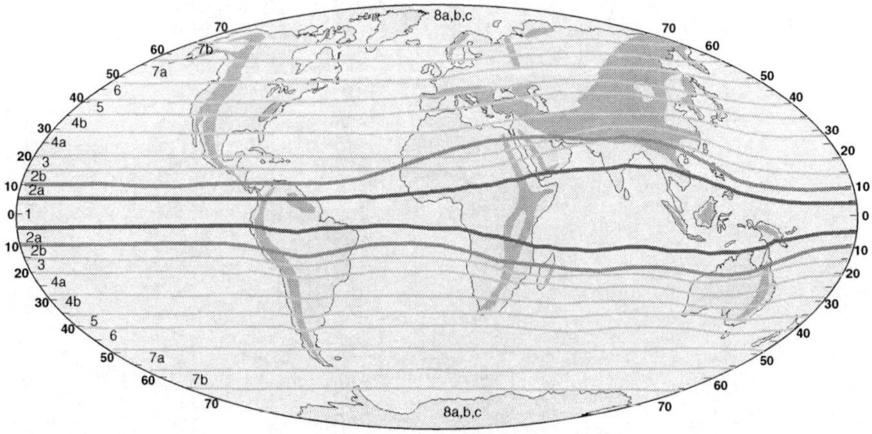


**0–3 MYBP**

(b) **(Perlmutter *et al.*, 1998)**

| Phase \ Belt | 1 | 2a | 2b | 3 | 4a | 4b | 5 | 6 | 7a | 7b | 8 |
|---|---|---|---|---|---|---|---|---|---|---|---|
| A | Tr/VH (6 mo.) | Tr/VH (6 mo.) | Tr/H (5 mo.) | Tr/H (4 mo.) | Tr/SH | Tr/D | Te/A | Te/D | Te/SH | Te/H | P/H-A |
| B1 | Tr/VH (6 mo.) | Tr/H (5 mo.) | Tr/H (4 mo.) | Tr/SH | Tr/D | Te/A | Te/D | Te/SH | Te/H | P/H | P/SH-A |
| B2 | Tr/H (4 mo.) | Tr/SH | Tr/SH | Tr/D | Te/A | Te/D | Te/SH | Te/H | P/H | P/SH | P/D-A |
| C | Tr/H (3 mo.) | Tr/SH | Tr/D | Te/A | Te/D | Te/SH | Te/H | P/H | P/SH | P/D | P/A |
| D1 | Tr/H (4 mo.) | Tr/SH | Tr/SH | Te/D | Te/A | Te/D | Te/SH | Te/H | P/H | P/SH | P/D-A |
| D2 | Tr/VH (5 mo.) | Tr/H (5 mo.) | Tr/H (4 mo.) | Tr/SH | Tr/D | Te/A | Te/D | Te/SH | Te/H | P/H | P/SH-A |

*Figure 14.21 Cyclostratigraphic belt map for the Pleistocene/Holocene and associated climatic successions. (a) Belt map showing areas of the earth's surface that to a first approximation have similar climatic shifts (Perlmutter and Matthews, 1989; Matthews and Perlmutter, 1994) – see colour version of this figure in Appendix. (b) Idealized, end-member and climatic*

when hemispheric insolation values are highest (Kerr, 1986; Park and Oglesby, 1994). Therefore, a rigorous, rainy monsoon could exist in one hemisphere, while a sluggish, dry monsoon occurs in the other, even if the geographic conditions are identical (latitude, land area, proximity to an ocean, sea surface temperature, elevation, prevailing wind direction, etc.) because of the differences in summer heating. These variations require that a paleoclimate evaluation for a specific time period include at least four seasonal maps, including the July and January for climatic maximum and minimum that account for the range of paleoclimatic indicators in an interval (Perlmutter *et al.*, 1995, 1998). When equatorial areas are being evaluated, an equinox map is interpreted for maximum and minimum conditions as well. Figure 14.22 shows a climatic analysis for the early Permian. High and low sea level conditions are a function of Southern Hemispheric glaciation (Perlmutter and Plotnick, 2002, 2003).

### 14.7.3 The Relationship of Climate and Sea Level

As noted above, climatic response to a particular insolation cycle is regional. That is, not all regions become warmer or cooler, wetter or drier, at the same time point in an insolation cycle. However, the glacioeustatic response to a particular insolation cycle is global, with synchronous sea level highs or lows everywhere on Earth. This means that regional climate cycles of different geographic areas may have different phase relationships with glacioeustatic cycles (Perlmutter *et al.*, 1995, 1998; Perlmutter and Plotnick, 2002, 2003). Because climate directly impacts weathering, sediment transport and environments of deposition, these differences in the phase relationships of regional climate cycles and global sea level cycles can produce very different stratigraphic patterns during the same insolation cycle. Recognizing that an insolation cycle can cause stratigraphic cycles to vary depending on the specific conditions of climate and deposition is extremely important for correlating intervals and developing a geochronologic framework. In order to appreciate this phenomenon fully, the impact of climate and sedimentation must be evaluated.

---

*Figure 14.21 Continued*

*sequences by cyclostratigraphic belt for the Pleistocene/Holocene. The climate cycle is divided into six phases: A – the climatic maximum, B1 and B2 – the cooling transition, C – the climatic minimum, and D1 and D2 – the warming transition. Uniform box size does not necessarily indicate equal lengths of time for each phase. Equatorial areas are affected by a wet monsoon during some climate phases, with the number of rainy months per year indicated in parentheses. Note that mid-latitudes become more humid toward the climatic minimum, while equatorial and polar latitudes become less humid. Abbreviations are as follows: Tr = tropical; Te = temperate; P = polar; VH = very humid; H = humid; SH = subhumid; D = dry; A = arid. See text for explanation.*

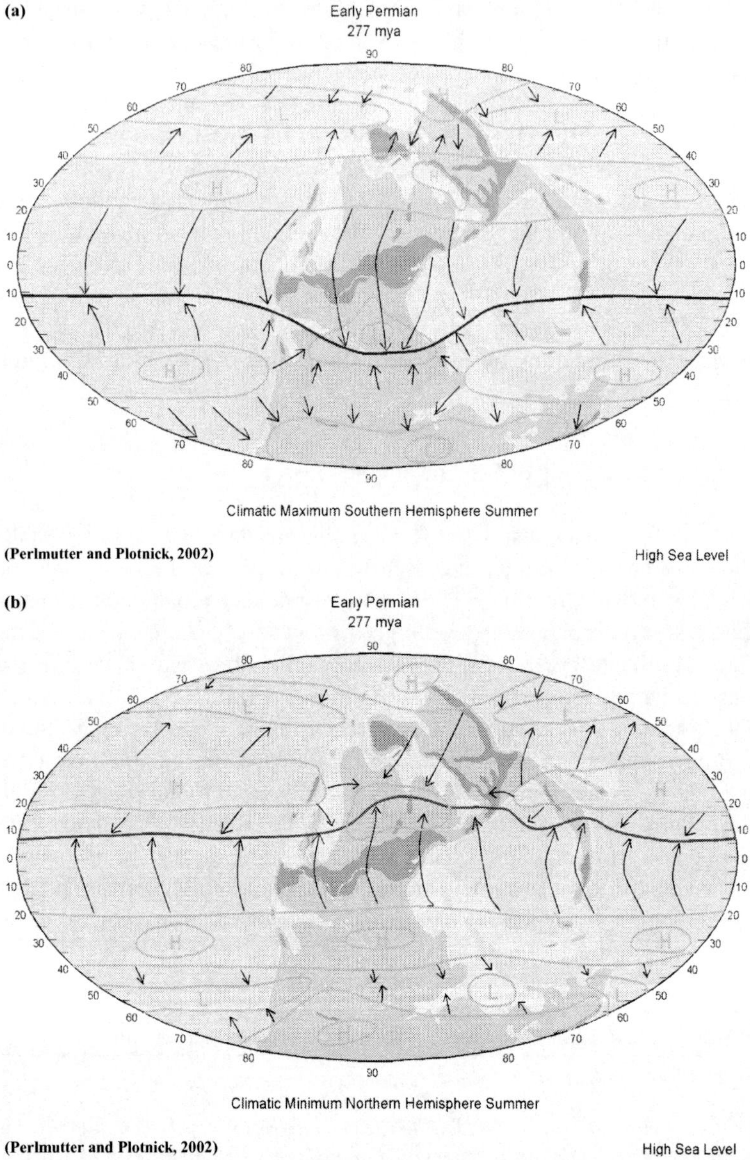


*Figure 14.22 Climatic variability in the early Permian. (a). Early Permian climatic maximum, Southern Hemisphere summer. High eccentricity. Hot Southern Hemisphere summers occur during perihelion. Cold Southern Hemisphere winters occur during aphelion. The Northern Hemisphere experiences mild summers and winters. Hot Southern Hemisphere summers are unfavorable for icecap formation in the Southern Hemisphere. This is a time of high sea level. (b) Early Permian climatic minimum, Northern Hemisphere summer. High eccentricity. Mild Northern Hemisphere summers occur during aphelion. Mild Northern Hemisphere winters occur during perihelion. The Southern Hemisphere experiences extreme summer (hot) and winter (cold) conditions. Hot summers are unfavorable for icecap formation in the Southern Hemisphere. This is a time of high sea level.*

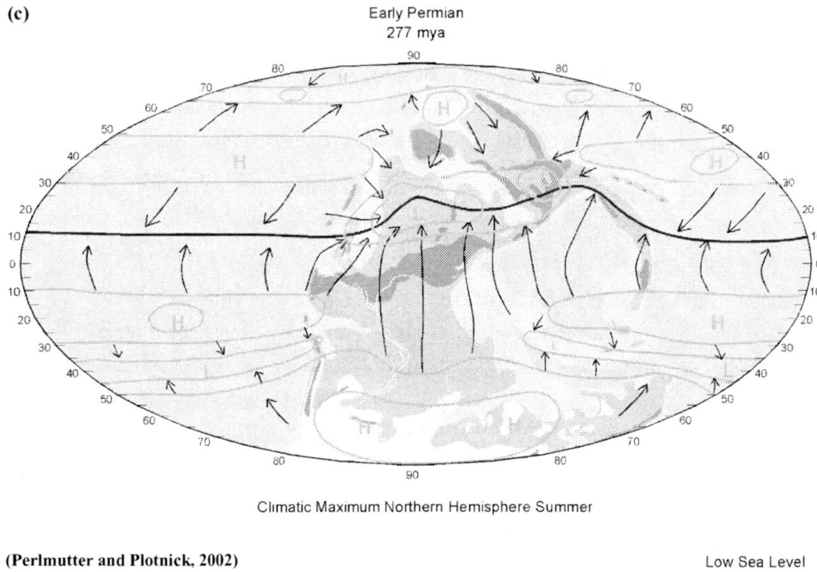


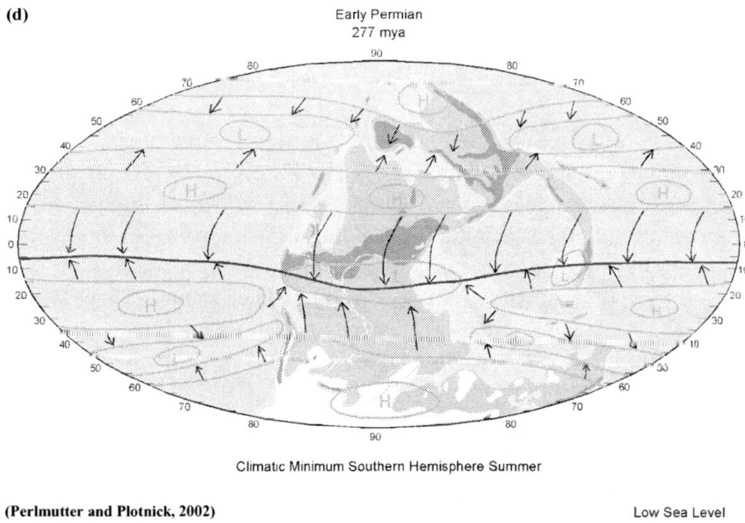


*Figure 14.22 Continued. (c) Early Permian climatic maximum, Northern Hemisphere summer. High eccentricity. Hot Northern Hemisphere summers occur during perihelion. Cold Northern Hemisphere winters occur during aphelion. The Southern Hemisphere experiences mild summers and winters. Low Southern Hemisphere summer temperatures are likely to be favorable for icecap formation in the Southern Hemisphere. This is time of low sea level. (d) Early Permian climatic minimum, Southern Hemisphere summer. High eccentricity. Mild Southern Hemisphere summers occur during aphelion. Mild Southern Hemisphere winters occur during perihelion. The Northern Hemisphere experiences extreme summer (hot) and winter (cold) conditions. Low Southern Hemisphere summer temperatures are likely to be favorable for icecap formation in the Southern Hemisphere. This is time of low sea level. Maps were interpreted from the geographic distribution*

## 14.8 The Impact of Climate on Sediment Yield

Sediment yield is a complex interaction of many processes, including elevation, slope, climate, provenance and vegetation. However, elevation is commonly considered the primary factor that determines yield (e.g. Milliman and Syvitski, 1992), with climate having only a secondary effect, despite a relatively large body of information indicating the importance of climate (e.g. Garner, 1959; Leopold *et al.*, 1964; Wilson, 1973; Schumm, 1977; Schumm and Brakenridge, 1987). Clearly, the total potential volume of sediment that can be produced from an area is a function of the volume of material above base level. However, the manner and rate that the material is weathered, eroded, transported and deposited is a function of the climate and slope. A mountain produces a very different sedimentary deposit depending on whether it erodes in a desert or a rainforest. The generalization that elevation is the primary control occurs because the detailed quantitative information needed to evaluate the effect of climate, and the interaction of climate and elevation, on sediment yield are very limited. In addition, time resolution of the geologic record at the scales necessary to differentiate rapid (sub-Milankovitch) variations in yield is difficult.

An initial attempt to resolve the effects of elevation and climate was made by plotting sediment yield/unit area vs. the maximum elevation of the largest 29 modern river systems (Fig. 14.23; Perlmutter *et al.*, 1998; Summerfield and Hulton, 1994). When plotted as an undifferentiated group, the analysis shows that elevation accounts for only about half the variance in the data. However, when the data were grouped by the climate of the headlands of each drainage basin projected down to sea level, differences in the load/unit area become apparent. This evaluation shows that in general, subhumid climates have the greatest yield, followed by monsoonal and dry Mediterranean climates, and finally humid climates. Arid climates are assumed to have the lowest yield. What this plot demonstrates is that for a given elevation sediment yield may vary by over an order of magnitude, depending on the climate (Perlmutter and Matthews, 1994; Perlmutter *et al.*, 1995, 1998). There is an ongoing effort to increase the total number of drainage basins used to determine the impact of climate and elevation on yield (Perlmutter *et al.*, 1998).

### 14.8.1 The Impact of Climate Change on Sediment Yield

As indicated above, the actual phases of climate cycles vary significantly depending on location. To demonstrate how climatic succession can affect sediment flux, yield

*Figure 14.22 Continued*
*and temporal range of paleoclimate indicators (after Perlmutter and Matthews, 1989). Base maps were generated using the program PGIS (Paleogeographic Information System; Ross, 1992) based on Paleomap paleogeographic reconstructions (Scotese et al., 1992). Conditions favorable for ice-cap formation were estimated using criteria from Kukla and Gavin (1992) – see colour version of this figure in Appendix.*

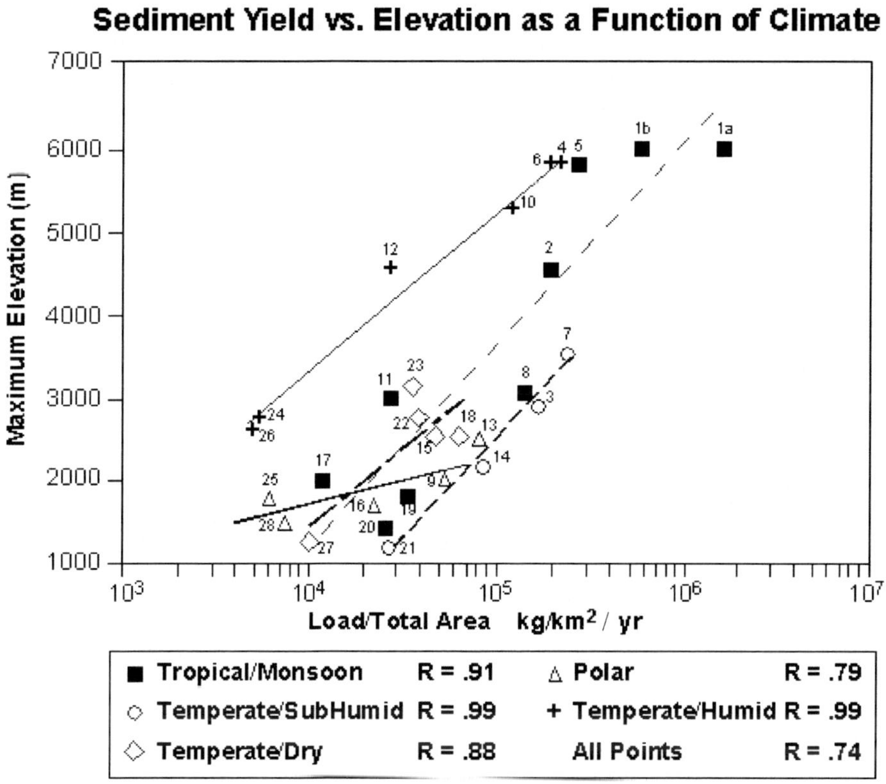


**(Perlmutter *et al.*,1998)**

*Figure 14.23 Sediment yield/unit area vs. elevation subdivided by climate (Perlmutter et al., 1998). Numbers refer to rivers: (1a) Brahamaputra, (1b) Ganges, (2) Amazon, (3) Mississippi, (4) Chiang Jiang, (5) Indus, (6) Mekong, (7) Colorado, (8) Orinoco, (9) MacKenzie, (10) Huang He, (11) Nile, (12) La Plata, (13) Yukon, (14) Danube, (15) Orange, (16) Amur, (17) Zaire, (18) Shatt-El-Arab, (19) Zambezi, (20) Niger, (21) Murray, (22) Columbia, (23) Rio Grande, (24) Ob, (25) Lena, (26) Yenisei, (27) São Francisco, and (28) Kolyma. Estimates of yield for headlands with a climatic range were made by averaging values.*

cycles were estimated for each cyclostratigraphic belt, assuming similar provenance areas and a maximum elevation of 2 km (Fig. 14.24). From this analysis, the greatest change in yield is shown to occur in a climatic succession that shifts from arid to subhumid. This climate shift occurs in mid-latitude belts 3, 4a, 4b, and 5.

Figure 14.24 highlights the fact that yield cycles are not globally synchronous. For example, in belt 1, yield is highest at the climatic maximum and lowest at the climatic minimum; in belt 4b, yield is highest at the climatic minimum and lowest in the climatic transitions, just before and after the climatic maximum. In general, belts show more than one peak episode of sediment yield over a climate cycle. As a consequence,

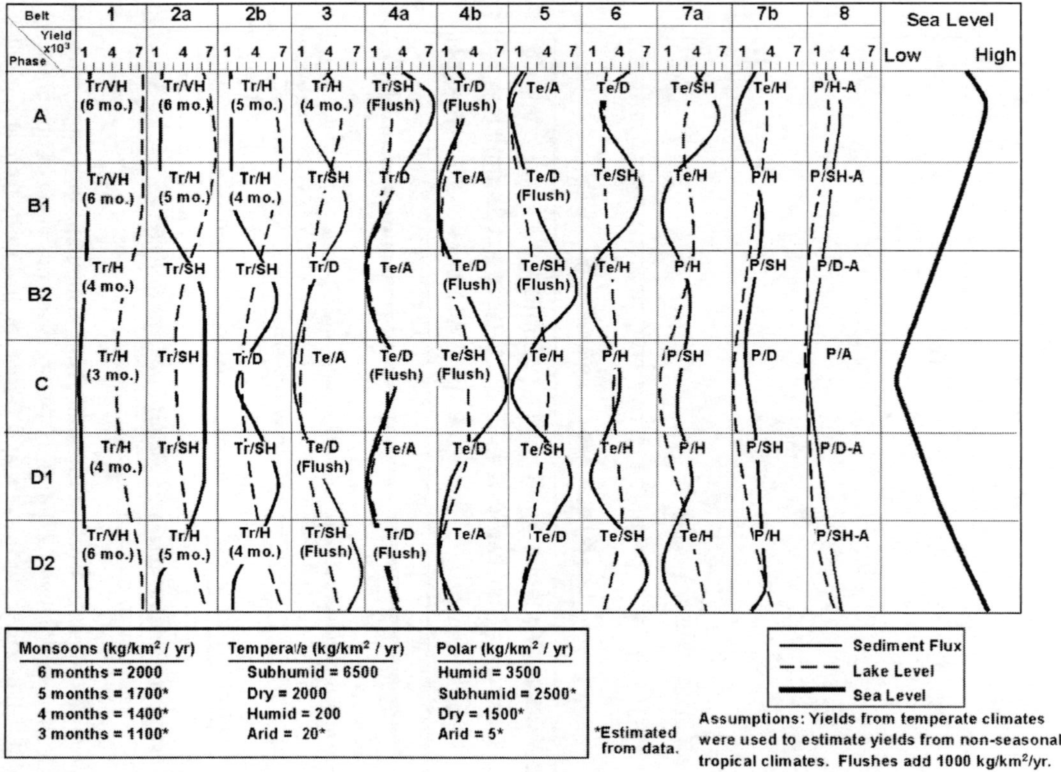


**(Perlmutter *et al.*,1998)**

*Figure 14.24* *Sediment yield for each cyclostratigraphic belt by climate phase, sea level, and lake level (Perlmutter et al., 1998). Solid curves are sediment yield and dashed curves are lake level; sea level is shown on the right. Estimates from data are indicated with asterisks. Abbreviations are as follows: Tr = tropical; Te = temperate; P = polar; VH = very humid; H = humid; SH = subhumid; D = dry; A = arid. See text for explanation.*

sedimentation cycles produced by river basins in different parts of the world do not have to be in phase with each other or with sea level.

## 14.9 Effects of Interaction of Climate Sediment and Sea Level Cycles on Stratigraphy

Sequence stratigraphic convention assumes that: (1) as sea or lake levels fall, the associated base-level fall increases sediment yield by increasing erosion of river valleys and newly exposed shoreline and shelf; and (2) during a sea or lake level rise, the associated base-level rise traps sediment in the fluvial drainage basins and the newly flooded areas, decreasing yield to distal regions of the depositional basin. It is clearly understood however, that this convention must be modified by climatic considerations (Vail *et al.*, 1977a). Figure 14.24 indicates that only in the climatic successions represented by belts 2a, 4b, and 5 can sediment yield actually increase at some point during falling and lowstand sea level and decrease during the rise. Other belts show maximum supply during the rise, low or highstand, depending on the location.

To investigate the potential effects of climate and sediment cycling on stratigraphy Perlmutter *et al.* (1998) simulated four different phase relationships between sediment yield and eustasy using the 2D stratigraphic computer model Sedpak (Kendall *et al.*, 1991). Sediment yield and sea level cycles were simulated as simple sinusoidal curves. The phase conditions simulated in these experiments were: (1) maximum yield during the fall, minimum during the rise; 2) maximum yield during lowstand and minimum during highstand; (3) maximum yield during the rise, minimum during the fall; and (4) maximum yield during highstand, minimum during lowstand. Model results were then interpreted using a sequence stratigraphic framework with well-established criteria (Vail *et al.*, 1991). Model input, output, and the interpretations of stratigraphic surfaces were then compared to determine possible effect.

Results of the comparisons are shown in Table 14.5. Even using the simplified conditions of the models, variations are evident in the timing of surfaces and systems tracts relative to the eustatic cycle. This is important information to consider because the common assumption made by most interpreters is that specific stratigraphic surfaces and systems tracts are associated with certain phases of sea level. This set of experiments suggests that the phase relationships of high frequency sedimentation cycles can impact the timing relative to sea level by delaying or accelerating the conditions that produce regressive and transgressive surfaces, and enhancing or inhibiting the development of systems tracts. Perlmutter (1985) and Kolla and Perlmutter (1993) demonstrate this phenomenon by re-evaluating the Pleistocene stratigraphy of the Gulf of Mexico in relationship to the sediment yield cycle of the Mississippi River. They show that the deposition of submarine fans extended well into the mid-rise of sea-level during the deposition of the transgressive systems tract. The phase relationship may also be critical if a specific surface, such as a maximum flooding surface, is used for correlation across a basin that is being filled by multiple river systems that drain

*Table 14.5 The effect of the interaction of sediment and glacioeustatic cycles on the interpretation of key stratigraphic surfaces and selected strata (Perlmutter et al., 1998). This table was compiled from interpretations of output from the stratigraphic computer Sedpak. A phase shift of 90° (fall) indicates the maximum amplitude of the sediment supply cycle coincides with sea level fall. A phase shift of 180° (low) indicates the maximum amplitude of the sediment supply cycle coincides with sea level lowstand. A phase shift of 270° (rise) indicates the maximum amplitude of the sediment supply cycle coincides with sea level transgression. A phase shift of 360° (high) indicates the maximum amplitude of the sediment supply cycle coincides with sea level highstand.*

| Surfaces & Strata / Phase Shift of Yield & Eustasy (Max. yield) | Sequence Boundary | Transgressive Surface | Maximum Flooding Surface | Condensed Section | Lowstand Prograding Complex |
|---|---|---|---|---|---|
| 90° (Fall) | Mid Fall – just prior to the inflection point | Very Early Rise – just after lowstand | Very Late Rise – just prior to highstand | Rise | Very Late Fall and Lowstand |
| 180° (Low) | Very Early Fall – just after highstand | Early Rise | Highstand | Highstand | Low to Early Rise |
| 270° (Rise) | Lowstand | Mid Rise – at the inflection point | Very Late Rise – just prior to highstand | Fall | Mid Rise |
| 360° (High) | Early Fall – midway between highstand and the inflection point | Lowstand | Late Rise – midway between the inflection point and highstand | Lowstand | Late Fall |

**Perlmutter *et al.*, 1998**

regions with different climates. The same kind of surface in different portions of the basin may actually represent different phases of the eustatic curve. The effects of the phase relationship of high frequency sediment supply cycles and lake level cycles on lacustrine depositional systems may be more profound than for marine systems because lakes and their associated drainage basins can react more rapidly and to a greater extent to local to regional climate changes (Perlmutter and Matthews, 1989; Matthews and Perlmutter, 1994).

In addition to changing the rate of the sediment supply, climate also affects river hydrodynamics, grain size, vegetation, turbidity, and nearshore salinity etc. For example, variations in weathering over a climate cycle can alter the sand/shale ratio (Perlmutter and Matthews, 1989). Additionally, for some climatic successions, the percent sand increases as total sediment volume decreases. Thus, for some regions, the timing of optimal reservoir potential will also be a function of the balance between the total volume of coarse material and the coarse fraction of sediment. The sand/shale ratio also impacts the potential for seals, which can effect hydrocarbon migration and entrapment. Vegetation and freshwater discharge can affect the type and content of organic matter in potential source rocks.

In summary, sea level cycles caused by changes in glacial mass at the poles tend to track the insolation cycle (Matthews and Frohlich, 1991; Kukla and Gavin, 1992;

Berger, 1992): sea level is high around the insolation maximum and low around the insolation minimum. In addition, the climatic succession of each belt relative to the insolation cycle is different, causing phase relationships of yield and sea level cycles to vary. Depending on location, maximum and minimum sediment yield can occur at any phase of glacioeustatic cycle. The same is true for the sediment yield and lake level curves.

## 14.10 Examples of the Effect of Climatic Succession and Glacioeustacy on Reservoir Distribution

### 14.10.1 Surma Basin, Bangladesh

The late Miocene section of the Surma Basin, Bangladesh, demonstrates that reservoir distribution can be controlled by a combination of climatic succession and glacioeustacy (Perlmutter *et al.*, 1998). At that time, the climatic maximum conditions in the fluvial drainage area providing sediment to the Surma basin were interpreted as tropical wet, with a short dry season. A wet tropical climate causes vigorous biochemical weathering and produces mostly very fine-grained sediments. The climatic minimum was interpreted as cooler, with a wet summer, and a dry winter monsoon, increasing both the volume and coarseness of sediment yield at this point in the cycle. Alpine glaciers may have also existed in the higher elevations of the headlands at this time, with seasonal melting of ice adding to the yield. As the climate cycle progressed and conditions warmed, the area became wetter and sediment yield and coarseness decreased.

The phase relationship of sediment yield and eustatic cycles for the Surma basin was affected by unipolar glaciation. In the late Miocene, the Southern Hemisphere had a polar icecap but the Northern Hemisphere did not. At the scale of precession, the Northern and Southern Hemispheres are out of phase so the climatic maximum in the Northern Hemisphere corresponds to the climatic minimum in the Southern Hemisphere. Consequently, the highest volume and coarsest sediment yield to the Surma basin (Northern Hemisphere climatic minimum) corresponded to high sea level (Southern Hemisphere climatic maximum, glacial minimum). The lowest volume, finest-grained yield (Northern Hemisphere climatic maximum) therefore corresponded to low sea level (Southern Hemisphere climatic minimum, glacial maximum). Sediment yield would have been highest in and around high sea level, peaking just as sea level fall began and the alpine glaciers in the headlands melted. Yield and grain size decreased as tropical wet conditions redeveloped in the drainage basin near the time of low sea level. High levels of river discharge would have caused erosion of the shelf as base-level fell.

Based on this correlation of yield and sea level, the potential for deposition of reservoir-forming strata in the Surma basin should have increased toward the eustatic high, with the highest reservoir potential in the later transgressive and highstand system tracts. A sediment pulse early in the sea-level fall would have increased this potential and increased the rate of highstand progradation. Data from the Surma basin indicate that the stratigraphy is indeed cyclic. Shallow shelf transgressive and highstand systems tracts are sandy, with marginal marine lowstand wedges and prograding complexes predominantly muddy and silty (Radovich *et al.*, 1995). Condensed sections are thin and muddy, overlying transgressive shoreline sands. These data and interpretations are completely consistent with the forecasts made using cyclostratigraphy.

### 14.10.2 Gulf of Mexico

The phase relationship of the sediment yield cycle of the Mississippi River and the glacioeustatic cycle in the late Pliocene and Pleistocene was previously evaluated by Perlmutter (1985) and Kolla and Perlmutter (1993). Maximum total yield occurred during the early to mid-rise of eustasy, and was over an order of magnitude greater than the minimum yield. A maximum coarse yield, 20 times larger than the minimum coarse yield, also occurred during sea level rise. Minimum total and coarse yields occurred during low runoff around the eustatic low. Sea level highstands were characterized by warmer and wetter conditions and fine-grained sediments. These analyses indicated that the highest potential for reservoirs should have occurred in intervals deposited during the early to mid-rise of sea level, in lowstand prograding complexes, thick sandy fans and early transgressive systems tracts (Kolla and Perlmutter, 1993).

Prior to the glaciation of North America in the late Pliocene (~2.5–3 ma; Galloway *et al.*, 1991), climate conditions in the western headlands of the Mississippi drainage basin, where most of the sediment was produced, were warm temperate/subhumid to dry during the climatic maximum and cool temperate/humid to subhumid at the climatic minimum (Perlmutter and Matthews, 1989; Matthews and Perlmutter, 1994). Alpine glaciers may have existed in the higher elevations of headlands during the climatic minimum. These climates were used to estimate early Pliocene sediment yield by using Figure 14.23; yield was highest during the climatic maximum (~9,000 $kg/km^2/yr$) and lowest during the climatic minimum (~3,000 $kg/km^2/yr$). Seasonal melt from Aalpine glaciers during the climatic minimum worked to counterbalance the effect of the reduced amount of water in the hydrologic cycle due to lower temperatures. These estimates indicate that maximum yield was likely to have been two to three times the minimum yield. The shift to a cooler but more humid climate in the western portion of the drainage basin would have tended to reduce grain size.

Again, prior to the build up of a glacier in North America, only the Southern Hemisphere had a large polar icecap and, therefore, controlled glacioeustasy. High eustatic sea level (Southern Hemisphere climatic maximum, glacial minimum) corresponded to the time of the lowest volume yield (Northern Hemisphere climatic minimum). As sea level fell, the climate in the Mississippi drainage basin warmed, runoff increased, alpine glaciers melted, and sediment load increased. The associated fall in base level, combined with an active river system, caused erosion of the shelf, possibly making the total volume of sediment delivered to the Gulf of Mexico during the fall about equal to that of low eustatic sea level (the time of the Northern Hemisphere climatic maximum and highest yield from the Mississippi drainage basin). As sea level rose during the next phase of the cycle, climate in the drainage area became cooler and more humid, decreasing yield and grain size. This, combined with the addition of accommodation space caused by rising sea level, would have significantly reduced sediment volume reaching the Gulf of Mexico. The resulting yield cycle has had its highest amplitude around the mid-fall and low sea level. Recall for comparison, that in the late Pliocene–Pleistocene section, maximum supply and coarse yield occurred during the early to mid-rise of eustasy (Perlmutter, 1985). In early Pliocene time, therefore, the highest potential for sandy reservoirs in the Gulf of Mexico should have been in systems tracts deposited in relative lowstand position, fans and prograding complexes.

A detailed sequence stratigraphic analysis of a regional line on the Louisiana shelf, Gulf of Mexico, was made by Radovich *et al.* (1990). These authors integrated seismic interpretation, electric logs, and biostratigraphy to interpret systems tracts in five prograding sequences that ranged in age from early Pliocene to early Pleistocene. Biostratigraphy indicated that water depths in the area of study were predominantly outer shelf and slope during deposition of transgressive and highstand systems tracts. This location permitted excellent definition of lowstand prograding complexes, which were expanded and sand-rich compared to the transgressive and highstand systems tracts, which are mostly thin and predominantly shaly. Thus, the interpretation of the distribution of potential reservoir-forming sands, made using sequence stratigraphy, does coincide with the forecasts made using cyclostratigraphy. Overall, the predominant lowstand locations of high potential reservoirs in the Gulf of Mexico are very different than the sandy reservoirs in the Surma basin, which occur in more updip positions, in transgressive and highstand systems tracts.

## 14.11 Summary and Discussion

High frequency climatic succession is a function of global position and is therefore interpretable and mappable. Climatic succession may cause sediment flux to vary by over an order of magnitude through the span of an insolation cycle. Climate succession is a function of geographic location, while glacioeustatic sea level is global and tends to track the insolation cycle. The phase relationship between the sediment yield cycle and glacioeustasy therefore varies as a function of paleogeography. Maximum or minimum sediment supply is not necessarily related to any specific stage of sea-level change. The impact of this variability is that the distribution of potential hydrocarbon reservoirs, seal and source rocks; (1) may be affected by the relative timing of the climate, sediment yield and base-level (lake and sea level) cycles, and (2) has predictable regional differences. This information should be incorporated into exploration programs to focus on those basins and time intervals with higher probabilities of hydrocarbon occurrence. More specifically, it can be used to focus prospect assessment and development toward those intervals with systems tracts with the highest probability of reservoir, capped by a competent sealing shale in proximity to a source rock.

Simplified simulations using the computer model Sedpak have been used to investigate the impact of the phase relationship of sediment cycles and sea (or lake) level cycles. Even in these relatively uncomplicated simulations, variation occurred in the timings of surfaces and systems tracts relative to the timing of the eustatic cycle. This suggests that the timing of the sediment supply cycle with respect to sea level may affect the interpretation of stratigraphy and reservoir potential.

The distribution of reservoirs in the late Miocene section of the Surma basin, Bangladesh, was compared with the Pliocene sections of the Gulf of Mexico in the area of the Mississippi River to demonstrate the effect climatic succession has on lithostratigraphy. Potential reservoirs in the Gulf were associated with thick, sandy, lowstand prograding complexes, while reservoirs in the Surma basin were associated with transgressive and highstand systems tracts. In both cases, actual distributions closely matched the distribution forecasts obtained by using cyclostratigraphic methods.

## 14.12 Acknowledgements

We thank ChevronTexaco Inc. and Petrobras for permission to publish and gratefully acknowledge Julia S. Wellner of Rice University and Kelly Dempster of ChevronTexaco for critical review of the manuscript. Drafting was provided by George Griffith, Linda Lovell and Heriberto V. dos Santos.

Chapter 15

# The Role and Value of "Biosteering" in Hydrocarbon Reservoir Exploitation

ROBERT W. JONES[1], STEPHEN LOWE[1], PAUL MILNER[2], PETER HEAVEY[2], SIMON PAYNE[3] and DAVID EWEN[3]

[1] British Petroleum (BP), Building H, Chertsey Road, Sunbury-on-Thames, Middlesex, TW16 7LN, UK.

[2] British Petroleum (BP), Godesetdalen 8, 4065 Stavanger, Norway.

[3] British Petroleum (BP), Farburn Industrial Estate, Dyce, Aberdeen AB21 7PB, UK.

With contributions from Nick HOLMES, Ichron, Northwich, Cheshire; Haydon BAILEY, and Liam GALLAGHER, Network Stratigraphic Consulting Ltd., Potters Bar, Hertfordshire; Patrice BRENAC, BSI, Llanddulas, North Wales; Keith RICHARDS, KRA, Llandudno, North Wales; and Richard ATTEWELL, Robertson Research International Ltd., Llandudno, North Wales.



## 15.1 Introduction

Case histories are presented of applications of biosteering in British Petroleum (BP). They have been selected so as to give a reasonable representation of geological settings, reservoir ages and facies, and analytical techniques. They are:

> *Andrew Field, North Sea* (Upper Paleocene deep marine clastic reservoir – biosteered using micropaleontology and microfacies);
> *Valhall Field, North Sea* (Upper Cretaceous deep marine carbonate reservoir – biosteered using micropaleontology and nannopaleontology);
> *Cusiana Field, Colombia* (Upper Cretaceous shallow marine clastic reservoir – biosteered using palynology);
> *Sajaa Field, Sharjah* (Lower Cretaceous shallow marine carbonate reservoir – biosteered using thin-section micropaleontology and microfacies).



*E.A.M. Koutsoukos (ed.), Applied Stratigraphy,* 339-355.

The value added to date by the application of biosteering throughout the BP portfolio runs into hundreds of millions of dollars. It is anticipated that this figure will further increase in the future, as the technology is transferred to fields in areas only now entering into production.

## 15.2 Biosteering

Biosteering involves real-time monitoring of stratigraphic position relative to reservoir in a (deviated) well by biostratigraphic techniques applied at well-site. It also involves, as necessary, for instance when encountering a sub-seismic fault or a problem with seismic depth conversion or survey data, issuing instructions to redirect the well trajectory to ensure optimal reservoir penetration. The high resolution of the technique (its "window"), usually established by analyses of closely spaced samples from offset wells or a pilot hole, is often of the order of only a few feet or metres.

The technique is, and will remain, critical to the exploitation of many petroleum reservoirs. It was first applied on Maersk's Dan Field in the Danish Sector of the North Sea as long ago as 1987 (see, for instance, Shipp and Marshall, 1995; Shipp, 1999).

## 15.3 Discussion of Results

### 15.3.1 Andrew Field, North Sea

*Reservoir* = Upper Paleocene deep marine clastics (Andrew Formation).
*Biosteering technology* = Micropaleontology and microfacies.
*Contractor* = Ichron Ltd., Northwich, Cheshire.

The Andrew Field (see also Payne *et al.*, 1999; Holmes, 1999) is situated in the UK Sector of the North Sea (Fig. 15.1). The reservoir comprises deep marine submarine fan sandstones and subordinate mudstones of Late Paleocene age (Andrew Formation). Reserves are estimated at 118 million barrels. The pay interval is comparatively thin. The field is currently (1999) being produced using horizontal wells. Production commenced in June, 1996, and rates are currently at 64000 barrels/day. Understanding of the reservoir facies and heterogeneities and consequences for fluid flow, and optimal placement of wells with respect to fluid contacts are critical to the maximization of oil production prior to the inevitable early gas and/or water breakthrough.

Micropaleontological biostratigraphy in conjunction with petrophysics and core sedimentology has been used to establish a subdivision of the reservoir into seven zones of the order of a few metres or tens of feet thick (Fig. 15.2). Microfacies in conjunction with core sedimentology has been used to identify facies and heterogeneities within each zone, thereby establishing the spatio-temporal distribution of reservoir and non-reservoir units and potential consequences for fluid flow (Fig. 15.2).

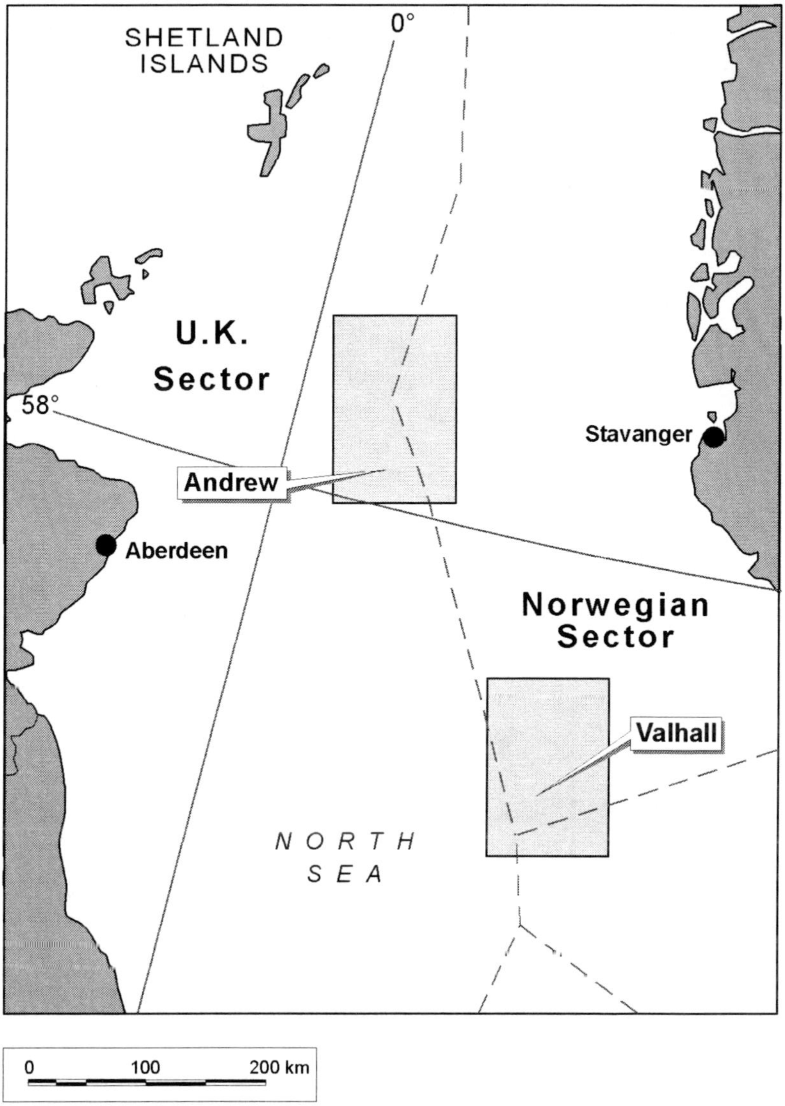


*Figure 15.1 Location Map, Andrew and Valhall Fields.*

Mudstones A3 and A1 have been interpreted on the basis of high abundance and diversity, low dominance (stress) "tranquil" assemblages of agglutinating foraminifera characterized by comparatively high incidences of complex infaunal "morphogroup" C as hemipelagic. They have therefore also been interpreted as potentially of field-wide extent and constituting barriers to fluid flow. This has been confirmed by pressure data.

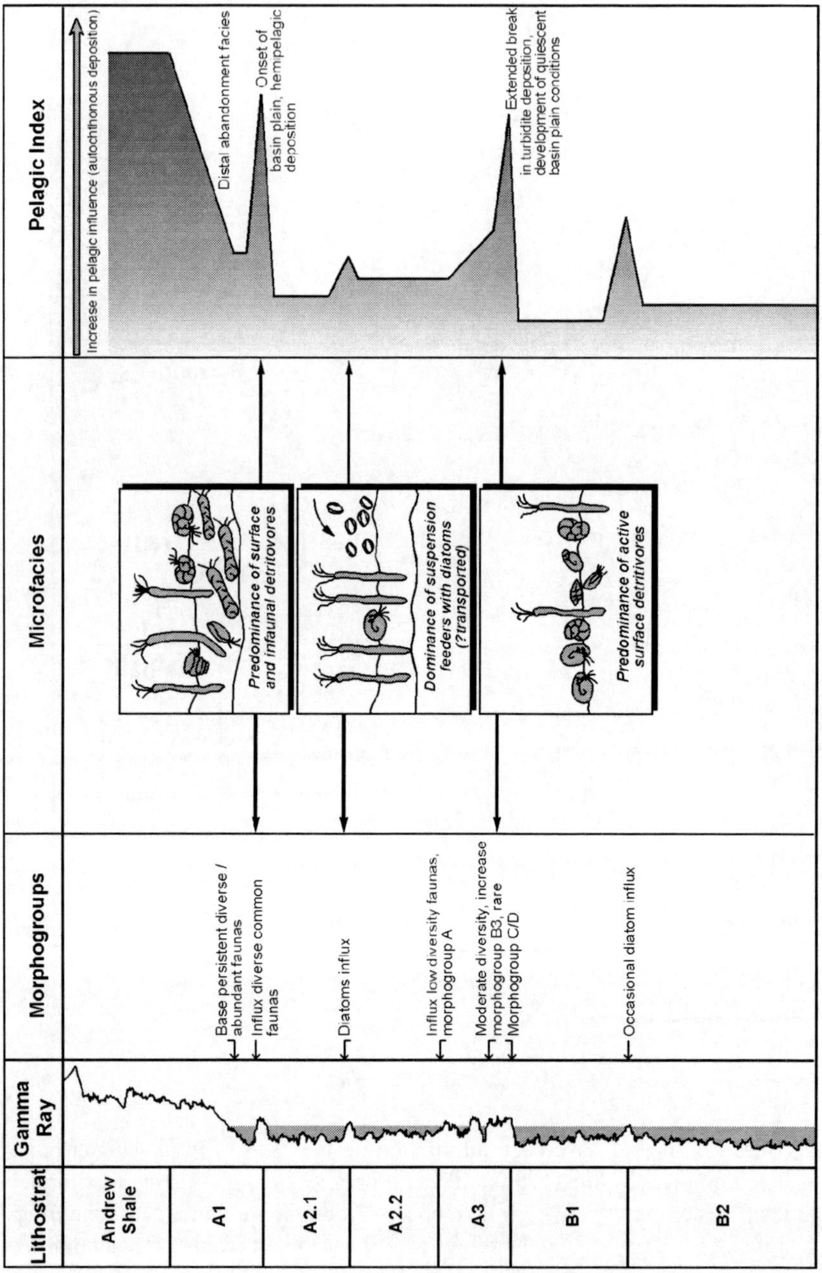


*Figure 15.2 Reservoir Stratigraphy and Facies, Andrew Field. After Payne et al. (1999).*

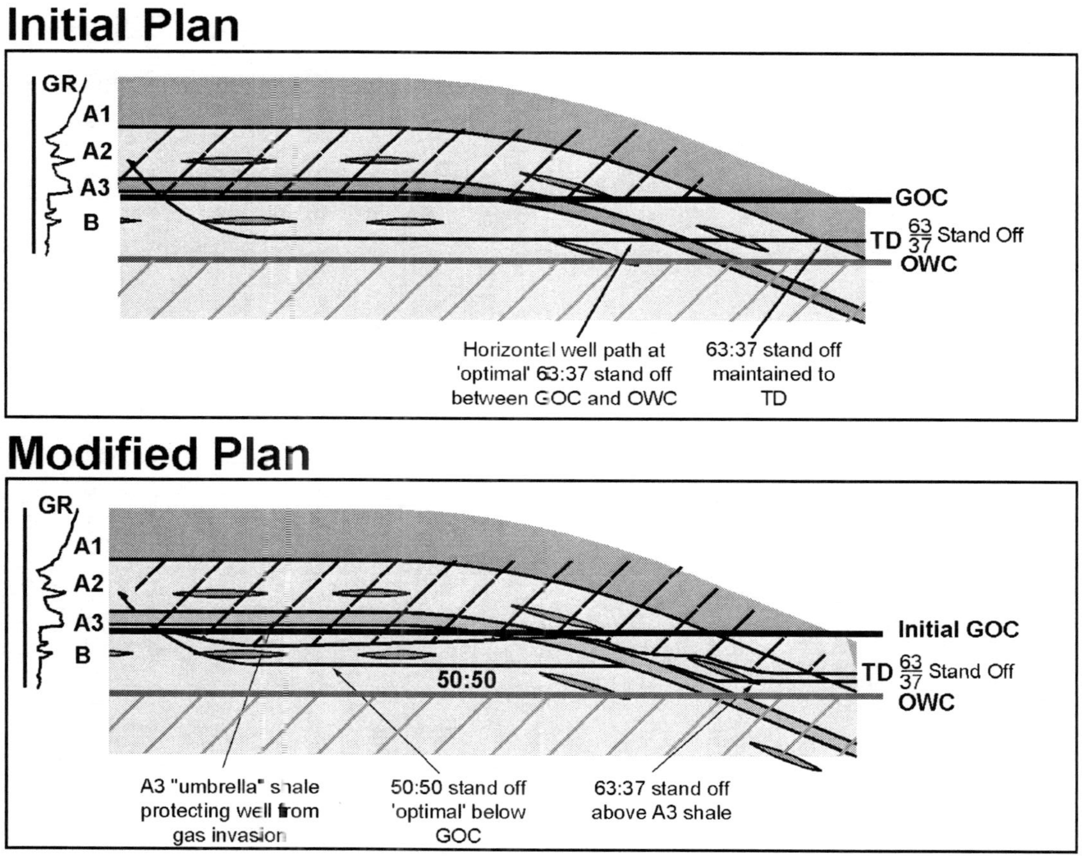


*Figure 15.3* *Impact of Biosteering, Andrew Field. After Payne et al. (1999). Well kept within best reservoir (Zone B) throughout trajectory by biosteering.*

Mudstone A2 has been interpreted on the basis of low abundance and diversity, high dominance (stress) "disturbed" assemblages dominated by simple epifaunal "morphogroups" A (suspension-feeders) and B as interturbiditic. It has therefore also been interpreted as potentially of limited areal extent, and constituting only a baffle to fluid flow. The "morphogroup" nomenclature alluded to above follows Jones and Charnock (1985), Jones (1996) and Jones (1999) and additional references cited therein; key stratigraphically and paleoenvironmentally useful species have been illustrated by Charnock and Jones (1990).

Biosteering – by means of agglutinating foraminiferal micropaleontology and microfacies – has targeted Reservoir Unit B (Fig. 15.2). Fortuitously, the top of this unit is effectively coincident with the gas-oil contact over the crest of the field. Realization of this fact has allowed the biosteered well-bore to be run more medially through the reservoir than in the initial well plan (with an initial 50 : 50 stand-off from the gas-oil contact as against a standard 63 : 37), with the overlying Mudstone A3 acting as a barrier to fluid flow (see above) and hence protecting it from gas invasion (Fig. 15.3).

It is estimated by the Andrew Business Unit that the optimal well placement enabled by the biosteering has added 10 million barrels of reserves to the books.

### 15.3.2 Valhall Field, North Sea

*Reservoir* = Upper Cretaceous deep marine carbonates (Tor Formation).
*Biosteering technology* = Micropaleontology and nannopaleontology.
*Contractor* = Network Stratigraphic Consulting Ltd., Potters Bar, Hertfordshire.

The Valhall Field (see also Bergen and Sikora, 1999; Sikora *et al.*, 1999) is situated in the Norwegian Sector of the North Sea (Fig. 15.1). The reservoir comprises deep marine carbonates (allochthonous chalks and chalky turbidites) of Late Cretaceous, essentially Maastrichtian, age (Tor Formation). Recoverable reserves are estimated at 705 million barrels. Offtake is by means of high-angle wells, of which 50 have already been drilled. Production rates are currently (1999) at 105,000 barrels/day.

Biostratigraphy in conjunction with petrophysics and core sedimentology has been used to establish a subdivision of the reservoir into seven zones of the order of a few metres of tens of feet thick (Fig. 15.4).

Biosteering – by means of calcareous benthic and planktonic foraminiferal micropaleontology and nannopaleontology – has targeted Zones C and D (Zones A and B possessing better reservoir properties in terms of porosity and permeability, but being unstable and prone to collapse under drawdown).

On particularly successful well that was kept within Zone D by biosteering is currently (1999) producing 12,000 barrels/day (Fig. 15.5).

Other well-site applications of biostratigraphy include setting casing close to the base of the overburden without drilling overbalanced into the underpressured zone at the top of the reservoir, thereby causing formation damage, and "biostopping" (making the TD call) at the base of the reservoir. Another application is identifying the

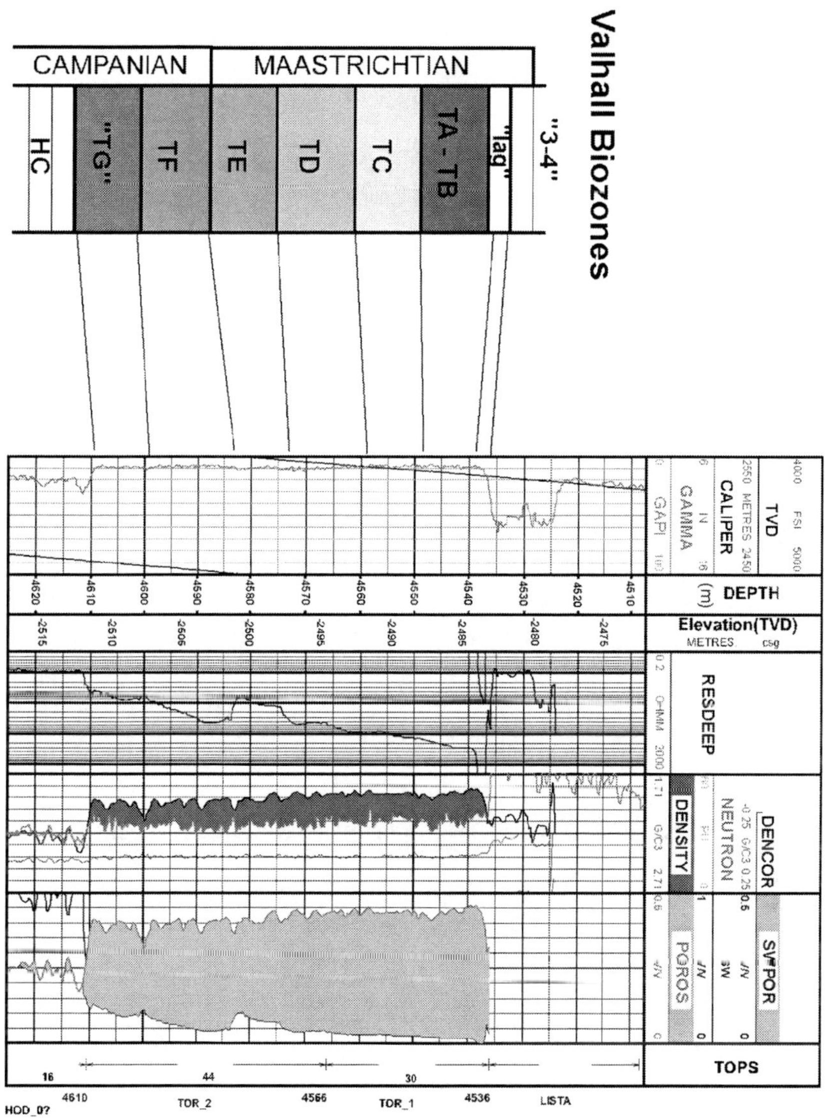


*Figure 15.4* *Reservoir Stratigraphy, Valhall Field.*

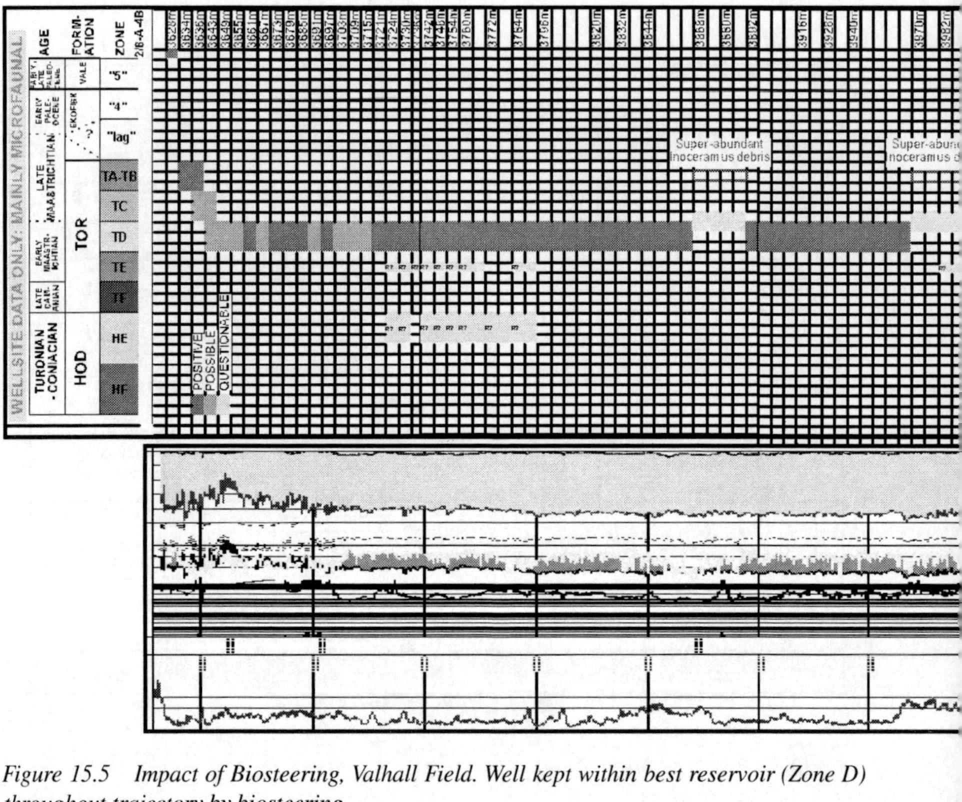


*Figure 15.5 Impact of Biosteering, Valhall Field. Well kept within best reservoir (Zone D) throughout trajectory by biosteering.*

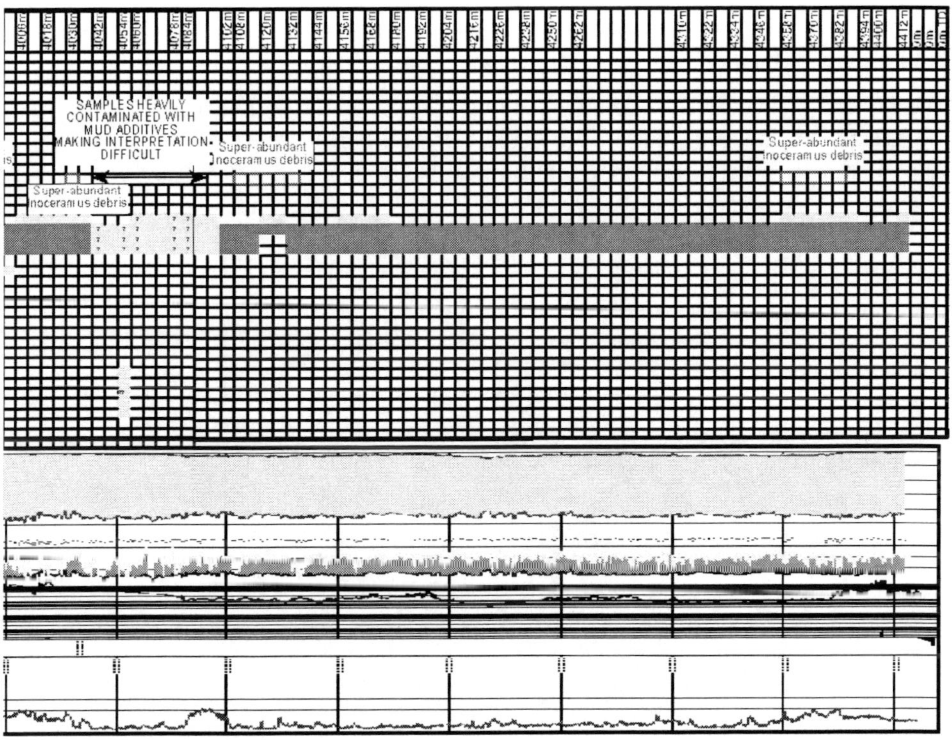


*Figure 15.5 Continued*

origin of caved material and hence unstable zones in the tophole, impacting well design and drilling mud requirements.

In terms of value added, 30% of the current field production of 105,000 barrels/day is attributed by the Valhall Business Unit to optimal reservoir placement enabled by biosteering. Moreover, a minimum of $1 million per well (7 days drilling, about $150,000 per day) is saved by being able to set casing in the correct place by means of biostratigraphy (see above).

### 15.3.3 Cusiana Field, Colombia

*Reservoir* = Upper Cretaceous shallow marine clastics (Guadalupe Formation).
*Biosteering technology* = Palynology.
*Contractors* = BSI (Brenac Stratigraphy International), Llanddulas, North Wales; KRA (Keith Richards & Associates), Llandudno, North Wales.

The Cusiana Field (see also Cazier *et al.*, 1995; Cooper *et al.*, 1995a–b) is situated in the Llanos Basin, more specifically in the frontal thrust-sheets of the Eastern Cordillera, some 150 miles north-east of Bogota in Colombia (Fig. 15.6). The reservoir

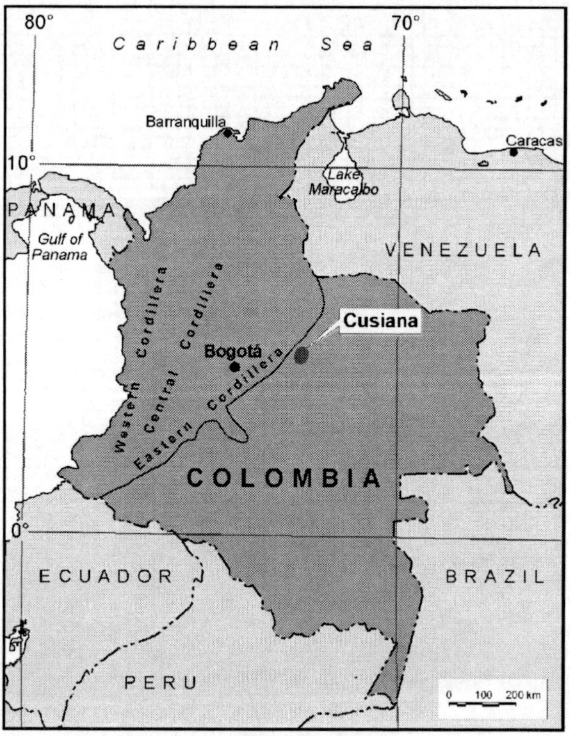


*Figure 15.6 Location Map, Cusiana Field.*

comprises shallow marine sandstones of Late Cretaceous, Santonian-Campanian, age (Guadalupe Formation). Recoverable reserves are estimated at 1.5 billion barrels of light oil and condensate and 3.4 trillion cubic feet of gas. The field is currently being produced using horizontal wells. Production rates are currently (1999) in excess of 300,000 barrels/day.

Biostratigraphy in conjunction with petrophysics and core sedimentology has been used to establish a subdivision of the reservoir into seven zones of the order of a few tens of feet thick (Fig. 15.7).

Biosteering – by means of palynology – has targetted Zones GR3–GR7, which are developed within the best ("Upper Phosphate") reservoir (Fig. 15.8). A BP designed portable unit has enabled safe handling at well-site of the dangerous chemicals used in

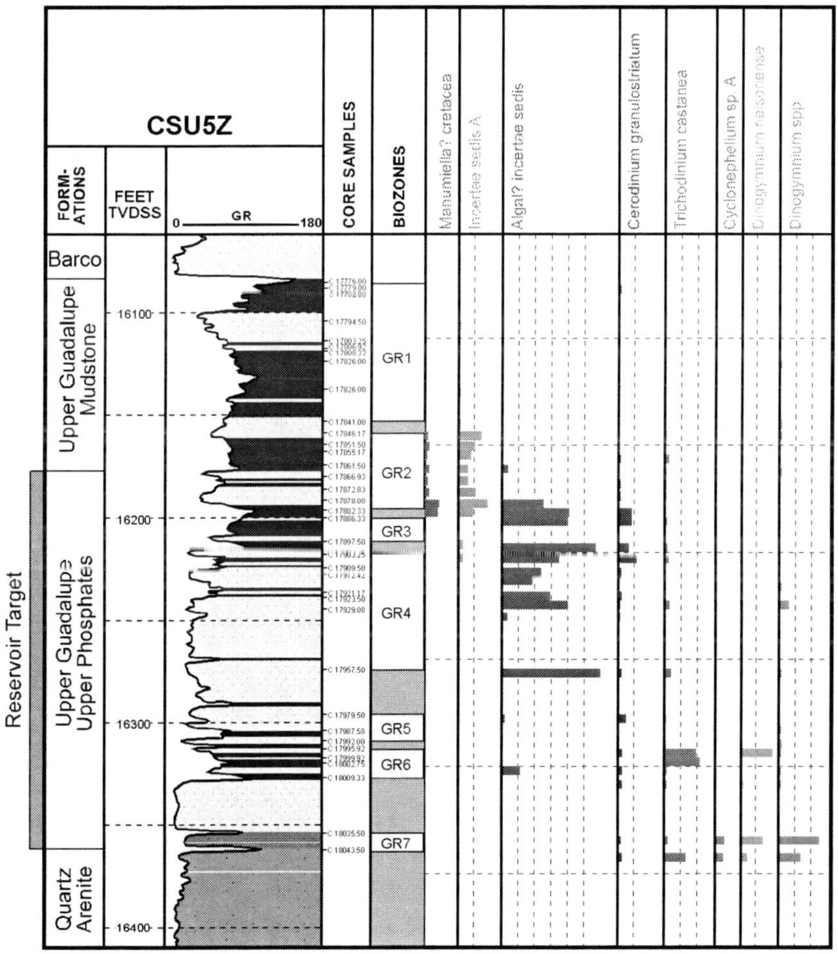


*Figure 15.7 Reservoir Stratigraphy, Cusiana Field.*

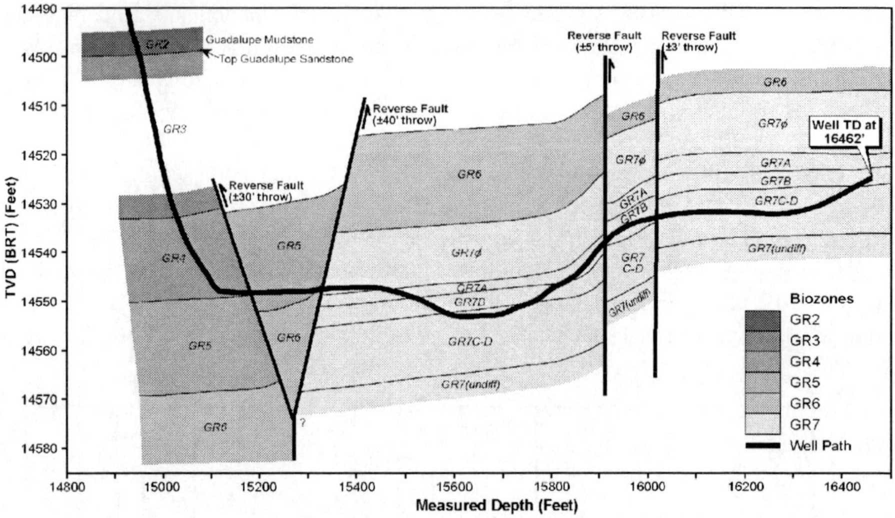


*Figure 15.8 Impact of Biosteering, Cusiana Field. Well kept within best reservoir (Zones GR3-7) throughout trajectory by biosteering.*

palynological processing. A BP sponsored research programme is currently underway with the objective of eliminating altogether the need for the use of such chemicals.

In terms of value added, the production potential from the first biosteered horizontal well of the field was approximately 30,000 barrels/day, as against 12,000 barrels/day from the best conventional vertical well. Moreover, one biosteered well costing $26 million effectively does the work of 3–4 conventional wells costing $15–18 million each. Thus, the three that have been drilled to date have resulted in savings in drilling costs of $57–138 million.

### 15.3.4 Sajaa Field, Sharjah

*Reservoir* = Lower Cretaceous shallow marine carbonates (Kharaib and Shuaiba Formations).
*Biosteering technology* = Thin-section micropaleontology and microfacies.
*Contractor* = Robertson Research International Ltd., Llandudno, North Wales.

The Sajaa Field (see also Blinten and Wahid, 1983; Beydoun, 1988) is situated in the frontal thrust-sheets of the Oman Mountains, some 25 miles east of Sharjah town in Sharjah in the United Arab Emirates (Fig. 15.9). The reservoir comprises shallow marine carbonates of Early Cretaceous, Barremian–Aptian, age (Kharaib and Shuaiba Formations). Recoverable reserves are estimated at between 100–400 million barrels of condensate and 1.5–6 trillion cubic feet of gas. Production commenced in 1982. 1985 rates were 60,000 barrels/day. The field is currently (1999) being produced using horizontal wells.

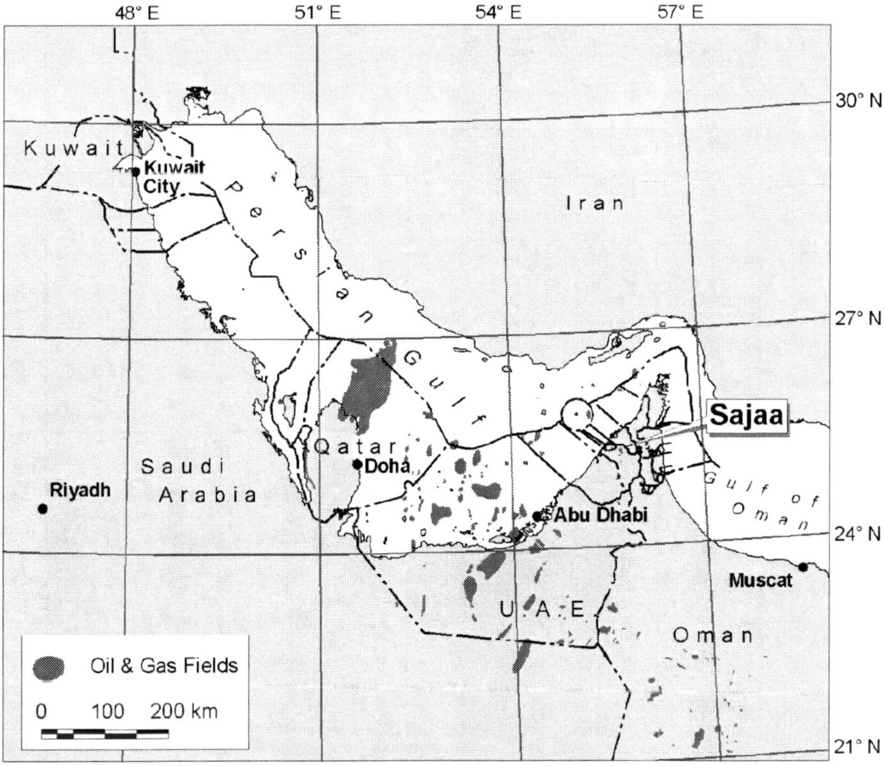


*Figure 15.9 Location Map, Sajaa Field.*

Integration of biostratigraphic, petrophysical and core sedimentological data has enabled a 21-fold subdivision of the Kharaib and Shuaiba reservoirs into zones of the order of a few tens of feet thick (Figs. 15.10a–b). Incidentally, detailed paleobathymetric interpretation of closely-spaced samples from equivalent reservoir sections elsewhere in the United Arab Emirates (in Abu Dhabi) and in Kuwait has contributed significantly to the understanding of parasequence-scale reservoir faces and architecture in these areas, and to the identification of reservoir "sweet spots" (reservoir quality being related primarily to depositional facies, with little diagenetic overprint). The methodology for the paleobathymetric interpretation is described by Banner and Simmons (1994) and Jones (1996); key stratigraphically and paleoenvironmentally useful species are illustrated by Whittaker *et al.* (1998).

Biosteering – by means of thin-section larger benthic foraminiferal micropaleontology and microfacies – has been achieved over distances of several thousands of feet (a world record of 10,228' in the equivalent reservoir section in Al-Shaheen Field, offshore Qatar; Shipp and Marshall, 1995). It has effectively replaced geosteering using coherency, which worked well in unfaulted but not in faulted sections (the approach taken

(a)

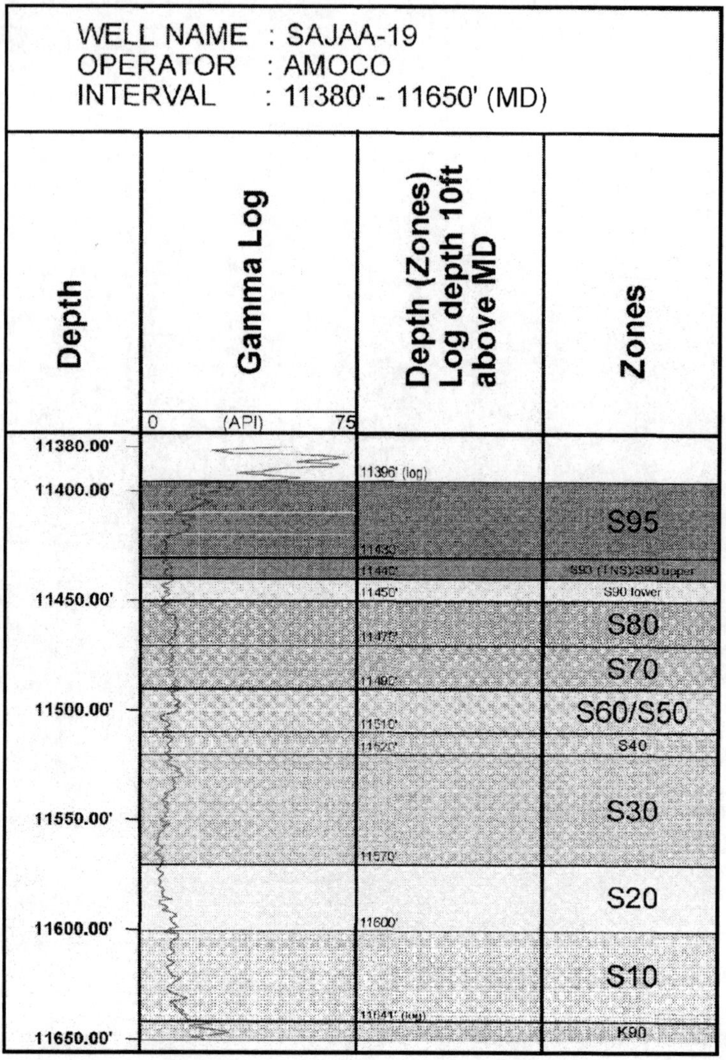


*Figure 15.10 Reservoir stratigraphy, Sajaa Field. (a) Shuaiba reservoir; (b) Kharaib reservoir.*

(b)

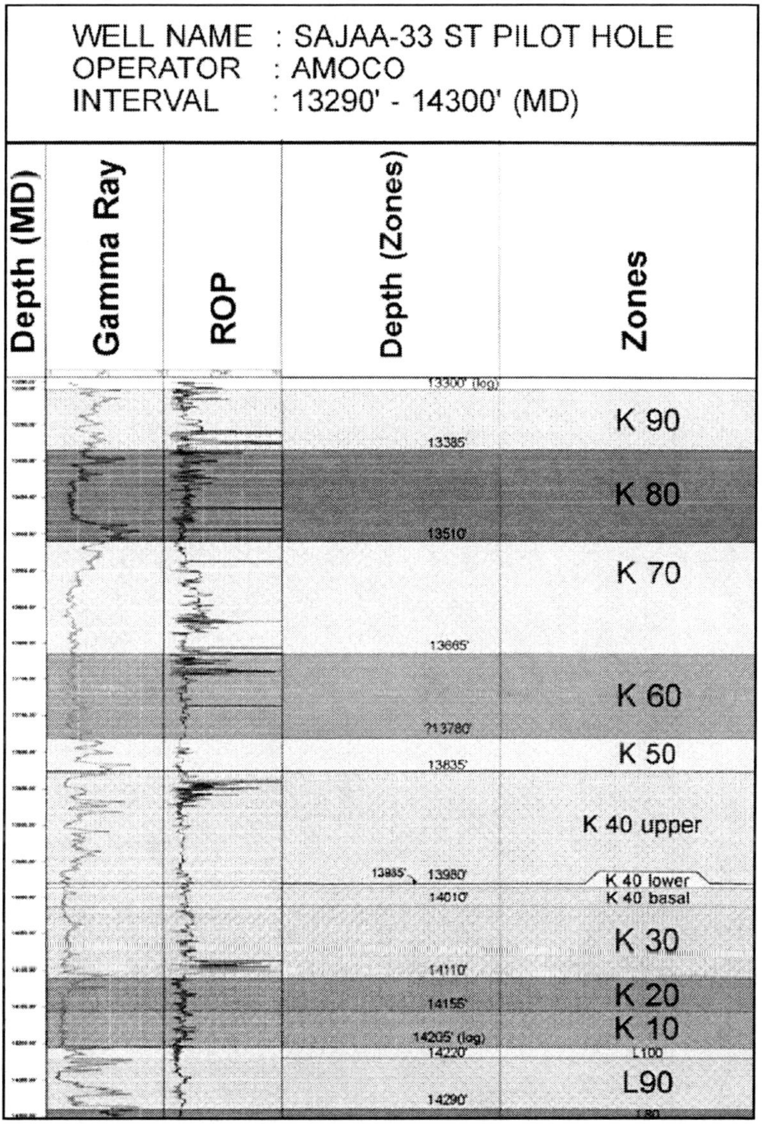


*Figure 15.10 Continued*

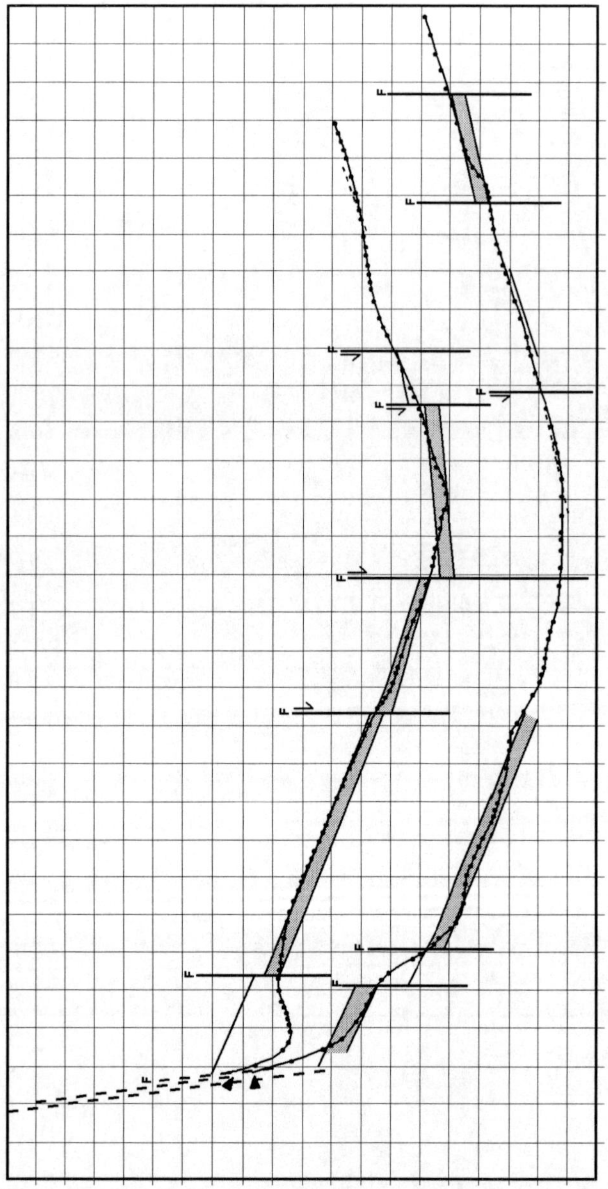

*Figure 15.11* Impact of Biosteering, Sajaa Field. Multilateral wells kept within best reservoirs throughout trajectory by biosteering (upper = Shuaiba reservoir; lower = Kharaib reservoir).

on encountering a fault being to steer upwards to a known point in the stratigraphy – usually the base of the Nahr Umr Formation – and then back down again, which could take up to 24 hours). Biosteering of multilateral wells has also been possible (Fig. 15.11).

## 15.4 Concluding Remarks

Benefits accruing from biosteering and optimal reservoir placement in the cited examples include:

- Savings of tens of millions of dollars in drilling costs;
- Addition of tens of millions of barrels of reserves;
- Addition of tens of thousands of barrels per day of production (sustainable throughout field life).

Costs incurred are typically less than 0.5% of an individual well budget. The value added to date by the application of biosteering on over 200 wells throughout the BP portfolio runs into hundreds of millions of dollars.

It is anticipated that this figure will further increase in the future, as the technology is transferred to fields only now entering into production, for example deep-water Angola and the Gulf of Mexico.

## 15.5 Acknowledgements

We would like to thank BP for providing permission to publish and word-processing facilities, Pat Randell for the draftwork, and Tom Dignes and Jamie Powell for their constructive reviews of the draft manuscript.

Part IV

# Modelling the Record

Chapter 16

# Quantitative Methods for Applied Microfossil Biostratigraphy

FELIX M. GRADSTEIN
Geology Museum, University of Oslo, N-0318 Oslo, Norway.



> "Quantitative stratigraphy uses relatively simple or complex mathematical-statistical methods to calculate stratigraphic models that with a minimum of data provide a maximum of predictive potency, and include formulation of confidence limits", (F.P. Agterberg, 1990).

## 16.1 Introduction

Modern biostratigraphy frequently copes with occurrence data from hundreds of fossil taxa, in thousands of samples, derived from many wells or sections in many different basins. Two challenges, particularly on a regional, basinwide scale, are:

(a) to increase stratigraphic resolution in biozonations, using a combination of eventls derived from several different micro- and/or macrofossil groups, and



*E.A.M. Koutsoukos (ed.), Applied Stratigraphy,* 359-386.

(b) to calibrate these zonations to modern time scales, and extract sequence stratigraphic signals that assist with seismic mapping of prospects, and burial history for basin modelling.

This study focuses on the first challenge. Conventional stratigraphic resolution places considerable emphasis on the end pioints of a few taxa in a few sections. Since a limited number of sections are likely to have the uniform presence or consistent order of all "zonal or index" taxa, there always is a fair amount of subjective judgments as to the perceived "true" order.

New tools in stratigraphy, using semi-quantitative or quantitative methods, make it easier to build integrated zonations, and individual wells or outcrop sections may be tested for 'stratigraphic normality'. Advantages of these tools in stratigraphy, particularly in frontier regions, are:

1. Standardization during digitization of the fossil record and execution of (semi-) objective stratigraphic methods gives easy access to all data and interpretations.
2. Data sets and results are easy to communicate and are rapidly updated with new information.
3. Integration of all fossil and also physical (e.g. isotope, well-log) events in one stratigraphic solution increases resolution and practical use.
4. Methods and results (zonation + correlation) are more objective than 'hand-made' solutions.
5. Zones, events and their correlations may have error bars attached, which improve insight into true stratigraphic resolution and reliability of event correlation.
6. Interpolation of missing event positions in sections increases detail in correlations.
7. Unlike subjective stratigraphy, the new methods provide more than one possible solution of the data, depending on run conditions (multiple working hypothesis).
8. Sequence stratigraphic levels or trends may be detected and defined.
9. The new methods handle large and complex data sets, and calculate reliable solutions quickly.

That is not to say that there are no limiations to quantitative biostratigraphy. For example, the fossil record cannot be modelled a priori for spatial and temporal distributions, and it is difficult to directly weigh records (observations) in terms of stratigraphic quality. In addition, the methods are time consuming because of demands on data organization and data formating. On the other hand, modern studies are showing that benefits of the quantitative approach outweigh limitations, and enhance the quality of geologic interpretations.

## 16.2 Properties of Stratigraphic Data

A paleontological record is the position of a fossil taxon in a rock sequence. The stratigraphic range of a fossil is a composite of all its records. The end-points of the range are biostratigraphic events, which includes the first appearance in time, and disappearance from the geologic record. A biostratigraphic event is the presence of a taxon in its time context, derived from its position in a rock sequence. Fossil events are the result of the continuing evolutionary trends of life on earth; they differ from physical events in that they are unique, non-recurrent, and that their order is irreversible.

Often the first and last occurrences of fossil taxa are relatively poorly defined records, based on few specimens in scattered samples. Particularly with time-wise scattered last occurrences, reworking may have locally extended the record, which may be distinguished by differentiating between the last occurrence (LO), and the last common or last consistent occurrence (LCO) of taxa.

The spacing in relative time between successive fossil events is called resolution. The greater the probability that such events follow each other in time, the greater the likelihood that correlation of the event record models isochrons. Most industrial data sets make use of sets of LO and LCO events. In an attempt to increase resolution in stratigraphy, particuarly when many sidewall cores are available, efforts are made to recognize a half dozen events along the stratigraphic range of a fossil taxon (Fig. 16.1), including last stratigraphic occurrence ('top' or LO event), last common or consistent occurrence (LCO event), last abundant occurrence (LAO event), first abundant occurrence (FAO event), first common or consistent occurrence (FCO event) and first occurrence (FO event). Unfortunately, such practice may not yield the desired increase in biostratigraphic resolution sought after, for reason of poor event traceability.

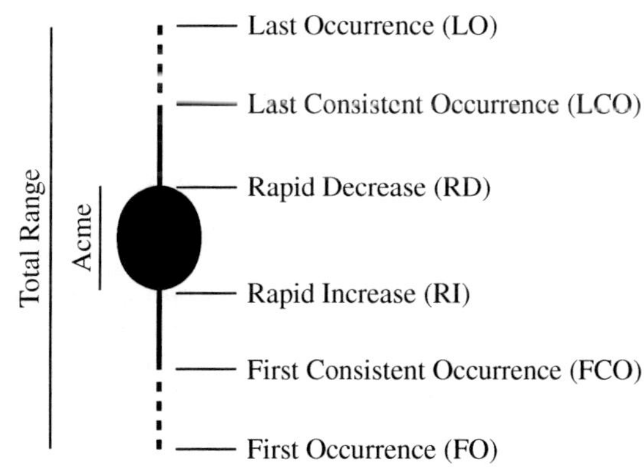


*Figure 16.1 Terminology of biostratigraphic events along the total stratigraphic range of a single taxon. The first and last consistent occurrence frequently coincide with the first and last common occurrence; the rapid increase and rapid decrease encapsulate the acme of the taxon.*

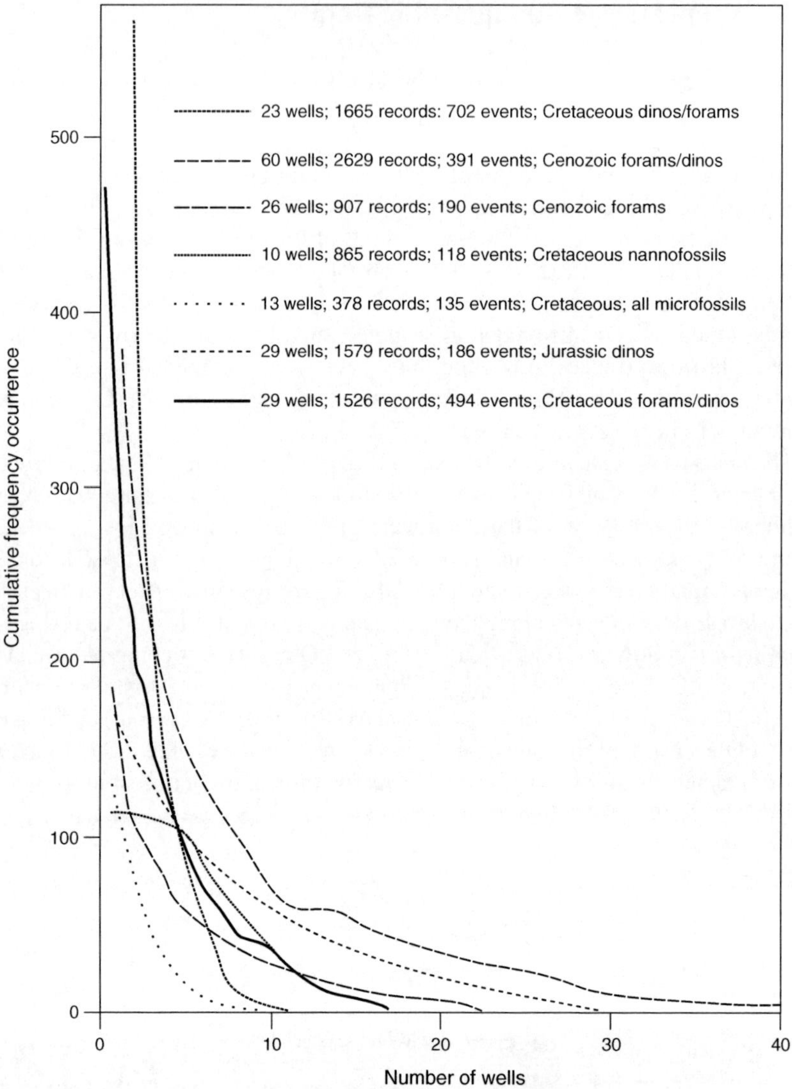


*Figure 16.2 Cumulative frequency of microfossil event occurrences versus number of wells for seven subsurface data sets. Dots – 23 wells/702 events, Cretaceous of North Sea and offshore Norway, various consultants, unpublished; dashes – 60 wells/391 events (Gradstein et al., 1994; Gradstein and Bäckström, 1996, and unpublished); dot-dash – 26 wells/190 events, Cenozoic offshore eastern Canada (Gradstein et al., 1994); closely spaced dots – Upper Cretaceous nannofossils (with relatively high traceability, as seen from the hump in the curve), offshore East Canada (Doeven et al., 1982); pairs of dots – 13 wells/135 events, Indian and Atlantic Oceans Deep Sea Drilling Sites and Ocean Drilling Sites with relatively low fossil diversities and frequent hiatusses (Gradstein et al., 1992); small dashes – Jurassic dinoflagellates, Troll area, offshore southwest Norway (R. Woollam, pers. comm., 1994); solid line – 29 wells/494 events, Cretaceous foraminifers and dinocysts, offshore Norway (Gradstein et al., 1999).*

Poor event traceability is illustrated in Figure 16.2, where cumulative event distributions are plotted using a wide variety of microfossils from different stratigraphic intervals in different basins. All curves are asymptotic, showing an inverse relation between event distribution and the number of wells. None of the events occur in all wells, and far fewer events occur in 5 or 6 wells than only in 1 or 2 wells; hence, the cumulative frequency drops quite dramatically with a small increase in the number of wells. Obviously, the majority of fossil events have poor traceability, which is true for most data sets, either from wells or from outcrops. Groups of microfossils with higher local species diversity, on average have lower event traceability.

Data sets with above average traceability of events are those where one or more dedicated observers have spent above average time examining the fossil record, verifying taxonomic consistency between wells or outcrop sections, and searching for 'missing' data. In general, routine examination of wells by consultants for drilling completion reports yields only half or (much) less of the taxa and events than may be detected with a slightly more dedicated approach.

There are other reasons than lack of details from analysis for why event traceability is relatively low. For example, lateral variations in sedimentation rate change the diversity and relative abundance of taxa in coeval samples between wells, particularly if sampling is not exhaustive, as with well cuttings or sidewall cores. Because chances of detection depend on many factors, stratigraphical, mechanical, and statistical in nature, increasing sampling and studying more than one microfossil group in detail is beneficial.

Although not always explicit, biostratigraphy relies almost as much on the absence, as on the presence of certain markers. This remark is particularly apt for microfossils that generally are widespread and relatively abundant, and compose many stratigraphically useful events. Only if non-existence of events is recognized in many, well-sampled sections, may absences be construed as affirmative for stratigraphic interpretations. If few samples are available over long stratigraphic intervals, the chance to find long-ranging taxa considerably exceeds the chance to find short-ranging forms. In actual practice, index fossils have a short stratigraphic range, are generally uncommon and hence easily escape detection. Therfore, interpretations based on absences should be used with caution.

## 16.3 Data Bookkeeping

An important aspect of quantitative stratigraphy is microfossil event input, and efficient bookkeeping of such records for many wells. The creation of datasets that provide meaningful stratigraphic answers is dependent on such. In the process, detailed checks are advisable to eradicate taxonomic errors and to remove gross outliers representing caved (in exploration wells) or geological recycled (reworked) events. There is no doubt that data input, data bookkeeping and data checking take most of the time in a project. Without a suitable computer program to digitize, organise, pre-digest and filter data of many wells or outcrop sections, such tasks can not be executed, and quantitative biostratigraphic methods cannot be applied. A key property of such a methodology is that all wells in a dataset are accessible simultaneously for queries, corrections, and modifications, so-called multi-well tasking. Both standard spreadsheet or relational database programs can be adapted to the task, and should offer some or all of the following options:

(a) bookkeeping and organisation of fossil events such that they can be queried simultaneously in all wells;
(b) calculation of simple census-type statistics;
(c) tracing of events over all well sites;
(d) finding of co-occurring events, synonyms and geographic substitutes;
(e) cross-plot events from two wells, or of well versus zonation to eliminate outliers;
(f) reformat files for direct input in quantitative stratigraphy programs;
(g) create subsets of the original data to verify local biozonal trends with selected taxa; and
(h) provide a complete printed record of all data, suitable for reports.

The curious anomaly in data processing is the lack of agreement on a standard format in which biostratigraphic data are processed and stored. This is a challenge that needs an urgent consensus.

## 16.4 Stratigraphic Methods

### 16.4.1 Introduction

Traditionally, biostratigraphic zonations are executed 'by hand' through a painstaking process of (mental) stacking in relative geologic time of numerous fossil events from many different outcrop or well sections. Subtle stratigraphic order relationships are evaluated, and frequent gaps are bridged by superpositional hypothesis, where data are scarce. The human mind is good at evaluating observed and virtual superpositional data and bridging data gaps.

Quantitative methods of biostratigraphy, like graphic zonation and correlation, or ranking and scaling cannot easily match the subtilities of very detailed subjective zonations, based on many, often incomplete stratigraphic sections, using much information on missing data. As mentioned above, the experienced biostratigrapher uses almost as much information on absence as on presence of data, and the former cannot be evaluated by a method. To produce a data set that is detailed and informative enough to yield quantifiable, high-resolution zonations is a considerable task. However, once such a dataset and its derived quantitative zonation is accomplished, and made available together with its raw data and data processing details, it serves as a more reliable model for correlation and chronostratigraphic calibration than a poorly documented, subjective zonation. True stratigraphic resolution improves if event spacing in relative time is assessed with standard deviations, that create an understanding as to the chance that two events are superpositional.

### 16.4.2 Deterministic and Probabilistic Methods

There are two principal families of quantitative stratigraphic methods: (a) deterministic, and (b) probabilistic. Deterministic methods seek the total or maximum stratigraphic range of taxa, whereas probabilistic methods calculate the most probable or average range (Fig. 16.3), accompanied by an estimate of stratigraphic uncertainty. Deterministic methods assume that inconsistencies in the stratigraphic range of a taxon from well to well are due to missing data. On the other hand, probabilistic methods assume that the inconsistencies are the result of random deviations from the most commonly occurring or average stratigraphic range. Athough this concept is relatively foreign to conventional biostratigraphy, the large (and often noisy) body of local range data for microfossils in many different basins makes this concept attractive for exploration biostratigraphers in particular.

Hood (1995) evaluated the use of average composite sections from graphic correlation, showing a model of taxon first occurrence with a localized speciation event, with delayed migration into different environments in a basin (Fig. 16.4; see also figure 49 in Thierry, 1997). The shape of the 'average' first appearance emphasises the difference between the use of maximum and average event positions for realistic zonations and

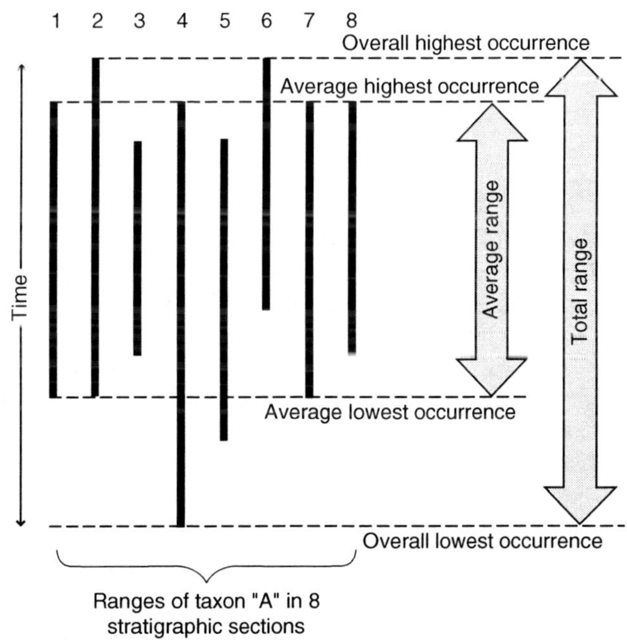


*Figure 16.3 Deterministic biostratigraphy tries to find the total range of a taxon, whereas probabilitistic methods seek the average stratigraphic range. The latter may have an estimate of uncertainty attached that is a function of the spread in local ranges of the taxon in the eight sections examined (after Cooper et al., in press).*

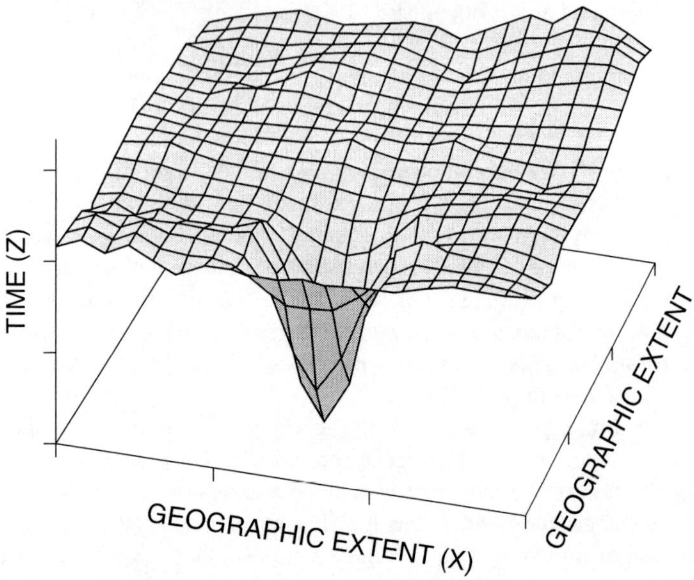


*Figure 16.4 Model of the first occurrence of a taxon with a localized speciation event, with delayed migration into different environments in a basin, that may be modelled with an average first occurrence surface (redrawn after Hood, 1995).*

correlations. Similar figures may be drawn for average last occurrence distributions of fossils.

Deterministic methods are traditionally simple, and most suitable for data sets of few wells or outcrop sections; they lack error analysis, are slow to execute, and sensitive to geological reworking and poor sampling. Probabilistic methods have a mathematical basis, and may be more complex, but have detailed error analysis, execute quickly, and are less sensitive to reworking of taxa or incomplete sampling. A prerequisite for both methods is good data input, and good data organisation, with probabilistic methods more effective with larger data sets. Characteristics of programs in both categories of methods are in Table 16.1.

### 16.4.3 Graphic Correlation

Among deterministic methods, Graphic Correlation is best known. Graphic Zonation and Correlation, also called Shaw's Method (Shaw, 1964) has become accepted and used by academic and industrial biostratigraphers as a simple, semi-objective tool to assess the fossil record for zonation and correlation purposes. This is actually a type of "crossplot" method, where a comparison is made between order and spacing of stratigraphic events in pairs of sections, using bivariate scatterplots. First, one of the stratigraphically more complete sections is selected as the reference, and a second section is crossplotted with it, using the events common to both. On one axis the events are in the order (and spacing if so desired) they have in the reference section, and along the other axis they are in the order (and spacing) that they have in the compared section. The line of correlation (LOC) between two sections is derived either from subjectively connecting points (events) in common in the two-way scattergrams, or from a statistically modeled best fit line between the scattered points. Traditionally, stratigraphers have used either single straight lines, or segmented straight lines as LOC. Using the LOC, the original section is interpolated with the second section, using a few simple rules, to produce a composite sequence. The composite of the two is then crossplotted with a third section, and again interpolated. This procedure is repeated for all (well) sections, and the final solution is called the composite standard. This composite standard is a handy and detailed zonation, expressed in composite standard units that may be used to interpret the sequence history of wells, not unlike conventional geohistory.

For theory and applications with graphic zonation and correlation the excellent book by Mann and Lane (eds., 1995) is recommended, as well as the elegant study with Ordovician graptolite data of Cooper and Lindholm (1991). The latter gives a clear illustration of the principle and utility of the method for closely controlled, high-resolution datasets (Fig. 16.5). The authors created a standard graptolite zonation with 45 FA (first appearance) and 45 LA (last appearance) events of 90 taxa, using fourteen closely sampled and closely studied sections in Australia, Texas, NW Canada, Newfoundland, S. Sweden and S.E. Norway. The Australian sequence, the richest and best known was taken as the initial reference section to be composited with the order of the other thirteen ones. Figure 16.5A shows a plot of the chosen reference section against the S. Sweden section. Figure 16.5B is a plot of the final composite standard sequence against the original Australian section, revealing the extent to which the Australian section has been modified by incorporation of the order of events in other sections. In positioning the LOC, the authors gave weight to events based on species that are relatively distinctive, relatively abundant, and relatively short ranging. The latter minimizes awkward 'unfilled range' situations, and allows plotting the data (by hand) on straight line segments.

### 16.4.4 Constrained Optimization

Recently, a new method has appeared that overcomes some disadvantages of graphic zonation and correlation. It is called CONOP (*con*strained *op*timization), and was designed by Kemple *et al.* (1995), with P.M. Sadler (University of California, Riverside) doing further development. A recent stratigraphic application is by Cooper *et al.* (2001; see below). As in graphic correlation, order and thickness spacing of events in sections are used, but the method is multi-dimensional in the sense that it treats the observations in all sections simultaneously. Like in RASC (see below) it can complete the task of sequencing (the ranking problem), before the task of scaling (the spacing problem).

*Table 16.1* *Properties of principal methods in quantitative biostratigraphy used in exploration biostratigraphy (modified after Cooper et al., in press). Together with their manuals, the programs GRAPHCOR, STRATCOR, CONOP, RASC and CASC provide professional and academic biostratigraphers with rapid and versatile tools to organise, explore and interpret biostratigraphic data for zonation and geological correlation, with estimates of uncertainty*

| *Graphic correlation* | *Constrained optimization* | *Ranking & scaling* |
|---|---|---|
| Programs GRAPHCOR, STRATCOR | Program CONOP | Programs RASC & CASC |
| Deterministic method: graphic correlation in bivariate plots (note: STRATCOR program can operates in a probabilistic or deterministic manner) | Mostly a deterministic method; can also simulate probabilistic solutions<br>Constrained optimisation with simulated annealing and penalty score | Probabilistic method: ranking, scaling normality testing, and automated, most likely graphic correlation with error analysis |
| Uses event order and thickness spacing; works best with datasets having both first and last occurrences of taxa | Uses event order, event cross-over, and thickness spacing; datasets best have both first and last occurrences of taxa | Uses event order from well to well, and scores of cross-over from well to well for all event pairs |
| Best suited for small data sets, but can operate also on large datasets | Processes medium to large data sets | Processes large data sets fast; has data management and data input module |
| Requires selection of an initial standard section, then section by section comparison with the intermediate composite in repeated rounds | Treats all sections and events simultaneously, and works inverse through iteratively improved 'guesses' about the solution | Treats all sections and events simultaneously |
| Line of Correlation (LOC) fitting, in section by section plots, can be partially automated | Multidimensional LOC; automated fitting; can generate several different composites depending on the many run | Automated execution; generates several scaled optimum sequences per dataset depending on run parameters, and tests |

| | | |
|---|---|---|
| Attempts to find maximum stratigraphic range of taxa among the sections | Attempts to find maximum or most common stratigraphic ranges of taxa | Finds average stratigraphic position of first and last occurrence events |
| Builds a composite of events by interpolation of missing events in successive section by section plots, via the LOC | Uses simulated annealing to find either the 'best' or a good multidimensional LOC and composite sequence of events | Uses scores of order relationships to find the most likely order of events, which represents the stratigraphic order found on average among the sections |
| Relative spacing of events is a composite of original event spacing in meters in the sections | Relative spacing of events in the composite is derived from original event spacing in meters or sample levels | Relative spacing of the events in the scaled optimum sequence derived from pair-wise cross-over frequency |
| **No automatic correlation of sections; can be used to build time scales** | **Correlates sections automatically; can be used to build a standard time scale** | **Automated correlation of sections using isochrones** |
| No error analysis; sensitive to geological reworking and other 'stratigraphic noise', and sensitive to order in which sections are composited during analysis | Numerous numerical tests and graphical analysis of stratigraphic results; finds best break points for assemblage zones | Three tests of stratigraphic normality of sections and events; calculates standard deviation of each event as a function of its stratigraphic scatter in wells |
| Interactive operation under DOS; graphic displays of scattergrams and best fit lines | Batch operation under windows; colour graphics displays shows progress of runs | Button operated under windows, fast batch runs; colour graphics of output and options for interactive graphics editing |

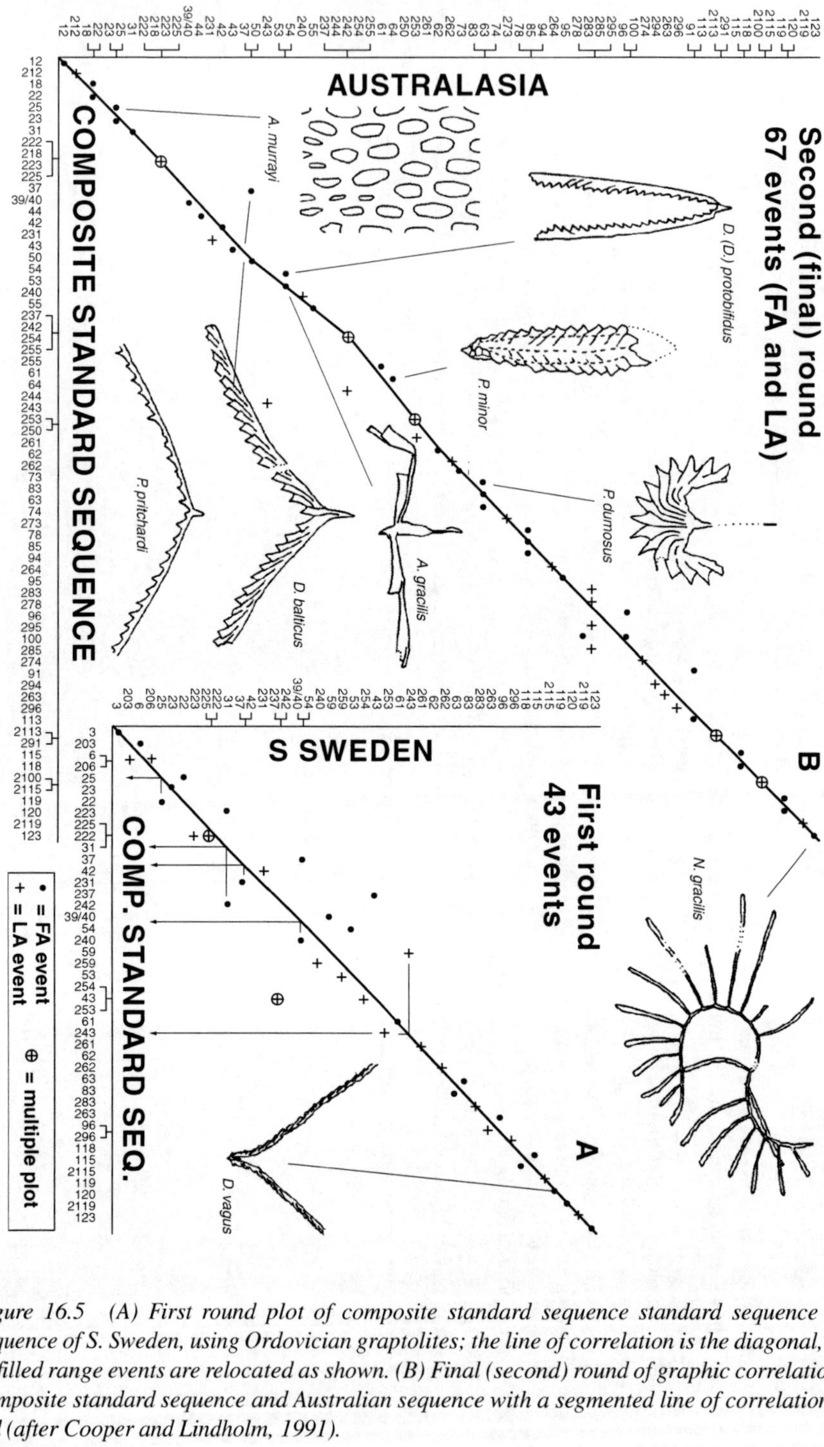


*Figure 16.5 (A) First round plot of composite standard sequence standard sequence and sequence of S. Sweden, using Ordovician graptolites; the line of correlation is the diagonal, and unfilled range events are relocated as shown. (B) Final (second) round of graphic correlation of composite standard sequence and Australian sequence with a segmented line of correlation fitted (after Cooper and Lindholm, 1991).*

Instead of building a solution from the data, CONOP works through a series of iteratively improved guesses about the solution. Each guess is compared with the data; the misfit between the solution and data guides the next guess. Geophysicists like to call the process 'inversion'. Unfortunately, solution time increases as $2^N$, where N is the number of events. This means that for an exhaustive search of e.g. 124 events (LO and FO of 62 taxa), the searching time becomes impossible. In order to find a good solution without waiting 'forever' CONOP uses a version of the simulated annealing algorithm, using heuristic search techniques; such a technique is incapable of proof, but serves to guide to acceptable solutions.

The method is constrained in that it eliminates impossible solutions (constraint), and then searches for the best of all the possible ones (optimization). The method may be thought of as fitting a multidimensional line of correlation (LOC) simultaneously to all points in all sections. The composite 'true' section of events is that hypothetical sequence of ordered and spaced events that causes the least net disruption or penalty when the ranges of taxa in each of the well sections are adjusted to match it. Like graphic correlation, the observed tops of species in individual well sections are extended stratigraphically upwards, and bases downward to achieve a best fit. In this sense, penalty represents a measure of inconsistency of individual tops or bases among the well sections, and is expressed in meters. This penalty resembles that used in method STRATCOR (Gradstein, 1996) that keeps track of the cumulative amount (distance) over all wells that events shift from their observed position to their interpolated one. CONOP has a host of other features, and presently builds a Lower Paleozoic conodont-graptolite composite that assists with the construction of a detailed geologic time scale (R. Cooper, pers. comm., 2000).

### 16.4.5 Ranking and Scaling

The principal method of probabilistic biostratigraphy is called Ranking and Scaling (RASC; Agterberg and Gradstein, 1999; Gradstein *et al.*, 1985, 1999). The many options in RASC method of biostratigraphy are listed in Table 16.2. During the last two decades, RASC has been applied to a wide variety of datasets involving many tyoes of microfossils. A majority of applications are with well data sets from industry or scientific ocean drilling. Published literature on and with the method is extensive, and is listed in the literature cited.

Unlike graphic correlation, the RASC method considers the stratigraphic order of all fossil events in all wells simultaneously, and calculates the most likely (optimum) sequence of events. In this sequence, each event position is an average of all individual positions encountered in the wells.

Ranking is based on superpositional relationships between events. In general, there are three possible types of superpositional relationships for a pair of events co-occurring in the same section. An event can be observed to occur above or below another event, or the two events coexist in the same sample. In the ranked optimum sequence, which is based on a large number of sections, two events can be coeval on

*Table 16.2 Products of the RASC and CASC programs for probabilistic stratigraphy*

***PROGRAM RASC – zonation, variance analysis and normality testing***

Bookkeeping
- Value of input parameters
- Sequence of wells
- Tabulation of event records, using frequency and cumulative frequency
- Summary of RASC run results (vital statistics)
- Dictionaries of events – numerical and alphabetic listings
- Occurrence table of events in all wells

Ranking and Scaling (= probabilistic zonation)
- Optimum sequence of events, with option to insert variances and unique (rare) events
- Final scaling of optimum sequence of events, with option to insert unique events

Normality 'testing' of event record in wells
- Graphical correlation of well sequence record and optimum sequence, with estimation 'how far' events are off best fit line (cubic spline)
- Step model per well, with penalty points for out of place events
- Rank correlation of event well sequences with (scaled) optimum sequence
- Normality test per well, with second order difference statistics for all events
- Comparison of observed and expected second order difference values

Variance analysis
- Standard deviations of events per well
- Event variance analysis (difference in each well between observed and stratigraphically expected event position + frequency distribution)
- Summary of event variance analysis results
- Estimation of event ranges – numerical and graphical representation of probable minimum–, probable maximum –, and average observed stratigraphic event positions
- Estimation of event cross-over ranges

***PROGRAM CASC – correlation of RASC zonation, with flattening option in graphics displays***
- Probable position in wells of optimum sequence events, with 95% confidence limits
- Observed event positions with 95% confidence limits

the average when one of them occurs exactly as many times above the other one as it occurs below it. If an event is observed above another event in some sections but below it in others, a stratigraphic inconsistency involving these two events is indicated. The purpose of constructing the optimum sequence is to utilize and resolve such inconsistencies. In fact, there is no point in applying RASC if inconsistencies are missing when sections are compared with one another. Lines of correlation connecting observed positions of events in sections show cross-overs when there are inconsistencies, which is a normal event feature (see Fig. 16.6).

Scaling of the optimum sequence in relative time provides information on the stratigraphic clustering of events, and is a function of the frequency with which events in each each pair in the RASC optimum sequence cross-over their relative positions (observed records) from well to well. The more often any two events cross-over, the smaller their interfossil distance. Final distance estimates are expressed in dendrogram

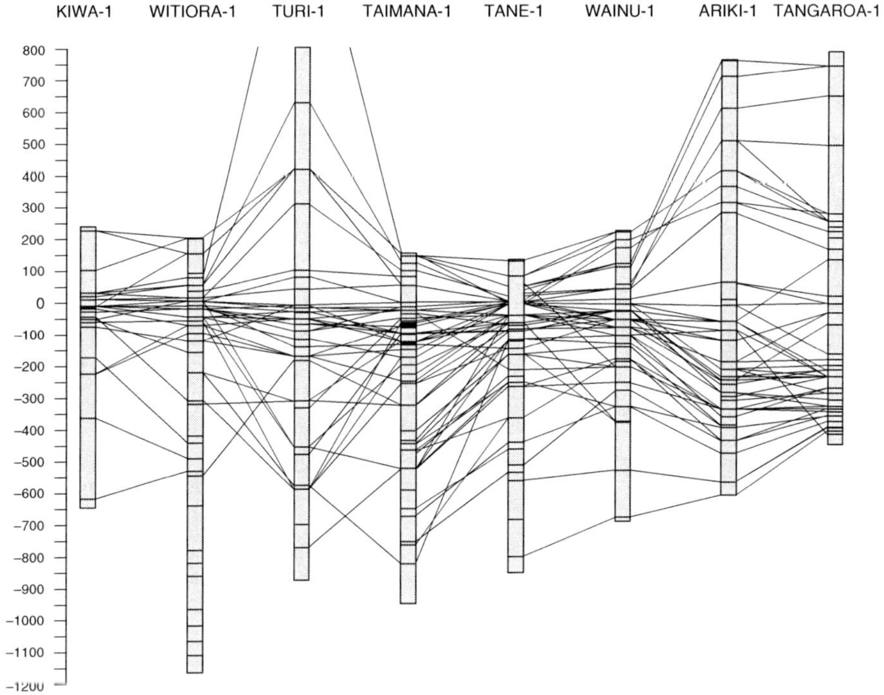


*Figure 16.6 Direct correlation of the events in the CONOP Composite Sequence of Figure 16.7, based on the observed depth of the range end events, produces an intricate network with the usual cross-overs and mismatches of conventional well correlations.*

format, where tightness of clustering is a measure of nearness of events along a stratigraphic scale. The scaled version of the optimum sequence features time successive clusters, each of which bundles distinctive events. Individual bundles of events are assigned zonal status. The process of zone assignment in the scaled optimum sequence is subjective, as guided by the stratigraphic experience of the users. Large interfossil distances between successive dendrogram clusters agree with zonal boundaries, reflecting breaks in the fossil record due to average grouping of event extinctions. Such extinctions occur for a variety of reasons, and may reflect sequence boundaries. From a practical point of view it suffices to say that taxa in a RASC zone on average group close together in relative time.

### 16.4.6 Variance Analysis

Anyone that tries to apply an event correlation framework from a zonation, quickly notices that closely spaced events tend to cross-over between wells (Fig. 16.6), indicative of some kind of uncertainty in event sampling and stratigraphic position. One way to rationalize such uncertainty is to use RASC zonations with variances, that identify more reliable stratigraphic markers (D'Iorio and Agterberg, 1989; Gradstein and Agterberg, 1998). The principle of variance calculation is straightforward. Individual well sequences are compared to the scaled optimum sequence using bivariate curve fitting to obtain the sum of differences between the observed and expected values of the events. The expected values are on the best fit line, meaning that an event would have zero variance. Once the differences between the observed and expected values are compiled for all RASC optimum sequence events in all wells, frequency distributions can be estimated. If an event is close to all lines of correlation considered it has a relatively small standard deviation; this means that it is a relatively good marker, as will be demonstrated for *Subbotina patagonica* in the lower Eocene, North Sea.

Graphical representation of differences between observed and expected positions at the well locations, such as on a map, may show that the large variance of an event is due to transgressive behaviour. Thus, variance analysis can be useful for tracking the time transgressive behaviour of events, as illustrated below.

RASC features two more tests to determine if the individual well record differs from the most likely zonation, the Stepmodel and the Normality tests (Table 16.2), both described in the literature cited. These tests, like variance analysis guide the user to outliers due to event misidentification, reworking or sample contamination, and to 'good and bad' wells.

### 16.4.7 Correlation and Standard Error Calculation

The preceding chapters dealt with zonation of fossil events, and tests to discern "good and bad" events and "good and bad" wells, using the RASC method. A companion method performs geologic correlation of RASC events and standard error calculation, and is named CASC, for *C*orrelation *a*nd *S*tandard Error *C*alculation. The technique is an extension of so-called "graphic correlation", as discussed in more detail in Gradstein *et al.* (1985), and in Agterberg (1990). Uncertainty limits (error bars) are calculated for both the most likely–, and the observed event positions in the wells or outcrop sections. Well correlation diagrams are both displayed in numerical and in colour graphics format. A detailed application in the Lower Cretaceous subsurface of the Grand Banks, eastern Canada, that also converts the RASC optimum sequence based on foraminifers to a RASC timescale and correlates isochrons, is in Williamson (1987); a recent application of CASC on a large Cretaceous dataset with many events using dinoflagellate cysts and foraminifers, offshore Norway is in Gradstein *et al.* (1999).

## 16.5 Stratigraphic Applications

### 16.5.1 Constrained Optimization: Taranaki Basin

The Taranaki Basin, New Zealand's producing hydrocarbon province, contains a highly fossiliferous Upper Cretaceous–Cenozoic sedimentary succession, resting unconformably on an erosional surface of varied relief that cuts across a 'basement' of Paleozoic and Mesozoic rocks. The basin has a complex depositional history with depositional breaks, condensed intervals, contemporaneous faults and folds, and lateral facies changes. Biostratigraphy from over 80 wells is an essential tool to interpret depositional history. Cooper *et al.* (2001) developed a detailed statistically based biozonation scheme for the purpose of increasing stratigraphic resolution, and assessing depositional rates across the basin. In addition, the team of stratigraphers wanted to evaluate the relative merits of deterministic and probabilistic approaches to quantitative biostratigraphic subdivision and correlation, using the methods outlined above: CONOP, RASC, and GRAPHCOR.

In all, the dataset of choice comprises 8 wells, from which 351 usable range tops of 351 foraminifers, nannofossils and palynomorphs were extracted from early Paleocene through early Miocene ages. In order to emphasize events that have correlation potential, and to calculate event variances, those events that occur in fewer than four wells were removed from the dataset. However, the methods allow unique events back in the analysis, such as index fossils or local marker horizons found in fewer than 4 wells. The dataset was thus reduced to 178 events in the 8 wells, with first stages of analysis leading to removal of 91 more events as being highly inconsistent in position from well to well, or having tops elsewhere above the youngest level sampled in the wells. The final dataset hence comprises 87 events, with 508 records. Direct correlation of the events in the CONOP composite, based on the observed depth of the range-end tops, produced an intricate network with the usual cross-overs and mismatches (Fig. 16.6) of conventional well correlations. In Figure 16.7, the CONOP ordinal and scaled composite sequences are displayed; the same figure also shows correlation of the calculated zonation to the regional scheme of stages using key markers, and correlation of the zonation through three wells with interpolated depths, like in graphic correlation. Note that the scaled composite section (right column of Fig. 16.7) is in arbitrary units based upon interpolation and extrapolation of the stratigraphic thicknesses of the interevent-units in all well sections.

The main conclusion of the study is that the RASC probable sequence (not shown) and the CONOP composite sequence are remarkably similar, and the two compare well with classical graphic correlation that cannot be executed in automated batch mode. Several stratigraphically promising species events were detected, not generally used for conventional biostratigraphy. The CONOP composite section gives the best estimate of the 'true' stratigraphic tops of taxa, based on the 8 well sections. It is consistent with the aims of conventional biostratigraphy based on range-end events, which are to establish zonations and correlations schemes based on the (maximum) ranges of species. Hence, it relates best to the conventional regional stratigraphy of New Zealand. The RASC scaled optimum sequence, on the other hand, gives the most probable order of events and its spacing and is particularly useful as a predictive zonation and correlation tool for future exploration drilling in the basin. The probabilistic and deterministic techniques are experienced as complimentary in order to best understand biostratigraphic potential of a dataset.

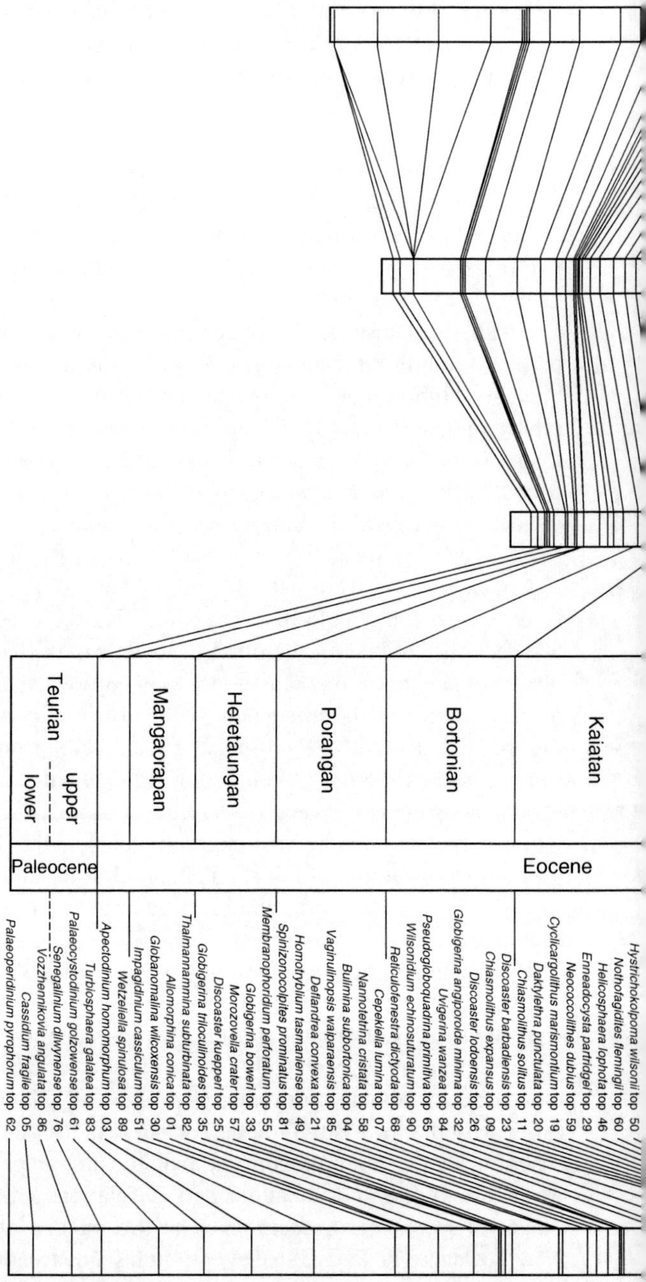


*Figure 16.7 The CONOP Composite Sequence, both ordinal and scaled, its correlation to the regional scheme of stages using key markers, and correlation of the zonation through*

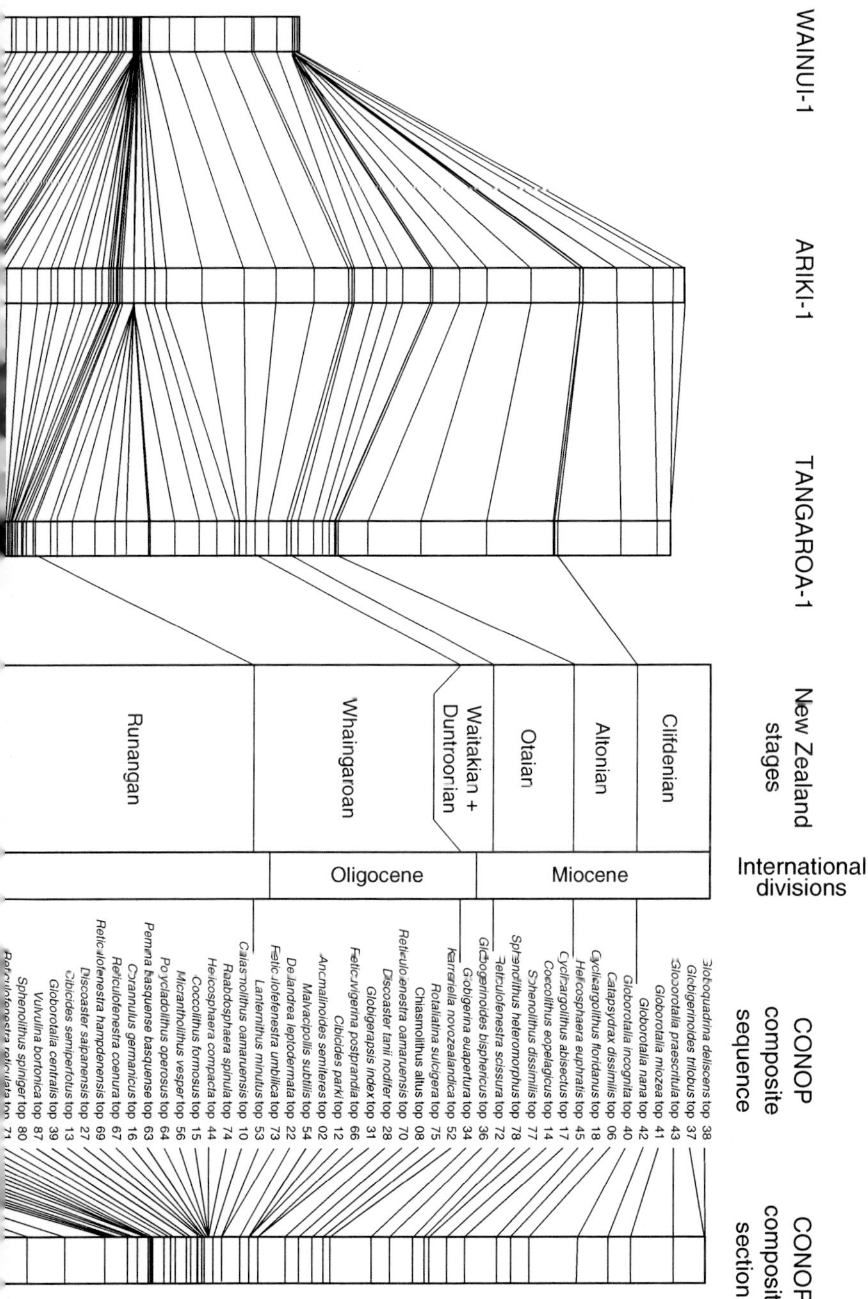


*Figure 16.7 Continued*

*three wells with interpolated depths, Paleocene to Miocene of the Taranaki Basin, New Zealand. Each event has a dictionary number behind it.*

### 16.5.2 Ranked and Scaled Optimum Sequences with Variances: North Sea

The North Sea region, a prolific petroleum province, contains remnants of stratigraphically superimposed sedimentary basins of Late Paleozoic through Cenozoic age, like stacks of half eaten pancakes. The regional history is complex; differential subsidence and uplifts are related to extensive mobilization of the North Atlantic rift systems. Deeper water, bathyal sediments, including minor and major gravity flow, siliciclast wedges, of middle Cretaceous through Paleogene age, are widespread and contain diversified agglutinated benthic foraminifera assemblages (Jones, 1988; Gradstein and Bäckström, 1996). In the southern part of the central North Sea, where deep water conditions prevailed into Miocene, the (DWAF) assemblage accordingly extends stratigraphically upwards (see below). The assemblages assist with biostratigraphy and paleobathymetry in exploration and exploitation wells, and DWAF taxa are prominent in the regional RASC zonation, as shown below.

The large scale deposition of basaltic ash (Balder unit) during earliest Eocene coincides with the eruption of major flood basalts in eastern Greenland and Rockal, at the onset of seafloor spreading in the Norwegian Sea. The ash is a prominent North Sea seismic reflector. Due to the flood-basalt outpourings, the North Sea became restricted, as reflected in the widespread distribution of diatoms, including the pyritized pillbox *Fenestrella antiqua* (Rank position 69 in Fig. 16.8, and Zone NSR3 in Fig. 16.9), and virtual absence of bottom fauna in the severely dysaerobic basin. Surface water salinity may have been abnormally low.

Correlations between the onshore NW Europe and North Sea Basin succession in the Paleocene and Eocene is achieved by dinoflagellate cyst biostratigraphy, integrated with the biostratigraphy provided by the calcareous plankton (foraminifera and nannoplankton) benthic foraminifera, magnetostratigraphy and volcanic ash stratigraphy. In this way a correlation network has been established over NW Europe, which serves as the background against which the probabilistic zonation was developed shown below. This probabilistic zonation serves as a template for range charts of DWAF in the petroleum basin.

Figure 16.8 shows the RASC optimum sequence with standard deviations, calculated with the variance analysis method (see section on Variance Analysis), using 1430 event records in 30 wells, based on the LO and LCO occurrences of 289 benthic and some planktonic foraminifera and dinoflagellates, most or them analysed 'in-house', which greatly enhances taxonomic consistency. In addition, North Sea log markers were incorporated (NS Log B–G) for which precise well depths data are available (see Gradstein *et al.*, 1994 for details). Each of the 88 events in the zonation occurs in at least 7 wells, except for 16 unique events (marked with two asterisks in Figs. 16.8 and 16.9) that occur in fewer than seven wells, and are inserted to complement the zonation, and/or assist with age calibration. Forty-two of the events in the optimum sequence, including many DWAF taxa, have standard deviation below average, which is quite a good number for an industrial type dataset.

The RASC scaled optimum sequence with zones assigned is shown in Figure 16.9. There are 18 zones and subzones assigned, named NSR1–13 (NSR = North Sea

RASC), of early Paleocene through early Pleistocene age. Large breaks (at events 129, 50, 206, 6, 266 and 23) indicated transitions between natural microfossil sequences, and/or hiatuses, and are candidates for sequence stratigraphic breaks if corroborated by regional seismic analysis. The zones contain 33 DWAF events (32 LO and 1 LCO events) for 32 taxa. On average, event observation in the wells may be closer to the average stratigraphic position than the last occurrence end points in regional range charts.

The average range end of DWAF in the North Sea (coarse agglutinated spp. in Fig. 16.9) is generally in mid-Cenozoic, and falls in zone NSR8A, late Oligocene. It occurs in 27 out of 30 wells in the dataset, and has an sd of 2.402, which is above average (Fig. 16.8). With variance analysis technique in RASC it is possible to display the well deviations (Fig. 16.10). Since the wells in this figure are arranged from north (left) to south (right), the method confirms that the DWAF LO is time-transgressive, with younger occurrences southward, as can be readily observed in southern (Central Graben) wells. Since in subregions of the dataset the DWAF LO is a reasonable good marker, a histogram of stratigraphic deviations (Fig. 16.10a), although rather wide still shows a reasonable central (normal Gaussian) distribution.

A more marked central tendency, and a much lower sd of 1.287, despite a sample size of only 14 and not 27 as for DWAF, is calculated for the distinctive planktonic foraminifer *Subbotina patagonica* of Zone NSR4, Lower Eocene (Fig. 16.10).

Given a large enough dataset (10 or more wells is a reasonable rule), the variance analysis technique allows to rapidly rank events in the RASC solutions on stratigraphic fidelity.

## 16.6 Computer Programs

A PC desktop computer version under DOS of the original method of graphic correlation is program GRAPHCOR (Hood, 1995; K.C. Hood, 9707 Arrowgrass Dr., Houston, TX 77064, USA). It was converted by Amoco Oil Co for interactive operation on a Unix workstation. An example of adaptation of graphic correlation to probabilistic stratigraphy is found in program STRATCOR (Gradstein, 1990), which has hybrid features to RASC.

Desktop PC program CONOP under DOS, has many features to analyse medium-size stratigraphic datasets. It has colour graphics displays, and the progress of search for the optimal range chart may be watched on screen, which is an instructive option. The program is actively being developed by Peter M. Sadler (Department of Earth Sciences, University of California, Riverside, CA 92521, USA).

The probabilistic stratigraphy programs that perform ranking, scaling, correlation and standard error calculation operate as a single module under MS Windows, and are called RASC & CASC. One windowing master menu controls the operation of the programs and their results that also include the data input and re-organising module called MAKEDAT, and correspondence analysis program COR (Hill, 1979; Bonham-Carter

| Nº & UI | | | | N | SD |
|---|---|---|---|---|---|
| 1 | | 77 | *Elphidium* spp. | 20 | 0.675 |
| 2 | | 228 | *Cassidulina teretis* | 20 | 0.559 |
| 3 | | 31 | *Cibicidoides scaldisensis* | 8 | 0.724 |
| 4 | | 270 | *Cibicidoides grossa* | 17 | 0.656 |
| 5 | | 1 | *Neogloboquadrina pachyderma* | 12 | 1.068 |
| 6 | | 171 | *Buccella frigida* | 7 | 1.421 |
| 7 | | 318 ** | *Monspeliensis pseudotepida* | | |
| 8 | | 316 | *Trifarina fluens* | 7 | 1.909 |
| 9 | 0 1 | 23 | *Sigmoilopsis schlumbergeri* | 18 | 1.219 |
| 10 | −10 | 4 | *Globorotalia inflata* | 9 | 1.123 |
| 11 | 0 1 | 269 | *Neogloboquadrina atlantica* | 14 | 1.162 |
| 12 | | 74 ** | *Neogloboquadrina humerosa* | | |
| 13 | −10 | 266 | *Globorotalia puncticulata* | 8 | 0.914 |
| 14 | 0 2 | 130 | *Siphonina advena* | 9 | 1.083 |
| 15 | −10 | 6 | *Neogloboquadrina acostaensis* | 7 | 0.929 |
| 16 | 0 1 | 219 | *Martinotiella cylindrica* | 11 | 2.347 |
| 17 | | 301 ** | *Bolboforma metzmacheri* | | |
| 18 | −11 | 91 | Diatoms/radiolarians LCO | 12 | 1.475 |
| 19 | −11 | 207 | NS Log G | 15 | 1.653 |
| 20 | −10 | 282 | *Uvigerina* ex.gr. *semiornata* | 7 | 1.777 |
| 21 | | 58 ** | *Neogloboquadrina atlantica* (dext) | | |
| 22 | | 411 ** | *Globocassidulina subglobosa* var. | | |
| 23 | | 123 | *Globigerinoides trilobus* | 8 | 1.022 |
| 24 | | 236 | *G.* ex.gr. *praescitula zealandica* | 16 | 1.281 |
| 25 | | 285 | *Caucasina elongata* | 10 | 1.478 |
| 26 | | 362 ** | *Cannosphaeropsis utinensis* | | |
| 27 | | 125 | *Neogloboquadrina continuosa* | 8 | 2.150 |
| 28 | | 17 | *Asterigerina gurichi* | 21 | 1.625 |
| 29 | | 18 | *Spiroplectammina carinata* | 7 | 1.867 |
| 30 | | 15 | *Globigerina praebulloides* | 18 | 2.369 |
| 31 | | 528 ** | *Catapsydrax parvulus* | | |
| 32 | | 348 ** | *Sphaeroidinella disjuncta* | | |
| 33 | | 138 | *Tenuitella angustiumbilicata* | 7 | 3.568 |
| 34 | | 111 | *Globigerina ex.gr. officinalis* | 7 | 3.855 |
| 35 | | 413 | *Silicious biofacies* | 8 | 1.026 |
| 36 | | 20 | *Gyroidina girardana* | 12 | 2.633 |
| 37 | | 25 | *Coarse agglutinated* spp. | 28 | 2.512 |
| 38 | | 369 ** | *Rhombodinium draco* | | |
| 39 | | 368 ** | *Chiropteridium mespilanum* | | |
| 40 | | 97 | *Cyclammina placenta* | 24 | 2.968 |
| 41 | | 182 | *Spirosigmoilinella compressa* | 22 | 1.052 |
| 42 | | 262 | *Karrerulina horrida* | 12 | 3.780 |
| 43 | | 259 | *Ammodiscus latus* | 9 | 2.407 |

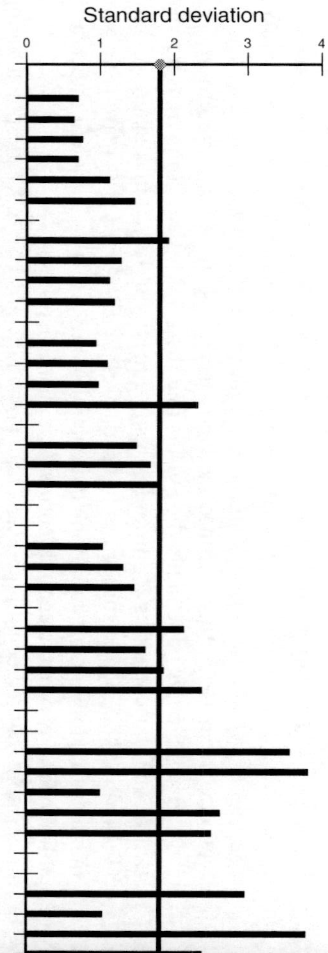

| | | | | | |
|---|---|---|---|---|---|
| 49 | | 289 | *Adercotryma agterbergi* | 13 | 1.255 |
| 50 | | 148 *– | *Globigerapsis index* | | |
| 51 | | 12 | *Areosphaeridium diktyoplokus* | 8 | 1.864 |
| 52 | | 535 | *Annectina biedai* | 8 | 3.061 |
| 53 | | 229 | *Recurvoides* ex.gr. *walteri* | 8 | 2.960 |
| 54 | | 206 | NS Log F | 14 | 2.066 |
| 55 | | 29 | *Reticulophragmium amplectens* | 28 | 1.511 |
| 56 | | 32 | *Ammosph.pseudopauciloculata* | 17 | 2.085 |
| 57 | | 263 | *Ammomarginulina aubertae* | 14 | 2.073 |
| 58 | | 260 | *Haplophragmoides kirki* | 14 | 2.113 |
| 59 | | 68 | *Spiroplectammina spectabilis* LO | 24 | 1.932 |
| 60 | | 37 ** | *Acarinina pentacamerata* | | |
| 61 | | 264 | *Karrerulina conversa* | 19 | 3.433 |
| 62 | | 205 | NS Log E | 15 | 1.337 |
| 63 | | 14 ** | *Eatonicysta ursulae* LCO | | |
| 64 | | 47 ** | *Planorotalites planoconicus* | | |
| 65 | | 54 | *Spiroplectammina navarroana* | 19 | 1.721 |
| 66 | | 279 | *Recurvoidella lamella* | 9 | 1.464 |
| 67 | | 50 | *Subbotina patagonica* | 14 | 1.287 |
| 68 | 0 1 | 277 | *Tuff* | 7 | 1.089 |
| 69 | –10 | 22 | *Fenestrella antiqua* | 12 | 1.323 |
| 70 | | 78 | *Trochamminoides coronatus* | 10 | 3.165 |
| 71 | | 204 | NS Log D | 15 | 0.983 |
| 72 | | 203 | NS Log C | 15 | 1.111 |
| 73 | | 105 | *Rzehakina minima* | 8 | 0.894 |
| 74 | | 343 ** | *Cenosphaera* sp. (disc or flat) | | |
| 75 | | 310 | *Saccammina placenta* | 11 | 2.162 |
| 76 | | 76 | *Reticulophragmium pauperum* | 19 | 0.897 |
| 77 | 0 2 | 283 | *Reticulophragmium garcilassoi* | 7 | 2.083 |
| 78 | –10 | 57 | *Spiroplectammina spectabilis* LCO | 20 | 0.972 |
| 79 | | 129 | *Ammoanita ruthvenmurrayi* | 11 | 1.839 |
| 80 | | 134 | *Caudammina excelsa* | 11 | 2.512 |
| 81 | | 532 | *Cystammina sveni* | 12 | 2.860 |
| 82 | | 531 ** | *Ammoanita ingerlisae* | | |
| 83 | | 55 | *Gavelinella beccariiformis* | 15 | 1.245 |
| 84 | | 107 | *Remesella varians* | 8 | 1.670 |
| 85 | | 202 | NS Log B | 12 | 1.143 |
| 86 | 0 1 | 61 | *Subbotina pseudobulloides* | 18 | 1.046 |
| 87 | –10 | 60 | *Planorotalites compressus* | 12 | 0.554 |
| 88 | | 253 | *Subbotina triloculinoides* | 10 | 0.593 |

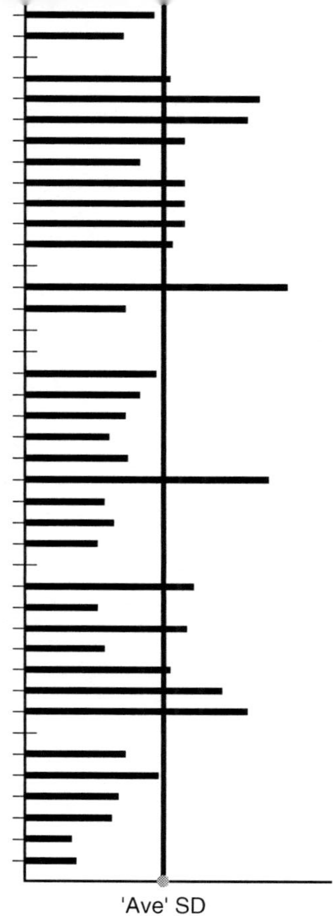


*Figure 16.8 RASC Optimum Sequence for the Cenozoic of the North Sea; low standard deviations are an indication of good stratigraphic markers; ave SD = average event standard deviation; N = event occurrence in wells; UI = uncertainty interval on event position (in this case +/– 1 or 2 positions); each event has a dictionary number in front.*

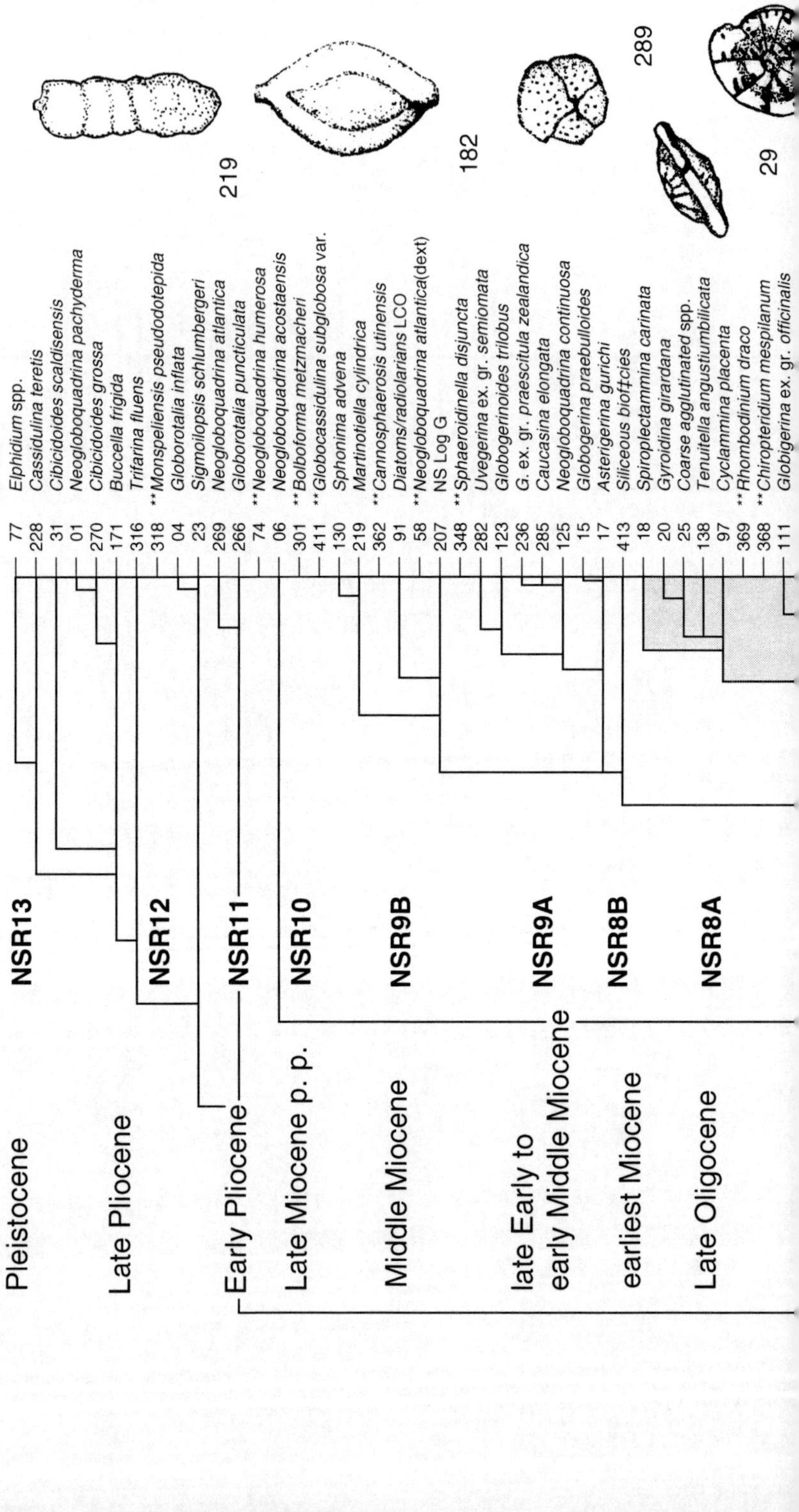
Pleistocene
Late Pliocene
Early Pliocene
Late Miocene p. p.
Middle Miocene
late Early to early Middle Miocene
earliest Miocene
Late Oligocene
NSR13
NSR12
NSR11
NSR10
NSR9B
NSR9A
NSR8B
NSR8A
77 Elphidium spp.
228 Cassidulina teretis
31 Cibicidoides scaldisensis
01 Neogloboquadrina pachyderma
270 Cibicidoides grossa
171 Buccella frigida
316 Trifarina fluens
318 **Monspeliensis pseudotepida
04 Globorotalia inflata
23 Sigmoilopsis schlumbergeri
269 Neogloboquadrina atlantica
266 Globorotalia puncticulata
74 **Neogloboquadrina humerosa
06 Neogloboquadrina acostaensis
301 **Bolboforma metzmacheri
411 **Globocassidulina subglobosa var.
130 Sphonima advena
219 Martinotiella cylindrica
362 **Cannosphaerosis utinensis
91 Diatoms/radiolarians LCO
58 **Neogloboquadrina atlantica(dext)
207 NS Log G
348 **Sphaeroidinella disjuncta
282 Uvegerina ex. gr. semiomata
123 Globogerinoides trilobus
236 G. ex. gr. praescitula zealandica
285 Caucasina elongata
125 Neogloboquadrina continuosa
15 Globogerina praebulloides
17 Asterigerina gurichi
413 Siliceous biof‡cies
18 Spiroplectammina carinata
20 Gyroidina girardana
25 Coarse agglutinated spp.
138 Tenuitella angustiumbilicata
97 Cyclammina placenta
369 **Rhombodinium draco
368 **Chiropteridium mespilanum
111 Globigerina ex. gr. officinalis
219
182
289
29

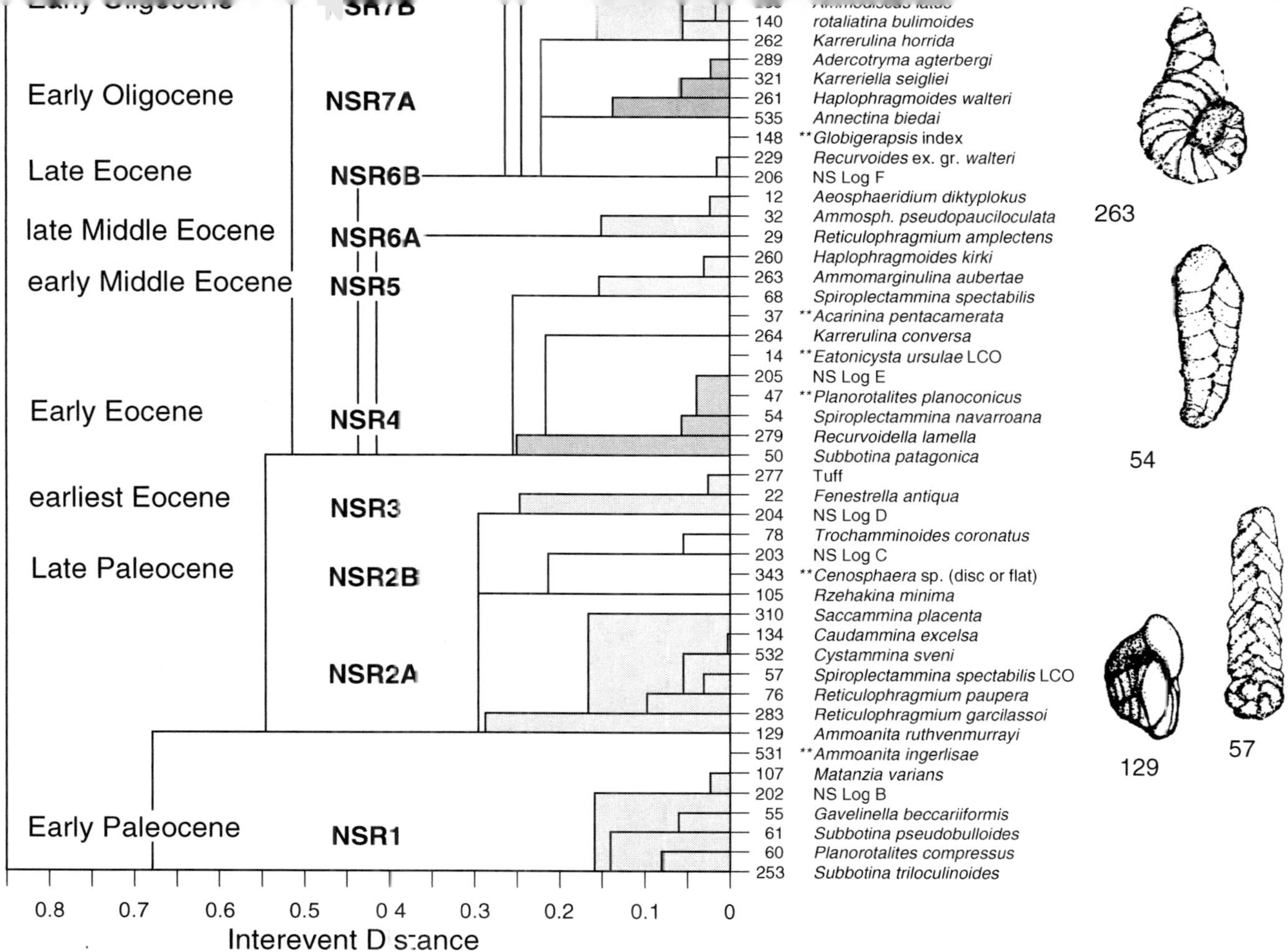


*Figure 16.9 Scaled version of the RASC Optimum Sequence of Figure 16.8, with eighteen interval zones assigned (NSR1 – 13) of early Paleocene through Pleistocene age.*

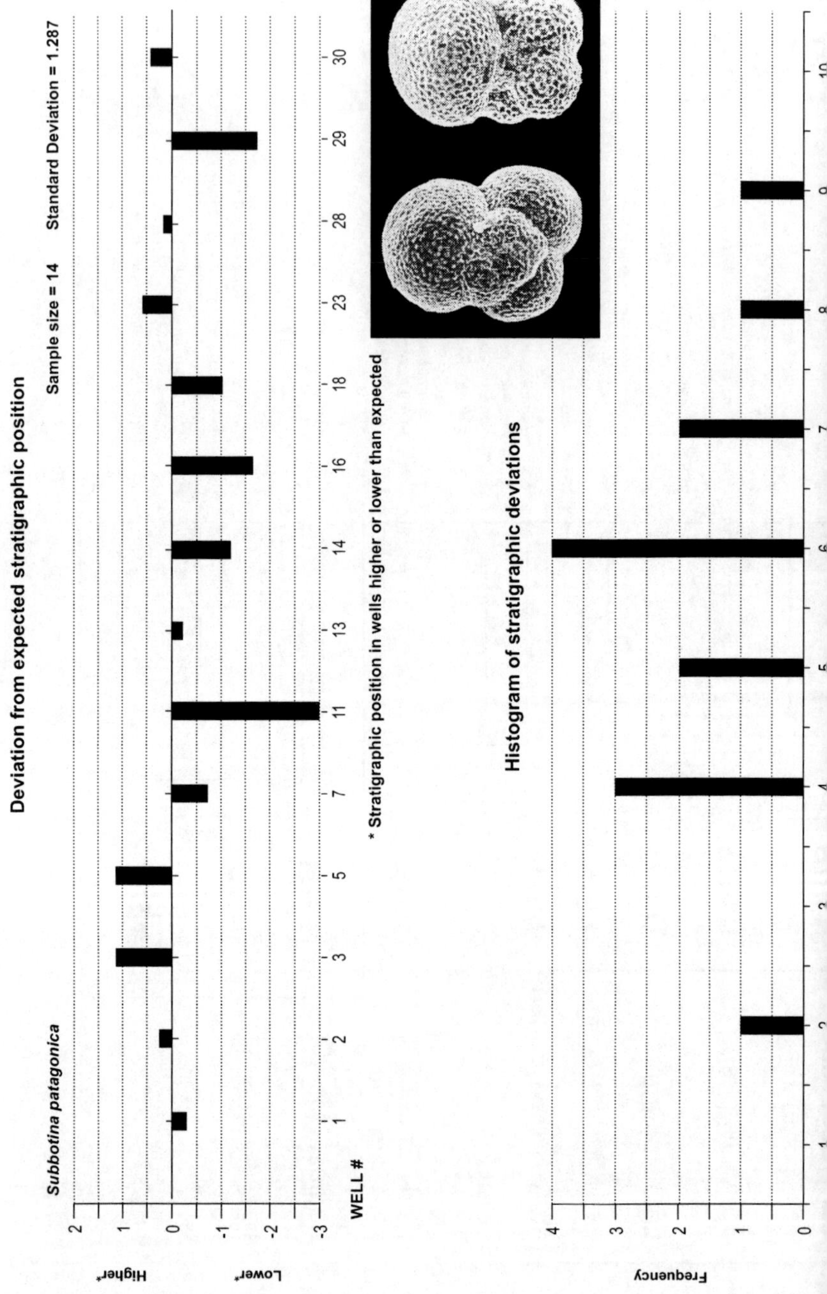
Subbotina patagonica
Deviation from expected stratigraphic position
Sample size = 14
Standard Deviation = 1.287
Higher*
Lower*
2
1
0
-1
-2
-3
WELL #
1
2
3
5
7
11
13
14
16
18
23
28
29
30
* Stratigraphic position in wells higher or lower than expected
Histogram of stratigraphic deviations
Frequency
4
3
2
1
0

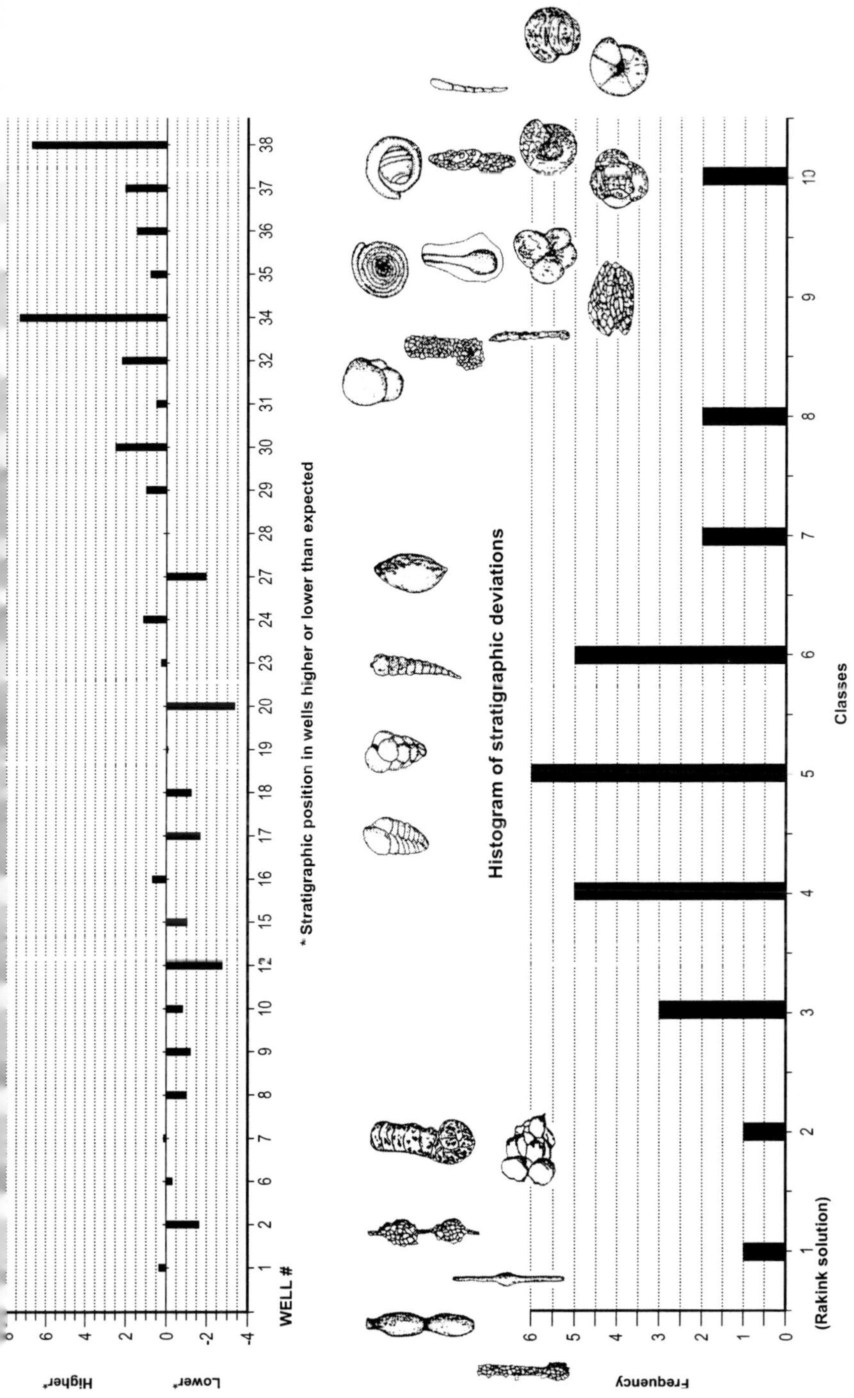


*Figure 16.10* *Deviations from expected stratigraphic position and histograms of stratigraphic deviations for the average LO of Deep Water Agglutinated Foraminifera, and of the planktonic foraminifer* Subbotina patagonica, *North Sea.*

*et al.*, 1986). Graphics results are displayed in colour, and may be modified and edited with a build-in 2D chart control program, and colour printed or plotted from the screen displays. The program is actively being developed by F.P. Agterberg and F.M. Gradstein. More details may be found on websites *www.rasc.uio.no*, and *www.stratigraphy.org*.

## 16.7 Acknowledgements

Thanks are due to Frits Agterberg, Peter Thompson and an anonymous reviewer for constructive comments that improved this manuscript, and to H.V. dos Santos, Petrobras-Cenpes, for excellent assistance with drafting of several figures. Roger Cooper and Peter Sadler provided valuable scientific advice and permission to utilize two figures.

# References


Abdala, N. F., Ribeiro, A. M., and Schultz, C. L., 2001, A rich cynodont fauna of Santa Cruz do Sul, Santa Maria Formation (Middle-Late Triassic), southern Brazil, *Neues Jahrbuch für Geologie und Paläontologie*, Monatshefte, Tübingen, **11**:669–687.

Achab, A., Asselin, E., and Soufiane, A., 1993, New morphological characters observed in the Order Operculatifera and their implications for the suprageneric chitinozoan classification, *Palynology* **17**:1–9.

Ager, D. V., 1973, *The Nature of the Stratigraphical Record*, New York, John Wiley & Sons.

Ager, D. V., 1981, *The Nature of the Stratigraphical Record*, Halsted Press, New York, John Wiley and Sons.

Ager, D. V., 1984, The stratigraphic code and what it implies, in: *Catastrophes and Earth History – the New Uniformitarism* (W. A. Berggren and J. A. van Couvering, eds.), Princeton, Princeton University Press, pp. 91–100.

Agterberg, F. P., 1990, *Automated Stratigraphic Correlation*, Amsterdam, Elsevier.

Agterberg, F. P., and Gradstein, F. M., 1999, The RASC method for ranking and scaling of biostratigraphic events, in: Proceedings, Conference on the 75th Birthday of C. W. Drogger, Utrecht, *Earth Science Review* **46**(1–4):1–25.

Aigner, T., 1985, *Storm Depositional Systems*, Berlin, Springer Verlag.

Aigner, T., and Bachmann, G. H., 1998, Sequence stratigraphy of the germanic triassic: a short overview, *Hallesches Jahrbuch der Geowissenschaften, Reihe B* **(6)**:23–26.

Albertão, G. A., Koutsoukos, E. A. M., Regali, M. P. S., Attrep Jr., M., and Martins, P. P., Jr., 1994, The Cretaceous-Tertiary boundary in southern low-latitude regions: preliminary study in Pernambuco, northeastern Brazil, *Terra Nova* **6**(4):366–375.

Ali, J. R., and Hailwood, E. A., 1998, Resolving possible problems associated with the magnetostratigraphy of the Paleocene/Eocene boundary in Holes 549, 550 (Goban Spur) and 690B (Maud Rise). Abstracts of the special meeting of the Geological Society of France, *Paleocene/Eocene Boundary Events in Europe*, Strata (in press).

Ali, J. R., Kent, D. V., and Hailwood, E. A., 2000, Magnetostratigraphic reinvestigation of the Palaeocene/Eocene boundary interval in Hole 690B, Maud Rise, Antarctica, *Geophysical Journal International* **141**:639–646.

Alpern, B., 1963, Méthode d'extraction des spores des roches du Houiller, *Pollen et spores* **1**:169–177.

Alpern, B., and Streel, M., 1972, Palynologie et stratigraphie du Paléozoïque moyen et supérieur, *Mémoires du Bureau de Recherches Géologiques et Minières, Paris* **77**:(1–1971): 217–241.

Alvarez, W., and Asaro, F., 1990, Die Kreide-Tertiär-Wende: ein Meteoriteneinschlag? *Spektrum der Wissenschaften* **12**:52–59.

Alvarez, W., and Muller, R. A., 1984, Evidence from crater ages for periodic impacts on the Earth, *Nature* **308**:718–720.

Alvarez, L. W., Alvarez, W., Asaro, F., and Michel H. V., 1980, Extraterrestrial cause for the Cretaceous-Tertiary extinction, *Science* **208**:1095–1108.

Alvarez, W., Alvarez, L. W., Asaro, F., and Michel, H. V., 1984a, The end of the Cretaceous: sharp boundary or gradual transition, *Science* **223**:1183–1186.

Alvarez, W., Kauffmann, E. G., Surlyk, F., Alvarez, L. W., Asaro, F., and Michel, H. V., 1984b, Impact theory of mass extinctions and the invertebrate fossil record, *Science* **223**:1135–1141.

Alvarez, W., Smit, J., Lowrie, W., Asaro, F., Margolis, S. V., Claeys, P., Kastner, M., and Hildebrand, A. R., 1992, Proximal impact deposits at the Cretaceous-Tertiary boundary in the Gulf of Mexico: A restudy of DSDP Leg 77 Sites 536 and 540, *Geology* **20**:697–700.

Alves, L. S. R., 1994, Integração entre a lignoflora e a palinologia no afloramento Passo São Borja (Fm. Irati–Bc. do Paraná), Dissertação (Mestrado em Geociências), Instituto de Geociências, Universidade Federal do Rio Grande do Sul, Porto Alegre, 174 p. (unpublished).

Alves, L. S. R., 1999, Anéis de Crescimento em Lenhos Fósseis como Indicadores Paleoclimáticos no Neopermiano da Bacia do Paraná (Fm. Irati e Serra Alta), Tese (Doutorado em Ciências), Instituto de Geociências, Universidade Federal do Rio Grande do Sul, Porto Alegre, 279 p. (unpublished).

Anan-Yorke, R., 1974, Devonian Chitinozoa and Acritarcha from exploratory oil wells on the shelf and coastal region of Ghana, West Africa, *Ghana Geology Survey Bulletin* **37**:1–217.

Anderson, J. M., 1977, The biostratigraphy of the Permian and Triassic. Part 3. A review of Gondwana palynology with particular reference to the northern Karoo Basin, South Africa, *Botanical Survey of South Africa Memoirs* **41**:133.

Anderson, J. M., 1980, World Permo-Triassic correlations: Their biostratigraphic basis. *Fifth International Gondwana Symposium*, Wellington, New Zealand, pp. 11–16.

Anderson, J. M., and Cruickshank, A. R. I., 1978, The biostratigraphy of the Permian and the Triassic. Part. 5. A review of the classification and distribution of Permo-Triassic tetrapods, *Palaeontologia Africana* **12**:15–44.

Archangelsky, S., and Baldoni, A., 1972, Revisión de las Bennettitales de la Formación Baqueró (Cretácico inferior), Província de Santa Cruz. I. Hojas, *Revista del Museo de La Plata, Paleontologia* **744**:195–265.

Archibald, J. D., 1991, Surivivorship patterns of non-marine vertebrates across the Cretaceous-Tertiary (K-T) boundary in the Western U.S., in: *Fifth Symposium on Mesozoic Terrestrial Ecosystems and Biota*, Contr. Paleont. Mus. Univers 364 (Z. Kielan-Jaworowska, N. Heintz, and H. A. Nakrem, eds.), Oslo, pp. 1–2.

Archibald, J. D., 1992, Dinosaur extinction: how much and how fast?, *Journal of Vertebrate Paleontology* **12**(2):263–264.

Archibald, J. D., and Brayant, L. J., 1990, Differential Cretaceous-Tertiary extinctions of non-marine vertebrates: evidence from northeastern Montana, in: *Global Catastrophes in Earth History*, Special Paper 247 (V. L. Sharpton and P. D. Ward, eds.), Geological Society of America, pp. 549–562.

Argast S., Farlow J. O., Gabet, R. M., and Brinkmann, D. L., 1987, Transport-induced abrasion of fossil reptilian teeth: Implications for the existence of Tertiary dinosaurs in the Hell Creek Formation, Montana, *Geology* **15**:927–930.

Armentrout, J . M., 1996, High resolution sequence biostratigraphy: examples from the Gulf of Mexico Plio-Pleistocene, in: *High Resolution Sequence Stratigraphy: Innovations and Applications*, Special Publication 104 (J. A. Howell and J. F. Aitken, eds.), London, Geological Society, pp. 65–86.

Armentrout, J. M., and Clement, J. F., 1990, Biostratigraphic calibration of depositional cycles: A case study in High Island-Galveston-East Breaks Areas, offshore Texas, in: *Sequence Stratigraphy as an Exploration Tool: Concepts and Practices in the Gulf Coast*, Eleventh Annual Research Conference (J. M. Armentrout and B. F. Perkins, eds.), Houston, Gulf Coast Section Society of Economic Paleontologists and Mineralogists Foundation, pp. 21–51.

Armesto, J. J., Villagrán, C., Aravena, J. C., Pérez, C., Smith-Ramirez, C., Cortés, M., and Hedin, L., 1995, Conifer forest of the Chilean Range, in: *Ecology of the Southern Conifers* (N. J. Enright and R. S. Hill, eds.), Washington, Smithsonian Institute Press, pp. 156–170.

Arnold, C. A., 1947, *An Introduction to Palaeobotany*, New York, McGraw Hill.

Arthur, M. A., Dean, W. E., and Schlanger, S. O., 1985, Variations in the global carbon cycle during Cretaceous related to climate, volcanism, and changes in atmospheric $CO_2$, in: *The Carbon Cycle and Atmospheric $CO_2$, Natural Variations Archean to Present*, Monograph **32** (E. T. Sundquist and W. S. Broecker, eds.), Washington, American Geophysical Union, pp. 504–529.

Ash, J., 1983, Growth rings in *Agathis robusta* and *Araucaria cunninghamii* from tropical Australia, *Australian Journal of Botany* **31**:269–275.

Askin, R. A., 1990, Campanian to Paleocene spore and pollen assemblages of Seymour Island, Antarctica, *Review of Palaeobotany and Palynology* **65**:105–113.

Askin, R. A., 1992, Late Cretaceous-Early Tertiary Antarctic outcrop evidence for past vegetation and climates, in: *The Antarctic Paleoenvironment: A Perspective on Global Change*, Volume 56 (J. P. Kennett and D. A. Warnke, eds.), American Geophysical Unions, Antarctic Resource Service, pp. 61–73.

Askin, R. A., and Jacobson, S. R., 1996, Palynological change across the Cretaceous-Tertiary boundary on Seymour Island, Antarctica: environmental and depositional factors, in: *The Cretaceous-Tertiary Mass Extinctions: Biotic and Environmental Changes* (N. MacLeod and G. Keller, eds.), New York, W. W. Norton and Co., pp. 7–26.

Aslan, A., and Behrensmeyer, A. K., 1996, Taphonomy and time resolution of bone assemblages in a contemporary fluvial system: the East Fork River Wyoming, *Palaios* **11**(5):411–421.

Attar, A., Candilier, A. M., Coquel, R., and Fournier, J., 1980, Étude palynologique du Dévonien terminal et du Carbonifère inférieur du bassin d'Illizi (Fort-Polignac), Algérie, *Bulletin du Centres Recherches et Exploration – Production Elf Aquitaine* **17**(1):79–147.

Aubry, M. P., 1991, Sequence stratigraphy: Eustasy or tectonic imprint?, *Journal of Geophysical Research* **96**:6641–6679.

Aubry, M.-P., 1993a, Neogene allostratigraphy and depositional history of the De Soto Canyon area, northern Gulf of Mexico, *Micropaleontology* **39**:327–366.

Aubry, M.-P., 1993b., Calcareous nannofossil stratigraphy of the Neogene formations of eastern Jamaica, in: *Biostratigraphy of Jamaica*, Memoir 182 (R. M. Wright, J. B. Saunders, and E. Robinson, eds.), Geological Society of America, pp. 131–178.

Aubry, M.-P., 1995, From chronology to stratigraphy: Interpreting the lower and middle Eocene stratigraphic record in the Atlantic Ocean, in: *Geochronology Time Scales and Global Stratigraphic Correlation*, Volume 54 (W. A. Berggren, D. V. Kent, M.-P. Aubry, and J. Hardenbol, eds.), Society of Economic Paleontologists and Mineralogists Special Publication, pp. 213–274.

Aubry, M.-P., 1997. Interpreting the (marine) stratigraphic record. In Aguilar, J. P., Michaux, J. and Legendre, S., (eds.), *Actes du Congrès BiochroM' 97, Mémoires et Travaux de l'Ecole pratique des Hautes Etudes*, Institut de Montpellier, v. 21, p. 15–32.

Aubry, M.-P., 1998, Stratigraphic (dis)continuity and temporal resolution of geological events in the upper Paleocene-lower Eocene deep sea record, in: *Late Paleocene–Early Eocene Climatic and Biotic Events* (M.-P. Aubry, S. Lucas, and W. A. Berggren, eds.), New York, Columbia University Press, pp. 37–66.

Aubry, M.-P., 2000, Where should the global stratotype section and point (GSSP) for the Paleocene/Eocene boundary be located?, *Bulletin de la Sociéte Géologique de France* **171**(4):461–476.

Aubry, M.-P., and Berggren, W. A., 2000a, When conflict arises between principles and rules: the case of the GSSP for the Paleocene/Eocene boundary, *Geologiska Föreningens i Stockholm Förhandlingar* **122**:13–14.

Aubry, M.-P., and Berggren, W. A., 2000b, The homeless GSSP: the dilemma of the Palaeocene/Eocene boundary, *Tertiary Research* **20**(1–4):107–112.

Aubry, M.-P., Berggren, W. A., Kent, D. V., Flynn, J. J., Klitgord, K. D., Obradovich, J. D., and Prothero, D. W., 1988, Paleogene geochronology: An integrated approach, *Paleoceanography* **3**(6):707–742.

Aubry, M.-P., Berggren, W. A., Stott, L., and Sinha, A., 1996, The upper Paleocene-lower Eocene stratigraphic record and the Paleocene-Eocene boundary carbon isotope excursion: Implications for geochronology, in: *Correlation of the Early Paleogene in Northwest Europe*, No. 101 (R. W. O'B. Knox, R. M. Corfield, and R. E. Dunay, eds.), Geological Society Special Publication, pp. 353–380.

Aubry, M.-P., Cramer, B., Miller, K. G., Wright, J., Kent, D. V., and Olsson, R. D., 2000, Late Paleocene event chronology: Unconformities not diachrony. *Journal of the Geological Society of France* **171**:367–378.

Axelrod, D. I., 1984, An interpretation of Cretaceous and Tertiary biota in polar regions, *Palaeogeography, Palaeoclimatology, Palaeoecology* **45**:105–147.

Azambuja Filho, N. C. de, 1990, The Oligo-Miocene turbidites and associated facies of Campos basin, offshore Brazil, Ph.D. thesis, Imperial College, London, 456 pp. (unpublished).

Azambuja Filho, N. C. de, and Azevedo, R. L. M. de, 1995, Cyclostratigraphy and stable isotopes in the Oligocene-Miocene succession of the Campos basin, in: *1st Workshop of Paleoceanography with Emphasis on Micropaleontology and Stratigraphy*, Gramado, Brazil, pp. 27–33.

Azcuy, C. L., 1975, Miosporas del Namuriano y Westfaliano de la comarca Malanzán-Loma Larga, provincia de La Rioja, Argentina. I. Localización geográfica y geológica de la comarca y descripciones sistematicás, *Ameghiniana* **XII**:1–69.

Azevedo-Soares, H.L.C., and Grahn, Y., in press, The Silurian-Devonian boundary in the Amazonas Basin, northern Brazil, *Neues Jahrbuch für Geologie und Palaeontologie.*

Backhouse, J., 1991, Permian palynostratigraphy of the Collie Basin, Western Australia, *Review of Palaeobotany and Palynology* **67**:237–314.

Backhouse, J., 1993, Palynology and correlation of Permian sediments in the Perth, Collie, and Officer Basins, Western Australia, *Geological Survey of Western Australia Report* **34**:111–128.

Backhouse, J., 1998, Palynological correlation of the Western Australian Permian, *Proceedings of the Royal Society of Victoria* **110**:107–114.

Backman, J., and Raffi, I., 1997, Calibration of Miocene events to orbitally tuned cyclostratigraphies from Ceara Rise, in: *Proceedings ODP, Scientific Results*, Volume 154 (N. J. Shackleton *et al.*, eds.), College Station, TX (Ocean Drilling Program), pp. 83–99.

Bailey, R. G., 1998, *Ecoregions: the Ecosystem Geography of the Oceans and Continents*, New York, Springer.

Bailey, R. G., and Cushwa, C. T., 1981, Ecoregions of North America. FWS/OBS-81/29. Washington, DC, U. S. Fish and Wildlife Service. 1: 12.000.000.

Baldoni, A. M., and Barreda, V., 1986, Estudio palinológico de las Formaciónes López de Bertodano y Sobral, Isla Vicecomodoro Marambio, Antartida. *Boletim IG-USP, Série Científica* **17**:89–98.

Baldoni, A., and Medina, F., 1989, Fauna y microflora del Cretácico en bahía Brandy, Isla James Ross, Antártica, *Serie Científica Instituto Antártico Chileno* **39**:43–58.

Balme, B. E., 1952, The principal microspores of the Permian coals of Collie, in: J. H. Lord, ed., Collie Mineral Field, *Geological Survey of Western Australia Bulletin*, **105**, Appendix **2**:164–201.

Balme, B. E., 1970, Palynology of Permian and Triassic strata in the Salt Range and Surghar Range, West Pakistan, in: *Stratigraphic Boundary Problems: Permian and Triassic of West Pakistan* (B. Kummel and C. Teichert, eds.), Lawrence, Kansas, University of Kansas Press, pp. 306–453.

Balme, B. E., 1980a, Palynology and the Carboniferous-Permian boundary in Australia and other Gondwana continents, *Palynology* **4**:43–55.

Balme, B. E., 1980b, Palynology of Permian-Triassic boundary beds at Kap Stosch, East Greenland, *Meddeleser om Grønland* **200**(6):1–37.

Balme, B. E., and Hennelly, J. P. F., 1955, Bisaccate sporomorphs from Australian Permian coals, *Australian Journal of Botany* **3**:89–98.

Balme, B. E., and Hennelly, J. P. F., 1956a, Monolete, monocolpate, and alete sporomorphs from Australian Permian sediments, *Australian Journal of Botany* **4**:54–67.

Balme, B. E., and Hennelly, J. P. F., 1956b, Trilete sporomorphs from Australian Permian sediments, *Australian Journal of Botany* **4**:240–260.

Balme, B. E., and Playford, G., 1967, Late Permian plant microfossils from the Prince Charles Mountains, Antarctica, *Revue de Micropaleontologie* **10**:179–192.

Bannan, M. W., 1954a, The wood structure of some Arizoan and Californian species of *Cupressus*, *Canadian Journal of Botany* **32**:285–307.

Bannan, M. W., 1954b, Ring width tracheid size and ray volume in stem wood of *Thuja occidentalis* L., *Canadian Journal of Botany* **32**:466–479.

Banner, F. T., and Simmons, M. D., 1994, Calcareous algae and foraminifera as water-depth indicators: an example from the Early Cretaceous sediments of northeast Arabia, in: *Micropalaeontology in Hydrocarbon Exploration in the Middle East* (M. D. Simmons, ed.), London, Chapman and Hall, pp. 243–252.

Banerjee, I., and Kidwell, S. M., 1991, Significance of molluscan shell beds in sequence stratigraphy: An example from the lower Cretaceous Manville Group of Canada, *Sedimentology* **38**:913–934.

Barberena, M. C., 1977, Bioestratigrafia preliminar da formação Santa Maria. *Pesquisas* **7**:111–129.

Barberena, M. C., Araújo, D. C., and Lavina, E. L. C., 1985a, Late Permian and Triassic tetrapods of Southern Brazil, *National Geographic Research* **1**:5–20.

Barberena, M. C., Araújo D. C., Lavina, E. L. C., and Azevedo, S. A. K., 1985b, O estado atual do conhecimento sobre os tetrápodes permianos e triássicos do Brasil meridional, in: *Coletânea de Trabalhos Paleontológicos. Série Geologia* **27**(2): 21–28.

Barberena, M. C., Araújo D. C., Lavina, E. L. C., and Faccini, U. F., 1991. Evidences for close paleofaunistic affinity between South America and Africa, *Proceedings of the International Gondwana Symposium*, **7**:455–467.

Barberena, M. C., Faccini, U. F., and Scherer, C. M. S., 1993, Situação atual e perspectivas da bioestratigrafia da Formação Santa Maria, in: 1° Simpósio sobre a Cronoestratigrafia da Bacia do Paraná, *Resumos*, pp. 100–101.

Barefoot, A. C., and Hankins, F. W., 1982, *Identification of Modern and Tertiary Woods*, Oxford, Clarendon.

Barreda, V. D., 1986, Acritarcos givetiano-frasnianos de la Cuenca del Noroeste, Provincia de Salta, Argentina, *Revue Español de Micropaleontology* **18**:229–245.

Barrera, E., 1994, Global environmental changes preceding the Cretaceous-Tertiary boundary: early-late Maastrichtian transition, *Geology* **22**:877–880.

Barrera, E., and Savin, S., 1999, Evolution of late Maastrichtian marine climates and oceans, *in: Evolution of the Cretaceous Ocean-Climate System* (E. Barrera and C. Johnson, eds.), Geological Society of America Special Papers, **332**, pp. 245–282.

Barrera, E., Huber, B. T., Savin, S. M., and Webb, P. N., 1987, Antarctic marine temperatures: late Campanian through early Palaeocene, *Paleoceanography* **2**:21–47.

Barron, J. A., Keller, G., and Dunn, D. A., 1985, A multiple microfossil biochronology for the Miocene, in: *The Miocene Ocean*, Memoir 163 (J. P. Kennet, ed.), Geological Society of America, pp. 21–48.

Barrón López, E., 1995, *Estudio Tafonómico y Análisis Paleoecológico de la Macro y Microflora Miocena de la Cuenca de La Cerdaña*, Tesis doctoral, Universidad Complutense de Madrid, Madrid, Facultad de Ciencias Biológicas, 773 pp.

Basinger, J. F., 1991, The fossil forests of the Buchanan Lake Formation (Early Tertiary), Axel Heiberg Island, Canadian Arctic Archipelago: Preliminary floristics and paleoclimate, in: *Tertiary Fossil Forests of the Geodetic Hills, Axel Heiberg Island, Antarctic Archipelago*, Volume 403 (R. L. Christir and N. J. McMillan, eds.), *Geological Survey of Canada Bulletin*, pp. 39–65.

Battail, B., 1993, On the biostratigraphy of triassic therapsid-bearing formations, in: *The Nonmarine Triassic, Bulletin of the New Mexico Museum of Natural History and Sciences* **3**:31–35.

Batten, D. J., 1973, Use of palynologic assemblage-types in Wealden correlation, *Palaeontology* **16**:1–40.

Batten, D. J., 1975, Wealden palaeoecology from the distribution of plant fossils, *Proceedings of the Geological Association*, **85**(for 1974):433–458.

Batten, D. J., 1996a, Palynofacies and palaeoenvironmental interpretation, in: *Palynology: Principles and Applications*, Volume 3 (J. Jansonius and D. C. McGregor, eds.), Dallas, American Association of Stratigraphic Palynologists' Foundation, pp. 1011–1064.

Batten, D. J., 1996b, Palynofacies and petroleum potential, in: *Palynology: Principles and Applications*, Volume 3 (J. Jansonius and D. C. McGregor, eds.), Dallas, American Association of Stratigraphic Palynologists' Foundation, pp. 1065–1084.

Batten, D. J., 1999a, Palynofacies analysis, in: *Fossil Plants and Spores: Modern Techniques* (T. P. Jones and N. P. Rowe, eds.), The Geological Society, London, pp. 194–198.

Batten, D. J., 1999b, Small palynomorphs, in: *Fossil Plants and Spores: Modern Techniques* (T. P. Jones and N. P. Rowe, eds.), London, The Geological Society, pp. 15–19.

Batten, D. J., and Grenfell, H. R., 1996, *Botryococcus*, in: *Palynology: Principles and Applications*, Volume 1 (J. Jansonius and D. C. McGregor, eds.), Dallas, American Association of Stratigraphic Palynologists' Foundation, pp. 205–214.

Batten, D. J., and Morrison, L., 1983, Methods of palynological preparation for palaeoenvironmental, source potential and organic maturation studies, in: *Palynology – Micropalaeontology: Laboratories, Equipment and Methods*, Bull. 2 (L. I. Costa, ed.), Norwegian Petroleum Directorate, pp. 35–53.

Baum, J. S., Baum, G. R., Thompson, P. R., and Humphrey, J. D., 1994, Stable isotopic evidence for relative and eustatic sea level changes in Eocene to Oligocene carbonates, Baldwin County, Alabama, *Geological Society of America Bulletin* **106**(6):824–839.

Beck, C. N., 1953, A new root species of *Callixylon*, *American Journal of Botany* **40**:226–233.

Becker, L., Poreda, R. J., Hunt, A. G., Bunch, T. E., and Rampino, M., 2001, An extraterrestrial impact at the Permian-Triassic boundary?, *Science* **291**:1530.

Beerling, D. J., McElwain, J. C., and Osborne, C. P., 1998, Stomatal response of the "living fossil" *Ginkgo biloba* L. to changes in atmospheric $CO_2$ concentrations, *Journal of Experimental Botany* **49**:1603–1607.

Behrensmeyer, A. K., 1978, Taphonomic and ecologic information from bone weathering, *Paleobiology* **4**(2):150–162.

Behrensmeyer, A. K., 1988, Vertebrate preservation in fluvial channels, *Palaeogeography, Palaeoclimatology, Palaeoecology* **63**:183–99.

Behrensmeyer, A. K., and Hook, R. W., 1992, Paleoenvironmental context and taphonomic modes, in: *Terrestrial Ecosystems through Time* (A. K. Behrensmeyer, J. D. Damuth, W. A. DiMichele, R. Potts, H. D. Sues, and S. L. Wing, eds.), Chicago, University of Chicago Press, pp. 15–136.

Behrensmeyer, A. K., and Kidwell, S. M., 1985, Taphonomy's contributions to paleobiology, *Paleobiology* **11**:105–119.

Behrensmeyer, A. K., Kidwell, S. M., and Gastaldo, R. A., 2000, Taphonomy and palebiology, in: *Deep Time – Paleobiology's Perspective* (D. H. Erwin and S. L. Wing, eds.), Supplement to *Palaeobiology* **26**(4):103–147.

Benton, M. J. 1983, Dinosaur success in the Triassic: a non-competitive ecological model, *Quarterly Review of Biology* **58**(1):29–55.

Benton, M. J., 1986, The Late Triassic tetrapod extinction events, in: *The Beginning of the Age of Dinosaurs: Faunal Change Across the Triassic-Jurassic Boundary* (K. Padian, ed.), Cambridge, New York, Cambridge University Press, pp. 303–320.

Benton, M. J., 1988, New data on dinosaur extinctions, *Geology Today*, pp. 159–161.

Benton, M. J., 1989, Mass extinction among tetrapods and the quality of the fossil record, *Philosophical Transactions of the Royal Society of London* B **325**:369–386.

Benton, M. J., 1992, Les Dinosaures polaires, *La Recherche* **239**(23):62–70.

Benton, M. J. 1994, Late Triassic to Middle Jurassic extinctions among continental tetrapods: testing the pattern, in: *In the Shadow of the Dinosaurs, Early Mesozoic tetrapods* (N. C. Fraser and H.-D. Sues, eds.), Cambridge, Cambridge University Press, pp. 366–397.

Bergamaschi, S., 1999, Análise estratigráfica do Siluro-Devoniano (Formação Furnas e Ponta Grossa) da sub-bacia de Apucarana, Bacia do Paraná, Brasil, USP, São Paulo, SP (Tese de Doutoramento, Instituto de Geociências), 167 p. (unpublished).

Bergen, J. A., and Sikora, P. J., 1999, Microfossil diachronism in southern Norwegian North Sea chalks: Valhall and Hod fields, in: *Biostratigraphy in Production and Development Geology*, No. 152 (R. W. Jones and M. D. Simmons, eds.), London, Geological Society, Special Publications, pp. 85–112.

Berger, A. L., 1978, Long term variations of caloric insolation resulting from the earth's orbital elements, *Quaternary Research* **9**:139–167.

Berger, A. L., 1980, Milankovitch astronomical theory of paleoclimates: a modern review, *Vistas in Astronomy* **24**(2):103–122.

Berger, A. L., 1992, Astronomical theory of paleoclimates and the last glacial-interglacial cycle, *Quaternary Science Reviews* **11**:571–581.

Berger, A. L., and Loutre, M. F., 1988, New insolation values for the climate of the last 10 million years, *Université Catholique de Louvain Scientific Reports* **1988**(13).

Berger, A. L., Loutre, M. F., and Laskar, J., 1992, Stability of the astronomical frequencies over the earth's history for paleoclimate studies, *Science* **255**:560–566.

Berggren, W. A., 1993, Neogene planktonic foraminiferal biostratigraphy of eastern Jamaica, in: *Biostratigraphy of Jamaica*, Memoir 182 (R. M. Wright, J. B. Saunders, and E. Robinson, eds.), Geological Society of America, pp. 179–217.

Berggren, W. A., and Aubert, J., 1975, Paleocene benthonic foraminiferal biostratigraphy, paleobiogeography and paleoecology of Atlantic-Tethyan regions: Midway-type fauna, *Palaeogeography, Palaeoclimatology, Palaeoecology* **18**:73–192.

Berggren W. A., and Aubry, M.-P., 1996, A late Paleocene–early Eocene NW European and North Sea magnetobiochronological correlation network, in: *Correlation of the Early*

*Paleogene in Northwest Europe*, No. 101 (R. W. O'B. Knox, R. M. Corfield, and R. E. Dunay, eds.), Geological Society Special Publication, pp. 309–352.

Berggren, W. A., and Hollister, C. D., 1974, Paleogeography, paleobiogeography and the history of circulation in the Atlantic Ocean, Society of Economic Paleontologists and Mineralogists, Special Publication **29**:129–186.

Berggren, W. A., and Van Couvering, J. A., 1978, Biochronology, in: *Contributions to the Geologic Time Scale*, Volume 6 (G. V. Cohee, M. F. Glaessner, and H. D. Hedberg, eds.), American Association of Petroleum Geologists, *Studies in Geology*, pp. 39–55.

Berggren, W. A., Kent, D. V., Flynn, J. J., and Van Couvering, J. A., 1985a, Cenozoic geochronology, *Geological Society of America Bulletin* **96**:1407–1418.

Berggren, W. A., Kent, D. V., and Flynn, J. J., 1985b, Paleogene geochronology and chronostratigraphy, in: *The Chronology of the Geological Record*, Memoir 10 (N. J. Snelling, ed.), London, Geological Society, pp. 141–195.

Berggren, W. A., Kent, D. V., Flynn, J. J., and Van Couvering, J. A., 1985c, Neogene geochronology and chronostratigraphy, in: *The Chronology of the Geological Record*, Memoir 10 (N. J. Snelling, ed.), London, Geological Society, pp. 211–260.

Berggren, W. A., Hilgen, F. J., Langereis, C. G., Kent, D. V., Obradovich, J. D., Raffi, I., Raymo, M. E., and Shackleton, N. J., 1995a, Late Neogene chronology: New perspectives in high-resolution stratigraphy, *Geological Society of America Bulletin* **107**:1272–1287.

Berggren, W. A., Kent, D. V., Aubry, M.-P., and Hardenbol, J. (eds.), 1995b, *Geochronology, Time Scales, and Global Stratigraphic Correlation: Framework for an Historical Geology*, Special Publication of The Society of Economic Geologists and Paleontologists **54**.

Berggren, W. A., Kent, D. V., Swisher, C., III, and M.-P. Aubry, 1995c, A revised Cenozoic geochronology and chronostratigraphy, in: *Geochronology, Time Scales and Global Stratigraphic Correlation: Framework for an Historical Geology*, Special Publication of The Society of Economic Geologists and Paleontologists **54** (W. A. Berggren, D. V. Kent, and J. Hardenbol, eds.), pp. 129–213.

Best, M. M. R., and Kidwell, S. M., 2000a, Bivalve taphonomy in tropical mixed siliciclastic-carbonate settings. I. Environmental variation in shell condition, *Paleobiology* **26**:80–102.

Best, M. M. R., and Kidwell, S. M., 2000b, Bivalve taphonomy in tropical mixed siliciclastic-carbonate settings. II. Effect of bivalve life habits and shell types, *Paleobiology* **26**:103–115.

Beydoun, Z. R., 1988, *The Middle East: Regional Geology and Petroleum Resources*, Beaconsfield, Bucks, Scientific Press Ltd.

Bharadwaj, D. C., 1962, The miospore genera in the coals of Raniganj stage (Upper Permian), India, *The Palaeobotanist* **9**:68–106.

Bhattacharya, J., and Walker, R. G., 1991, Allostratigraphic subdivision of the Upper Cretaceous Dunvegan, Shaftesbury and Kaskapau formations, northwestern Alberta subsurface, *Bulletin of Canadian Petrologists and Geologists* **39**:145–164.

Birkenmajer, K., 1980, Tertiary volcanic-sedimentary succession at Admiralty Bay, King George Island (South Shetland Islands, Antarctica), *Studia Geologica Polonica* **64**:7–65.

Birkenmajer, K., 1988, Tertiary glacial and interglacial deposits (South Shetland Islands, Antarctica: Geochronology *versus* Biostratigraphy (A Progress Report), *Bulletin of the Polish Academy of Sciences, Earth Sciences* **36**(2):133–145.

Birkenmajer, K., and Zastawniak, E., 1989, Late Cretaceous-Early Neogene vegetation history of the Antarctic peninsula sector, Gondwana break-up and tertiary glaciations, *Bulletin of the Polish Academy of Sciences, Earth Sciences* **37**(1–2):63–88.

Birkenmajer, K., Delitala, M. C., Narebski, W., Nicoletti, M., and Petrucciani, C., 1986a, Geochronology and migration of Cretaceous through tertiary plutonic centers, South

Shetland Islands (West Antarctica): Subduction and hot spot magmatism, *Bulletin of the Polish Academy of Sciences, Earth Sciences* **34**(3):243–255.

Birkenmajer, K., Delitala, M. C., Narebski, W., Nicoletti, M., and Petrucciani, C. 1986b, Geochronology of Tertiary island-arc volcanics and glacigenic deposits, King George Island South Shetland Islands (West Antarctica), *Bulletin of the Polish Academy of Sciences, Earth Sciences* **34**(3):257–273.

Bishop, W. W., 1980, Paleogeomorphology and continental taphonomy, in: *Fossils in the Making* (A. K. Behrensmeyer and A. P. Hill, eds.), Chicago, University of Chicago Press, pp. 20–40.

Blinten, J. S., and Wahid, I. A., 1983, A review of Sajaa Field development, Sharjah, UAE, *Society of Petroleum Engineers* 11516 (Middle East oil exhibition; 14 Mar 1983; Manama, Bahrain): 601–606.

Blair, T. C., 1986, Tectonic and hydrologic controls on cyclic alluvial fan, fluvial and lacustrine rift basin sedimentation, Jurassic-lowermost Cretaceous Todos Santos Formation, Chiapas, Mexico, *Journal of Sedimentary Petrology* **57**:845–862.

Blatt, H., Berry, W. B. N., and Brande, S., 1991, *Principles of Stratigraphic Analysis*, Boston, Blackwell Scientific Publications.

Blodgett, R. B., Fryda, J., and Lenz, A. C., 2001, *Semitubina yukonensis* new species, first occurrence of this biogeographically distinctive Old World Realm gastropod genus in the Lower Devonian of the western hemisphere, *Journal of Paleontology* **75**:466–470.

Blow, W. H., 1979, *The Cainozoic Globigerinada: A Study of the Morphology, Taxonomy, Evolutionary Relationships and the Stratigraphical Distribution of some Globigerinada (Mainly Globigerinacea)*, Leiden, E. J. Brill, 3 vols.

Blum, M. D., and Price, D. M., 1998, Quaternary alluvial plain construction in response to interacting glacio-eustatic and climatic controls, Texas Gulf Coastal plain, in: *Relative Role of Eustasy, Climate and Tectonism in Continental Rocks*, Special Publication 59 (K. W. Shanely and P. J. McCabe, eds.), SEPM, pp. 31–48.

Boersma, A., Shackleton, N., Hall, M., and Given, Q., 1979, Carbon and oxygen isotope records at DSDP site 384 (North Atlantic) and some Paleocene paleotemperatures and carbon isotope variations in the Atlantic Ocean, in: *Initial Reports of the Deep Sea Drilling Project*, 34, Washington, DC, pp. 695–717.

Bohor, B. F., 1990, Shock-induced microdeformations in quartz and other mineralogical indications of an impact event at the Cretaceous-Tertiary boundary, *Tectonophysics* **171**(1–4):359–372.

Bohor, B. F., Foord, D. D., Modreski, P. J., and Triplehorn, D. M., 1984, Mineralogic evidence of an impact event at the Cretaceous-Tertiary boundary, *Science* **224**:867–869.

Bohor, B. F., Modreski, P. J., and Foord, E. E., 1987, Shocked quartz in the Cretaceous-Tertiary boundary clay: evidence for a global distribution, *Science* **236**:705–709.

Bolli, H. M., 1966a, Zonation of Cretaceous to Pliocene marine sediments based on planktonic foraminifera. Boletin Informativo, *Asociacion Venezolana de Geologia, Mineria y Petroleo* **9**:3–32.

Bolli, H. M., 1966b, The planktonic foraminifera in Well Bodjonegoro-I of Java, *Eclogae Geological Helvetiae* **59**:450–464.

Bonaparte, J. F., 1966a, Cronologia de algunas formaciones Triásicas de Argentina basada en restos de tetrápodos, *Revista de la Asociación Geologica Argentina* **21**:20–38.

Bonaparte, J. F., 1966b, Chronological survey of the tetrapod-bearing Triassic of Argentina, *Breviora* **251**:1–13.

Bonaparte, J. F., 1973, Edades/Reptil para el Triásico de Argentina y Brasil, V Congreso Geológico Argentino, Villa Carlos Paz, Buenos Aires, *Actas* **III**:93–129.

Bonaparte, J. F., 1982, Faunal replacement in the Triassic of South America, *Journal of Vertebrate Palaeontology* **2**:362–371.

Bonaparte, J. F., and Kielan-Jaworowska, Z., 1987, Late Cretaceous dinosaur and mammal faunas of Laurasia and Gondwana, in: *Symposium on Mesozoic Terrestrial Ecosystems*, Fourth Occasional Paper, Tyrrell Museum of Palaeontology, 3 (P. J. Currie and E. H. Koster, eds.), pp. 24–29.

Bonham-Carter, G. F., Gradstein, F. M., and D'Iorio, M. A., 1986, Distribution of Cenozoic Foraminifera from the Northwestern Atlantic Margin analyzed by correspondence analysis, *Computers and Geosciences* **12**:621–635.

Bott, M. H. P., and Johnson, G. A. L., 1967, The controlling mechanism of Carboniferous cyclic sedimentation, *Quarterly Journal of the Geological Society of London* **22**:421–441.

Bottjer, D. J., Arthur, M. A., Dean, W. E., Hattin, D. E., and Savrda, C. E., 1986, Rhythmic bedding produced in Cretaceous pelagic carbonate environments: sensitive recorders of climatic cycles, *Paleoceanography* **1**(4):467–481.

Boucot, A. J., 1975, *Evolution and Extinction Rate Controls*, Amsterdam, Elsevier.

Boucot, A. J., 1978, Community evolution and rates of cladogenesis, *Evolutionary Biology* **11**:545–655.

Boucot, A. J., 1983, Does evolution take place in an ecological vacuum?, II. *Journal of Paleontology* **57**:1–30.

Boucot, A. J., 1984a, Ecostratigraphy, in: *Stratigraphy, Quo Vadis?, AAPG Studies in Geology*, No. 16 (E. Seibold and J. D. Meulenkamp, eds.), IUGS Special Publication No. 14, pp. 55–60.

Boucot, A. J., 1984b, The pattern of Phanerozoic community evolution, in: *Proceedings, 27th International Geological Congress*, Volume 2, *Palaeontology*, VNU Science Press, pp. 13–21.

Boucot, A. J., 1988, Devonian biogeography: an update, in: *Proceedings of the 2nd International Symposium on the Devonian System*, Volume III, Canadian Society of Petroleum Geologists Memoir 14, pp. 211–227.

Boucot, A. J., 1990, Silurian biogeography, in: *Palaeozoic Palaeogeography and Biostratigraphy*, Memoir 121 (W. S. McKerrow and C. R. Scotese, eds.), London, Geological Society, pp. 191–196.

Boucot, A. J., 1994, The episodic, rather than periodic nature of extinction events. *Revista de la Sociedad Mexicana de Paleontologia*, **7**:15–36.

Boucot, A. J., 1996, Epilogue, *Palaeogeography, Palaeoclimatology, Palaeoecology* **127**:339–359.

Boucot, A. J., 1997, Community evolution, in: *Evolutsia zhizin na Zemle*, Materiali I-go Mezhdunarodi Simpozuma, Tomsk, pp. 13–15.

Boucot, A. J., 1999a, Some Wenlockian-Gedinnian, chiefly brachiopod dominated, communities of North America, in: *Paleocommunities – A Case Study from the Silurian and Lower Devonian* (A. J. Boucot and J. D. Lawson, eds.), Cambridge, Cambridge University Press, pp. 549–591.

Boucot, A. J., 1999b, Southern African Phanerozoic marine invertebrates: biogeography, palaeoecology, climatology and comments on adjacent regions, *Journal of African Earth Sciences* **28**:129–143.

Boucot, A. J., and Blodgett, R. B., 2001, Silurian-Devonian Biogeography, in: *Brachiopods*, Special Series 63 (L. R. M. Cocks, H. Brunton, and S. L. Long, eds.), Proceedings of the Fourth International Brachiopod Congress, Systematics Association, London, Taylor & Francis, pp. 335–344.

Boucot, A. J., and Lawson, J. D. (eds.), 1999, *Paleocommunities – A Case Study from the Silurian and Lower Devonian*, Cambridge, Cambridge University Press.

Boucot, A. J., and Racheboeuf, P. R., 1993, Biogeographic summary of the Malvinokaffric Realm Silurian and Devonian fossils, in: *Fosiles y Facies de Bolivia, 2. Invertebrados y Paleobotanica*, 13–14 (R. Suarez Soruco, ed.), Revista Tecnica Y.P.F.B., pp. 71–75.

Boucot, A. J., Johnson, J. G., and Shagam, R., 1972, Braquiopodos Siluricos de los Andes Merideños de Venezuela. Memoria IV Congreso Geologico Venezolano, 2. *Boletin de Geologia*. Publicacion Especial 5, Ministerio de Minas e Hidrocarburos. Caracas, pp. 585–727.

Boucot, A. J., Bahlburg, B., Breitkreuz, C., Isaacson, P. E., Niemeyer, H., and Urzua, F., 1995, Devonian brachiopods from northern Chile, *Journal of Paleontology* **69**:257–263.

Boucot, A. J., Blodgett, R. B., and Stewart, J. H., 1997, European Province Late Silurian brachiopods from the Cuidad Victoria area, Tamaulipas, northeastern Mexico. Special Paper 321, Geological Society of America, pp. 273–293.

Boucot, A. J., Chen Xu, and Scotese, C. R., in press. Preliminary compilation of Cambrian through Miocene Climatically Sensitive Deposits, SEPM.

Boumendjel, K., 1987, Les Chitinozoaires du Silurien supérieur et du Dévonien du Sahara algérien (cadre géologique, systématique, biostratigraphie), Université de Rennes I, D.Sc. thesis (unpublished).

Boumendjel, K., Loboziak, S., Paris, F., Steemans, P., and Streel, M., 1988, Biostratigraphie des miospores et des chitinozoaires du Silurien supérieur et du Dévonien dans le Bassin d'Illizi (S. E. du Sahara algérien), *Geobios* **32**(3):329–357.

Bourgeois, J., Hansen, T. A., Wiberg, P. L., and Kauffman, E. G., 1988, A tsunami deposit at the Cretaceous-Tertiary boundary in Texas, *Science* **241**:567–570.

Bowen, G. J., Koch, P., Gingerich, P. D., Norris, R. C., Bains, S., and Corfield, R. M., 2001, Refined isotope stratigraphy across the continental Paleocene-Eocene boundary on Polecat Bench in the northern Bighorn Basin, in: *Paleocene-Eocene Stratigraphy and Biotic Change in the Bighorn and Clarks Fork Basins, Wyoming*, Volume 33 (P. R. Gingerich, ed.), University of Michigan Papers on Paleontology, pp. 73–88.

Bown, P. R. (ed.), 1999, *Calcareous Nannofossil Biostratigraphy*, British Micropalaeontological Society Publication Series, Dordrecht, Kluwer Academic Publishers.

Boy, J. A., 1977, Typen und Genese Jungpaläozoischer Tetrapoden-Lagerstätten, *Paleontographica Abteilung A* **156**(4–6):111–167.

Bradley, W. H., 1929, The varves and climate of the Green River Epoch, *United States Geological Survey Professional Papers* **154**:86–110.

Brain, C. K., 1969, The contribution of Namib Desert Hottentotts to an understanding of Australopithecine bone accumulations, *Scientific Papers of the Namibia Research Station* **39**:13–22.

Bralower, T. J., Paull, C. K., and Leckie, R. M., 1998, The Cretaceous-Tertiary boundary cocktail: Chicxulub impact triggers margin collapse and extensive sediment gravity flows, *Geology* **26**(4):331–334.

Bramlette, M. N., and Martini, E., 1964, The great change in calcareous nannoplankton fossils between the Maestrichtian and Danian, *Micropaleontology* **10**(3):291–322.

Brandt, D. S., 1989, Taphonomic grades as a classification for fossiliferous assemblages and implications for paleoecology, *Palaios* **4**:303–309.

Brenckle, P. L., and Manger, W. L. (eds.), 1991, Intercontinental correlation and division of the Carboniferous system, *Courier Forschungsinstitut Senckenberg* **1990**:130:1–350.

Brett, C. E., 1995, Sequence stratigraphy, biostratigraphy, and taphonomy in shallow marine environments, *Palaios* **10**:597–616.

Brett, C. E., 1998, Sequence stratigraphy, paleoecology and evolution: biotic clues and responses to sea-level fluctuations, *Palaios* **13**:241–262.

Brett, C. E., and Baird, G. C., 1986, Comparative taphonomy: a key to paleoenvironmental interpretation based on fossil preservation, *Palaios* **4**:303–309.

Brett, C. E., and Baird, G. C., 1992, Coordinated stasis and evolutionary ecology of Silurian-Devonian marine biotas in the Appalachian basin, Geological Society of America, *Abstracts with Program* **24**:A139.

Brett, C. E., and Baird, G. C., 1995, Coordinated stasis and evolutionary ecology of Silurian to Middle Devonian faunas in the Appalachian Basin, in: *New Approaches to Speciation in the Fossil Record* (D. H. Erwin and R. L. Anstey, eds.), New York, Columbia University Press, pp. 285–315.

Brett, C. E., and Baird, G. C., 1997, *Paleontological Events: Stratigraphic, Ecological, and Evolutionary Implications*, New York, Columbia University Press.

Brett, C. E., and Seilacher, A., 1991, Fossil Lagerstätten: a taphonomic consequence of event sedimentation, in: *Cycles and Events in Stratigraphy* (G. Einsele, W. Ricken, and A. Seiacher, eds.), New York, Berlin, Heidelberg, Springer-Verlag, pp. 283–297.

Brett, C. E., and Speyer, S. E., 1990, Taphofacies, in: *Paleobiology: a Synthesis* (B. E. K. Briggs and P. R. Crowther, eds.), Oxford, Blackwell, pp. 258–263.

Brett, C. E., Ivany, L. C., and Schopf, K. M., 1996, Coordinated stasis: An overview, *Palaeogeography, Palaeoclimatology, Palaeoecology* **127**:1–20.

Briggs, D. E .G., 1995, Experimental taphonomy, *Palaios* **10**:539–550.

Brinkhuis, H., and Visscher, H., 1995, The upper boundary of the Eocene Series: a reappraisal based on dinoflagellate cyst biostratigraphy and sequence stratigraphy, in: *Geochronology, Time Scales and Global Stratigraphic Correlations*, Special Publication No. 54 (W. A. Berggren, D. V. Kent, M.-P. Aubry, and J. Hardenbol, eds.), Society of Economic Paleontologists and Mineralogists, pp. 295–304.

Brinkhuis, H., and Zachariasse, W. J., 1988, Dinoflagellate cysts, sea level changes and planktonic foraminifers across the Cretaceous-Tertiary boundary at El Haria, northwest Tunisia, *Marine Micropaleontology* **13**:153–191.

Brinkhuis, H., Romein, A. J. T., Smit, J., and Zachariasse, W. J., 1994, Danian-Selandian dinoflagellate cysts from lower latitudes with special reference to the El Kef section, NW Tunisia, *Geologiska Föreningens i Stockholm Förhandlingar* **116**:46–48.

Brinkhuis, H., Bujak, J. P., Smit, J., Versteegh, G. J. M., and Visscher, H., 1998, Dinoflagellate-based sea surface temperature reconstructions across the Cretaceous-Tertiary boundary, *Palaeogeography, Palaeoclimatology, Palaeoecology* **141**(1–2):67–83.

Brito, I. M., 1971, Contribuição ao conhecimento dos microfósseis silurianos e devonianos da Bacia do Maranhão. V – Acritarcha Herkomorphitae e Prismatomorphitae, *Anais de Academia Brasileira de Ciências* **43**:201–208.

Bromley, R. G., 1996, *Trace Fossils, Biology and Taphonomy*, 2nd ed., London, Chapman & Hall.

Bromley, R. G., and Asgaard, U., 1991, Ichnofacies: a mixture of taphofacies and biofacies, *Lethaia* **24**:153–163.

Bromley, R. G., Pemberton, S. G., and Rahmani, R. A., 1984, A Cretaceous woodground: the *Teredolites* ichnofacies, *Journal of Palaeontology* **58**:488–498.

Broom, R., 1906, On the Permian and Triassic Faunas of South Africa, *Geological Magazine* **5**:29–30.

Broom, R., 1907, On the geological horizons of the vertebrate genera of the Karoo Formation. *Records of the Albany Museum* **2**:156–163.

Broom, R., 1909, An attempt to determine the horizons of the fossil vertebrates of the Karoo, *Annals of the South African Museum* **7**:285–289.

Brouwers, E. M., and De Deckker, P., 1996, Earliest origins of northern hemisphere temperate nonmarine ostracode taxa: Evolutionary development and survival through the Cretaceous-Tertiary boundary mass-extinction event, in: *Cretaceous-Tertiary Mass Extinctions: Biotic and Environmental Changes* (N. MacLeod and G. Keller, eds.), New York, W. W. Norton & Company, pp. 205–229.

Brouwers, E. M., Clemens, W. A., Spicer, R. A., Ager, T. A., Carter, L. D., and Sliter, W. V., 1987, Dinosaurs on the North Slope, Alaska: high latitude, latest cretaceous environments, *Science* **237**(2):1608–1610.

Brown, F. H., Sarna-Wojcicki, A. M., Meyer, C. E., and Haileab, B., 1992, Correlation of Pliocene and Pleistocene tephra layers between the Turkana Basin of East Africa and the Gulf of Aden, *Quaternary International* **13/14**:55–67.

Bruns, P., Rakoczy, H., Pernicka, E., and Dullo, W.-Ch., 1997, Slow sedimentation and Ir anomalies at the Cretaceous/Tertiary boundary, *Geologische Rundschau* **86**:168–177.

Buatois, L. A., and Mángano, M. G., 1995, The palaeoenvironmental and palaeoecological significance of the *Mermia* ichnofacies: an archetypal subaqueous non-marine trace fossil assemblage, *Ichnos* **4**:151–161.

Buck, K., Olson, H. C., and Austin, J. A., 1999, Paleoenvironmental evidence for latest Pleistocene sea-level fluctuations on the New Jersey outer continental shelf: Combining high-resolution sequence stratigraphy and foraminiferal analysis, *Marine Geology* **154**:287–304.

Burjack, M. I. A., Loboziak, S., and Streel, M., 1987, Quelques données nouvelles sur les miospores dévoniennes du Bassin du Paraná, *Sciences Géologiques Bulletin* **40**(4):381–391.

Burnham, R. J., 1993, Reconstructing richness in the plant fossil record, *Palaios* **8**(4):376–384.

Buzas, M. A., and Hayek, L. C., 1998, SHE Analysis for Biofacies Identification, *Journal of Foraminiferal Research* **28**(3):233–239.

Byvscheva, T. V., 1974, Zonation of the Tournaisian and lower and middle Visean deposits of the Volga-Urals region on the basis of spore analysis, in: *Palynology of Proterophyte and Palaeophyte* (S. N. Naumova, ed.), Proceedings, 3rd International *Palynology Conference*, pp. 100–105 [in Russian].

Cadée, G. C., 1991, The history of taphonomy, in: *The Processes of Fossilization* (S. K. Donovan, ed.), London, Belhaven Press, pp. 3–21.

Cande, S. C., and Kent, D. V., 1992, A new geomagnetic polarity time scale for the Late Cretaceous and Cenozoic, *Journal of Geophysical Research* **97**(B10):13917–13951.

Cande, S. C., and Kent, D. V., 1995, Revised calibration of the geomagnetic polarity time scale for the Late Cretaceous and Cenozoic, *Journal of Geophysical Research* **100**(B4).6093–6095.

Cantrill, D. J., and Nichols, G. J., 1996, Taxonomy and paleoecology of Early Cretaceous (Late Albian) angiosperm leaves from Alexander Island, Antarctica, *Review of Palaeobotany and Palynology* **92**:1–28.

Carlsquist, S., 1975, *Ecological Strategies of Xylem Evolution*, Berkeley, CA, University of California Press.

Carpenter, R. J., Hill, R., and Jordan, G. J., 1994, Cenozoic vegetation in Tasmania, in: *History of the Australian Vegetation: Cretaceous to Recent* (R. Hill, ed.), Cambridge, Cambridge University Press, pp. 276–298.

Case, J. A., 1988, Paleogene Floras from Seymour Island, Antarctic Peninsula, in: *Geology and Paleontology of Seymour Island, Antarctic Peninsula*, Volume 169 (R. Feldmann and M. O. Woodburne, eds.), Memoir of the Geological Society of America, Boulder, Colorado, pp. 489–498.

Case, J. A., 1995, Biostratigraphic patterns of Antarctic marine vertebrates across the Cretaceous-Tertiary boundary, *Journal of Vertebrate Paleontology* **15**(3 Suppl.):22A.

Castro, A. H. A. de, 1999, Estudo de ciclicidade orbital com base em dados sísmicos exemplo para a bacia de campos, Brasil, Dissertação (Mestrado em Geociências), Instituto de Geociências, Universidade Federal do Rio Grande do Sul, Porto Alegre, 147 pp. (unpublished material).

Cazier, E. C., Hayward, A. B., Espinosa, G., Velanda, J., Mugniot, J. F., and Leel, W. G., Jr., 1995, Petroleum geology of the Cusiana Field, Llanos Basin foothills, Colombia, *American Association of Petroleum Geologists Bulletin* **79**(10):1444–1469.

Césari, S., and Gutiérrez, P. R., 2001, Palynostratigraphy of upper Paleozoic sequences in central-western Argentina, *Palynology* **24**:113–146. [imprinted 2000; published February 8, 2001, *vide* R. T. Clarke].

Chaloner, W. G., and Creber, G. T., 1973, Growth rings in fossil woods as evidence of past climates, in: *Implications of Continental Drift to the Earth Sciences* (D. H. Tarling and S. K. Runcorn, eds.), London, Academic Press, pp. 425–437.

Chaloner, W. G., and Creber, G. T., 1988, Fossil plants as indicators of late Palaeozoic plate positions, in: *Gondwana and Tethys*, Special Publication 37 (M. G. Audley-Charles and A. Hallam, eds.), London, Geological Society, pp. 201–210.

Chaloner, W. G., and Creber, G. T., 1990, Do fossil plants give a climatic signal?, *Journal of the Geological Society* **147**:343–350.

Chaloner, W. G., and Lacey, W. S., 1973, The distribution of late Palaeozoic floras, in: *Organisms and Continents through Time*, Special Paper in Palaeontology 12 (N. F. Hughes, ed.), pp. 271–289.

Chaloner, W. G., and McElwain, J., 1997, The fossil plant record and global climatic change, *Review of Palaeobotany and Palynology* **95**:73–82.

Chaloner, W. G., and Meyen, S. V., 1973, Carboniferous and Permian floras of the northern continents, in: *Atlas of Palaeogeography* (A. Hallam, ed.), Amsterdam, Elsevier, pp. 169–186.

Channell, J. E. T., Erba, E., Nakanishi, M., and Tamaki, K., 1995, Late Jurassic-early Cretaceous time scales and oceanic magnetic anomaly block models, in: *Geochronology, Time Scales and Global Stratigraphic Correlation*, Special Publication No. 54 (W. A. Berggren, D. V. Kent, M.-P. Aubry, and J. Hardenbol, eds.), Society of Economic Paleontologists and Mineralogists, pp. 51–63.

Chapman, J. L., 1994, Distinguishing internal developmental characteristics from external palaeoenvironmental effects in fossil wood, *Review of Palaeobotany and Palynology* **81**:19–32.

Charig, A. J., 1989, The Cretaceous-Tertiary boundary and the last of the dinosaurs, *Philosophical Transactions of the Royal Society of London* **B 325**:387–400.

Charnock, M. A., and Jones, R. W., 1990, Agglutinated Foraminifera from the Palaeogene of the North Sea, in: *Paleoecology, Biostratigraphy, Paleoceanography and Taxonomy of Agglutinated Foraminifera* (C. Hemleben *et al.*, eds.), Dordrecht, Kluwer Academic Publishers, pp. 139–244.

Chatfield, C., 1989, *The Analysis of Time Series – An Introduction*, London, Chapman & Hall.

Chiappe, L. M., 1995, The first 85 million years of avian evolution, *Nature* **378**:349–355.

Clarke, L. J., and Jenkyns, H. G., 1999, New oxygen isotope evidence for long-term Cretaceous climatic change in the Southern Hemisphere, *Geology* **27**:699–702.

Clayton, G., 1995, Carboniferous miospore and pollen assemblages from the Kingdom of Saudi Arabia, *Review of Palaeobotany and Palynology* **89**:115–123.

Clayton, G., 1996, Mississippian miospores, in: *Palynology: Principles and Applications*, Volume 2, Chapter 18C (J. Jansonius and D. C. McGregor, eds.), Dallas, American Association of Stratigraphic Palynologists Foundation, pp. 589–596.

Clayton, G., Coquel, R., Doubinger, J., Gueinn, K.J., Loboziak, S., Owens, B., and Streel, M., 1977, Carboniferous miospores of Western Europe: illustration and zonation, *Mededelingen Rijks Geologische Dienst* **29**:1–71.

Clayton, G., Owens, B., Al-Hajri, S., and Filatoff, J., 2000, Latest Devonian and Early Carboniferous miospore assemblages from Saudi Arabia, in: *Stratigraphic Palynology of the Palaeozoic of Saudi Arabia*, GeoArabia Special Publication **1** (S. Al-Hajri and B. Owens, eds.), Manama, Bahrain, Saudi Arabia, Gulf PetroLink, pp. 146–153.

Clemens, W. A., and Nelms, L. G., 1993, Paleoecological implications of Alaskan terrestrial vertebrate fauna in latest Cretaceous time at high paleolatitudes, *Geology* **21**:503–506.

Clement, B. M., and Robinson, F., 1987, The magnetostratigraphy of Leg 94 sediments, in: *Initial Reports of the Deep Sea Drilling Project*, Volume 30 (W. F. Ruddiman, R. B. Kidd, E. Thomas *et al.*, eds.), Washington, DC, Government Printing Office, pp. 635–650.

CLIMAP Project Members, 1976, The surface of the Ice-Age earth, *Science* **191**(4232):1131–1137.

Cline, R. M., and Hays, J. D. (eds.), 1976, *Investigation of Late Quaternary Paleoceanography and Paleoclimatology*, Memoir 145, Boulder, Geological Society of America.

Coccioni, R., and Galeotti, S., 1994, K-T boundary extinction: Geologically instantaneous or gradual event? Evidence from deep-sea benthic foraminifera, *Geology* **22**(9):779–782.

Coccioni, R., and Galeotti, S., 1998, What happened to small benthic foraminifera at the Cretaceous/Tertiary boundary?, *Bulletin de la Societé Géologique de France* **169**(2):271–279.

Cohen, A., Lockley, M., Halfpenny, J., and Michel, A. E., 1991, Modern vertebrate track taphonomy at Lake Manyara, Tanzania, *Palaios* **6**:371–389.

Colbath, G. K., and Grenfell, H. R., 1995, Review of biological affinities of Paleozoic acid-resistant, organic-walled eukaryotic algal microfossils (including "acritarchs"), *Review of Palaeobotany and Palynology* **86**:287–314.

Colbert, E. H., 1984, Triassic tetrapod faunas, in: *Short Papers*, Third Symposium on Mesozoic Terrestrial Ecosystems (W.-E. Reif and F. Westphal, eds.), Tübingen, Attempto Verlag, pp. 55 59.

Coles, G., 1990, A comparison of the evolution, diversity, and composition of the Cainozoic Ostracoda in the deep water North Atlantic and shallow water environments of North America and Europe, in: *Ostracoda and Global Events* (R. Whatley and C. Maybury, eds.), Chapman Hall, pp. 71–86.

Combaz, A., 1964, Les palynofaciès, *Revue de Micropaléontologie* **7**:205–213, 4 pls.

Combaz, A., 1980, Les kérogènes vus au microscope, in: *Kerogen: Insoluble Organic Matter from Sedimentary Rocks* (B. Durand, ed.), Paris, Éditions Technip, pp. 55–111, 10 pls.

Combaz, A., Calandra, F., Jansonius, J., Millepied, P., Poumot, C., and Van Oyen, F. H., 1967, *Microfossiles organiques du Paléozoïque, 2, Les chitinozoaires, morphographie*, C. N. R. S., Editions du C. N. R.S., pp. 7–43.

Contreras, V. H., 1999, Rhynchosaurs from the Ischigualasto Formation (Upper Triassic, late Carnian), San Juan, Argentina, VII International Symposium on Mesozoic Terrestrial Ecosystems, Buenos Aires, Argentina, *Abstracts*, p. 19.

Cook, E., 1995, Taphonomy of two non-marine Lower Cretaceous bone accumulations from southeastern England, *Palaeogeography, Palaeoclimatology, Palaeoecology* **116**:263–270.

Cooper, M. A., Addison, F. T., Alvarez, R., Coral, M., Graham, R. H., Hayward, A. B., Howe, S., Martinez, J., Naar, J., Penas, R., Pulham, A. J., and Taborda, A., 1995a, Basin development and tectonic history of the Llanos Basin, eastern Cordillera and Middle Magdalena Valley, Colombia, *American Association of Petroleum Geologists Bulletin* **79**(10):1421–1443.

Cooper, M. A., Addison, F. T., Alvarez, R., Hayward, A. B., Howe, S., Pulham, A. J., and Taborda, A., 1995b, Basin development and tectonic history of the Llanos Basin, in: *Petroleum Basins of South America*, Memoir, No. 62 (A. Tankard *et al.*, eds.), Tulsa, Oklahoma, American Association of Petroleum Geologists, pp. 659–666.

Cooper, M. R., 1982, A mid-Permian to earliest Jurassic tetrapod biostratigraphy and its significance, *Arnoldia Zimbabwe* **9**:77–104.

Cooper, R. A., and Lindholm, K., 1991, A precise worldwide correlation of early Ordovician graptolite sequences, *Geological Magazine* **127**(6):497–525.

Cooper, R. A., Crampton, J. S., Raine, J. I., Gradstein, F. M., Morgans, H. E. G., Sadler, P. M., Strong, C. P., Waghorn, D., and Wilson, G. J., 2001, Quantitative biostratigraphy of the Taranaki Basin, New Zealand – a deterministic and probabilistic approach, *American Association of Petroleum Geologists Bulletin* **85**(8):1469–1498.

Coquel, R., and Latrèche, S., 1989, Étude palynologique de la Formation d'Illerène (Dévono-Carbonifère) du Bassin d'Illizi (Sahara algérien oriental), *Palaeontographica Abteilung B* **212**:47–70.

Coquel, R., Loboziak, S., Stampfli, G., and Stampfli-Vuille, B., 1977, Palynologie du Dévonien supérieur et du Carbonifère inférieur dans l'Elburz oriental (Iran Nord-Est), *Review of Micropalaeontology* **20**(2):59–71.

Coquel, R., Doubinger, J., and Massa, D., 1988, Nouvelles donnees palynologiques sur l'intervalle CarbonifèreViseen/Namurien Bassin de Rhadames (Libye), *Revue de l'Institut Français du Pétrole* **43**:1–16.

Costa, R. H. C., 1981, Madeiras Gimnospérmicas da Formação Irati no Rio Grande do Sul – Contribuição ao estudo anatômico e paleoecológico, Dissertação (Mestrado em Geociências), Instituto de Geociências, Universidade Federal do Rio Grande do Sul, Porto Alegre, 99 pp. (unpublished).

Courtillot, V. E., 1990a, Deccan volcanism at the Cretaceous-Tertiary boundary: past climatic crises as a key to the future?, *Paleogeography, Paleoclimatology, Paleoecology* **189**:291–299.

Courtillot, V. E., 1990b, Die Kreide-Tertiär-Wende: verheerender Vulkanismus?, *Spektrum der Wissenschaften* **12**:60–69.

Courtillot, V., 2000, *Evolutionary Catastrophes: The Science of Mass Extinction*, New York, Cambridge University Press.

Courtillot, V. E., Féraud, G., Maluski, H., Vandamme, D., Moreau, M. G., and Besse, J., 1988, The Deccan flood basalts and the Cretaceous/Tertiary boundary, *Nature* **333**(6176): 843–846.

Courtillot, V., Féraud, G., Aluski, H., Vandamme, D., Moreau, M. G., and Besse, J., 1988, Deccan flood basalts and the Cretaceous/Tertiary boundary, *Nature* **333**:843–846.

Courtillot, V., Jaeger, J.-J., Yand, Z., Féraud, G., and Hofmann, C., 1996, The influence of continental flood basalts on mass extinctions: where do we stand? in: *The Cretaceous-Tertiary Event and Other Catastrophes in Earth History*, Special Paper 307 (G. Ryder, D. Fastovsky, and S. Gartner, eds.), Boulder, Geological Society of America, pp. 513–525.

Cowie, J. W., and Basset, M. G., 1989, *1989 Global Stratigraphic Chart*, International Union of Geological Sciences (IUGS).

Cowie, J. W., Ziegler, W., and Remane J., 1989, Stratigraphic commission accelerates progress 1984–1989, *Episodes* **112**:79–83.

Cox, C. B., 1973, Triassic tetrapods, in: *Atlas of Palaeobiogeography* (A. Hallam, ed.), Amsterdam, Elsevier, pp. 213–223.

Cox, C. B., 1991, The Pangaea Dicynodont Rechnisaurus and the comparative biostratigraphy of Triassic dicynodont faunas, *Palaeontology* **34**(4):767–784.

Cox, C. B., and Moore, P. D., 2000, *Biogeography: An Ecological and Evolutionary Approach*, 6th ed., Oxford, Blackwell Scientific.

Crabtree, D. R., 1987, Angiosperms of the Northern Rocky Mountains: Albian to Campanian (Cretaceous) megafossil floras, *Annals of the Missouri Botanical Garden* **74**:707–747.

Craig, H., 1965, The measurement of oxygen isotopes paleotemperatures, in: *Proceedings Spoleto Conference on Stable Isotopes in Oceanographic Studies and Paleotemperatures*, Volume 3 (E. Tongiorgi, ed.), Pisa, pp. 3–24.

Crame, J. A., 1992, Late Cretaceous palaeoenvironments and biotas: An Antarctic perspective, *Antarctic Science* **4**:371–382.

Cramer, B. S., 2001, Latest Paleocene-earliest Eocene cyclostratigraphy: using core photographs for reconnaissance geophysical logging, *Earth and Planetary Science Letters* **186**:231–244.

Cramer, F., 1964, Microplankton from three Palaeozoic formations in the province of León, NW Spain, *Leidse Geologische Mededelingen* **30**:253–361.

Creber, G. T., 1977, Tree rings: a natural data-storage system, *Biological Review* **52**:349–383.

Creber, G. T., and Chaloner, W. G., 1984, Climatic indications from growth rings in fossil woods, in: *Fossil and Climate* (P. J. Brenchley, ed.), Liverpool, John Wiley & Sons, pp. 49–74.

Crook, K. A. W., 1967, Tectonics, climate and sedimentation, in: *Proceedings*, Volume 1 (7th International Sedimentologic Congress), pp. 1–3.

Cross, T. A., and Homewood, P. W., 1997, Amanz Gressly's role in founding modern stratigraphy, *Geological Society of America Bulletin* **109**:1617–1630.

Cross, T. A., Baker, M. R., Chapin, M. A., Clark, M. S., Gardner, M. H., Hanson, M. S., Lessenger, M. A., Little, L. D., McDonough, K. J., Sonnefeld, M. D., Valasek, D. W., Williams, M. R., and Writter, D. N., 1993, Application of high-resolution sequence stratigraphy to reservoir analysis, in: *Subsurface Reservoir Characterization from Outcrop Observation* (R. Eschard and B. Doligez, eds.), Paris, Technip, pp. 11–33.

Culver, S. J., 2003, Benthic foraminifera across the Cretaceous-Tertiary (K-T) boundary: a review, *Marine Micropaleontology* **47**:177–226.

Cunha, A. A. S., 2000, Cicloestratigrafia no Cenomaniano Superior e Turoniano do Atlântico Sul, Tese (Doutorado em Ciências), Instituto de Geociências, Universidade Federal do Rio Grande do Sul, Porto Alegre, 317 pp. (unpublished material).

Cutler, A. H., and Behrensmeyer, A. K., 1996, Models of vertebrate mass mortality events at the K/T boundary, in: *The Cretaceous-Tertiary Event and Other Catastrophes in Earth History*, Special Paper 307 (G. Ryder, D. Fastovsky, and S. Gartner, eds.), Boulder, Geological Society of America, pp. 375–379.

Daemon, R. F., 1974, Palinomorfos-guias do Devoniano Superior e Carbonífero Inferior das bacias do Amazonas e Parnaíba, *Anais da Academia Brasileira de Ciências* **46**(3/4):549–587.

Daemon, R. F., 1976, Correlação bioestratigráfica entre os sedimentos do Siluriano, Devoniano e Carbonífero Inferior das bacias do Amazonas, Parnaíba e Paraná, in: *29° Congr. Brasil Geol.*, Volume 2, Ouro Preto, Anais, pp. 189–194.

Daemon, R. F., and Contreiras, C. J. A., 1971, Zoneamento palinológico da Bacia do Amazonas, in: *Anais XXV Congr. Bras. Geol.*, Volume 3, São Paulo, pp. 79–88.

Daemon, R. F., Quadros, L. P., and Silva, L. C., 1967, Devonian palynology and biostratigraphy of the Paraná Basin, in: *Problems in Brazilian Devonian Geology*, Bol. Parana. Geoc., 21/22 (J. J. Bigarella, ed.), pp. 99–132.

Dagys, A. S., and Weitschat, W., 1993, Correlation of the Boreal Triassic, *Mitteilungen aus dem Geologisch Paläontologischen Institut der Universität Hamburg*, **75**:193–209.

Damuth, J. E., 1975, Echo character of the western Equatorial Atlantic floor and its relationship to the dispersal and distribution of terrigeneous sediments, *Marine Geology* **18**:17–45.

Damuth, J. E., 1980, Use of high-frequency (3.5–12 kHz) echograms in the study of near-bottom sedimentation processes in the deep sea: A review, *Marine Geology* **38**:1–75.

Damuth, J. E., and Olson, H. C., 2001, Evaluation of Latest Quaternary depositional proesses and geohazards in Texas-Louisiana intraslope basins based on high-resolution seismic facies and shallow piston cores, Abstract, American Association of Petroleum Geologists Annual Meeting, Denver, Colorado.

Davies, D. J., Powell, E. N., and Stanton, R. J., 1989, Taphonomic signature as a function of environmental processes: Shells and shell beds in a hurricane-influenced inlet on the Texas coast, *Palaeogeography, Palaeoclimatology, Palaeoecology* **72**:317–356.

Davies, S. D., and Walker, R. G., 1993, Reservoir geometry influenced by high-frequency forced regressions within an overall transgression: Caroline and Garrington fields, Viking Formation (Lower Cretaceous), Alberta, *Canadian Petroleum Geology* **41**: 407–421.

Davis, J. C., 1986, *Statistics and Data Analysis in Geology*, Singapore, John Wiley & Sons.

Davydov, V. I., Snyder, W. S., and Spinosa, C., 1997, Permian foraminiferal biostratigraphy and sequence stratigraphy of Nevada, in: *Late Paleozoic Foraminifera, their Biostratigraphy, Evolution and Paleoecology and the Mid-Carboniferous Boundary*, Special Publication 36 (C. A. Ross, J. R. P. Ross, and P. L. Brenckle, eds.), Cushman Foundation for Foraminiferal Research, pp. 31–34.

Dean, W. E., and Gardner, J. V., 1986, Milankovitch cycles in Neogene deep sea sediment, *Paleoceanography* **1**:539–553.

Del Valle, R. A., Elliot, D. H., and MacDonald, D. I. M., 1992, Sedimentary basins on the east flank of the Antarctic Peninsula: proposed nomenclature, *Antarctic Science* **4**(4): 477–478

Demaison, G. J., and Moore, G. T., 1980, Anoxic environment and oil source bed genesis, *Bulletin of the American Association of Petroleum Geologists* **64**:1179–1209.

Demangeot, J., 1987, *Les milieux "naturels" du globe*, Paris.

Denne, R., and Sen Gupta, B., 1989, Effects of taphonomy and habitat on the record of benthic foraminifera in modern sediments, *Palaios* **4**:414–423.

Dettmann, M. E., 1986, Significance of the spore genus *Cyatheacidites* in tracing the origin and migration of *Lophosoria* (Filicopsida), *Special Papers in Paleontology* **35**:63–94.

D'Hondt, S. L., 1991, Phylogenetic and stratigraphic analysis of earliest Paleocene biserial and triserial planktonic foraminifera, *Journal of Foraminiferal Research* **21**(2):168–181.

Díaz, E., Isaacson, P., Lema, J. C., Christensen, K. L., Vavrdová, M., and Antelo, B., 1993a, Significance of a Late Devonian (Famennian) glacial marine diamictite, northern Altiplano, Bolivia, in: *12th International Conference on the Carboniferous and Permian*, Volume 1, Buenos Aires, Compte Rendu, pp. 293–304.

Díaz, E., Palmer, B. A., and Lema, J. C., 1993b, The Carboniferous sequence of the northern Altiplano of Bolivia: from glacial-marine to carbonate deposition, in: *12th International Conference on the Carboniferous and Permian,* Volume 2, Buenos Aires, Compte Rendu, pp. 203–222.

Díaz, E., Vavrdová, M., Bek, J., and Isaacson, P. E., 1999, Late Devonian (Famennian) glaciation in Western Gondwana: Evidence from the Central Andes, in: *Mid-Paleozoic Terranes, Stratigraphy and Biota*, (R. Feist, J. A. Talent, and A. Daurer, eds.), Abh. Geol. Bundesanstalt, Volume 54, pp. 213–237.

Díez, M. (del) C. R., and Cramer, F. H., 1978, Iberian chitinozoans. – II. Lower Devonian forms (La Vid shales and equivalents), *Palinología* núm. extr. **1**:149–201.

Dingus, L., 1984, Effects of stratigraphic completeness on interpretations of extinction rates across the Cretaceous-Tertiary boundary, *Paleobiology* **10**(4):420–438.

Dino, R., and Rodrigues, M. A., 1995, Palinomorfos eodevonianos da Formação Furnas – Bacia do Paraná, *Anais da Academia Brasileira de Ciências* **67**(1):107–116.

Dino, R., Bergamaschi, S., Pereira, E., Melo, J. H. G., Loboziak, S., and Steemans, P., 1995, Biochronostratigraphic investigations of the Pragian and Emsian stages on the southeastern border of the Paraná Basin, in: *2° Simp. Cronoestr. B. Paraná*, Porto Alegre, Bol. Res. Expand., pp. 19–25.

D'Iorio, M. A., and Agterberg, F. P., 1989, Marker event identification technique and correlation of Cenozoic biozones on the Labrador Shelf and Grand Bank, *Canadian Petroleum Geology Bulletin* **37**:346–357.

Di Pasquo, M., and Azcuy, C. L., 1997, Palinomorfos retrabajados en el Carbonífero tardío de la cuenca Tarija (Argentina) y su aplicación a la datación de eventos diastróficos, *Revista de la Universidad du Guarulhos (Geociências)* **II** (n° esp.), 28–42.

Ditchfield, P. W., Marshall, J. D., and Pirrie, D., 1993, High latitude palaeotemperature variation: new data from the Tithonian to Eocene of James Ross Island, Antarctica. *Palaeogeography, Palaeoclimatology, Palaeoecology* **107**(1–2):79–101.

Détienne, P., 1989, Appearance and periodicity of growth rings in some tropical woods, *IAWA Bulletin* n.s., **10**(2):123–132.

Dodson, P., and Tatarinov, L. P., 1990, Dinosaur extinction, in: *The Dinosauria* (D. B. Weishampel, P. Dodson, and H. Osmolska, eds.), University of California Press, pp. 55–62.

Doeven, P. H., Gradstein, F. M., Jackson, A., Agterberg, F. P., and Nel, L. D., 1982, A quantitative nannofossil range chart, *Micropaleontology* **28**:85–92.

Doktor, M., Gazdzicki, A., Marenssi, S., Porebski, S. J., Santillana, S. N., and Vrba, A. V., 1988, Argentine-Polish geological investigations on Seymour (Marambio) Island, Antarctica, *Polish Polar Research* **9**(4):521–541.

Donohoo, L. L., 2001, CLAMP analysis and magnetostratigraphy of Early Tertiary Chuckanut Formation, Northwest Washington, in: *Abstracts of the Geological Society of America Annual Meeting*, Washington, **2002**:abst.28383.

Dorf, E., 1959, Climate changes of the past and present, *Contributions to the Museum of Palaeontology* **13**(8):181–210.

Dott, R. H. Jr., 1983, SEPM Presidential Address: Episodic sedimentation – How normal is average? Does it matter?, *Journal of Sedimentary Petrography* 1982:**53**(1):05–23.

Downie, C., Evitt, W. R., and Sarjeant, W. A. S., 1963, Dinoflagellates, hystrichospheres and the classification of the acritarchs, *Stanford University Publications in Geological Sciences* **7**:1–16.

Downing, K. P., and Walker, R. G., 1988, Viking Formation, Joffre Field, Alberta: shoreface origin of long, narrow sand body encased in marine mudstones, *American Association of Petroleum Geologists Bulletin* **72**:1212–1228.

Dowsett, H., 1989, Application of the graphic correlation method to Pliocene marine sequences, *Marine Micropaleontology* **14**.3–32.

Dulhunty, J. A., 1945, Principal microspore types in the Permian coals of New South Wales, *Proceedings of the Linnean Society, New South Wales* **70**:147–157.

Dulhunty, J. A., 1946, Distribution of microspore types in the New South Wales Permian coalfields, *Proceedings of the Linnean Society, New South Wales* **71**:239–251.

Dumait, P., 1962, L'action des ultrasons sur les pollens. Note préliminaire, *Pollen et Spores*, **IV**(1):175–180.

Durden, C. J., 1974, Biomerization: an ecological theory of provincial differentiation, in: *Paleogeographic Provinces and Provinciality*, Volume 21 (C. Ross, ed.), Special Publication of the Society of Economic Paleontology and Mineralogy, pp. 18–73.

Du Toit, A. L., 1918, The Zones of the Karroo System and their distribution, *Proceedings of the Geological Society of South Africa* **21**:17–37.

Dutra, T. L., 1997a, Primitive leaves of *Nothofagus* (Nothofagaceae) in Antarctic Peninsula: an upper Campanian record and a betulaceous more than fagaceous morphological character, in: *Actas del VIII Congresso Geologico Chileno, Antofagasta, Chile, Sesión Temática* 3, pp. 511–514.

Dutra, T. L., 1997b, Climate and vegetation in the Antarctic Peninsula during the Cretaceous and Tertiary, *Revista da Universidade de Guarulhos, Geociências* **2**:230.

Dutra, T. L., 1999, Life and death near volcanoes: taphonomic constraints over the upper Cretaceous to lower Tertiary floras of King George Island, Northern Antarctic Peninsula, in: *Boletim de Resumos*, 1° Simpósio sobre Vulcanismo e Ambientes Associados, Gramado, RS, Brasil, pp. 105.

Dutra, T. L., 2000, *Nothofagus* no noroeste da Península Antártica. I. Cretáceo superior (*Nothofagus* in northestern Antarctic Peninsula. I. upper Cretaceous), *Revista da Universidade de Guarulhos* **5**:102–106.

Dutra, T. L., 2001, Paleoflora da ilha 25 de Mayo (King George Island), Península Antártica: contribuição à paleogeografia, paleoclima e para a evolução de Nothofagus, *Asociación Paleontológica Argentina*, Publicación Especial **8**:29–37.

Dutra, T. L., and Batten, D., 2000, The Upper Cretaceous flora from King George Island, an update of information and the paleobiogeographic value, *Cretaceous Research* **21**:181–209.

Dutra, T. L., Leipnitz, B., Faccini, U. F., and Lindenmayer, Z., 1996, A non marine Upper Cretaceous interval in West Antarctica (King George Island, Northern Antarctic Peninsula, *SAMC (South Atlantic Mesozoic Correlations) News* **5**:21–22.

Dymond, J., 1981, Geochemistry of Nazca Plate surface sediments: an evaluation of hydrothermal, biogenic, detrital, and hydrogeneous sources, *Geological Society of America Memoir* **154**: 133–174.

Eaton, J. E., Kirkland, J. I., and Doi, K., 1989, Evidence of reworked Cretaceous fossils and their bearing on the existence of tertiary dinosaurs, *Palaios* **5**:281–286.

Edwards, L. E., 1984, Insights on why graphic correlation (Shaw's method) works, *Journal of Geology* **92**:583–597.

Efremov, J. A., 1940, Taphonomy: new branch of paleontology, *Pan-American Geologists* **74**:81–93.

Ehrmann, W. U., 1991, Implications of sediment composition on the southern Kerguelen Plateau for paleoclimate and depositional environment, in: *Proceedings of the Ocean Drilling Program. Scientific results. Kerguelen Plateau-Prydz Bay*, Volume 119 (E. K. Mazzullo and N. J. Stewart, eds.), pp. 185–210.

Ehrenberg, S. N., and Siring, E., 1992, Use of bulk chemical analyses in stratigraphic correlations of sandstones: An example from the Statfjord Nord Field, Norwegian Continental Shelf, *Journal of Sedimentary Petrology* **62**:318–330.

Einsele, G., 1998, Event stratigraphy: recognition and interpretation of sedimentary event horizons, in: *Unlocking the Stratigraphic Record: Advances in Modern Stratigraphy* (P. Doyle and M. R. Bennett, eds.), Chichester, John Wiley & Sons, pp. 145–193.

Einsele, G., and Seilacher, A. (eds.), 1982, *Cyclic and Event Stratification*, Berlin, Heidelberg, New York, Springer.

Einsele, G., Ricken, W., and Seilacher, A., 1991, Cycles and events in stratigraphy, basic concepts and terms, in: *Cycles and Events in Stratigraphy* (G. Einsele, W. Ricken, and A. Seilacher, eds.), Berlin, Springer-Verlag, pp. 1–19.

Eisenack, A., 1931, Neue Mikrofossilien des baltischen Silurs, I, *Paläontologische Zeitschrift* **13**:74–118.

Ekdale, A. A., 1985, Paleoecology of the marine endobenthos, *Palaeogeography, Palaeoclimatology, Palaeoecology* **50**:63–81.

Ekdale, A. A., Bromley, R. G., and Pemberton, S. G., 1984, *Ichnology: Trace Fossils in Sedimentology and Stratigraphy*, Soc. Econ. Paleo. Mineralol., Short Course Notes 15.

Ekman, S., 1953, *Zoogeography of the Sea*, London, Sidgwick and Jackson.

Elkins, M. G., and Wieland, G. R., 1914. Cordaitean wood from the Indiana Black Shale, Ibid. **4**:38–67.

Elliot, D., 1988, The James Ross Basin, northern Antartic Peninsula, *Comunicaciones* **39**:226.

Elliot, D. H., 1991, Triassic-Early Cretaceous evolution of Antarctica, in: *Geological Evolution of Antarctic* (M. R. A. Thomson, J. A. Crame, and J. W. Thomson, eds.), Cambridge, Cambridge University Press, pp. 541–547.

Elliot, D. H., and Trautman, T. A., 1982, Lower Tertiary strata on Seymour Island, Antarctic Peninsula, in: *Antarctic Geoscience-Symposium on Antarctic Geology and Geophysics*, Volume **B**(4) (C. Craddock, ed.), Madison, University of Wisconsin Press, International Union of Geological Sciences, pp. 287–297.

Elliot, D. H., Askin, R. A., Kyte, F. T., and Zinsmeister, W. J., 1994, Iridium and dinocysts at the Cretaceous-Tertiary boundary on Seymour Island, Antarctica: implications for the K–T event, *Geology* **22**:347–355.

Embry, A. S., 1988, Triassic sea-level changes: evidence from the Canadian arctic archipelago, in: *Sea-level Changes: An Integrated Approach*, SEPM Special Publication, **42**:249–259.

Emerson, S., and Hedges, J. I., 1988, Processes controlling the organic carbon content of open ocean sediments, *Paleoceanography*, **3**:621–634.

Emery, D., and Myers, K., 1996, *Sequence Stratigraphy*, Oxford, Blackwell Science.

Emiliani, C., 1955, Pleistocene paleotemperatures, *Journal of Geology* **63**: 538–578.

Emiliani, C., 1972, Quaternary paleotemperatures and the duration of the high-temperature intervals, *Science* **178**:398–401.

Epstein, S., Buchbaum, R., Lowenstam, H., and Urey, H. C., 1951, Carbonate-water isotopic paleotemperature scale, *Bulletin of the Geological Society of America* **63**: 417–426.

Epstein, S., Buchbaum, R., Lowenstam, H., and Urey, H. C., 1953, Revised carbonate-water isotopic paleotemperature scale, *Bulletin of the Geological Society of America* **64**: 1315–1325.

Erez, J., and Luz, B., 1983, Experimental paleotemperature equation for planktonic foraminifera, *Geochimica Cosmochimica Acta* **47**: 1025–1031.

Ericson, D. B., and Wollin, G., 1968, Pleistocene climates and chronology in deep-sea sediments, *Science* **162**(3859):1227–1234.

Esau, K., 1976, *Anatomia das plantas com sementes*, São Paulo, Ed. Edgar Blucher.

Eshet, Y., Druckman, Y., Cousminer, H. L., Habib, D., and Drugg, W. S., 1988, Reworked palynomorphs and their use in the determination of sedimentary cycles, *Geology* **16**:662–665.

Espitalié, J., Laporte, J. L, Madec, M., Marquis, F., Leplat, P., and Paulet, J. 1997, Méthode rapide de caractérization dês roches mères, de leur potenciel pétrolier et de leur degree d'évolution, *Revue Institute Français de Pétrole* **32**: 23–43.

Estrella, G. O., Gaglianone, P. C., and Melo, M. R., 1984, The Espirito Santo Basin-Brazil: source rock characterization and petroleum habitat, in: *Petroleum Geochemistry and Basin Evaluation*, AAPG Memoir 35, pp. 253–271.

Evernden, J. F., Savage, D. E., Curtis, G. H., and James, G., 1964, K-Ar dates and Cenozoic mammalian chronology of North America, *American Journal of Science* **262**:145–198.

Evitt, W. R., 1963, A discussion and proposals concerning fossil dinoflagellates, hystrichospheres, and acritarchs, I II, *Proceedings of the National Academy of Sciences* **49**:158–164, 298–302.

Eyles, N., Mory, A. J., and Backhouse, J., 2002, Carboniferous-Permian palynostratigraphy of West Australian marine rift basins: resolving tectonic and eustatic controls during Gondwanan glaciations, *Palaeogeography, Palaeoclimatology, Palaeoecology* **184**:305–319.

Fahn, A., 1974, *Anatomia Vegetal*, Madrid, H. Blume.

Fairbridge, R. W., 1976, Effects of Holocene climatic change on some tropical geomorphic processes, *Quaternary Research* **6**:529–556.

Falcon-Lang, H. J., 1999a, The Early carboniferous (Asbian Brigatian) seasonal tropical climate of northern Britain, *Palaios* **14**:116–126.

Falcon-Lang, H. J., 1999b, The Early carboniferous (Courceyan Arudian) monsoonal climate of the British isles, *Geology Magazine* **136**:177–187.

Falcon-Lang, H. J., 2000, The relationship between leaf longevity and growth ring markedness in modern conifer woods and its implications for palaeoclimatic studies, *Palaeogeography, Palaeoclimatology, Palaeoecology* **160**:317–328.

Fara, E., and Hungerbühler, A., 2000, *Paleorhinus magnoculus* from the Upper Triassic of Morocco: A juvenile primitive phytosaur (Archosauria), *Comptes Rendus de l'Académie des Sciences de Paris, Sciences de la Terre e des planètes/Earth and Planetary Sciences* **331**:831–836.

Faure, G., 1986, *Principles of Isotope Geology*, 2nd ed. New York, John Wiley and Sons.

Fauth, G., 2000, Cretaceous-Tertiary (K-T) boundary ostracodes from the Poty quarry, Pernambuco-Paraíba Basin, northeastern Brazil: systematics, biostratigraphy, palaeoecology and palaeobiogeography, Ph.D. thesis, University of Heidelberg, Geologisch-Paläontologisches Institut, Heidelberg, Germany, 170 pp. (unpublished).

Fauth, G., Colin, J. P., Koutsoukos, E. A. M., and Bengtson, P., in press, Taxonomic study of the Cretaceous-Tertiary boundary ostracodes from the Poty Quarry, Pernambuco, northeastern Brazil, in: *Mesozoic Palaeontology and Stratigraphy of South America and the South Atlantic* (P. Bengtson, ed.), *Journal of South American Earth Sciences*, Special Issue (Proceedings of the 2nd European Meeting on the Palaeontology and Stratigraphy of South America and the Regional Meeting of IGCP 381 "South Atlantic Mesozoic Correlations", 2–4 September 1997, Heidelberg, Germany), Part 2.

Fensome, R. A., Williams, G. L., Barss, M. S., Freeman, J. M., and Hill, J. M., 1990, Acritarchs and fossil prasinophytes: an index to genera, species and infraspecific taxa, *American Association of Stratigraphists and Palynologists, Contribution Series* **25**:1–771.

Fensome, R. A., Williams, G. L., Barss, M. S., Freeman, J. M., and Hill, J. M., 1991, Alphabetical listing of acritarch and fossil prasinophyte species, *American Association of Stratigraphists and Palynologists, Contribution Series* **26**:1–111.

Fernández-López, S., 1999, Tafonomía y fosilización, in: *Tratado de Paleontologia. Tomo 1*, (B. Meléndez, ed.), Consejo Superior de Investigaciones Científicas, pp. 51–107, 438–441.

Fernández-López, S., 2000, Ammonite taphocycles in carbonate epicontinental platforms, *GeoResearch Forum* **6**:293–300.

Fernández-López, S., Duarte, L. V. and Henriques, M. H. P., 2000, Ammonites from lumpy limestones in the Lower Pliensbachian of Portugual: taphonomic analysis and palaeoenvironmental implications, *Revista Sociedad Geológica de España* **13**:3–15.

Fertl, W. H., Chilingar, G. V., and Yen, T. F., 1986, Organic carbon content and source-rock identification based on geophysical well logs, *Energy Sources* **8**:381–439.

Fischer, A. G., 1986, Climatic rhythms recorded in strata, *Annual Review of Earth Planetary Science Letters* **14**:351–376.

Fischer, A. G., 1991, Orbital cyclicity in Mesozoic strata, in: *Cycles and Events in Stratigraphy* (G. Einsele, W. Ricken, and A. Seilacher, eds.), Berlin, Springer-Verlag, pp. 48–62.

Fischer, A. G., 1993, Cyclostratigraphy of Cretaceous chalk-marl sequences, in: *Evolution of the Western Interior Basin*, Special Paper 39 (W. G. E. Caldwell and E. G. Kauffman, eds.), Geological Association of Canada, pp. 283–295.

Fischer, A. G., 1995, Cyclostratigraphy, *quo vadis?*, in: *Orbital Forcing Timescales and Cyclostratigraphy*, Geological Society special publication 85 (M. R. House and A. S. Gale, eds.), pp. 199–204.

Fischer, A. G., and Arthur, M. A., 1977, Secular variations in the pelagic realm, in: *Deep Water Carbonate Environments*, special publication 25 (H. E. Cook and P. Enos, eds.), Soc. Con. Paleontol. Mineral., pp. 351–376.

Fischer, A. G., and Bottjer, D. J., 1991, Orbital forcing and sedimentary sequences, *Journal of Sedimentary Petrology* **61**(7):1063–1069.

Flessa, K. W., 1986, Causes and consequences of extinctions, in: *Patterns and Processes in the History of Life* (D. M. Raup and D. Jablonski, eds.), Dahlem Konferezem, Springer-Verlag, pp. 235–257.

Flessa, K. W., 1987, *Paleoecology and Taphonomy of Pleistocene to Recent Intertidal Deposits, Gulf of California*, special publications No. 2, Paleontological Society.

Flessa, K. W., Kowalewski, M., and Walker, S. E., 1992, Post-collection taphonomy: shell destruction and the Chevrolet, *Palaios* **7**:553–554.

Florentin, J. M., Maurrasse, R., and Sen, G., 1991, Impacts, tsunamis, and the Haitian Cretaceous-Tertiary boundary layer, *Science* **252**:1690–1693.

Fluteau, F., Besse, J., Broutin, G., and Ramstein, G., 2001, The Late Permian climate. What can be inferred from climate modelling concerning Pangea scenarios and Hercynian range altitude, *Palaeogeography, Palaeoclimatology, Palaeoecology* **167**:39–71.

Flynn, J. J., and Tauxe, L., 1998, Magnetostratigraphy of upper Paleocene-lower Eocene marine and terrestrial sequences, in: *Late Paleocene-Early Eocene Climatic and Biotic Events* (M.-P. Aubry, S. Lucas, and W. A. Berggren, eds.), New York, Columbia University Press, pp. 67–90

Flynn, J. J., Parrish, J. M., Rakotosamimanana, B., Simpson, W. F., Whatley, R. L., and Wyss, A. R., 1999, A Triassic fauna from Madagascar, including early dinosaur, *Science* **286**:763 765.

Flynn, J. J., Parrish, J. M., Rakotosamimanana, B., Ranivoharimanana, L., Simpson, W. F., and Wyss, A. R., 2000, New traversodontids (Synapsida, Eucynodontia) from the Triassic of Madagascar, *Journal of Vertebrate Palaeontology* **20**(3):422–427.

Foster, C. B., 1975, Permian plant microfossils from the Blair Athol Coal Measures, central Queensland, Australia, *Palaeontographica Abteilung B* **154**:121–171.

Foster, C. B., 1978, Aspects of palynological correlation between the Late Permian of the Euramerican and Gondwanan (eastern Australian) provinces, *Queensland Government Mineral Journal* **79**:376–379.

Foster, C. B., 1979, Permian plant microfossils of the Blair Athol Coal Measures, Baralaba Coal Measures, and basal Rewan Formation of Queensland, Geol. Surv. Qld. Publ. 372, *Palaeontology Papers* **45**:1–244.

Foster, C. B., 1982, Spore-pollen assemblages of the Bowen Basin, Queensland (Australia): their relationship to the Permian/Triassic boundary, *Review of Palaeobotany and Palynology* **36**:165–183.

Foster, C. B., Logan, G. A., and Summons, R. E., 1998, The Permian-Triassic boundary in Australia: where is it and how is it expressed?, *Proceedings of the Royal Society of Victoria* **110**:247–266.

Frakes, L. A., 1979, *Climates Throughout Geologic Time*, New York, Elsevier.

Frakes, L. A., and Bolton, B. R., 1984, Origin of manganese giants: sea-level change and anoxic-oxic history, *Geology* **12**:83–86.

Frakes, L. A., Francis, J. E., and Syktus, J. I., 1992, *Climate Modes of the Phanerozoic: the History of the Earth's Climate over the Past 600 Million Years*, Cambridge, Cambridge University Press.

Francis, J. E., 1991, Paleoclimatic significance of Cretaceous-Early Tertiary fossil forests of the Antarctic Peninsula, in: *Geological Evolution of Antarctic* (M. R. A. Thomson, J. A. Crame, and J. W. Thomson, eds.), Cambridge, Cambridge University Press, pp. 623–628.

Francis, J. E., 1994, Paleoclimates of Pangea – Geological Evidence, in: *Pangea: Global Environments and Resources. Canadian Society of Petroleum Geologists, Memoir* **17**:265–274.

Francis, J., and Pole, I., 2002, Cretaceous and early tertiary climates of Antarctica: evidence from fossil wood, *Palaeogeography, Palaeoclimatology, Palaeoecology* **182**:47–64.

Frest, T. J., Witzke, B. J., and Brett, C. E., 1999, Caradocian-Gedinnian echinoderm associations of central and eastern North America, in: *Paleocommunities – A Case Study from the Silurian and Lower Devonian* (A. J. Boucot and J. D. Lawson, eds.), Cambridge, Cambridge University Press, pp. 638–783.

Frey, R. W., Pemberton, S. G., and Fagerstrom, J. A., 1984, Morphological, ethological, and environmental significance of the ichnogenera *Scoyenia* and *Ancorichnus, Journal of Paleontology* **58**:511–528.

Frey, R. W., Howard, J. D., and Hong, J. S., 1987, Prevalent lebensspuren on a modern macrotidal flat, Inchon, Korea: Ethological and environmental significance, *Palaios* **2**:571–593.

Frey, R. W., Pemberton, S. G., and Saunders, T. D. A., 1990, Ichnofacies and bathymetry: a passive relationship, *Journal of Paleontology* **64**:155–158.

Funkhouser, J. W., and Evitt, W. R., 1959, Preparation techniques for acid-insoluble microfossils, *Micropaleontology* **5**:369–375.

Fürsich, F. T., and Aberhan, M., 1990, Significance of time-averaging for palaeocommunity analysis, *Lethaia* **23**:143–152.

Fürsich, F. T., and Mayr, H., 1981, Non marine *Rhizocorallium* (trace fossil) from the Upper Freshwater Molasse (Upper Miocene) of southern Germany. *Neues Jarbuch für Geologie und Paläontologie Monatshefte* **6**:321–333.

Fürsich, F. T., and Pandey, D. K., 2003, Sequence stratigraphic significance of sedimentary cycles and shell concentrations in the Upper Jurassic–Lower Cretaceous of Kachchh, western India. *Palaeogeography, Palaeoclimatology, Palaeoecology* **193**:285–309.

Galbrun, M. B., 1984, Magnetostratigraphie de la limite Jurassique-Crétacé. Proposition d'une échelle de polarité à partir du Stratotype du Berriasien (Berrias, Ardèche, France) et de la sierra de Lugar (Province de Murcie, Espagne), Paris, Thèse de Doctorat 3e Cycle, L'Université Pierre et Marie Curie (Paris VI), 95 pp.

Gale, A. S., 1998, Cyclostratigraphy, in: *Unlocking the Stratigraphic Record: Advances in Modern Stratigraphy* (P. Doyle and M. R. Bennett, eds.), Chichester, John Wiley & Sons, pp. 195–220.

Gale, A. S., and Smith, A. B., 1982, The paleobiology of the Cretaceous irregular echinoids *Infulaster* and *Hagenowia*, *Palaeontology* **25**:11–42.

Gale, A. S., Jenkyns, H. C., Kennedy, W. J., and Corfield, R., 1993, Chemostratigraphy versus biostratigraphy: data from around the Cenomanian-Turonian boundary, *Journal of the Geological Society of London* **150**:29–32.

Galloway, W. E., 1989, Genetic stratigraphic sequences in basin analysis: I, Architecture and genesis of flooding-surface bounded depositional units, *American Association of Petroleum Geologists Bulletin* **73**(2):125–142.

Galloway, W. E., Bebout, D. G., Fisher, W. L., Dunlap, J. B., Jr., Cabrera-Castro, R., Lugo-Rivera, J. E., and Scott, T. M, 1991, Cenozoic, in: *The Gulf of Mexico Basin, The Geology of North America* (A. Salvador, ed.), V. J, Geol. Soc. Amer. Inc., Boulder, Colo., pp. 245–324.

Gao Lianda, 1985, Carboniferous spore assemblages in China, *Comptes Rendues 9th Congr. Internat. Strat. Géol. Carb., Washington and Champaign-Urbana*, **1979**:2:103–108.

Gardin, S., and Monechi, S., 1998, Palaeoecological change in middle to low latitude calcareous nannoplankton at the Cretaceous/Tertiary boundary, *Bulletin de la Societé Geologique de France* **169**(5):709–723.

Garner, H. F., 1959, Stratigraphic-sedimentary significance of contemporary climate and relief in four regions of the Andes Mountains, *Geological Society of America Bulletin* **70**:1327–1368.

Gartner, S., 1992, Miocene nannofossil chronology in the North Atlantic, DSDP Site 608, *Marine Micropaleontology* **18**:307–331.

Gastaldo, R. A., 1986, Implications on the paleoecology of authocthonous carboniferous lycopods in clastic sedimentary environments, *Palaeogeography, Palaeoclimatology, Palaeoecology* **53**:191–212.

Gauffre, F.-X., 1993, Biochronostratigraphy of the lower Elliot Formation (southern Africa) and preliminary results on the Maphutseng dinosaur (Saurischia: Prosauropoda) from the same formation of Lesotho, *New Mexico Museum of Natural History and Science Bulletin* **3**:147–149.

Gázdzicki, A., and Stolarski, J., 1992, An Oligocene record of the coral *Flabellum* from Antarctica, *Polish Polar Research* **13**(3–4):265–272.

Geiger, R., 1965, *The Climate Near the Ground*, Cambridge, MA, Harvard University Press.

Gilbert, G. K., 1895, Sedimentary measurement of Cretaceous time, *Journal of Geology* **3**:121–127.

Gingras, M. K., Pemberton, S. G., Mendoza, and Henk, B., 1999, Modeling fluid flow in trace fossils; assessing the anisotropic permeability of *Glossifungites* surfaces, *Petroleum Geoscience* **5**:349–357.

Ginsburg, R., Keller, G., Smit, J., Orue-Etxebarria, X., Canudo, I., Olsson, R. D., and Masters, B. M., 1997, Blind test of the foraminiferal extinctions at the K/T boundary stratotype section, El Kef (Tunisia), *Marine Micropaleontology* **29**:76.

Glasby, G. P., and Kunzendorf, H., 1996, Multiple factors in the origin of the Cretaceous/Tertiary boundary: the role of environmental stress and Deccan Trap volcanism, *Geologische Rundschau* **85**(2):191–210.

Glennie, K. W., 1984, Early Permian-Rotliegend, in: *Introduction to the Petroleum Geology of the North Sea* (K. W. Glennie, ed.), Boston, MA, Blackwell Scientific Publications, pp. 41–60.

Goldhammer, R. K., Dunn, P. A., and Hardie, L. A., 1987, High frequency glacio-eustatic sea level oscillations with Milankovitch characteristics recorded in Middle Triassic platform carbonates in northern Italy, *American Journal of Sciences* **287**:853–892.

Golonka, J., Ross, M., and Scotese, C. R., 1994, Phanerozoic palaeographic and paleoclimatic modeling maps. I, in: *Pangea: Global Environments and Resorces*, Memoir 17 (A. F. Embry, B. Beauchamp, and D. J. B. Glass, eds.), Canadian Society of Petroleum Geologists, pp. 1–48.

Gomankov, A. V., 1992, The interregional correlation of the Tatarian and the problem of the Permian upper boundary, *International Geology Review* **34**:1015–1020.

Gomankov, A. V., Balme, B. E., and Foster, C. B., 1998, Tatarian palynology of the Russian Platform: a review, *Proceedings of the Royal Society of Victoria* **110**:115–135.

Gómez, J. J., and Fernández-López, S., 1994, Condensation processes in shallow platforms, *Sedimentary Geology* **92**:147–59.

Goodwin, P. W., and Anderson, E. J., 1985, Punctuated agradational cycles, *Jounal of Geology* **93**:515–533.

Goolaerts, S., Kennedy, W. J., Dupuis, C., and Steurbaut, E., 2004, Terminal Maastrichtian ammonites from the Cretaceous–Paleogene Global Stratotype Section and Point, El Kef, Tunisia, *Cretaceous Research* **25**:313–328.

Gordon, W. A., 1973, Marine life and ocean surface currents in the Cretaceous, *Journal of Geology* **81**:269–284.

Gould, S. J., 1984, Toward the vindication of punctuational change, in: *Catastrophes and Earth History* (W. A. Berggren and J. A. Van Couvering, eds.), Princeton, Princeton University Press, pp. 464.

Gould, S. J., and Eldredge, N., 1977, Punctuated equilibria: the tempo and mode of evolution reconsidered, *Paleobiology* **3**:115–151.

Graciansky, P. de, Hardenbol, J., Jacquin T., and Vail, P. R. (eds.), 1998, *The Mesozoic-Cenozoic Sequence Stratigraphy of European Basins*, Tulsa: Society of Economic Paleontologists and Mineralogists Special Publication 60.

Gradstein, F. M., 1990, Program Stratcor for zonation and correlation of fossil events, Geological Survey of Canada, Open File Report 2285 (includes source code on diskettes).

Gradstein, F. M., 1996, Stratcor – Graphic Zonation and Correlation software. Users Guide, Authors Edition 4.

Gradstein, F. M., and Agterberg, F. P., 1998, Uncertainty in stratigraphic correlation, in: *Sequence Stratigraphy – Concepts and Applications* (F. M. Gradstein *et al.*, eds.), Amsterdam, Elsevier, pp. 9–29.

Gradstein, F. M., and Bäckström, S. A., 1996, Cainozoic biostratigraphy and palaeobathymetry, northern North Sea and Haltenbanken, *Norsk Geologisk Tidsskrift* **76**:3–32.

Gradstein, F. M., and Ogg, J. G., 1996, A phanerozoic timescale, *Episodes* **19**:3–5.

Gradstein, F. M., Agterberg, F. P., Bower, J. C., and Schwarzacher, W. S., 1985, *Quantitative Stratigraphy*, Dordrecht, Reidel.

Gradstein, F. M., Kaminski, M. A., Berggren, W. A., Kristiansen, I. L., and D'Iorio, M. A., 1994, Cainozoic biostratigraphy of the North Sea and Labrador shelf, *Micropaleontology*, **40** (Supplement 1994).

Gradstein, F. M., Agterberg, F. P., Ogg, J. G., Hardenbol, J., Van Veen, P., Thierry, J., and Huang, Z., 1995, A Triassic, Jurassic and Cretaceous time scale, in: *Geochronology, Time Scales and Global Stratigraphic Correlation*, Special Publication No. 54 (W. A. Berggren, D. V. Kent, M.-P. Aubry, and J. Hardenbol, eds.), Society of Economic Paleontologists and Mineralogists, pp. 95–126.

Gradstein, F. M., Agterberg, F. P., Ogg, J. G., Hardenbol, J., and Backstrom, S., 1999, On the Cretaceous time scale, *Neues Jahrbuch für Geologie und Paläontologie, Abhandlungen* **212**(1–3):3–14.

Gradstein, F. M., Kaminski, M. A., and Agterberg, F. P., 1999, Biostratigraphy and paleoceanography of the Cretaceous Seaway between Norway and Greenland, *Earth-Science Reviews* **46**:27–98.

Graefe, K.-U., 1999, Foraminiferal evidence for Cenomanian Sequence Stratigraphy and Palaeoceanography of the Boulonnais (Paris Basin, northern France), *Palaeogeography, Palaeoclimatology, Palaeoecology* **153**:41–70.

Grahn, Y., 1981, Ordovician Chitinozoa from the Stora Åsbotorp boring in Västergötland south-central Sweden, *Sveriges Geologiska Undersoekning Serie C* **787**:1–40.

Grahn, Y., 1992, Revision of Silurian and Devonian strata of Brazil, *Palynology* **16**:35–61.

Grahn, Y., 1997, Chitinozoan biostratigraphy in the Devonian outcrop belt of the east part of the Paraná Basin, and in older formations from the Paraná Basin, Brazil, Universidade do Estado do Rio de Janeiro, unpubl. Uerj-Padct/Finep report (Nov. 1997).

Grahn, Y., 1998a, Devonian chitinozoan biostratigraphy in the north and northwest part of the Paraná Basin, Brazil, Universidade do Estado do Rio de Janeiro, unpubl. Uerj-Padct/Finep report (Apr. 1998).

Grahn, Y., 1998b, Silurian and Early Devonian chitinozoan biostratigraphy of the Paraná Basin, Paraguay, Universidade do Estado do Rio de Janeiro, unpubl. Uerj-Padct/Finep report (Jun. 1998).

Grahn, Y., 1999a, Recent progress in Silurian and Devonian chitinozoan biostratigraphy of the Paraná Basin in Brazil and Paraguay, in: *Ordovician-Devonian Palynostratigraphy in Western Gondwana: Update, Problems and Perspectives* (M. A. C. Rodrigues and E. Pereira, eds.), Rio de Janeiro, Uerj, pp. 147–163.

Grahn, Y., 1999b, Silurian and Devonian chitinozoan biostratigraphy of the Tarija Basin, Argentina, and the stratigraphic position of the Devonian rocks in the Paraná Basin of Uruguay, Universidade do Estado do Rio de Janeiro, unpublished Uerj-Padct/Finep report (Sept. 1999).

Grahn, Y., 2002, Upper Silurian and Devonian Chitinozoa from central and southern Bolivia, central Andes, *Journal of South American Earth Sciences* **15**:315–326.

Grahn, Y., 2003a, Silurian and Devonian chitinozoan assemblages from the Chaco-Paraná Basin, northeastern Argentina and central Uruguai, *Revista Española de Micropaleontologica* **35**:1–8.

Grahn, Y., 2003b, Chitinozoan biostratigraphy in the north-northwestern and eastern Devonian outcrop belts of the Paraná Basin, Universidade do Estado do Rio de Janeiro, unpublished Uerj-Padct/Finep and Petrobras/Profex report (Jun. 2003).

Grahn, Y., and Melo, J. H. G., 2002, Chitinozoan biostratigraphy of the Late Devonian formations in well Caima PH-2, Tapajós River area, Amazonas Basin, northern Brazil, *Review of Palaeobotany and Palynology* **118**:115–139.

Grahn, Y., and Melo, J. H. G., (in press, a) Middle and Late Devonian Chitinozoa and biostratigraphy of the Parnaíba and Jatobá Basins, northeastern Brazil, *Palaeontographica Abteilung A*.

Grahn, Y., and Melo, J. H. G., (in press, b), Integrated correlation of Middle Devonian chitinozoans and miospores in the Amazonas Basin, northern Brazil, *Journal of Micropalaeontology*.

Grahn, Y., and Melo, J. H. G., 2003, Silurian–Devonian chitinozoan biostratigraphy along the Urubu, Uatumã and Abacate Rivers in the western part of the Amazonas Basin, northern Brazil, *Bulletin of Geosciences (Prague)* **78**(4):373–391.

Grahn, Y., Melo, J. H. G., and Loboziak, S., 1997, Preliminary biostratigraphic integration of chitinozoan and miospore successions in the type Barreirinha Formation, Curuá Group (Upper Devonian, Amazonas Basin), *Revista de la Universidad du Guarulhos (Geociências)* **II** (No. Esp.), 218.

Grahn, Y., Pereira, E., and Bergamaschi, S., 2000, Silurian and Lower Devonian chitinozoan biostratigraphy of the Paraná Basin in Brazil and Paraguay, *Palynology* **24**:143–172.

Grahn, Y., Loboziak, S., and Melo, J. H. G., 2001, Integrated miospore-chitinozoan biozonation of the Parnaíba Basin and its correlation with Petrobras (Müller 1962) Silurian–Lower Carboniferous palynozones, in: *Correlação de seqüências paleozóicas sul-americanas*, (J. H. G. Melo and G. J. S. Terra, eds.), *Ciência e Técnica Petróleo (Seção: Exploração de Petróleo)* **20**:81–89. (CD Rom, ISSN-0100–1515.)

Grahn, Y., Pereira, E., and Bergamaschi, S., 2002, Middle and Upper Devonian chitinozoan biostratigraphy of the Paraná Basin in Brazil and Paraguay, *Palynology* **26**:135–165.

Grahn, Y., Loboziak, S., and Melo, J. H. G., 2003, Integrated correlation of Late Silurian (Pridoli s.l.)–Devonian chitinozoans and miospores in the Solimões Basin, northern Brazil, *Acta Geologica Polonica* **53**(4):283–300.

Grassi, A. A., 2000, O limite Cretáceo-terciário nas bacias de Pernambuco-Paraíba e Campos: um estudo multidisciplinar com ênfase na bioestratigrafia de nanofósseis, Dissertação (Mestrado), Universidade Federal do Rio Grande do Sul, Instituto de Geociências, Porto Alegre, RS, Brazil, 158 pp. (unpublished).

Gregory, W. A. and Hart, G. F., 1992, Towards a predictive model for the palynologic response to sea-level changes, *Palaios* **7**:3–33.

Greguss, P., 1972, *Xylotomy of the Living Conifers*, Budapest, Akademiai kiado.

Greenwood, D. R., 1991, The taphonomy of plant macrofossils, in: *The Processes of Fossilization* (S. K. Donovan, ed.), London, Belhaven Press, pp. 141–169.

Greenwood, D. R., 1992, Taphonomic constraints on foliar physiognomy interpretations of late Cretaceous and tertiary paleoclimates, *Review of Palaeobotany and Palynology* **71**:149–190.

Greenwood, D. R., 1994, Palaeobotanical evidence for Tertiary climates, in: *History of the Australian Vegetation, Cretaceous to Recent* (R. S. Hill, ed.), Cambridge, Cambridge University Press, pp. 44–59.

Grignani, D., Lanzoni, E., and Elatrash, H., 1991, Palaeozoic and Mesozoic Subsurface Palynostratigraphy in the Al Kufrah Basin, Libya, in: *The Geology of Libya*, Volume 4 (M. J. Salem, O. S. Hammuda, and B. A. Eliagoubi, eds.), Amsterdam, Elsevier, pp. 1159–1227.

Grimsdale, T. F., and van Morkhoven, F. P. C. M., 1955, The ratio between pelagic and benthonic foraminifera as a means of estimating depth of deposition of sedimentary rocks, *Proceedings of the Fourth World Petroleum Congress* **1/D**:473–491.

Groenwald, G. H., and Kitching, J. W., 1995, Biostratigraphy of the *Lystrosaurus* Assemblage Zone, in: *Biostratigraphy of the Beaufort Group (Karoo Supergroup). South African Comitte for Stratigraphy, Biostratigraphic Series*, Volume 1 (B. S. Rubidge, ed.), pp. 35–39.

Guennel, G. K., 1952, Fossil spores of the Alleghenian coals in Indiana, *Indiana Geological Survey, Report of Progress*, **4**:1–40.

Guerra, M., 1976, Uma nova madeira petrificada na Formação Irati do Rio Grande do Sul, *Ameghiniana* **13**(3/4):254–267.

Guerra-Sommer, M., 1977, *Damudoxylon* (Maheshwari) Maheshwari, 1972. Um gênero ocorrente no Gondwana do Brasil, *Pesquisas* **7**:131–144.

Guerra-Sommer, M., 1978, Ocorrência de gênero nórdico de conífera no Gondwana sul-brasileiro, *Pesquisas* **10**:65–76.

Guy-Ohlsen, D., 1996, Prasinophycean algae, in: *Palynology: Principles and Applications*, Volume 1, Chapter 7B (J. Jansonius and D. C. McGregor, eds.), Dallas, American Association of Stratigraphic Palynologists Foundation, pp. 181–189.

Hachiro, J., 1991, Litotipos, associações faciológicas e sistemas deposicionais da Formação Irati no Estado de São Paulo, Dissertação (Mestrado em Geologia Sedimentar), São Paulo, Instituto de Geociências, Universidade de São Paulo, p. 175 (unpublished material).

Håkansson, E., and Thomsen, E., 1979, Distribution and types of bryozoan communities at the boundary in Denmark, in: *Cretaceous-Tertiary Boundary Events. I. The Maastrichtian and Danian of Denmark* (T. Birkelund and R. G. Bromley, eds.), University of Copenhagen, pp. 78–91.

Håkansson, E., and Thomsen, E., 1999, Benthic extinction and recovery patterns at the K/T boundary in shallow water carbonates, Denmark, *Palaeogeography, Palaeoclimatology, Palaeoecology* **154**(1–2):67–85.

Hallam, A., ed., 1973, *Atlas of Palaeobiogeography*, Elsevier, Amsterdam.

Hallam, A., 1985, A review of Mesozoic climates, *Journal of the Geological Society of London* **142**:433–445.

Hallam, A., 1987, End-Cretaceous Mass Extinction Event: Argument for Terrestrial Causation, *Science* **238**:1237–1242.

Hallam, A., 1989a, Catastrophism in geology, in: *Catastrophes and Evolution: Astronomical Foundations* (S. W. M. Clube, ed.), Cambridge, Cambridge University Press, pp. 25–55.

Hallam, A., 1989b, The case for sea-level change as a dominant causal factor in mass extinction of marine invertebrates, *Royal Society of London Philosophical Transactions* ser. B, **325**:437–455.

Hallam, A., 1992, *Phanerozoic Sea-level Changes*, New York, Columbia University Press.

Hallam, A., and Wignall, P. B., 1999, Mass extinctions and sea-level changes, *Earth-Science Reviews* **48**:217–250.

Hallam, A. J. M., 1985, A review of Mesozoic climate, *Journal of the Geological Society of London* **142**:433–445.

Hancock, J. M., and Kauffman, E. G., 1979, The great transgressions of the Late Cretaceous, *Journal of the Geological Society* **136**:175–186.

Hansen, H. J., Gwozdz, R., Bromley, R. G., Rasmussen, K. L., Wogesen, E. W., and Pedersen, K. R., 1986, Cretaceous/Tertiary spherules from Denmark, New Zealand, and Spain, *Geological Society of Denmark Bulletin* **35**:75–82.

Hansen, H. J., Mohabey, D. M., and Toft, P., 2001, No K/T boundary at Anjar, Gujarat, India: Evidence from magnetic susceptibility and carbon isotopes, *Proceedings of the Indian Academy of Sciences* **100**(2):133–142.

Hanson, C. B., 1980, Fluvial taphonomic processes: models and experiments, in: *Fossils in the Making* (A. K. Behrensmeyer and A. P. Hill, eds.), Chicago, University of Chicago Press, pp.156–181.

Haq, B. U., Hardenbol, J., and Vail, P. R., 1987, Chronology of fluctuating sea levels since the Triassic (250 million years ago to present), *Science* **235**:1156–1167.

Haq, B. U., Hardenbol, J., and Vail, P. R., 1988, Mesozoic and Cenozoic chronostratigraphy and cycles of sea-level change, in: *Sea-level Changes: an Integrated Approach*, Special Publication 42 (C. K. Wilgus, B. S. Hastings, G. S. St. C. Kendall, H. W. Posamentier, C. A. Ross, and J. C. van Wagoner, eds.), Soc. Econ. Paleontol. Mineral., pp. 71–108.

Harbaugh, J. W., Watney, W. L., Rankey, E. C., Slingerland, R., Goldstein, R. H., and Franseen, E. K. (eds.), 1999, *Numerical Experiments in Stratigraphy: Recent Advances in Stratigraphic and Sedimentologic Computer Simulations*, SEPM (Society for Sedimentary Geology), Special Publication **62**.

Hardenbol, J., and Berggren, W. A., 1978, A new Paleogene mumerical time scale, in: *The Geologic Time Scale*, Studies in Geology 6 (G. V. Cohee, M. F. Glaessner, and H. D. Hedberg, eds.), Tulsa, American Association of Petroleum Geologists, pp. 213–234.

Hardenbol, J., Thierry, J., Farley, M. B., Jacquin, T., Graciansky, P. C. de, and Vail, P. R., 1998, Mesozoic and Cenozoic sequence chronostratigraphic framework of European basins, in: *Mesozoic and Cenozoic Sequence Stratigraphy of European Basins*, Special Paper **60** (P. C. de Graciansky, J. Hardenbol, T. Jacquin, and P. R. Vail, eds.), SEPM (Society of Sedimentary Geology), pp. 3–14.

Harland, W. B., Cox, A. V., Llewellyn, P. G., Pickton, C. A. G., Smith, A. G., and Walters, R., 1982, *A Geologic Time Scale*, Cambridge, Cambridge University Press.

Harland, W. B., Armstrong, R. L., Cox, A. V., Craig, L. E., Smith, A. G., and Smith, D. G., 1990, *A Geologic Time Scale*, Cambridge, Cambridge University Press.

Harrison, C. G. A., and Funnel, B. M., 1964, Relationship of palaeomagnetic reversals and micropalaeontology in two Late Caenozoic cores from the Pacific Ocean, *Nature* **204**:566.

Hart, G. F., 1960, Microfloral investigation of the Lower Coal Measures (K2); Ketewaka-Mchuchuma Coalfield, Tanganyika, *Geological Survey of Tanganyika, Bulletin* **30**:1–18.

Hart, G. F., 1965a, Microflora from the Ketewaka-Mchuchuma coalfield, Tanganyika, *Geological Survey of Tanganyika, Bulletin* **36**:1–27 [imprinted 1963].

Hart, G. F., 1965b, Miospore zones in Karroo sediments of Tanzania, *Palaeontologica Africana* **IX**:139–150.

Hart, G. F., 1969, Palynology of the Permian period, in: *Aspects of Palynology* (R. H. Tschudy and R. A. Scott, eds.), New York, Wiley, pp. 271–289.

Hashemi, H., and Playford, G., 1998, Upper Devonian palynomorphs of the Shishtu Formation, Central Iran Basin, east-central Iran, *Palaeontographica Abteilung B* **246**:115–212.

Hathway, B., 2000, Continental rift to back-arc basin: Jurassic-Cretaceous stratigraphical and structural evolution of the Larsen basin, Antarctic Peninsula, *Journal of the Geological Society of London* **157**:417–432.

Hayes, J. D., Imbrie, J., and Shackleton, N. J., 1976, Variations in the earth's orbit: pacemaker of the Ice Ages, *Science* **194**:1121–1132.

Hayward, B. W., 1976, Lower Miocene bathyal and submarine canyon ichnocoenoses from Northland, New Zealand, *Lethaia* **9**:149–162.

Heckel, P. H., and Witzke, B. J., 1979, Devonian world palaeogeography determined from distribution of carbonates and related lithic palaeoclimatic indicators, in: *The Devonian System*, Special Papers in Palaeontology 23 (M. R. House, C. T. Scutton, and M. G. Bassett, eds.), pp. 99–123.

Hedberg, H. D. (ed.), 1976, *International Stratigraphic Guide, A Guide to Stratigraphic Classification, Terminology, and Procedure*, International Subcommission on Stratigraphic Classification (ISSC), New York, John Wiley & Sons.

Heirtzler, J. R., Dickson, G. O., Herron, E. M., Pittman III, W. C., and LePichon, X., 1968, Marine magnetic anomalies, geomagnetic field reversals, and motions of the ocean floor and continents, *Journal of Geophysics Research* **73**:2119–2136.

Helby, R., 1966, Sporologische Untersuchungen an der Karbon/Perm Grenze im Pfälzer Bergland, *Geologisches Landesamt Rheinland und Westfalen Fortschritte* **13**:645–704.

Helenes, J., and Somoza, D., 1999, Palynology and sequence stratigraphy of the Cretaceous of eastern Venezuela, *Cretaceous Research* **20**:447–463.

Henriksson, A. S., 1996, Calcareous nannoplankton productivity and succession across the Cretaceous-Tertiary boundary in the Pacific (DSDP Site 465) and Atlantic (DSDP Site 527) Oceans, *Cretaceous Research* **17**:451–477.

Herbert, T. D., and Fischer, A. G., 1986, Milankovitch climatic origin of the mid-Cretaceous black shale rhythms in central Italy, *Nature* **321**:739–743.

Herrera, Z. A., and Rachboeuf, P. R., 1997, Affinidades paleobiogeográficàs de la fauna de braquiópodos devónicos de la Precordillera Argentina, XIII Jornadas de Paleontologia, V Reunión Internacional Proyecto 351, Programa Internacional de Correlación Geológica, Paleozoico Inferior del Noroeste de Gondwana, pp. 83–86.

Hess, H. H., 1959, Nature of the great oceanic ridges: International Oceanography Congress Preprints, Washington, DC, AAAS, pp. 33–34.

Hess, H. H., 1962, History of ocean basins, in: *Petrologic Studies: A Volume in Honor of A. F. Buddington Bolder* (A. E. J. Engel, H. L. James, and B. F. Leonar, eds.), Geological Society of America, pp. 599–620.

Hess, J., Bender, M. L., and Schilling, J. G., 1986, Sea-water $^{87}Sr/^{86}Sr$ evolution from Cretaceous to Present – Applications to Paleoceanography, *Science* **231**:979–984.

Hickey, L. J., 1984, Changes in angiosperm flora across the Cretaceous-Tertiary boundary, in: *Catastrophes and Earth History – The new Uniformitarism* (W. A. Berggren and J. Van Couvering, eds.), Princeton, NJ: Princeton University Press, pp. 279–314.

Hickey, L. J., and McWeeney, L. J., 1992, Plants at the K/T boundary, *Nature* **356**:295–296.

Higgs, K. T., 1996, Paleozoic spores and pollen – introduction, in: *Palynology: Principles and Applications*, Volume 2, Chapter 18 (J. Jansonius and D. C. McGregor, eds.), Dallas, American Association of Stratigraphic Palynologists Foundation, pp. 553–554.

Hildebrand, A. R., and Boynton, W. V., 1990, Proximal Cretaceous-Tertiary boundary impact deposits in the Caribbean, *Science* **248**:843–847.

Hildebrand, A. R., Penfield, G. T., Kring, D. A., Pilkington, M., Camargo, Z. A., Jacobsen, S. B., and Boynton, W. V., 1991, Chicxulub crater: a possible Cretaceous/Tertiary boundary impact crater on the Yucatán Peninsula, Mexico, *Geology* **19**:867–871.

Hildebrand, A. R., Pilkington, M., Connors, M., Ortiz-Aleman, C., and Chavez, R. E., 1995, Size and structure of the Chicxulub crater revealed by horizontal gravity gradients and cenotes, *Nature* **376**:415–417.

Hilgen, F. J., 1991a, Astronomical calibration of Gauss to Matuyama sapropels in the Mediterranean and implication for the Geomagnetic Polarity Time Scale, *Earth and Planetary Science Letters* **104**:226–244.

Hilgen, F. J., 1991b, Extension of the astronomically calibrated (polarity) time scale to the Miocene/Pliocene boundary, *Earth and Planetary Science Letters* **107**:349–368.

Hilgen, F. J., and Krijgsman, W., 1999, Cyclostratigraphy and astrochronology of the Tripoli diatomite formation (pre-evaporite Messinian, sicily, Italy), *Terra Nova* **11**:16–22.

Hilgen, F. J., and Langereis, C. G., 1994, A critical re-evaluation of the Miocene/Pliocene boundary as defined in the Mediterranean, *Earth and Planetary Science Letters* **118**:167–179.

Hilgen, F. J., Krijgsman, W., Langereis, C. G., Lourens, L. J., Santarelli, A., and Zachariasse, W. J., 1995, Extending the astronomical (polarity) time scale into the Miocene, *Earth and Planetary Science Letters* **136**:495–510.

Hilgen, F. J., Krijgsman, W., Langereis, C. G., and Lourens, L. J., 1997, Breakthrough made in dating of the geological record, EOS, *Transactions of the American Geophysical Union* **78**:285, 288–289.

Hilgen, F. J., Krijgsman, W., Raffi, I., Turco, E., and Zachariasse, W. J., 2000, Integrated stratigraphy and astronomical calibration of the Serravallian/Tortonian boundary section at Monte Gibliscemi (Sicily, Italy), *Marine Micropaleontology* **38**:181–212.

Hill, A. P., 1979, Disarticulation and scattering of manual skeletons, *Paleobiology* **5**:261–274.

Hill, M. O., 1979, DECORANA – a FORTRAN program for detrended correspondence analysis and reciprocal averaging: Ecology and Systematics, Cornell University, Ithaca, New York.

Hill, R. S., 1983, Evolution of *Nothofagus cunninghamii* and its relationship to *N. moorei* as inferred from Tasmanian macrofossils, *Australian Journal of Botany* **31**:453–465.

Hill, R. S., 1990, Evolution of the modern high latitude Southern Hemisphere flora: evidence from the Australian macrofossil record, in: *Proceedings of the 3rd International Organization of Palaeobotany Conference* (J. G. Douglas and D. C. Christophel, eds.), Melbourne, A-Z Printers, pp. 31–42.

Hill, R. S., 1992, *Nothofagus*: evolution from a southern perspective, *Trends in Ecology and Evolution* **7**:190–194.

Hill, R. S., 1994, The history of selected Australian taxa, in: *Australian Vegetation History: Cretaceous to Recent* (R. S. Hill, ed.), Cambridge, Cambridge University Press, pp. 390–420.

Hill, R. S., and Brodribb, T. J., 2001, Macrofossil evidence for the onset of xeromorphy in Australian Casuarinaceae and tribe Banksia (Proteaceae), *Journal of Mediterranean Ecology* **2**:127–136.

Hill, R. S., Macphail, M. K., Jordan, G. J., 1999, Tertiary history and origins of the flora and vegetation, in: *Vegetation of Tasmania*, Flora of Australia Supplementary Series, 8 (J. B. Reid, R. S. Hill, H. J. Brown, and A. Hovenden, eds.), Canberra, ABRS, pp. 39–63.

Hinde, G. J., 1882, On annelid remains from the Silurian strata of the Isle of Gotland, *Bihang till Kungl. Vetenshapsakad. Handl.* **7**(5):3–28.

Hodell, D. A., and Woodruff, F., 1994, Variations in the strontium isotope ratio of seawater during the Miocene: Stratigraphic and geochemical implications, *Paleoceanography* **9**:405–426.

Hoganson, J. W., Erickson, J., Cvancara, A. M., and Holland, F. D., Jr., 1997, Terminal Cretaceous extinction event documented by marine cartilaginous fishes from the Fox Hills (Maastrichtian) and Cannonball (Danian/Thanetian) Formations, North Dakota, *Journal of Vertebrate Paleontology* **17**(3, Suppl.):53A.

Hole, M. J., 1998, Stratigraphical applications of radiogenic isotope geochemistry, in: *Unlocking the Stratigraphic Record: Advances in Modern Stratigraphy* (P. Doyle and M. R. Bennett, eds.), Chichester, John Wiley & Sons, pp. 351–382.

Holland, S. M. 1995a, The stratigraphic distribution of fossils, *Palaeobiology* **21**(1):92–109.

Holland, S. M., 1995b, Sequence stratigraphy, facies control, and their effects on the stratigraphic distribution of fossils, in: *Sequence Stratigraphy and Depositional Response to Eustatic, Tectonic and Climatic Forcing* (B. U. Haq, ed.), Dordrecht, Kluwer Academic, pp. 1–23.

Holland, S. M., 1999, The new stratigraphy and its promise for paleobiology, *Paleobiology* **25**(3):409–416.

Holland, S. M., 2000, The quality of the fossil record: a sequence stratigraphic perspective, in: *Deep Time – Paleobiology's Perspective*, Supplement to *Paleobiology* **26** (D. H. Erwin and S. L. Wing, eds.), pp. 148–168.

Holland, S. M., and Patzkowsky, M. E., 1999, Models for simulating the fossil record, *Geology* **27**:491–494.

Holmes, A., 1911, The association of lead with uranium in rock minerals, and its application to the measurrement of geologic time, *Proceedings of the Royal Society of London* A **85**:248–256.

Holmes, N. A., 1999, The Andrew Formation and "biosteering" – different reservoirs, different approaches, in: *Biostratigraphy in Production and Development Geology*, Special Publication No. 152 (R. W. Jones and M. D. Simmons, eds.), London, Geological Society, pp. 155–166.

Holser, W. T., 1984, Gradual and abrupt shifts in ocean chemistry during Phanerozoic time, in: *Patterns of Change in Earth Evolution* (H. D. Holland and A. F. Trendall, eds.), New York, Springer-Verlag, pp. 123–143.

Holser, W. T., and Margaritz, M., 1989, Application of isotopes in stratigraphic correlation, *Proceedings of the XXVIII International Geological Congress*, Washington.

Holz, M., 1997, Early Permian sequence stratigraphy and paleophysiography of the Paraná Basin in northeastern Rio Grande do Sul State, Brazil, *Anais da Academia Brasileira de Ciências* **69**:521–543.

Holz, M., and Barberena, M. C., 1994, Taphonomy of the Brazilian Triassic paleoherpetofauna: pattern of death, transport and burial, *Palaeogeography, Palaeoclimatology, Palaeoecology* **107**:179–197.

Holz, M., and Dias, M. E., 1998, Taphonomy of palynological records in a sequence stratigraphic framework: an example from the Early Permian Paraná Basin of Southern Brazil, *Review of Palaeobotany and Palynology* **99**:217–234.

Holz, M., and Scherer, C. M. S., 2000, Sedimentological and paleontological evidence of paleoclimatic change during the Southbrazilian Triassic, the register of a global trend towards a humid paleoclimate, *Zentralblatt für Geologie und Paläontologie* **1**(11–12): 1589–1609.

Holz, M., and Souto-Ribeiro, A. W., 2000, Taphonomy of the South-Brazilian Triassic vertebrates, *Revista Brasileira de Geociências* **30**(3):487–490.

Hood, K. C., 1995, Evaluating the use of average composite sections and derived correlations in the graphic correlation technique, in: *Graphic Correlation, Society for Sedimentary Geology*, Special Publication **53**:83–93.

Horner, J. R., Varricchio, D. J., and Goodwinn, M. B., 1992, Marine transgressions and the evolution of Cretaceous dinosaurs, *Nature* **358**:59–61.

Hsü, K. J., 1983, Actualistic Catastrophism – Address of the retiring President of the International Association of Sedimentologists, *Sedimentology* **30**:3–9.

Hsü, K. J., and McKenzie, J. A., 1985, A "Strangelove" ocean in the earliest Tertiary, in: *The Carbon Cycle and Atmospheric $CO_2$: Natural Variations Archean to Present* (E. T. Sundquist and W. S. Broecker, eds.), American Geophysical Monographs, Washington DC., **32**: 487–492.

Huber, B. T., 1996, Evidence for planktonic foraminifer reworking vs. survivorship across the Cretaceous-Tertiary boundary at high latitudes, in: *The Cretaceous-Tertiary Event and Other Catastrophes in Earth History*, Geological Society of America Special Paper 307 (G. Ryder, D. Fastovsky, and S. Gartner, eds.), Boulder, pp. 319–334.

Huber, B. T., MacLeod, K. G., and Norris, R. D., 2002, Abrupt extinction and subsequent reworking of Cretaceous planktonic foraminifera across the Cretaceous-Tertiary boundary: evidence from the subtropical North Atlantic, *Geological Society of America Special Paper* **356**:277–289.

Huber, P., Lucas, S. G., and Hunt, A. P., 1993, Vertebrate biochronology of the Newark Supergroup Triassic, eastern North America, *New Mexico Museum Natural History Science Bulletin* **3**:179–186.

Humphreys, B., Morton, A. C., Hallsworth, C. R., Gatliff, R. W., and Riding, J. B., 1991, An integrated approach to provenance studies: A case example from the Upper Jurassic of the Central graben, North Sea, in: *Developments in Sedimentary Provenance Studies*, Special Publication 57 (A. C. Morton, S. P. Todd, and P. D. W. Haughton, eds.), London, Geological Society, pp. 251–262.

Hungerbühler, A., 2000, Heterodonty in the European phytosaur *Nicrosaurus kapffi* and its implications for the taxonomic utility and functional morphology of phytosaur dentitions, *Journal of Vertebrate Paleontology* **20**(1):31–48.

Hunt, A. P., and Lucas S. G., 1991, The *Paleorhinus* biochron and the correlation of the nonmarine Upper Triassic of Pangaea, *Palaeontology* **34**:191–198.

Hunt, D., and Tucker, M. E., 1992, Stranded parasequences and the forced regressive wedge systems tract: Deposition during base level fall, *Sedimentary Geology* **81**:1–9.

Hurlbert, S. H., and Archibald, J. D., 1995, No statistical support for sudden (or gradual) extinction of dinosaurs, *Geology* **23**(10):881–884.

Hsü, K. J., and McKenzie, J. A., 1985, A "Strangelove" ocean in the earliest Tertiary, in: *The Carbon Cycle and Atmospheric $CO_2$: Natural Variations Archean to Present*, Volume 32 (E. T. Sundquist and W. S. Broecker, eds.), Washington, DC, American Geophysical Monographs, pp. 487–492.

Hut, P., Alvarez, W., Elder, W. P., Hansen, T., Kauffman, E. G., Keller, G., Shoemaker, E. M., and Weissman, P. R., 1987, Comet showers as a cause of mass extinctions, *Nature* **329**:118–126.

Imbrie, J., and Imbrie, K. P., 1979, *Ice Ages – Solving the Mystery*, Cambridge, MA, Harvard University Press.

International Subcommission on Stratigraphic Classification (ISSC), 1979. Magnetostratigraphy polarity units – a supplementary chapter of the ISSC International Stratigraphic Guide, *Geology* **7**:578–583.

Isaacson, P., and Díaz-Martinez, E., 1995, Evidence for a middle-late Paleozoic foreland basin and significant paleolatitudinal shift, central Andes, in: *Petroleum Basins of South America*, American Association of Petroleum Geologists, Memoire 62 (A. J. Tankard, R. Suárez, and H. J. Welsink, eds.), pp. 231–249.

Ivany, L. C., and Salawitch, R. J., 1993, Carbon isotopic evidence for biomass burning at the K-T boundary, *Geology* **21**:487–490.

Izett, G. A., 1987, The Cretaceous-Tertiary (K-T) boundary interval, Raton Basin, Colorado and New Mexico, and its content of shock-metamorphosed minerals: implications concerning the K-T boundary impact-extinction theory, Denver, USGS (Open File Report No. 87–606).

Izett, G. A., Dalrymple, G. B. and Snee, L. W., 1991, $^{40}Ar/^{39}Ar$ age of the Cretaceous-Tertiary boundary tektites from Haiti, *Science* **252**:1539–1542.

Jablonski, D., 1998, Geographic variation in the molluscan recovery from the end-Cretaceous extinction, *Science* **279**:1327–1330.

Jablonski, D., and Raup, D. M., 1995, Selectivity of end-Cretaceous marine bivalve extinctions, *Science* **268**:389–391.

Jacobs, B. F., 1999, Estimation of rainfall variables from leaf characters in tropical Africa, *Paleogeography, Paleoclimatology, Palaeoecology* **145**:231–250

Jacoby, G. C., 1989, Overview of tree-ring analysis in tropical regions, *IAWA Bulletin* n.s. **10**(2):99–108.

Jasonius, J., and McGregor, D. C., 1996, Introduction, in: *Palynology: Principles and Applications*, Volume 1, Chapter 1 (J. Jansonius and D. C. McGregor, eds.), Dallas, American Association of Stratigraphic Palynologists Foundation, pp. 1–10.

Jardiné, J., 1972, Microplankton (Acritarches) et limites stratigraphiques du Silurien terminal au Dévonien supérieur, in : *7ème Congr. Int. Strat. Géol. Carb.*, Volume 1, Krefeld, Compte Rendu, pp. 313–323.

Jardiné, S., and Yapaudjian, L., 1968, Lithostratigraphie et palynologie du Dévonien-Gothlandien gréseux du Bassin de Polignac (Sahara), *Revue de l'Institut Français du Pétrole et Annales de Combustible Liquides*, **23**(4):439–467.

Jardiné, S., Combaz A., Magloire, L., Peniguel, G., and Vachey, G., 1974, Distribution stratigraphique des Acritarches dans le Paléozoïque du Sahara, *Review of Palaeobotany and Palynology* **18**(1/2):99–129.

Jefferey, C. H., 1997, All change at the Cretaceous-Tertiary boundary? Echinoids from the Maastrichtian and Danian of the Mangyshlak Peninsula, Kazakhstan, *Palaeontology* **40**(3):659–712.

Jefferson, T. H., 1981, Paleobotanical contributions to the geology of Alexander Island, Antarctica, Ph.D. thesis, Cambridge University, pp. 149.

Jefferson, T. H., 1982, Fossil forests from the Lower Cretaceous of Alexander Island, Antarctica, *Palaeontology* **25**(4):681–708.

Jefferson, T. H., 1983, Paleoclimate significance of some Mesozoic Antarctic fossil floras, in: *Antarctic Earth Science*, Proceedings of the Fourth International Symposium on Antarctic Earth Sciences (R. L. Oliver, P. R. James, and J. B. Jago, eds.), Australian Academy of Science, Canberra. Cambridge, Cambridge University Press, pp. 593–598.

Jekhowsky, B., 1959, Une technique standard de préparation des roches pour l'étude des microfossiles organiques, *Revue de l'Institut Français de Pétrole* **XIV**(3):315–320.

Jenkins, W. A. M., 1970, Chitinozoa, *Geoscience and Man* **1**:1–21.

Jin Yugan, Wardlaw, B. R., Glenister, B. F., and Kotlyar, G. V., 1997, Permian chronostratigraphic subdivisions, *Episodes* **20**:10–15.

Johnson, C., and Kauffman, E. G., 1996, Maastrichtian extinction patterns of Caribbean province rudists, in: *Cretaceous-Tertiary Mass Extinctions: Biotic and Environmental Changes* (N. MacLeod and G. Keller, eds.), New York, W. W. Norton, pp. 231–273.

Johnson, D. A., and Nigrini, C. A., 1985, Time-transgressive late Cenozoic radiolarian events of the equatorial Indo-Pacific, *Science* **230**:538–540.

Johnson, K. R., and Hickey, L. J., 1990, Megafloral change across the Cretaceous/Tertiary boundary I the northern Great Plains and Rocky Mountains, U.S.A., in: *Global Catastrophes in Earth History* (V. L. Sharpton and P. D. Ward, eds.), Geological Society of America Special Paper **247:**433–444.

Jones, G. D., 1988, A paleoecological model of Late Paleocene "flysch-type" agglutinated foraminifera using the paleoslope transect approach, Viking Graben, North Sea, in: Proc. 2nd Workshop Agglutinated Foraminifera, Vienna 1986, *Abhandlungen der Geologischen Bundesanstalt*, **41**:143–153.

Jones, M. J., and Truswell, E. M., 1992, Late Carboniferous and Early Permian palynostratigraphy of the Joe Joe Group, southern Galilee Basin, Queensland, and implications for Gondwanan stratigraphy, *BMR Journal of Australian Geology and Geophysics* **13**:143–185.

Jones, R. A., 1994, The application of microwave technology to the oxidation of kerogen for use in palynology, *Review of Palaeobotany and Palynology* **80**:333–338.

Jones, R. W., 1996, *Micropalaeontology in Petroleum Exploration*, Oxford, Oxford University Press.

Jones, R. W., 1999, Forties Field (North Sea) revisited: a demonstration of the value of historical micropalaeontological data, in: *Biostratigraphy in Production and Development Geology*, Special Publication, No. 152 (R. W. Jones and M. D. Simmons, eds.), London, Geological Society, pp. 185–200.

Jones, R. W., and Charnock, M. A., 1985, "Morphogroups" of agglutinating foraminifera, their life positions and feeding habits and potential applicability in (paleo)ecological studies, *Revue de Paleobiologie* **4**(2):311–320.

Jones, T. P., and Rowe N. P. (eds.), 1999, *Fossil Plants and Spores: Modern Techniques*, London, Geological Society.

Kalmykova, M. A., Kashik, D. S., Miklukho-Maklay, K. V., and Ukharskaya, L. B., 1979, A complex utilization of biostratigraphic and cyclostratigraphic methods while studying the stratigraphy of marine Permian deposits of the Russian Platform. *Ezhegodnik Vsesoyuznogo Paleontologicheskogo Obshchestva* **22**:235–252.

Kardong, K. V., 1997/8, Squamates at geological boundaries: unbounded evolution, *Zoology* **100**:152–163.

Katz, M. E., Pak, D. K., Dickens, G. R., and Miller, K. G., 1999, The source and fate of massive carbon input during the latest Paleocene thermal maximum, *Science* **286**:1531–1533.

Kauffmann, E. G., 1984, The fabric of Cretaceous marine extinctions, in: *Catastrophes and Earth History* (W. A. Berggren and J. A. Van Couvering, eds.), Princeton, NJ, Princeton University Press, pp. 151–246.

Kauffman, E. G., 1987, High-resolution event stratigraphy: concepts, methods and Cretaceous examples. in: *A Field Trip Guidebook: High Resolution Event Stratigraphy, Greenhorn Cyclothem (Cretaceous: Cenomanian-Turonian), Western Interior of Colorado and Utah* (E. G. Kauffmann, B. B. Sageman, E. R. Gustason, and W. P. Elder, eds.), Boulder, Geological Society of America, Rocky Mountains Section, pp. 2–34.

Kauffman, E. G., 1988, Concepts and methods of high-resolution event stratigraphy, *Annual Review of Earth Planetary Sciences* **16**:605–654.

Kauffman, E. G., and Hazel, J. E., eds., 1977, *Concepts and methods in stratigraphy*, Stroudsburg, Pennsylvania, Dowden, Hutchinson and Ross.

Keller, A. M., and Hendrix, M. S., 1997, Paleoclimatological analysis of Late Jurassic petrified forest southeastern Mongolia, *Palaios* **12**:82–291.
Keller, G., 1983, Biochronology and paleoclimatic implications of middle Eocene to Oligocene planktic foraminiferal faunas, *Marine Micropaleontology* **7**:463–486.
Keller, G., 1988, Extinction, survivorship and evolution of planktic foraminifers across the Cretaceous/Tertiary boundary at El Kef, Tunisia, *Marine Micropaleontology* **13**:239–263.
Keller, G., 1989a, Extended Cretaceous/Tertiary boundary extinctions and delayed population change in planktonic foraminifera from Brazos River, Texas, *Paleoceanography* **4**:287–332.
Keller, G., 1989b, Extended period of extinctions across the Cretaceous/Tertiary boundary in planktonic foraminifera of continental-shelf sections: Implications for impact and volcanism theories, *Geological Society of America Bulletin* **101**:1408–1419.
Keller, G., 1993, The Cretaceous-Tertiary boundary transition in the Antarctic Ocean and its global implications, *Marine Micropaleontology* **21**:1–45.
Keller, G., and Lindinger, M., 1989, Stable isotope, TOC and $CaCO_3$ record across the Cretaceous/Tertiary Boundary at El Kef, Tunisia, *Palaeogeography, Palaeoclimatology, Palaeoecology* **73**(3/4):243–265.
Keller, G., Barrera, E., Schmitz, B., and Mattson, E., 1993, Gradual mass extinction, species survivorship, and long-term environmental changes across the Cretaceous-Tertiary boundary in high latitudes, *Geological Society of America Bulletin* **105**:979–997.
Keller, G., Li, L., and MacCleod, N., 1995, The Cretaceous/Tertiary boundary stratotype section at El Kef, Tunisia: How catastrophic was the mass extinction? *Palaeogeography, Palaeoclimatology, Palaeoecology* **119**(3/4):255–273.
Keller, G., Stinnesbeck, W., Adatte, T., and Stüben, D., 2003, Multiple impacts across the Cretaceous-Tertiary boundary, *Earth-Science Review* **62**:327–363.
Kelly, S. R. A., and Moncrieff, A. C. M., 1992, Marine molluscan constraints on the age of Cretaceous fossil forest of Alexander Island, Antarctica, *Geology Magazine* **129**(6): 771–778.
Kemp, E. M., Balme, B. E., Helby, R. J., Kyle, R. A., Playford, G., and Price, P. L., 1977, Carboniferous and Permian palynostratigraphy in Australia and Antarctica: a review, *BMR Journal of Australian Geology and Geophysics* **2**:177–208.
Kemple, W. G., Sadler, P. M., and Strauss, D. J., 1995, Extending Graphic Correlation to Many Dimensions: Stratigraphic Correlation as Constrained Optimization, Society for Sedimentary Geology, Special Publication **53**:65–82.
Kendall, C. G. St. C., Moore, P., Strobel, J., Cannon, R,. Perlmutter, M., Bezdek, J., and Biswas, G., 1991, Simulation of the sedimentary fill of basins, in: *Sedimentary Modeling*, Volume 233 (E. K. Franseen, W. K. Watney, C. G. St. C. Kendall, and W. Ross, eds.), *Kansas Geological Survey Bulletin*, pp. 9–30.
Kennedy, W., 1989, Thoughts on the evolution and extinction of Cretaceous ammonites, *Proceedings of the Geologists Association* **100**:251–279.
Kennedy, W. J., 1993, Ammonite faunas of the European Maastrichtian: diversity and extinction, in: *The Ammonoidea: Environment, Ecology and Evolutionary Change* (M. R. House, ed.), Systematics Association, Special Volume, pp. 285–326.
Kennett, J. P., and Srinavasan, M. S., 1984, Neogene planktonic foraminiferal data planes of the South Pacific: mid to equatorial latitudes, in: *Pacific Neogene Datum Planes Contribution to Biostratigraphy and Chronology* (R. Tsuchi and N. Ikebe, eds.), Tokyo, University of Tokyo Press, pp. 11–25.
Kennett, J. P., and Stott, L. D., 1990, Proteeus and proto-oceanus: ancestral Paleogene oceans as revealed from Antarctic stable isotopic results; ODP Leg 113, in *Proceedings of the Ocean Drilling, Scientific Results*, **113**:865–879.

Kennett, J. P., and Stott, L. D., 1991, Abrupt deep-sea warming, palaeoceanographic changes and benthic extinctions at the end of the Palaeocene, *Nature* **353**:225–229.

Kent, D. V., Olsen, P. E., and Witte, W. K., 1995, Late Triassic-earliest Jurassic geomagnetic polarity sequence and paleolatitudes from drill cores in the Newark rift basin, eastern North America, *Journal of Geophysical Research* **100**(B8):14,965–14,998.

Kerp, H., 2002, Atmospheric $CO_2$ from fossil plant cuticles, *Nature* **415**:38.

Kerr, R. A., 1986, Mapping orbital effects on climate, *Science* **234**:283–284.

Kerr, R. A., 1996, Ancient sea level swings confirmed, *Science* **272**:1097–1098.

Kidwell, S. M., 1989, Stratigraphic condensation of marine transgressive records: origin of major shell deposits in the Miocene of Maryland, *Journal of Geology* **97**:1–24.

Kidwell, S. M., 1991, Stratigraphy of shell concentrations, in: *Taphonomy, Releasing the Data Locked in the Fossil Record* (P. A. Allison and D. E. G. Briggs, eds.), New York, Plenum Press, pp. 211–290.

Kidwell, S. M., 1993, Patterns of time-averaging in the shallow marine fossil record, in: *Short Courses in Paleontology* (S. M. Kidwell and A. K. Behrensmeyer, eds.), Taphonomic approaches to time resolution in fossil assemblages, Paleontological Society No. 6, pp. 275–300.

Kidwell, S. M., 1998, Time-averaging in the marine fossil record: overview of strategies and uncertainties, *Geobios* **30**:977–995.

Kidwell, S. M., 2002, Mesh-size effects on the ecological fidelity of death assemblages: A meta-analysis of molluscan live-dead studies, *Geobios mémoire special* **24**:107–119.

Kidwell, S. M., and Behrensmeyer, A. K., 1993a, Summary estimates of time-averanging, in: *Taphonomic Approaches to Time Resolution in Fossil Assemblages*, Volume 6 (S. M. Kidwell and A. K. Behrensmeyer, eds.). Short Courses in Paleontology, The Paleontological Society, pp. 301–302.

Kidwell, S. M., and Behrensmeyer, A. K., 1993b, Taphonomic approaches to time resolution in fossil assemblages, Paleontological Society, *Short Courses in Paleontology* No. 6.

Kidwell, S. M., and Bosence, D. W. J., 1991, Taphonomy and time-averaging of marine shelly faunas, in: *Taphonomy: Releasing the Data Locked in the Fossil Record* (P. A. Allison and D. E. G. Briggs, eds.), New York, Plenum Press, pp. 115–209.

Kidwell, S. M., and Flessa, K. W., 1995, The quality of the fossil record: populations, species, and communities, *Annual Reviews of Ecology and Systematics* **26**:269–299.

Kidwell, S. M. and Labarbera, M., 1993, Experimental taphonomy, *Palaios* **8**:1–2.

Kidwell, S. M., Fürsich, F. T., and Aigner, T., 1986, Conceptual framework for the analysis of fossil concentrations, *Palaios* **1**:228–238.

Kidwell, S. M., Rothfus, T. A., and Best, M. M. R., 2001, Sensitivity of taphonomic signatures to sample size, sieve size, damage scoring system, and target taxa, *Palaios* **16**:26–52.

Kitching, J. W., 1995, Biostratigraphy of the *Cynognathus* Assemblage Zone, in: *Biostratigraphy of the Beaufort Group (Karoo Supergroup)*, Biostratigraphic Series 1 (B. S. Rubidge, ed.), South African Committee for Stratigraphy, pp. 40–46.

Kitching, J. W., and Raath, M. A., 1984, Fossils from the Elliot and Clarens formations (Karoo Sequence) of the northeastern Cape, Orange Free State and Lesotho, and a suggested biozonation based on tetrapods, *Palaeontologia Africana* **25**:111–125.

Klapper, G., 1997, Graphic correlation of Frasnaian (Upper Devonian) sequence in Montagne Noire, France, and western Canada, in: *Paleozoic Sequence Stratigraphy, Biostratigraphy, and Biogeography: Studies in Honor of J. Granville ("Jess") Johnson*, Special Paper **321** (G. Klapper, M. A. Murphy, and J. A. Talent, eds.), Geological Society of America, pp. 113–129.

de Klasz, I., and Koutsoukos, E. A. M., 1991, Late Cretaceous foraminifera biogeography in low latitude Atlantic regions: northeastern Brazilian shelf (Sergipe-Alagoas, Pernambuco-Paraíba and Potiguar basins) and central west African basins, in: *Resumos do XII Congresso Brasileiro de Paleontologia*, p. 60.

Knoll, A. H., 1996, Archean and Proterozoic paleontology, in: *Palynology: Principles and Applications*, Volume 1, Chapter 4 (J. Jansonius and D. C. McGregor, eds.), Dallas, American Association of Stratigraphic Palynologists Foundation, pp. 51–80.

Koch, P. L., Zachos, J. C., and Dettman, D. L., 1995, Stable isotope stratigraphy and paleoclimatology of the Paleogene Bighorn Basin (Wyoming, USA), *Palaeogeography, Palaeoclimatology, Palaeoecology* **115**:61–89.

Koch, P. L., Zachos, J. C., and Gingerich, P. D., 1992, Correlation between isotope records in marine and continental carbon reservoirs near the Palaeocene/Eocene boundary, *Nature* **358**:319–322.

Koeberl, C., 1993, Chicxulub Crater, Yucatan: Tektites, impact glasses, and the geochemistry of target rocks and breccias, *Geology* **21**:211–214.

Kolla, V., and Perlmutter, M. A., 1993, Timing of turbidite sedimentation – Mississippi fan, *Bulletin of the American Association of Petroleum Geologists* **77**(7):1129–1141.

Köppen, W., 1931, *Grundriss der Klimakunde*, Berlin, Walter de Gruyter.

Kosanke, R. M., 1969, Mississippian and Pennsylvanian palynology, in: *Aspects of Palynology* (R. H. Tschudy and R. A. Scott, eds.), New York, Wiley, pp. 223–269.

Koss, J. E., Ethridge, F. G., and Schumm, S. A., 1994, An experimental study of the effects of base-level change on fluvial, coastal plain and shelf systems, *Journal of Sedimentary Research* **B64**(2):90–98.

Koutsoukos, E. A. M., 1996a, Phenotypic experiments into new pelagic niches in early Danian planktonic foraminifera: aftermath of the K/T boundary event, in: *Biotic Recovery from Global Mass Extinction Events*, Geological Society Special Publication 102 (M. B. Hart, ed.), London, pp. 319–335.

Koutsoukos, E. A. M., 1996b, The Cretaceous-Tertiary boundary at Poty, NE Brazil: Event stratigraphy and palaeoenvironments, in: *Géologie de l'Afrique et de l'Atlantique Sud* (Compte-rendu des Colloques de géologie d'Angers, 16–20 juillet 1994), Mem. 16 (S. Jardiné, I. De Klasz, and J.-P. Debenay, eds.), *Bulletin de Centres de Recherches Exploration-Production Elf Aquitaine*, pp. 413–431.

Koutsoukos, E. A. M., 1998, An extraterrestrial impact in the early Danian: a secondary K/T boundary event?, *Terra-Nova* **10**:68–73.

Kowalewski, M., 1996, Time-averaging, overcompleteness, and the geological record, *Journal of Geology* **104**:317–326.

Kowalewski, M., 1997, The reciprocal taphonomic model, *Lethaia* **30**:86–88.

Kowalewski, M., Flessa, K. W., and Aggen, J. A., 1994, Taphofacies analysis of Recent shelly cheniers (beach ridges), northeastern Baja California, Mexico, *Facies* **31**:209–242.

Kowalewski, M., Goodfriend, G. A., and Flessa, K. W., 1998, High-resolution estimates of temporal mixing within shell beds: the evils and virtues of time-averaging, *Paleobiology* **24**:287–304.

Kramer, P. J., and Kozlowsky, T., 1972, *Fisiologia das Árvores*, Lisboa, Fundação Calouste Gulbenkian.

Krijgsman, W., Garcés, M., Langereis, C. G., Daams, R., van Dam, J., van der Meulen, A. J., Agustí, J., and Cabrera, L., 1996, A new chronology for the middle to late Miocene continental record in Spain, *Earth and Planetary Science Letters* **142**:367–380.

Krijgsman, W., Hilgen, F. J., Raffi, I., Sierro, F. J., and Wilson, D. S., 1999, Chronology, causes and progression of the Messinian salinity crisis, *Nature* **400**:652–655.

Kring, D. A., 1993, The Chicxulub impact event and possible causes of K/T boundary extinctions, in: *Proceedings of the First Annual Symposium of Fossils of Arizona* (D. Boaz and M. Dornan, eds.), Mesa, Arizona, Mesa Southwest Museum and Southwest Paleontological Society, pp. 63–79.

Kring, D. A., and Boynton, W. V., 1993, K/T melt glasses, *Nature* **363**:503–504.

Krynine, P. D., 1942, Differential sedimentation and its products during one complete geosynclinal cycle, *1st Congreso Panamericano de Ingeniera de Minas y Geologia* **2**:537–561.

Kuhnt, W., and Kaminski, M.A., 1993, Changes in the community structure of deep water agglutinated foraminifers across the K/T boundary in the Basque basin (northern Spain), *Revista Española de Micropaleontologia* **25**:57–92.

Kuhnt, W., and Kaminski, M. A., 1996, The response of benthic foraminifera to the K/T boundary event – A review, *Géologie de l'Afrique et de l'Atlantique Sud, Actes Colloques Angers* 1994:433–442.

Kuhnt, W., Herbin, J. P., Thurow, J., and Wiedman, J., 1990, Distribution of Cenomanian-Turonian organic facies in the Western Mediterranean and along the adjacent Atlantic margin, in: *Deposition of Organic Facies*, AAPG Studies in Geology 30 (A. Y. Huc, ed.), Tulsa, Oklahoma, pp. 133–160.

Kukal, Z., 1990, The rate of sedimentation, in Z. Kukal (ed.), 1990, The Rate of Geological Processes, *Earth-Science Reviews* **28**(1/2/3):94–133.

Kukla, G., and Gavin, J., 1992, Insolation regime of the warm to cold transitions, in*: NATO ASI Series*, Volume 13 (G. J. Kukla and E. Went, eds.), pp. 307–339.

Kulbicki, G., and Rumeau, J. L., 1967, Influence du milieu sur les teneurs en elements-traces des vase marines du Golf de Gascogne, *Bulletin de la Centre de Recherches Pau-SNPA* **1**:111–124.

Kulkarni, S., Maithy, P. K., and Surange, K. R., 1970, On *Barakaroxylon jhariense*, *The Palaeobotanist* **18**(3):305–308.

Kumangai, H., Sweda, T., Hayashi, K., Kojima, S., Basinger, J. F., Shibuya, M., and Fukaoa, Y., 1995, Growth-ring analysis of Early Tertiary conifer woods from the Canadian High Arctic and its palaeoclimatic implications, *Palaeogeography, Palaeoclimatology, Palaeocology* **116**:247–262.

Kyte, F. T., and Smit, J., 1986, Regional variations in spinel compositions: An important key to the Cretaceous/Tertiary event, *Geology* **14**:485–487.

Lagoe, M. B., Gary, A. C., and Zellers, S., 1992, Significance of mixed assemblages of benthic foraminifera in the late Cenozoic, glaciomarine diamictites of the Yakataga Formation, Gulf of Alaska, in: *Studies in Benthic Foraminifera, BENTHOS '90*, Tokyo, Tokyo University Press, pp. 265–275.

Lagoe, M. B., Davies, T., Austin, J. A., and Olson, H. C., 1997, Foraminiferal constraints on very high-resolution seismic stratigraphy and late Quaternary glacial history, New Jersey continental shelf, *Palaios* **12**:249–266.

LaMarche, V. C., 1982, Sampling strategics, in: *Climate from Tree Rings* (M. K. Hughes, P. M. Kelly, J. R. Pilcher, and V. C. LaMarche, eds.), Cambridge, Cambridge University Press, p. 4.

Lane, H. R., and Ziegler, W. (eds.), 1985, Toward a boundary in the middle of the Carboniferous: stratigraphy and paleontology, *Courier Forschungsinstitut Senckenberg* **74**:1–195.

Lange, F. W., 1967a, Subdivisão bioestratigráfica e revisão da coluna siluro-devoniana da Bacia do Baixo Amazonas, *Atas do simpósio sobre a Biota Amazônica (Geociências)* **1**:215–326.

Lange, F. W., 1967b, Biostratigraphic subdivision and correlation of the Devonian in the Paraná Basin, in: *Problems in Brazilian Devonian Geology*, Bol. Parana. Geoc. 21/22 (J. J. Bigarella, ed.), pp. 63–98.

Langer, M. C., and Schultz, C. L., 2000, A new species of the Late Triassic rhynchosaur *Hyperodapedon* from the Santa Maria Formation of South Brazil, *Paleontology* **43**(4): 633–652.

Langer, M. C., Boniface, M., Cuny, G., and Barbieri, L. 2000a, The phylogenetic position of *Isalorhynchus genovefae*, a Late Triassic rhynchosaur from Madagascar, *Annales de Paléontologie* **86**(2):101–127.

Langer, M. C., Ferigolo, J., and Schultz, C. L., 2000b, Heterochrony and tooth evolution in hyperodapedontine rhynchosaurs (Reptila, Diapsida), *Lethaia* **33**:119–128.

Lanzoni, E., and Magloire, L., 1969, Associations palynologiques et leurs applications stratigraphiques dans le Dévonien supérieur et Carbonifère inférieur du Grand Erg Occidental (Sahara algérien), *Revue de l'Institut Français du Pétrole et Annales Combustibles Liquides* **24**(4):441–469.

Larcher, W., 1986, *Ecofisiologia Vegetal*, São Paulo, Editora Pedagógica Universitária.

Lask, P. B., 1993, The hydrodynamic behaviour of sclerites from the trilobite *Flexicalymene meeki*, *Palaios* **8**:219–225.

Laufeld, S., 1967, Caradocian Chitinozoa from Dalarna, Sweden, *Geologiska Föreningens i Stockholm Förhandlingar* **89**:275–349.

Laufeld, S., 1974, Silurian Chitinozoa from Gotland, *Fossils and Strata* **5**:1–130.

Lavina, E. L., Araújo-Barberena, D. C., and Azevedo, S. A. K., 1991, Tempestades de Inverno e Altas Taxas de mortalidade de Répteis Mesossauros. Um exemplo a partir do Afloramento Passo São Borja, RS, *Pesquisas* **18**(1):64–70.

Lawrence, D. R., 1968, Taphonomy and information losses in fossil communities, *Geological Society of America Bulletin* **79**:1315–30.

Legaretta, L., and Ulina, M. A., 1998, Anatomy of hinterland depositional sequences: Upper Cretaceous fluvial strata, Neuquém Basin, West-Central Argentina, in: *Relative Role of Eustasy, Climate and Tectonism in Continental Rocks*, Special Publication of the SEPM 59 (K. W. Shanley and P. J. McCabe, eds.), pp. 83–92.

Le Hérissé, A., 2001, Evolution of Devonian phytoplanktonic assemblages in the upper Ererê Formation and Curuá Group (Barreirinha and lower Curiri Formations), Tapajós River area, Amazon Basin, northern Brazil, in: *Correlação de seqüências paleozóicas sul-americanas*, Ciênc.-Téc.-Petról., Seção: Expl. Petról. 20 (J. H. G. Melo and G. J. S. Terra, eds.), pp. 117–124. (CD Rom, ISSN-0100–1515.)

Le Hérissé, A., Gourvennec, R., and Wicander, R., 1997a, Biogeography of Late Silurian and Devonian acritarchs and prasinophytes, *Revue de Palaeobotanie et Palynologie* **98**:105–124.

Le Hérissé, A., Rubinstein, C., and Steemans, P., 1997b, Lower Devonian palynomorphs from the Talacasto Formation, Cerro del Fuerte section, San Juan Precordillera, Argentina, in: *Acritarcha in Praha*, Acta Universitatis Carolinae Geologica 40 (O. Fatka and T. Servais, eds.), pp. 497–515.

Lé Hérissé, A., Servais, T., and Wicander, R., 2000, Devonian acritarchs and related forms, *Courier Forschungsinstitut Senckenberg* **220**:195–205.

Leopold, L. B., Wolman, M. G., and Miller, J. P., 1964, *Fluvial Processes in Geomorphology*, San Francisco, Freeman and Co.

Le Riche, H. H., 1959, The distribution of certain trace elements in the Lower Lias of Southern England, *Geochimica et Cosmochimica et Acta* **16**:101–122.

Le Tourneau, P. M., 1985, Alluvial fan developments in the Lower Jurassic Portland Formation, central Connecticut – implications for tectonics and climate, in: *Proceedings of the 2nd Geological Survey Workshop on the Mesozoic Basins of Eastern United States* (G. P. Robison and A. J. Frohlich, eds.), US Geological Survey Circular 947, pp. 17–26.

Li, L., and Keller, G., 1998, Diversification and extinction in Campanian-Maastrichtian planktic foraminifera of northwestern Tunisia, *Eclogae Geologicae Helvetiae* **91**:75–102.

Lima, F. H. O., 2002, Estratigrafia integrada do Maastrichtiano (Formação Gramame) da Bacia de Pernambuco/Paraíba – Brasil: caracterização faciológica e evolução paleoambiental, Tese (Doutorado), Universidade Federal do Rio Grande do Sul, Instituto de Geociências, Porto Alegre, RS, Brazil, 2 vols, 295 pp.

Limachi, R., Goitia, V. H., Sarmiento, D., Arispe, O., Montecinos, R., Díaz-Martinez, E., Dalenz-Farjat, A., Liachenco, N., Pérez-Leyton, M., and Aguilera, E., 1996, Estratigrafía, geoquímica, correlaciones, ambientes sedimentários y bioestratigrafía del Silúrico-Devónico de Bolivia, in: *12° Congress of Geology Bolivia*, Tarija, Memorias, pp. 183–197.

Lindsay, E. H., Opdyke, N. D., and Johnson, N. M., 1984, Blancan-Hemphillian Land Mammal Ages and late Cenozoic mammal dispersal events, *Annual Review of Earth and Planetary Sciences* **12**:445–488.

Lipps, H. Jere (ed.), 1993, *Fossil Prokaryotes and Protists*, Cambridge, Blackwell Scientific Publications.

Liu, C., and Olsson, R. K., 1992, Evolutionary radiation of microperforate planktonic foraminifera following the K/T mass extinction event, *Journal of Foraminiferal Research* **22**:328–346.

Lobo-Boneta, J., 1975, Sobre algunos palinomorfos del Devónico superior y Carbónico inferior de la zona Subandina Sur de Bolivia, *Revista Técnica de Yacimientos Petroliferos Fiscales Bolivianos (YPFB)* **4**(3):159–176.

Lobo-Boneta, J., Suárez-Riglos, M., and Suárez-Soruco, R., 1977, Nuevas unidades cronoestratigráficas para el Paleozóico médio de la Provincia Austral Sudamericana, *Boletín Técnico de Asociacíon Regional de Empresas de Petroleo Gas Natural en Latinoamerica y el Caribe (ARPEL)* **6**(2):215–219.

Loboziak, S., 1999, Late Early to Late Devonian palynologic events in Western Gondwana: An application to Brazilian Devonian basins, in: *Ordovician-Devonian Palynostratigraphy in Western Gondwana: Update, Problems and Perspectives* (M. A. C. Rodrigues and E. Pereira, eds.), Rio de Janeiro, UERJ, pp. 63–75.

Loboziak, S., 2000, Middle to early Late Devonian miospore biostratigraphy of Saudi Arabia, in: *Stratigraphic Palynology of the Palaeozoic of Saudi Arabia*, Special GeoArabia Publication 1 (S. Al-Hajri and B. Owens, eds.), pp. 134–145.

Loboziak, S., and Clayton, G., 1988, The Carboniferous palynostratigraphy of northeast Libya, in: *Subsurface Palynostratigraphy of Northeast Libya* (A. El-Arnauti, B. Owens, and B. Thusu, eds.), Benghazi, Libya, Garyounis University Publications, pp. 129–149.

Loboziak, S., and Melo, J. H. G., 2000, Miospore events from Late Early to Late Devonian strata of Western Gondwana, *Geobios* **34**(3):399–407.

Loboziak, S., and Melo, J. H. G., 2002, Devonian miospore successions of Western Gondwana: update and correlation with Southern Euramerican miospore zones. *Revue de Palaeobotanie et Palynologie* **121**:133–148.

Loboziak, S., and Streel, M., 1989, Middle-Upper Devonian miospores from the Ghadamis Basin (Tunisia-Libya): systematics and stratigraphy, *Revue de Palaeobotanie et Palynologie* **58**:173–196.

Loboziak, S., and Streel, M., 1995a, West Gondwanian aspects of the Middle and Upper Devonian miospore zonation in North Africa and Brazil, *Revue de Palaeobotanie et Palynologie* **86**:147–155.

Loboziak, S., and Streel, M., 1995b, Late Lower and Middle Devonian miospores from Saudi Arabia, *Revue de Palaeobotanie et Palynologie* **89**:105–113.

Loboziak, S., Streel, M., and Burjack, M. I. A., 1988, Miospores du Dévonien moyen et supérieur du bassin du Paraná, Brésil: systématique et stratigraphie, *Sciences Géologiques Bulletin* **41**(3–4):351–377.

Loboziak, S., Streel, M., and Burjack, M. I. A., 1989, Déductions paléoclimatiques d'une comparaison entre les assemblages de miospores du Dévonien moyen et supérieur de Libye et du Brésil, *Geobios* **22**(2):247–351.

Loboziak, S., Streel, M., Caputo, M. V., and Melo, J. H. G., 1991, Evidence of West European defined miospore zones in the uppermost Devonian and Lower Carboniferous of the Amazonas Basin (Brazil), *Geobios* **24**(1):5–11.

Loboziak, S., Steemans, P., Streel, M., and Vachard, D., 1992a, Biostratigraphie par miospores du Dévonien inférieur à supérieur du sondage MG-1 (Bassin d'Hammadah, Tunisie) – Comparaison avec les données des faunes, *Revue de Palaeobotanie et Palynologie* **74**:193–205.

Loboziak, S., Streel, M., Caputo, M. V., and Melo, J. H. G., 1992b, Middle Devonian to Lower Carboniferous miospore stratigraphy in the Central Parnaíba Basin (Brazil), *Annales de la Societé Géologique Belgique* **115**(1):215–226.

Loboziak, S., Streel, M., Caputo, M. V., and Melo, J. H. G., 1993, Middle Devonian to Lower Carboniferous miospores from selected boreholes in Amazonas and Parnaíba basins (Brazil): additional data, synthesis, and correlation, *Documents des Laboratoires de Géologie de la Faculté des Sciences de Lyon* **125**:277–289.

Loboziak, S., Melo, J. H. G., Steemans, P., and Barrilari, I. M. R., 1995a, Miospore evidence for pre-Emsian and latest Famennian sedimentation in the Devonian of the Paraná Basin, South Brazil, *Anais da Academia Brasileira de Ciências* **67**(3):391–392.

Loboziak, S., Melo, J. H. G., Quadros, L, P., Daemon, R. F., Barrilari, I. M. R., and Streel, M., 1995b, Biocronoestratigrafia de miósporos do Devoniano Médio–Carbonífero Inferior das bacias do Solimões e Parnaíba (estado da arte), *Anais da Academia Brasileira de Ciências* **67**(3):394–395.

Loboziak, S., Melo, J. H. G., Rodrigues, R., Streel, M., Quadros, L. P., and Barrilari, I. M. R., 1996, Age and correlation of the Barreirinha Formation (Curuá Group, Amazon Basin): new evidence from the miospore biostratigraphy, *Anais da Academia Brasileira de Ciências* **68**(2):207–212.

Loboziak, S., Melo, J. H. G., Matsuda, N. S., and Quadros, L. P., 1997a, Miospore biostratigraphy of the type Barreirinha Formation (Curuá Group, Upper Devonian) in the Tapajós River area, Amazon Basin, North Brazil, *Bulletin de Centres de Recherches et Exploration – Production Elf Aquitaine* **21**(1):187–205.

Loboziak, S., Melo, J. H. G., Quadros, L. P., and Streel, M., 1997b, Palynological evaluation of the Famennian *Protosalvinia* (*Foerstia*) Zone in the Amazon Basin, northern Brazil: a preliminary study, *Revue de Palaeobotanie et Palynologie* **96**:31–45.

Loboziak, S., Melo, J. H. G., and Streel, M., 2000a, Latest Devonian and Early Carboniferous palynostratigraphy of northern Brazil and North Africa: a proposed integration of western European and Gondwanan miospore biozonations, *Bulletin de Centres de Recherches et Exploration – Production Elf Aquitaine* **22**(2): 241–259.

Loboziak, S., Caputo, M. V., and Melo, J. H. G., 2000b, Middle Devonian–Tournaisian miospore biostratigraphy in the southwestern outcrop belt of the Parnaíba Basin, north-central Brazil, *Revue de Micropaléontologie* **43**(4):301–318.

Lockwood, J. G., 1980, Milankovitch theory and ice ages, *Progress in Physical Geography* **4**:79–87.

Lohman, W. H., 1986, Calcareous nannoplankton biostratigraphy of the southern Coral Sea, Tasman, Sea, and southwestern Pacific Ocean, Deep Sea Drilling Project leg 90: Neogene and Quaternary, in*: Initial Reports of the Deep Sea Drilling Project*, Volume 90 (J. P. Kennett, C. C. von der Boch *et al.*, eds.*)*, Washington, DC, Government Printing Office, pp. 763–793.

Lourens, L. J., Antonarakou, A., Hilgen, F. J., Van Hoof, A. A. M., Vergnaud-Grazzini, C., and Zachariasse, W. J., 1996, Evaluation of a Plio-Pleistocene astronomical timescale, *Paleoceanography* **11**:391–413.

Lucas, S. G., 1990, Toward a vertebrate biochronology of the Triassic, *Albertiana* **8**:36–41.

Lucas, S. G., 1993, Vertebrate biochronology of the Triassic of China, *New Mexico Museum Natural History Science Bulletin* **3**:301–306.

Lucas, S. G., 1998, Global Triassic tetrapod biostratigraphy and biochronology, *Palaeogeography, Palaeoclimatology, Palaeoecology* **143**:345–382.

Lucas, S. G., 2001, Age and correlation of triassic tetrapod assemblages from Brazil, *Albertiana* **26**:13–20.

Lucas, S. G., and Hancox, P. J., 2001, Tetrapod-based correlation of the nonmarine Upper Triassic of Southern Africa, *Albertiana* **25**:5–9.

Lucas, S. G., and Heckert, A. B., 2002, The *Hyperodapedon* Biochron, Late Triassic of Pangea, *Albertiana* **27**:30–38.

Lucas, S. G., and Hunt, A. P., 1993, Tetrapod biochronology of the Chinle Group (Upper Triassic), western United States, *New Mexico Museum of Natural History and Science Bulletin* 3.

Lucas, S. G., Mateer, N. J., Hunt, A. P., and O'Neill, F. M., 1987, Dinosaurs, the age of the Fruitland and Kirkland Formations, and the Cretaceous-Tertiary boundary in the San Juan Basin, New Mexico, in: *The Cretaceous-Tertiary Boundary in the San Juan and Raton Basins, New Mexico and Colorado*, Geological Society of America Special Paper 209 (J. E. Fassett and J. K. Rigby Jr., eds.), pp. 35–50.

Lucas, S. G., Heckert, A. B., and Huber, P., 1998, *Aetosaurus* (Archosauromorpha) from the Upper Triassic of the Newark Supergroup, Eastern United States, and its biochronological significance, *Palaeontology* **41**(6):1215–1230.

Lyell, Sir Charles, 1830–3, *Principles of Geology*, 3 vols., London, John Murray.

MacEachern, J. A., and Pemberton, S. G., 1992, Ichnological aspects of Cretaceous shoreface successions and shoreface variability in the Western Interior Seaway of North America, in: *Applications of Ichnology to Petroleum Exploration, a Core Workshop*, Core Workshop 17 (S. G. Pemberton, ed.), Society of Economic Paleontologists and Mineralogists, pp. 57–84.

MacEachern, J. A., and Pemberton, S. G., 1994, Ichnological character of incised valley fill systems from the Viking Formation of the Western Canada Sedimentary Basin, Alberta, Canada, in: *Recognition and Facies of Incised-Valley Fills*, Special Publication 51 (R. Dalrymple, R. Boyd, and B. Zaitlin, eds.) Society of Economic Paleontologists and Mineralogists, pp. 129–157.

MacEachern, J. A., Zaitlin, B. A., and Pemberton, S. G., 1998, High resolution sequence stratigraphy of early transgressive incised shoreface and early transgressive valley/embayment deposits of the Viking Formation, Joffre Field, Alberta, Canada, *Bulletin of the American Association of Petroleum Geologists* **82**:729–756.

MacLeod, K. G., and Huber, B., 1996, Strontium isotopic evidence for extensive reworking in sediments spanning the Cretaceous/Tertiary boundary at ODP Site 738, *Geology* **24**:463–466.

MacLeod, N., 1996a, Nature of the Cretaceous-Tertiary (K-T) planktonic foraminiferal record: stratigraphic confidence intervals, Signor-Lipps effect, and patterns of survivorship, in: *The*

*Cretaceous–Tertiary Mass Extinction: Biotic and Environmental Changes* (N. MacLeod and G. Keller, eds.), New York, W. W. Norton and Co., pp. 85–138.

MacLeod, N., 1996b, Stratigraphic completeness and planktic foraminiferal survivorship across the Cretaceous–Tertiary (K/T) boundary, in: *Microfossils and Oceanic Environments* (A. Moguilevsky and R. Whatley, eds.), Aberystwyth, University of Wales, pp. 327–353.

MacLeod, N., 1996c, Resting patterns of Cretaceous–Tertiary planktonic foraminiferal extinctions at El Kef (Tunisia), in: *The Cretaceous–Tertiary Event and Other Catastrophes in Earth History*, Special Paper 307 (G. Ryder, D. Fastovsky *et al.*, eds.), Boulder, Geological Society of America, pp. 287–302.

MacLeod, K. G., and Huber, B., 1996, Strontium isotopic evidence for extensive reworking in sediments spanning the Cretaceous/Tertiary boundary at ODP Site 738, *Geology* **24**:463–466.

MacLeod, N., and Keller, G., 1991, How complete are Cretaceous/Tertiary boundary sections? A chronostratigraphic estimate based on graphic correlation, *Geological Society of America Bulletin* **103**:1439–1457.

MacLeod, N., and Keller, G., 1994, Comparative biogeographic analysis of planktic foraminiferal survivorship across the Cretaceous-Tertiary boundary, *Paleobiology* **20**:143–177.

MacLeod, N., Rawson, P. F., Forey, P. L., Banner, F. T., Boudagher-Fadel, M. K., Bown, P. R., Burnett, J. A., Chambers, P., Culver, S., Evans, S. E., Jeffery, C., Kaminski, M. A., Lord, A. R., Milner, A. C., Milner, A. R., Morris, N., Owen, E., Rosen, B. R., Smith, A. B., Taylor, P. D., Urquhart, E. and Young, J. R., 1997, The Cretaceous–Tertiary biotic transition, *Journal of the Geological Society of London* **154**:265–292.

Magloire, L., 1968, Étude stratigraphique par la palynologie des dépôts argilo-gréseux du Silurien et du Dévonien inférieur dans la région du Grand Erg Occidental (Sahara algérien), in: *International Symposium on the Devonian System*, Volume 3, Calgary, 1967, Proceedings, pp. 473–491.

Mai, H., 1999, Paleocene coccoliths and coccospheres in deposits of the Maastrichtian stage at the 'type locality' and type area in SE Limburg, The Netherlands, *Marine Micropaleontology* **36**:1–12.

Mann, K. O and Lane, R. H., (eds.), 1995, *Graphic Correlation*, Society for Sedimentary Geology, Special Publication 53.

Margaleff, R., 1983, *Limnologia*, Barcelona, Omega.

Marincovitch, Jr., L., 1996, Survivorship of Mesozoic mollusks in the Paleocene Arctic Ocean, in: *Cretaceous–Tertiary Mass Extinctions: Biotic and Environmental Changes* (N. MacLeod and G. Keller, eds.), New York, W. W. Norton & Company, pp. 275–287.

Marshall, C. R., and Ward, P. D., 1996, Sudden and gradual molluscan extinctions in the latest Cretaceous of western European Tethys, *Science* **274**:1360–1363.

Martin, F., 1993, Acritarchs: a review, *Biology Reviews* **68**:475–538.

Martin, R. J., 1999, *Taphonomy: A Process Approach*, Cambridge, Cambridge University Press.

Martini, E., 1971, Standard Tertiary and Quaternary calcareous nannoplankton zonation. In: *Proceedings of the Second Planktonic Conference* (A. Farinacci, ed.), Rome, Tecnoscienza, pp. 739–785.

Massa, D., and Moreau-Benoit, A., 1976, Essai de synthèse stratigraphique et palynologique du Système Dévonien en Libye occidentale, *Revue de l'Institut Français du Petrole* **31**(2):287–333.

Massa, D., Coquel, R., Loboziak, S., and Taugourdeau-Lantz, J., 1980, Essai de synthèse stratigraphique et palynologique du Carbonifère en Libye occidentale, *Annales de la Société géologique du Nord* **99**:429–442.

Matten, L. C., and Banks, H. P., 1967, Relationship between the Devonian progymnosperm genera *Sphenoxylon* and *Tetraxylopteris*, *Bulletin of the Torrey Botanical Club* **94**:321–333.

Matthews, M. D., and Perlmutter, M. A., 1994, Global cyclostratigraphic analysis of the Green River basin, in: *Orbital Forcing and Cyclic Sequence*, Special Publication No. 19 (P. L. de Boer and D. G. Smith, eds.), International Association of Sedimentologists, pp. 459–481.

Matthews, R. K., and Frohlich, C.,1991, Orbital forcing of low frequency glacioeustasy, *Journal of Geophysical Research* **96**(B4):6797–6803.

Maziane, N., Higgs, K., and Streel, M., 1999, Revision of the late Famennian miospore zonation scheme in eastern Belgium, *Journal of Micropaleontology* **18**(1):17–25.

McArthur, J. M., 1998, Strontium isotope stratigraphy, in: *Unlocking the Stratigraphic Record: Advances in Modern Stratigraphy* (P. Doyle and M. R. Bennett, eds.), Chichester, John Wiley & Sons, pp. 221–241.

McCarthy, F. M. G., and Gostlin, K. E., 2000, Correlating Pleistocene sequences across the New Jersey margin, *Sedimentary Geology* **134**:181–196.

McCartney, K., Huffman, A. R. and Tredoux, M., 1990, A paradigm for endogenous causation of mass extinctions, in: *Global Catastrophes in Earth History*, Special Paper 247 (V. L. Sharpton and P. D. Ward, eds.), Geological Society of America, pp. 125–138.

McCauley, S. E., and DePaolo, D. J., 1997, The marine 87Sr/86Sr and 18C records, Himalayan alkalinity fluxes, and Cenozoic climate models, in: *Tectonic Uplift and Climate Changes* (W. F. Ruddiman, ed.), New York, Plenum Press, pp. 427–467.

McClelland, E., Finegan, B., and Butler, R. W. H., 1996, A magnetostratigraphic study of the onset of the Mediterranean Messinian salinity crisis; Caltanissetta Basin, Sicily, in: *Paleomagnetism and Tectonics of the Mediterranean Region*, Special Publication No. 105 (A. Morris and D. H. Tarling, eds.), Geological Society, pp. 205–217.

McGregor, D. C., 1984, Late Silurian and Devonian spores from Bolivia, *Acad. Nac. Ci. Cordoba Miscelanea* **69**:1–43.

McGregor, D. C., and Playford, G., 1993, Canadian and Australian Devonian spores: zonation and correlation, *Bulletin of the Geological Survey of Canada* **438**:1–125 [imprinted 1992].

McGregor, D. C., Anderson, T. W., Davey, R. J., Dettmann, M. E., Dolby, G., Edwards, L. E., Hacquebard, P. A., Jansonius, J., Kaska, H. V., Mathewes, R. W., Mudie, P. J., Partridge, A. D., Piel, K. M., Riding, J. B., Schwab, K. W., Shane, J. D., Whitney, B. L., Wiggins, V. D., Williams, V. E., Witmer, R. J., and Zippi, P. A., 1996, Economic applications of palynology: examples, in: *Palynology: Principles and Applications*, Volume 3 (J. Jansonius and D. C. McGregor, eds.), American Association of Stratigraphic Palynologists Foundation, pp. 1277–1286.

McLaren, D. J., and Goodfellow, W. D., 1990, Geological and biological consequences of giant impacts, *Annual Review of Earth Planetary Sciences* **18**:123–171.

McLean, D. M., 1985, Deccan Traps mantle degassing in the terminal Cretaceous marine extinctions, *Cretaceous Research* **6**:235–259.

McLoughlin, S., and Hill, R. S., 1996, The succession of Western Australian Phanerozoic terrestrial floras, in: *Gondwanan Heritage: Past, Present and Future of the Western Australian Biota* (Hopper *et al.*, eds.), Chipping Norton, Surrey Beatty and Sons, pp. 61–80.

Mello, M. R., Telnaes, N., Gaglianone, P. C., Chicarelli, M. I., Brassel, S., and Maxwell, J. R., 1988, Organic geochemical characterization of depositional palaeoenvironments of sources rocks and oils in Brazilian marginal basins, *Organic Geochemistry* **13**:31–45.

Melo, J. H. G., 2000, Palynological evaluation and correlation of some Silurian–Devonian sections of southern Bolivia, in: *14° Congress of Geology, Bolivia*, La Paz, Memorias, pp. 136–141.

Melo, J. H. G., and Loboziak, S., 2001, New miospore zonation of Devonian–Early Carboniferous strata in the Amazon Basin: a preliminary account, in: *Correlação de Seqüências Paleozóicas Sul-Americanas*, Ciênc.-Téc.-Petról., Seção: Expl. Petról. 20 (J. H. G. Melo and G. J. S. Terra, eds.), pp. 99–107. (CD Rom, ISSN-0100–1515.)

Melo, J. H. G., and Loboziak, S., 2003, Devonian–Early Carboniferous miospore biostratigraphy of the Amazon Basin, northern Brazil, *Review of Palaeobotany and Palynology* **124**:131–202.

Melo, J. H. G., Grahn, Y., Loboziak, S., and Paris, F., 1996, First record of *Fungochitina fenestrata* (Taugourdeau and Jekhowsky 1960) (Chitinozoa) in the Devonian of Brazil and its biostratigraphic significance, in: Simpósio Sul-Americano Siluro-Devoniano: Estratigrafia e Paleontologia, Ponta Grossa, Anais, pp. 263–281.

Menegatti, A. P., Weissert, H., Brown, R. S., Tyson, R. V., Farrimond, P., Strasser, A., and Caron, M., 1998, High-resolution $\delta^{13}C$ stratigraphy through the early Aptian "Livello Selli" of the Alpine Tethys, *Paleoceanography* **13**:530–545.

Menéndez, C. A., 1965, Contenido palinológico en sedimentos con "*Rhacopteris ovata*" (McCoy) Walk. de la Sierra de Famatina, La Rioja, *Revista del Museo Argentino de Ciencias Naturales "Bernardino Rivadavia" – Paleontologia* **1**:45–80.

Menéndez, C. A., and Pöthe de Baldis, E. D., 1967, Devonian spores from Paraguay, *Review of Palaeobotany and Palynology* **1**:161–172.

Menéndez, C. A., Cacccavari de Filice, M. A., 1975, Distribución y frecuencia del polen fósil de *Nothofagus* en Depósitos Cretacicos y Terciarios del Norte de Tierra del Fuego, Argentina, in: 2° Congreso Argentino de Paleontología y Biostratigrafía, *Actas* **2**:241–255.

Meyen, S. V., 1987, *Fundamentals of Palaeobotany*, London, Chapman and Hall.

Miall, A. D., 1991, Stratigraphic sequences and their chronostratigraphic correlation, *Journal of Sedimentology and Petrology* **61**:497–505.

Miall, A. D., 1994, Sequence stratigraphy and chronostratigraphy: problems of definition and precision in correlation, and their implications for global eustasy, *Geosciences Canada* **21**:1–26.

Miall, A. D., 1996, *The Geology of Fluvial Deposits*, Berlin, Springer.

Miall, A. D., 1997, *The Geology of Stratigraphic Sequences*, Berlin, Heidelberg, New York, Springer-Verlag.

Miall, A. D., and Miall, C. E., 2001, Sequence stratigraphy as a scientific enterprise: the evolution and persistence of conflicting paradigms, *Earth-Science Reviews* **54**:321–348.

Milani, E. J., Faccini, U. F., Scherer, C. M., Araújo, L. M., and Cupertino, J. A., 1998, Sequences and stratigraphic hierarchy of the Paraná Basin (Ordovician to Cretaceous) Southern Brazil, *Boletim do Instituto de Geociências da Universidade de São Paulo* **29**:125–173.

Milankovitch, M., 1941, *Kanon der Erdbestranlung and seine anwendung auf das eiszeitenproblem* (Canon of insolation and the ice age problem): Transactions of the Royal Serbian Academy, Belgrade, Spec. Publication, v. 132, Section of Mathematical and Natural Sciences, v. 33, 674 pp.

Miller, A. R., Barrera, E., Olsson, R. K., Sugarman, P. J., and Savin, S. M., 1999, Does ice drive early Maastrichtian eustacy? *Geology* **27**:783–786.

Miller, F. X., 1977, The graphic correlation method in biostratigraphy, in: *Concepts and Methods in Stratigraphy* (E. G. Kauffman and J. E. Hazel, eds.), Stroudsburg, Pennsylvania, Dowden, Hutchinson and Ross, pp. 165–186.

Miller, K. G., Aubry, M. P., Kahn, M. J., Melillo, A. J., Kent, D. V., and Berggren, W. A., 1985, Oligocene-Miocene biostratigraphy, magnetostratigraphy, and isotopic stratigraphy of the Western North Atlantic, *Geology* **13**:257–261.

Miller, K. G., Feigenson, M. D., Kent, D. V., and Olsson, R. K., 1988, Upper Eocene to Oligocene isotope ($^{87}Sr/^{86}Sr$, $\delta^{18}O$, $\delta^{13}C$) standard section, Deep Sea Drilling Project Site 522, *Paleoceanography* **3**:223–233.

Miller, K. G., Wright, J. D., and Fairbanks, R. G., 1991, Unlocking the ice house: Oligocene-Miocene oxygen isotopes, eustacy, and margin erosion, *Journal of Geophysical Research* **96**:6829–6848.

Miller, K. G., Feigenson, M. D., Wright, J. D., and Clement, B. M., 1991, Miocene isotope reference section, Deep Sea Drilling Project Site 608: An evaluation of isotope and biostratigraphic resolution, *Paleoceanography* **6**:33–52.

Miller, K. G., Wright, J. D., Van Fossein, M. C., and Kent, D. V., 1994, Miocene stable isotopic stratigraphy and magnetostratigraphy of Buff Bay, Jamaica, *Geological Society of America Bulletin* **106**:1605–1620.

Miller, M. A., 1996, Chitinozoa, in: *Palynology: Principles and Applications*, Volume 1, Chapter 11 (J. Jansonius and D. C. McGregor, eds.), Dallas, American Association of Stratigraphic Palynologists Foundation, pp. 307–336.

Miller, R., 1983, *Continents in Collision*, Alexandria, Virginia, Time-Life Books (Planet Earth Series).

Milliman, J. D., and Syvitski, J. P. M., 1992, Geomorphic/tectonic control of sediment discharge to the ocean: the importance of small mountainous rivers, *Journal of Geology* **100**:525–544.

Miranda, A. P., Souza Cruz, C. E., and Oller, J., 2000, Sequence stratigraphy of the Late Silurian–Devonian Subandean Basin, southern Bolivia and northern Argentina, in: *14° Congress of Geology Bolivia*, La Paz, Memorias, pp. 46–50.

Mitchum, R. M., and Van Wagoner, J. C., 1991, High frequency sequences and their stacking patterns: sequence-stratigraphic evidence of high frequency eustatic cycles, *Sedimentary Geology* **70**(2–4):131–160.

Mohr, B. A. R., 1990, Early Cretaceous palynomorphs from ODP Sites 692 and 693, the Weddell Sea, Antarctica, in: *Proceedings of the Ocean Drilling Program Scientific Results*, Volume 113 (P. F. Barker and J. P. Kennett, eds.), Texas, College Station, pp. 449–464.

Molina, E., Arenillas, I., and Arz, J. A., 1998, Mass extinction in planktic foraminifera at the Cretaceous/Tertiary boundary in subtropical and temperate latitudes, *Bulletin de la Societé Geologique de France* **169**(3):351–363.

Molyneux, S. G., Le Hérissé, A., and Wicander, R., 1996, Paleozoic Phytoplankton, in: *Palynology: Principles and Applications*, Volume 2 (J. Jansonius and D. C. McGregor, eds.), American Association of Stratigraphic Palynologists Foundation, pp. 493–529.

Monechi, S., 1979, Variations in nannofossil assemblage at the Cretaceous/Tertiary boundary in the Bottaccione Section (Gubbio, Italy), in: *Cretaceous/Tertiary Boundary Events Symposium* (K. W. Christensen and T. Birkelund, eds.), Proceedings **2**, pp. 164–169.

Mongin, D., 1959, Study of some American Miocene lamellibranchs and comparison with related European species, *Bulletin of Mexican Paleontology* **39**.283–343.

Monteiro, S. M., 1979, Contribuição ao Estudo Paleoxilológico da Formação Irati do Rio Grande do Sul, Dissertação (Mestrado em Geociências), Instituto de Geociências, Universidade Federal do Rio Grande do Sul, Porto Alegre, 143 pp. (unpublished).

Montgomery, H., Pessagno, E., Soegaard, K., Smith, C., Muñoz, I., and Pessagno, J., 1992, Misconceptions concerning the Cretaceous/Tertiary boundary at the Brazos River, Falls County, Texas, *Earth Planetary Science Letters* **109**:593–600.

Moore, R. C., 1955, Invertebrates and the geological time scale, *Geological Society of America Special Paper* **62**:547–573.

Moore, T. C., Jr., van Andel, T. H., Sancetta, C., and Pisias, N., 1978, Cenozoic hiatuses in pelagic sediments, *Micropaleontology* **24**:113–138.

Moore, T. L., Oglesby, R. J., Perlmutter, M. A., Matthews, M. D., Plotnick, R. E., and Maasch, K. A., 2000, Deciphering orbital maximum and minimum states during the Pleistocene, *Proceedings of the Geological Society of America Annual Meeting*, Reno, Nevada.

Moreau-Benoit, A., 1984, Acritarches et Chitinozoaires du Dévonien moyen et supérieur de Lybie occidentale, *Review of Palaeobotany and Palynology* **43**:187–216.

Moreau-Benoit, A., Coquel, R., and Latrèche, S., 1993, Étude palynologique du Dévonien du Bassin d'Illizi (Sahara oriental algérien), approche biostratigraphique, *Geobios* **26**(1): 3–31.

Mosbrugger, V., and Utescher, T., 1997, The coexistence approach – a method for quantitative reconstructions of Tertiary terrestrial paleoclimate data using plant fossils, *Palaeogeography, Palaeoclimatology, Palaeoecology* **134**:61–86.

Moshkovitz, S., and Habib, D., 1993, Calcareous nannofossil and dinoflagellate stratigraphy of Cretaceous–Tertiary boundary, Alabama and Georgia, *Micropaleontology* **39**(2):167–191.

Mu En-Zhi, Boucot, A. J., Chen Xu, and Rong Jia-yu, 1986, Correlation of the Silurian Rocks of China, *Geological Society of America, Special Paper* **202**.

Murphy, M. A., and Salvador, A. (eds.), 2000, International Stratigraphic Guide – An abridged version, http://www.micropress.org/stratigraphy/guide.htm

Nadon, G. C., 1993, The association of anastomosed fluvial deposits and dinosaur tracks, eggs and nests: implication of floodplain environment and a possible survival strategy for ornithopods, *Palaios* **8**:31–44.

Nagy, J., Eli, K. F., Henning, D., and Bremer, M. G. A., 2001, Response of foraminiferal facies to transgressive–regressive cycles in the Callovian of northeast Scotland, *Journal of Foraminiferal Research* **31**(4):324–349.

Naumova, S. N., 1953, Spore-pollen assemblages of the Upper Devonian of the Russian Platform and their stratigraphic significance, *Transactions of Institute of Geological Sciences, Academy of Sciences of the USSR* **143** (Geological Series 60): 1–204. (in Russian).

Neal, J. E., Stein, J. A., and Gamber, J. H., 1995, Integration of the graphic correlation methodology in a sequence stratigraphic study: examples from North Sea Paleogene sections, in: *Graphic Correlation*, Special Publication 53 (K. O. Mann and H. R. Lane, eds.), Tulsa, SEPM Society for Sedimentary Geology, pp. 5–113.

Neveling, J., Rubidge, B. S., and Hancox, J. P., 1999, A lower *Cynognathus* Assemblage Zone fossil from the Katberg Formation (Beaufort Group (Karoo Group, South Africa), *South African Journal of Sciences* **95**:555–556.

Newberry, J. S., 1874, Circles of deposition in American sedimentary rocks, *American Association for the Advancement of Science Proceedings* **22**:185–96.

Nichols, D. J., 1992, Plants at the K/T boundary, *Nature* **356**:295.

Nichols, D. J., Jarzen, D. M., Orth, C. J., and Oliver, P. Q., 1986, Palynological and iridium anomalies at Cretaceous–Tertiary boundary, South-Central Saskatchewan, *Science* **231**:714–717.

Norris, R. D., and Röhl, U., 1999, Carbon cycling and chronology of climate warming during the Palaeocene/Eocene transition, *Nature* **401**:775–778.

Norris, R. D., Huber, B. T., and Self-Trail, J., 1999, Synchroneity of the K–T oceanic mass extinction and meteorite impact: Blake Nose, western North Atlantic, *Geology* **27**(5):419–422.

North American Commission on Stratigraphic Nomenclature, 1983, North American Stratigraphic Code, *Bulletin of the American Association of Petroleum Geologists* **67**(5):841–875.

O'Brien, N., Brett, C. E., and Taylor, W. L., 1994, Microfabric and taphonomic analysis in determining sedimentary processes in marine mudstones: example from Silurian of New York, *Journal of Sedimentary Research* **A64**(4):847–852.

Ochev, V. G., and Shishkin, M. A., 1989, On the principles of global correlation of the continental Triassic on the tetrapods, *Acta Palaeontologica Polonica* **34**:149–173.

Odin, G. S., 1992, New stratotypes for the Paleogene, the Cretaceous/Paleogene, Eocene/ Oligocene and Paleogene/Neogene boundaries, *Neues Jahrbuch für Geologie und Paläontologie, Abhandlungen* **186**(1):7–20.

Odin, G. S., and Odin, C., 1990, Echelle Numerique des Temps Geologiques, *Geochronologie* **35**:12–20.

Officer, C. B., Hallam, A., Drake, C. L., and Devine, J. D., 1987, Late Cretaceous and paroxysmal Cretaceous/Tertiary extinctions, *Nature* **326**:143–149.

Ogg, J. G., Steiner, M. B., Oloriz, F., and Tavera, J. M., 1984, Jurassic magentostratigraphy: 1. Kimmeridgian-Tithonian of Sierra Gorda and Carcabuey, southern Spain, *Earth and Planetary Science Letters* **71**:147–162.

Ogg, J. G., Hasenyager, R. W., Wimbledon, W. A., Channell, J. E. T., and Bralower, T. J., 1991, Magnetostratigraphy of the Jurassic–Cretaceous boundary interval, Tethyan and English faunal realms, *Cretaceous Research* **12**:455–482.

Okada, H., and Bukry, D., 1980, Supplementary modification and introduction of code numbers to the low-latitude coccolith biostratigraphic zonation (Bukry, 1973: 1975), *Marine Micropaleontology* **5**:321–325.

Olafsson, G., 1991, Late Oligocene through Late Miocene calcareous nannofossil biostratigraphy and biochronology, PhD thesis, Meddelanden fran Stockholms Universitets Institution för Geologi och Geokemi, No. 283.

Oliveira, S. F., 1997, Palinologia da seqüência devoniana da Bacia do Paraná no Brasil, Paraguai e Uruguai: implicações biocronoestratigráficas, paleoambientais e paleogeográficas, Universidade de São Paulo, D.Sc. thesis (unpublished).

Oliveira, S. F., and Burjack, M. I. A., 1996, *Petrovina connata* gen. n., sp. n., provavelmente relacionado a Chlorophyta, da Formação Ponta Grossa (Neodevoniano), Bacia do Paraná, MT, Brasil, *Revista de la Universidad Guarulhos (Geociencia)*, **I**(3):46–50.

Oliveira, S. F., Antonioli, L., and Dino, R., 1995, Análise palinológica de sedimentos da Formação Ponta Grossa, aflorantes no flanco NW da Bacia do Paraná, MT, Brasil, in: *2° Simp. Cronoestr. B. Paraná, Bol. Res.* Porto Alegre, Expand, pp. 1–9.

Olsen, P. E., 1980, Fossil great lakes of the Newark Supergroup in New Jersey, in: *Field Studies of New Jersey Geology and Guide to Field Trips* (W. Manspeizer, ed.), 52nd Annual Meeting of the N.Y. State Geological Association, Newark, N.J., Rutgers University, pp. 352–398.

Olsen, T., Steel, R., Hogseth, K., Skar, T., and Roe, S. E., 1995, Sequential architecture in a fluvial succession: sequence stratigraphy in the Upper Cretaceous Mesaverde Group, Price Canyon, Utah, *Journal of Sedimentary Research* **B65**(2):265–280.

Olson, E. C., 1958, Fauna of the Vale and Choza: Summary, review and integration of the geology and the faunas, Chicago, Ill., *Fieldiana Geology* **10**:397–448.

Olson, H. C., and Leckie, R. M. (eds.), 2003, *Micropaleontologic Proxies for Sea-Level Change and Stratigraphic Discontinuities*, Special Publication 75, Tulsa, SEPM, Society for Sedimentary Geology.

Olson, H. C., Damuth, J. E., Thompson, P., and Moss, G., 2001, Latest Quaternary studies of downslope transport within a stratigraphic framework, Gulf of Mexico intraslope basins area, abstract, American Association of Petroleum Geologists Annual Meeting, Denver, Colorado.

Olson, H. C., Gary, A., and Jones, G. D., 2003, Similarity curves as indicators of stratigraphic discontinuities, in: *Micropaleontologic Proxies for Sea-Level Change and Stratigraphic Discontinuities*, Special Publication 75 (H. C. Olson and R. M. Leckie, eds.), Tulsa, SEPM, Society for Sedimentary Geology, pp. 89–96.

Olsson, R. K., and Liu, C., 1993, Controversies on the placement of Cretaceous–Paleogene boundary and the K/P mass extinction of planktonic foraminifera, *Palaios* **8**:127–139.

Olsson, R. K., Hemleben, C., Berggren, W. A. and Liu, C., 1992, Wall texture classification of planktonic foraminifera genera in the lower Danian, *Journal of Foraminiferal Research* **22**:195–213.

Olszewski, T., 1999, Taking advantage of time-averaging, *Paleobiology* **25**:226–238.

Opdyke, N. D., 1989, Magnetic stratigraphy of Cenozoic terrestrial sediments and mammalian dispersal, *Journal of Geology* **98**:621–637.

Opdyke, N. D., 1995, Magnetostratigraphy of Permo-carboniferous time, in: *Geochronology, Time Scales and Global Stratigraphic Correlation*, Special Publication No. 54 (W. A. Berggren, D. V. Kent, M.-P. Aubry, and J. Hardenbol, eds.), Society of Economic Paleontologists and Mineralogists, pp. 41–47.

Opdyke, N. D., and Channell, J. E. T., 1997, *Magnetic Stratigraphy*, International Geophysical Series, v. 64, San Diego, Academic Press.

Opdyke, N. D., Lindsay, E., Johnson, G. D., Johnson, N., Tahirkheli, R. A. K., and Mirza, M. A., 1979, Magnetic polarity stratigraphy and vertebrate paleontology of the upper Siwalik Subgroup of northern Pakistan, *Palaeogeography, Palaeoclimatology, Palaeoecology* **27**:1–34.

Orbigny, A., d', 1849, *Cours élémentaire de Paléontologie et de Géologie stratigraphique*, Volume 1, Paris, Masson.

Orbigny, A. d', 1851, *Cours élémentaire de Paléontologie et de Géologie stratigraphique*, Volume 2, Paris, Masson.

Orth, C. J., Gilmore, J. S., Knight, J. D., Pillmore, C. L., Tschudy, R. H., and Fassett, R. H., 1981, An iridium abundance anomaly at the palynological Cretaceous-Teriary boundary in northern New Mexico, *Science* **214**:1341–1343.

Ottone, E. G., 1996, Devonian palynomorphs from the Los Monos Formation, Tarija Basin, Argentina, *Palynology* **20**:105–155.

Ottone, E. G., and Rossello, E. A., 1996, Palinomorfos devónicos de la Formación Tequeje, Angosto del Beu, Bolivia, *Ameghiniana* **33**(4):443–451.

Owens, B., 1996, Upper Carboniferous spores and pollen, in*: Palynology: Principles and Applications*, Volume 2, Chapter 18D (J. Jansonius and D. C. McGregor, eds.), Dallas, American Association of Stratigraphic Palynologists Foundation, pp. 597–606.

Owens, B., Clayton, G., Gao Lianda, and Loboziak, S., 1989, Miospore correlation of the Carboniferous deposits of Europe and China, *C. R. 11th. Congr. Internat. Strat. Géol. Carb. Beijing 1987*, **3**:189–210.

Owens, B., Loboziak, S., and Teteriuk, V. K., 1978, Palynological subdivision of the Dinantian to Westphalian deposits of northwest Europe and the Donetz Basin of the U.S.S.R. *Palynology* **2**:69–91.

Palmer, A. R., compiler, 1983, Decade of North American Geology (DNAG). Geologic Time Scale, *Geology* **11**:503–504.

Paola, C., Mullin, J., Ellis, C., Mohrig, D. C., Swenson, J. B., Parker, G., Hickson, T., Heller, P. L., Pratson, L., Syvitski, J., Sheets, B., and Sttrong, N., 2001, Experiment Paris, 1981, Les chitinozoaires dans le Paléozoïque du sud-ouest de l'Europe, *Memoires de la Societé Geologique et Mineralogique Bretagne* **26**:1–412.

Papavassiliou, C. T., and Cosgrove, M. E., 1982, The geochemistry of DSDP sediments from site 223 (Indian Ocean), *Chemical Geology* **37**:299–315.

Paproth, E., and Streel, M. (eds.), 1984, The Devonian-Carboniferous boundary, *Courier Forschungsinstitut Senckenberg* **67**:1–258.

Paris, F., 1990, Chitinozoans: a fascinating and mysterious microfossil-group (technical card and direction for use), Documents of Rennes 1 University (1990), pp. 1–67.

Paris, F., 1996, Chitinozoan biostratigraphy and palaeoecology, in: *Palynology: Principles and Applications*, Volume 2, Chapter 17 (J. Jansonius and D. C. McGregor, eds.), Dallas, American Association of Stratigraphic Palynologists Foundation, pp. 531–552.

Paris, F., and Nõlvak, J., 1999, Biological interpretation and paleobiodiversity of a cryptic fossil group: the "chitinozoan animal," *Geobios* **32**:315–324.

Paris, F., Richardson, J. B., Riegel, W., Streel, M., and Vanguestaine, M., 1985, Devonian (Emsian–Famennian) Palynomorphs, in: *Palynostratigraphy of North-east Libya* (B. Thusu and B. Owens, eds.), *Journal of Micropaleontology* **4**(1):49–82.

Paris, F., Grahn, Y., Nestor, V., and Lakova, I., 1999, A revised chitinozoan classification, *Journal of Paleontology* **73**(40):549–570.

Paris, F., Winchester-Seeto, T., Boumendjel, K., and Grahn, Y., 2000, Toward a global biozonation of Devonian chitinozoans, *Courier Forschungsinstitut Senckenberg* **220**:39–55.

Park, J., and Oglesby, R. J., 1994, The effect of orbital cycles on Late and Middle Cretaceous climate: a comparative general circulation model study, in: *Orbital Forcing and Cyclic*, No. 19 (P. L. de Boer and D. G. Smith, eds.), Sequences, International Association of Sedimentologists, Special Publication, pp. 509–530.

Parrish, J. M., Parrish, J. T., and Ziegler, A. M., 1986, Permian-Triassic paleogeography and paleoclimatology and implications for therapsid distribution, in: *The Ecology and Biology of Mammal-like Reptiles* (N. Hotton III, P. D. MacLean, J. J. Roth, and C. E. Roth, eds.), Washington, DC, Smithsonian Institution Press, pp. 109–131.

Parrish, J. T., 1990, Gondwanan paleogeography and paleoclimatology, in: *Antarctic Paleobiology: its Role in the Reconstruction of Gondwana* (T. N. Taylor and E. L. Taylor, eds.), New York, Springer-Verlag, pp. 15–26.

Parrish, J. T., and Spicer, R. A., 1988, Middle Cretaceous wood from the Nanushk Group, central north Slope, Alaska, *Palaeontology* **31**:19–34.

Payne, S. N. J., Ewen, D. F., and Bowman, M. J., 1999, The role and value of "high-impact biostratigraphy" in reservoir appraisal and development, in: *Biostratigraphy in Production and Development Geology*, Special Publication No. 152 (R. W. Jones and M. D. Simmons, eds.), London, Geological Society, pp. 5–22.

Payton, C. E. (ed.), 1977, Seismic stratigraphy – applications to hydrocarbon exploration, *American Association of Petroleum Geologists Memoir* **26**.

Pearce, T. J., and Jarvis, I., 1992, Applications of geochemical data to modelling sediment dispersal patterns in distal turbidites: Late Quaternary of the Madeira Abyssal Plain, *Journal of Sedimentary Petrology* **62**:1112–1129.

Pearce, T. J., and Jarvis, I., 1995, High-resolution chemostratigraphy of Quaternary distal turbidites: A case study of new methods for the Analysis and correlation of barren sequences, in: *Non-biostratigraphical Methods of Dating and Correlation: London, United Kingdom*, Special Publication No. 89 (R. E. Dunay and E. A. Hailwood, eds.), London, Geological Society, pp. 107–143.

Pearce, T. J., Besly, B. M., Wray, D. S., and Wright, D.K., 1999, Chemostratigraphy: A method to improve interwell correlation in barren sequences – A case study using onshore Duckmantian/ Stephanian sequences (West Midlands, U.K.), *Sedimentary Geology* **124**:197–220.

Pedersen, T. F., and Calvert, S. E., 1990, Anoxia vs. productivity: What controls the formation of organic-rich sediments and sedimentary rocks?, *Bulletin of the American Association of Petroleum and Geologists* **74**:454–466.

Pemberton, S. G., and Frey, R. W., 1985, The Glossifungites ichnofacies: modern examples from the Georgia coast, U.S.A., in: *Biogenic Structures: Their Use in Interpreting Depositional Environments*, Special Publication No. 35 (H. A. Curran, ed.), Society of Economic Paleontologists and Mineralogists, pp. 237–259.

Pemberton, S. G., and MacEachern, J. A., 1995, The sequence stratigraphic significance of trace fossils: examples from the Cretaceous foreland basin of Alberta, Canada, in: *Sequence Stratigraphy of Foreland Basin Deposits – Outcrop and Subsurface Examples from the Cretaceous of North America*, Memoir 64 (J. C. Van Wagoner and G. Bertram, eds.), American Association of Petroleum Geologists, pp. 429–475.

Pemberton, S. G., and MacEachern, J. A., 1997, The ichnological signature of storm deposits: the use of trace fossils in event stratigraphy, in: *Palaeontological Event Horizons Ecological and Evolutionary Implications* (C. E. Brett, ed.), Columbia University Press, pp. 73–109.

Pemberton, S. G., MacEachern, J. A., and Frey, R. W., 1992, Trace fossil facies models: environmental and allostratigraphic significance, in: *Facies Models: Response to Sea Level Change* (R. G. Walker and N. James, eds.), St. John's, Newfoundland, Geological Association of Canada, pp. 47–72.

Pemberton, S. G., Spila, M. V., Pulham, A. J., Saunders, T., MacEachern, J. A., Robbins, D., and Sinclair, I., 2002, *Ichnology and Sedimentology of Shallow and Marginal Marine Systems: Ben Nevis and Avalon Reservoirs, Jeanne D'Arc Basin.* St. John's, Newfoundland: Geological Association of Canada Short Course Notes 15.

Penck, A., 1914, The shifting of climate belts, *Scottish Geographical Magazine* **30**:281–293.

Peppers, R. A., 1996, Palynological correlation of major Pennsylvanian (Middle and Upper Carboniferous) chronostratigraphic boundaries in the Illinois and other coal basins, *Geological Society of America Memoirs* **188**:1–111.

Perch-Nielsen, K., 1979, Calcareous nannofossils in Cretaceous/Tertiary boundary sections in Denmark, in: *Cretaceous–Tertiary Boundary Events Symposium II, Proceedings* (W. K. Christensen and T. Birkelund, eds.), University of Copenhagen, pp. 120–126.

Perch-Nielsen, K., McKenzie, J. A., and He, Q., 1982, Bio- and isotope-stratigraphy and the 'catastrophic' extinction of calcareous nannoplankton at the Cretaceous/Tertiary boundary, in: *Geological Implications of Impacts of Large Asteroids and Comets on the Earth*, Special Paper 190 (L. T. Silver and P. H. Schultz, eds.), Geological Society of America, pp. 353–371.

Percival, S. F., and Fischer, A. G., 1977, Changes in calcareous nannoplankton in the Cretaceous–Tertiary biotic crisis at Zumaya, Spain, *Evolutionary Theory* **2**:1–35.

Pérez-Leyton, M., 1991, Miospores du Dévonien moyen et supérieur de la coupe de Bermejo-La Angostura (sud-est de la Bolivie), *Annales de la Societé Geologique Belgique* **113**(2) (1990):373–389.

Perlmutter, M. A., 1985, Deep water clastic reservoirs in the Gulf of Mexico: a depositional model, *Geo-Marine Letters* **5**:105–112.

Perlmutter, M. A., 1995, Using global cyclostratigraphy to predict lithofacies in continental and marine basins, *Norwegian Petroleum Society Conference on Predictive High Resolution Sequence Stratigraphy*, Stavanger, Norway, pp. 16 A-C.

Perlmutter, M. A., and Matthews, M. D., 1989, Global cyclostratigraphy – A model, in*:* *Quantitative Dynamic Stratigraphy* (T. A. Cross, ed.), Englewood Cliffs, New Jersey, Prentice-Hall, pp. 233–260.

Perlmutter, M. A., and Matthews, M. D., 1992, Global cyclostratigraphy, in: *Encyclopedia of Earth Systems Science* (W. A. Nierenberg, ed.), New York, Academic Press, pp. 379–393.

Perlmutter, M. A., and Matthews, M. D., 1994, Global cyclostratigraphy: predicting stratigraphic response to Milankovitch Cycle, in: *Predictive Stratigraphic Analysis – Concept and Applications*, Volume 2110 (C. B. Cecil and N. T. Edgar, eds.), *U.S.G.S. Survey Bulletin*, pp. 6–7.

Perlmutter, M. A., and Plotnick, R. E., 1997, Hemispheric asymmetry of the marine stratigraphic record; proof of the existence of a polar icecap, *A.A.P.G. Annual Meeting Abstracts with Programs*, p. A92.

Perlmutter, M. A., and Plotnick, R. E., 2002, Predicable variations in the marine stratigraphic record of the northern and southern hemispheres and reservoir potential, in: *Sequence Stratigraphic Models for Exploration and Production: Evolving Methodology, Emerging Models and Application Histories* (J. M. Armentrout and N. C. Rosen, eds.), 22nd Annual Gulf Coast SEPM Research Conference, Proceedings, pp. 231–256.

Perlmutter, M. A., and Plotnick, R. E., 2003, Hemispheric asymmetry of the marine stratigraphic record; conceptual proof of a unipolar icecap, in: *Climate Controls on Stratigraphy* (C. B. Cecil and N. T. Edgar, eds.), SEPM Special Publication 77, pp. 51–66.

Perlmutter, M. A., Brennan, P. A., Hook, S. C., Dempster, K., and Pasta, D., 1995, Global cyclostratigraphic analysis of the Seychelles southern shelf for potential reservoir, seal and source rocks, in: *Paleoceanography and Environments of Exploration of the Indian Ocean Basin and Margins*, Volume 96(1/2) (T. A. Davis and M. Coffin, eds.), *Sedimentary Geology* Special Issue, pp. 93–118.

Perlmutter, M. A., Radovich, B. J., Mathews, M. D., and Kendall, C. G. St. C., 1998, The impact of high-frequency sedimentation cycles on stratigraphic interpretation, in: *Sequence Stratigraphy – Concepts and Applications*, Special Publication 8 (F. M. Gradstein, K. O. Sandvik, and N. J. Milton, eds.), NPF, Amsterdam, Elsevier Science B.V., pp. 141–170.

Pessôa, R. H. C., 1985, Madeiras gimnospérmicas no Irati do Rio Grande do Sul: Observações sobre Anatomia, Sistemática e Paleoecologia, in: *Série Geologia*, Volume 27, n.2 (DNPM), Congresso Brasileiro de Paleontologia, 7, Salvador, 1983, Rio de Janeiro, pp. 623–639.

Peterson, L. N., 1999, *Atlas of Carboniferous Spores and Pollen of the Tychan Diamondiferous Region*, Krasnoyarsk, 80 pp. [in Russian].

Pfefferkorn, H. W., Fuchs, K., Hecht, C. A., Hofmann, C. H., Rabold, J. M., and Wagner, T., 1988, Recent geology and taphonomy of the Orinoco delta – overview and field observations, *Heidelberger Geowissenschaftlische Abhandlungen* **20**:21–56.

Philippe, M., Barale, G., Torres, T., and Covacevich, V., 1993, First study of *in situ* fossil woods from the Upper Cretaceous of Livington Island, South Shetland Islands, Antarctica: palaeoecological investigations, *Compte Rendu de l'Academie des Sciences, Paris* **317**(2):103–108.

Phillips, T. L., Pfefferkorn, H. W., and Peppers, R. A., 1973, Development of paleobotany in the Illinois Basin, *Illinois State Geological Survey Circular* **480**:1–86.

Phipps, D., and Playford, G., 1984, Laboratory techniques for extraction and concentration of palynomorphs from sediments, *Papers, Department of Geology, University of Queensland* **11**(1):1–23.

Piasecki, S., 1984, Preliminary palynostratigraphy of the Permian-Lower Triassic sediments in Jameson Land and Scorsesby Land, East Greenland, *Bulletin of the Geological Society of Denmark* **32**:139–144.

Pickford, M., Senut, B., Poupeau, G., Brown, F. H., and Haileab, B., 1991,Correlation of tephra layers from the Western Rift Valley (Uganda) to the Turkana Basin (Ethiopean/Kenya) and the Gulf of Aden, *Compte Rendu de l'Academie des Sciences, Paris* **313**(II):223–229.

Pirrie, D., 1998, Interpreting the record: facies analysis, in: *Unlocking the Stratigraphic Record: Advances in Modern Stratigraphy* (P. Doyle and M. R. Bennett, eds.), Chichester, John Wiley & Sons, pp. 395–420.

Pirrie, D., and Marshall, J. D., 1990, High-paleolatitude Late Cretaceous paleotemperatures: new data from James Ross Island, Antarctica, *Geology* **18**:31–34.

Plattner, G.-K., Joos, F., Stocker, T. F., and Marchal, O., 2001, Feedback mechanisms and sensitivities of ocean carbon uptake under global warming, *Tellus* **53B**:564–592.

Playford, G., 1978, Lower Carboniferous spores from the Drummond Basin, Queensland, *Palaeontographica Abteilung B* **167**:105–160.

Playford, G., 1985, Palynology of the Australian Lower Carboniferous: a review, *Compte Rendu 10th Congress of International Stratigraphists and Geologists Carb. Madrid* 1983 **4**:247–265.

Playford, G., 1991, Australian Lower Carboniferous miospores relevant to extra-Gondwanic correlations: an evaluation, in: *Intercontinental Correlation and Division of the Carboniferous System*, Volume 130 (P. L. Brenckle and W. L. Manger, eds.), *Courier Forschungsinstitut Senckenberg*, pp. 85–125 [imprinted 1990].

Playford, G., and Dettmann, M. E., 1996, Spores, in: *Palynology: Principles and Applications*, Volume 1, Chapter 8 (J. Jansonius and D. C. McGregor, eds.), Dallas, American Association of Stratigraphic Palynologists Foundation, pp. 227–260.

Playford, G., and Dino, R., 2000a, Palynostratigraphy of upper Palaeozoic strata (Tapajós Group), Amazonas Basin, Brazil: part one, *Palaeontographica Abteilung B* **255**:1–46.

Playford, G., and Dino, R., 2000b, Palynostratigraphy of upper Palaeozoic strata (Tapajós Group), Amazonas Basin, Brazil: part two, *Palaeontographica Abteilung B* **255**:87–145.

Playford, G., and Dring, R. S., 1981, Late Devonian acritarchs from the Carnarvon Basin, Western Australia, *Special Papers in Palaeontology* **27**:1–78.

Playford, G., and McGregor, D. C., 1993, Miospores and organic-walled microphytoplankton of Devonian-Carboniferous boundary beds (Bakken Formation), southern Saskatchewan: a systematic and stratigraphic appraisal, *Geological Survey of Canada Bulletin* **445**:1–107.

Plint, A. G., 1996, Recognition of marine and non-marine systems tracts in fourth order sequences in the mid-Cenomanian Dungevan Formation, northeastern British Columbia, Canada, in: *High Resolution Sequence Stratigraphy, Innovations and Applications*, Special Paper 104 (J. Howell and J. D. Aitken, eds.), London, Geological Society, pp.159–191.

Plint, A. G., Walker, R. G., and Duke, W. L., 1988, An outcrop of subsurface correlation of the Cardium Formation in Alberta, in: *Sequences, Stratigraphy, Sedimentology: Surface and Subsurface*, Memoir 15 (D. P. James and D. A. Leckie, eds.), Canadian Society of Petroleum Geologists, pp. 167–184.

Plotnick, R., Gardner, R., and O'Neill, R., 1993, Lacunarity indices as measures of landscape texture, *Landscape Ecology* **8**:201–211.

Plotnick, R. E., Gardner, R. H., Hargrove, W. H., Prestegaard, K., and Perlmutter, M. A., 1996, Lacunarity analysis: a general technique for the analysis of spatial patterns, *Physics Review E* **53**: 5461–5468.

Poag, C. W., and Commeau, J. A., 1995, Paleocene to middle Miocene planktic foraminifera of the southwestern Salisbury Embayment, Virginia and Maryland: Biostratigraphy, allostratigraphy, and sequence stratigraphy, *Journal of Foraminiferal Research* **25**(2):134–155.

Pole, M., 1994, An Eocene macroflora from the Taratu Formation at Livingstone, North Otago, New Zealand, *Australian Journal of Botany* **42**:341–367.

Pomerol, B., 1983, Geochemistry on the Late Cenomanian-Early Turonian chalks of the Paris Basin: manganese and carbon isotopes in carbonates as paleoceanographic indicators, *Cretaceous Research* **4**:85–93.

Poole, I., Hunt, R. J., and Cantrill, D. J., in press, A fossil flora from King George Island: ecological implications for an Antarctic Eocene vegetation, *Annals of Botany*.

Posamentier, H. W., and Allen, G. P., 1994, Siliciclastic sequence stratigraphy. Concepts and applications. AAPG Short Course.

Posamentier, H. W., and Vail, P. R., 1988, Eustatic controls on clastic deposition II – sequence and systems tract models, in: *Sea Level Changes – An Integrated Approach. Mineral*. Special Publication 42 (C. K. Wilgus, B. S. Hastings, C. G. St. C. Kendall, H. W. Posamentier, C. A. Ross, and J. C. Van Wagoner, eds.), *Society of Economic Paleontologists*, pp. 125–154.

Posamentier, H. W., Jervey, M. T., and Vail, P. R., 1988, Eustatic controls on clastic deposition I – conceptual framework, in: *Sea Level Changes – An Integrated Approach, Paleontol. Mineral.* Special Publication 42 (C. K. Wilgus, B. S. Hastings, C. G. St. C. Kendall, H. W. Posamentier, C. A. Ross, and J. C. Van Wagoner, eds.), *Society of Economic Paleontologists*, pp. 109–124.

Posamentier, H. W., Allen, G. P., James, D. P., and Tesson, M., 1992, Forced regressions in a sequence stratigraphic framework: concepts, examples, and exploration significance, *Bulletin of the American Association of Petroleum Geologists* **76**:1687–1709.

Pospichal, J. J., 1994, Calcareous nannofossils and the K/T boundary, El Kef: No evidence for stepwise, gradual, or sequential extinctions, *Geology* **22**:99–102.

Pospichal, J. J., and Wise, S. W., Jr., 1990, Paleocene to middle Eocene calcareous nannofossils of ODP Sites 689 and 960, Maud Rise, Weddell Sea, in: *Proceedings of the Ocean Drilling Program*, Volume 113 (P. F. Barker, J. P. Kennett *et al.*, eds.), Scientific Results, College Station, TX (Ocean Drilling Program), pp. 613–666.

Pöthe de Baldis, E. D., 1974, El microplancton del Devonico Medio de Paraguay, *Revista Española de Micropaleologia* **6**:367–379.

Pöthe de Baldis, E. D., 1978, Paleomicroplancton adicional del Devonico Inferior de Uruguay, *Revista Española de Micropaleologia* **9**:235–250.

Pöthe de Baldis, E. D., 1979, Acritarcos y quitinozoas del Devonico Superior de Paraguay, *Palinología* núm. extr. **1**:161–177.

Powell, A. J., 1992, Making the most of microfossils, *Geoscientist* **2**(1):12–16.

Pratt, L. M., Arthur, M. A., Dean, W. E., and Scholle, P. A., 1994 [for 1993], Paleo-oceanographic cycles and events during the Late Cretaceous in the western Interior seaway of North America, in: *Evolution of the Western Interior Basin*, Special Paper 39 (W. G. E. Caldwell and E. G. Kauffman, eds.), Geological Association of Canada, pp. 333–353.

Prauss, M., 1993, Sequence palynology – evidence from Mesozoic sections and conceptual framework, *Neues Jahrbuch für Geologie und Paläontologie, Abhandlungen* **190**:143–163.

Prell, W. L., and Hays, J. D., 1976, Late Pleistocene faunal and temperature patterns of the Colombia Basin, Caribbean Sea, in: *Investigations of Late Quaternary Paleoceanography and Paleoclimatology*, Memoir 145 (R. M. Cline and J. D. Hays, eds.), Geological Society of America, pp. 201–220.

Premoli-Silva, I., and Jenkins, D. G., 1988, Decision on the Eocene-Oligocene boundary, *Episodes* **16**:379–382.

Prentice, I. C., 1986, Vegetation responses to past climatic variation, *Vegetatio* **67**:131–141.

Price, P. L., 1997, Permian to Jurassic palynostratigraphic nomenclature of the Bowen and Surat Basins, in: *The Surat and Bowen Basins, South-East Queensland* (P. M. Green, ed.), Queensland Minerals & Energy Review Series, Queensland Department of Mines & Energy, Brisbane, pp. 137–178.

Price, P. L., Filatoff, J., Williams, A. J., Pickering, S. A., and Wood, G. R., 1985, Late Palaeozoic and Mesozoic palynostratigraphic units, C.S.R. Oil & Gas Division, *Palynology Facility, Report* **274/25**, 20 pp, Brisbane (unpublished).

Puri, G. S., Meher-Homji, V. M., Gupta, R. K., and Puri, S., 1983, Forest ecology, in: *Phytogeography and Forest Conservation*, Volume 1, Oxford and IBH, Oxford, 2nd ed.

Quadros, L. P., 1982, Distribuição bioestratigráfica dos Chitinozoa e Acritarchae na Bacia do Parnaíba, *Ciência e Técnica Petróleo (Seção: Exploração de Petróleo)* **12**:1–76.

Quadros, L. P., 1988, Zoneamento bioestratigráfico do Paleozóico inferior e médio (seção marinha) da Bacia do Solimões, *Boletin de la Geociências Petrobras* **2**:95–109.

Quadros, L. P., 1999, Silurian-Devonian acritarch assemblages from Paraná Basin: an update and correlation with northern Brazilian basins, in: *Ordovician-Devonian Palynostratigraphy in*

*Western Gondwana: Update, Problems and Perspectives* (M. A. C. Rodrigues and E. Pereira, eds.), Rio de Janeiro, UERJ, pp. 105–145.

Racey, A., Love, M. A., Bobolecki, R. M., and Walsh, J. N., 1995, The use of chemical element analyses in the study of biostratigraphically barren sequences: An example from the Triassic of the central North Sea (UKCS), in: *Non-biostratigraphical Methods of Dating and Correlation: London, United Kingdom*, Special Publication 89 (R. E. Dunay and E. A. Hailwood, eds.), London, Geological Society, pp. 69–105.

Racheboeuf, P. R., and Villeneuve, M., 1989, *Australostrophia senegalensis* n.sp.: first chonostrophiid brachiopod (Chonetacea) from NW Africa. Implications for the northwestern Gondwanaland margin, *Neues Jahrbuch für Geologie und Paläontologie Monatshefte*, **12**:737–748.

Racheboeuf, P. R., and Villeneuve, M., 1992, Une fauna Malvino-Cafre de brachiopodes Siluriens du Bassin Bove (Guinée, Ouest de l'Afrique), *Geologica et Palaeontologica* **26**:1–11.

Raddysh, H. K., 1988, Sedimentology and "geometry" of the Lower Cretaceous Viking Formation, Gilby A and B Fields, Albert, in: *Sequences, Stratigraphy, Sedimentology: Surface and Subsurface*, Memoir 15 (D. P. James and D. A. Leckie, eds.), Canadian Society of Petroleum Geologists, pp. 417–430.

Radovich, B. J., Powell, T., Lovell, M., and Mitchum, R. M., 1990, Sequence stratigraphy interpretation of the central shelf area, offshore Louisiana, Gulf Coast Section, Society of Economic Paleontologists and Mineralogists, Eleventh Annual Research Conference, *Program and Abstracts*, pp. 287–298.

Radovich, B. J., Hoffman, M. W., and Perlmutter, M. A., 1995, Upside-down sequence stratigraphy, sandy highstands, and muddy prograding complexes in the Surma basin, Bangladesh, *American Association of Petroleum Geologists Annual Meeting Program*, p. 79A.

Raffi, I., Rio, D., d'Afri, A., Fornaciari, E., and Rochetti, S., 1995, Quantitative distribution patterns and biomagnetostratigraphy of middle and late Miocene calcareous nannofossils from equatorial Indian and Pacific Oceans (legs 115, 130, and 138), in: *Proceedings of the Ocean Drilling Program*, Volume 138 (N. G. Pisias, L. A. Mayer, T. R. Janacek *et al.*, eds.), Scientific Results, College Station, TX (Ocean Drilling Program), pp. 479–502.

Raiswell, R., and Berner, R. A., 1985, Pyrite formation in euxinic and semi-euxinic sediments, *American Journal of Science* **285**:710–724.

Rampino, M. R., and Stothers, R. B., 1984, Terrestrial mass extinctions, cometary impacts and the Sun's motion perpendicular to the galactic plane, *Nature* **308**:709–711.

Rampino, M. R., and Stothers, R. B., 1988, Flood basalt volcanism during the past 250 million years, *Science* **241**:663–668.

Raunkiaer, C., 1934, The use of leaf size in biological plant geography, in: *The Life Forms of Plants and Statistical Plant Geography*, The Collected Papers. London, Clarendon Press, pp. 368–378.

Raup, D. M., and Jablonski, D., 1993, Geography of end-Cretaceous marine bivalve extinctions, *Science* **260**:971–973.

Raup, D. M., and Stanley, S. M., 1978, *Princípios de Paleontologia*, 1a ed., Barcelona, Espanha.

Raven, P. H., Evert, R. F., and Eichhorn, S. E., 1996, *Biologia Vegetal*, 5a ed., Rio de Janeiro, Guanabara, Ed. Koogan.

Ravn, R. L., 1986, Palynostratigraphy of the lower and middle Pennsylvanian coals of Iowa, *Iowa Geological Survey Technical Papers* **7**:1–245.

Raychaudhuri, I., Brekke, H. G., Pemberton, S. G., and MacEachern, J. A., 1992, Depositional facies and trace fossils of a low wave energy shoreface succession, Albian Viking

Formation, Chigwell Field, Alberta, Canada, in: *Applications of Ichnology to Petroleum Exploration*, Core Workshop 17 (S. G. Pemberton, ed.), Society of Economic Paleontologists and Mineralogists, pp. 319–337.

Reading, H. G., 1978, Facies, in: *Sedimentary Environments and Facies* (H. G. Reading, ed.), Oxford, Blackwell Scientific Publications, pp. 4–14.

Reading, H. G., 1996, *Sedimentary Environments: Processes, Facies and Stratigraphy*. Oxford, Blackwell Scientific Publications.

Reif, W. E.,1971, Zur Genese des Muschelkalk-Keuper Grenzbonebeds in Südwestdeutschland, *Neues Jahrbuch für Geologie und Paleontologie, Abhandlungen, Stuttgart*, **139**:369–404.

Reif, W. E., 1982, Muschelkalk/Keuper bone-beds (Middle Triassic, SW-Germany) – storm condensation in a regressive cycle, in: *Cyclic and Event Stratification* (G. Einsele and S. A. Seilacher, eds.), Berlin, Springer-Verlag, pp. 299–325.

Reissinger, A., 1939, Die "Pollenanalyse" ausgedehnt auf alle Sediment-gesteine der geologischen Vergangenheit. I. Teil, *Palaeontographica Abteilung B* **84**:1–19.

Remane, J. M., Kelller, G., Hardenbol, J., and Ben Haj Ali, M., 1999, International workshop on Cretaceous–Paleogene transitions in Tunisia: The El Kef stratotype for the Cretaceous–Paleogene boundary confirmed. Tunis, Tunisia, May 12–16, 1998, *Episodes* (March 1999) **22**(1):47–48.

Remmert, H., 1988, *Autoecologia, Ecologia de Populações e Estudo de Ecossistemas*, Barcelona, Editorial Blume.

Rich, F. J., 1989, A review of the taphonomy of plant remains in Lacustrine sediments, *Review of Palaeobotany and Palynology* **58**:33–46.

Richardson, J. B., and Ahmed, S., 1988, Miospore zonation and correlation of Upper Devonian sequences from western New York State and Pennsylvania, in: *Devonian of the World, Memoir* 14(3) (N. J. McMillan, A. F. Embry, and D. J. Glass, eds.), Canadian Society of Petroleum Geologists, pp. 541–558.

Richardson, J. B., and McGregor, D. C., 1986, Silurian and Devonian spore zones of the Old Red Sandstone Continent and adjacent regions, *Geological Survey of Canada Bulletin* **364**:1–79.

Richter, R., 1928, Aktuapaläontologie und Paläobiologie, eine Abgrenzung, *Senckenbergiana* **10**:285–292.

Riegel, W., 1996, The geological significance of the Late Paleozoic phytoplankton blackout, in: 9th International Palynology Conference, *Program and Abstracts*, Houston, pp. 133–134

Rigby, J. F., and Hekel, H., 1977, Palynology of the Permian sequence in the Springsure Anticline, central Queensland, *Geological Survey of Queensland Publication* 352, *Paleontology Papers* **37**:1–76.

Rigby, Jr., J. K., Newmann, K. R., Van der Kaars, J. S. S., Sloan, R. E., and Rigby, J. K., 1987, Dinosaurs from the Paleocene Part of the Hell Creek Formation, McCone County, Montana, *Palaios* **5**:296–302

Riley, N. J., 1993, Dinantian (Lower Carboniferous) biostratigraphy and chronostratigraphy in the British Isles, *Journal of the Geological Society of London* **150**:427–446.

Rio, D., Fornaciari, E., and Raffi, I., 1990, Late Oligocene through early Pleistocene calcareous nannofossils from western equatorial Indian Ocean (Leg 115), in: *Proceedings of Ocean Drilling Program*, Volume 115 (R. A. Duncan, J. Backman, L. C. Peterson *et al.,* eds.), Scientific Results, College Station, TX (Ocean Drilling Program), pp. 175–235.

Roberts, J., Claoué-Long, J. C., and Jones, P. J., 1995, Australian early Carboniferous time, in: *Geochronology Time Scales and Global Stratigraphic Correlation*, Special Publication No. 54 (W. A. Berggren, D. V. Kent, M. P. Aubry, and J. Hardenbol, eds.), Society of Economic Paleontologists and Mineralogists, pp. 23–40.

Robin, E., and Rocchia, R., 1998, Ni-rich spinel at the Cretaceous-Tertiary boundary of El Kef, Tunisia, *Bulletin de la Societé Geologique de France* **169**(3):365–372.

Robin, E., Boclet, D., Bonte, P., Froget, L., and Jehanno, C., 1991, The stratigraphic distribution of Ni-rich spinels in Cretaceous–Tertiary boundary rocks at El Kef (Tunisia), Caravaca (Spain) and Hole 761C (Leg 122), *Earth Planetary Science Letters* **107**:15–21.

Robinson, P. L., 1973, Paleoclimatology and continental drift, in: *Implications of Continental Drift to Earth Sciences*, Volume 1 (D. H. Tarling and S. K. Runcorn, eds.), London and New York, Academic Press, pp. 451–476.

Rodrigues, R., 1973, Aplicação da geoquímica orgânica ao problema de geração de hidrocarbonetos na Bacia Amazônica, in: *Anais do XXVII Congresso Brasileiro de Geologia*, Aracaju, Sergipe, **3**:53–66.

Rodrigues, R., 1995, A geoquímica orgânica na Bacia do Parnaíba, Tese (Doutorado em Ciências), Instituto de Geociências, Universidade Federal do Rio Grande do Sul, Porto Alegre, p. 225 (unpublished).

Rodrigues, R., 2001, Caracterização geoquímica do evento anóxico-disóxico do Frasniano nas bacias do Amazonas e Parnaíba, in: *Correlação de Seqüências Paleozóicas Sul-Americanas*, (J. H. G. Melo and G. J. S. Terra, eds.), Ciência-Técnica-Petróleo (Seção: Exploração de Petróleo) **20**, pp. 109–115. (CD Rom, ISSN-0100-1515.)

Rodrigues, R., Loboziak, S., Melo, J. H. G., and Alves, D. B., 1995, Geochemical characterization and miospore biochronostratigraphy of the Frasnian anoxic event in the Parnaíba Basin, northeast Brazil, *Bulletin des Centres de Recherches Exploration-Production Elf Aquitaine* **19**(2):319–327.

Rogers, R. R., Swisher, C. C. I., Sereno, P. C., Monetta, A. M., Forster, C. A., and Martínez, R. N., 1993, The Ischigualasto tetrapod assemblage (Late Triassic, Argentina) and 40Ar/39Ar dating of dinosaur origins, *Science* **260**:794–797.

Röhl, U., Bralower, T. J., Norris, R. D., and Wefer, G., 2000, New chronology for the late Paleocene thermal maximum and its environmental implications, *Geology* **28**:927–930.

Romer, A. S., 1966, The Chañares (Argentina) Triassic Reptile fauna. I. Introduction, *Breviora* **247**:1–14.

Romer, A. S., 1975, Intercontinental correlations of Triassic Gondwana vertebrate faunas, in: *Gondwana Geology* (K. S. N. Campbell, ed.), Canberra, Australian National University Press, pp. 469–473.

Romero, E. J., 1978, Paleoecologia y paleofitogeografia de las tafofloras del Cenofítico de Argentina y areas vecinas, *Ameghiniana* **15**(1–2):209–227.

Rong Jia-yu, Boucot, A. J., Su Yang-zheng, and Strusz, D. L., 1995, Biogeographical analysis of late Silurian brachiopod faunas from Asia and Australia, *Lethaia* **28**:39–60.

Ross, C. A., and Ross, J. R. P., 1988, Late Paleozoic transgressive–regressive deposition, in: *Sea-Level Changes: An Integrated Approach*, Special Publication 42 (C. K. Wilgus, B. S. Hastings, C. G. St. C. Kendall, H. W. Posamentier, C. A. Ross, and J. C. Van Wagoner, eds.), *Society of Economic Paleontologists and Mineralogists*, pp. 227–247.

Ross, C. A., Baud, A., and Menning, M., 1994, Project Pangea Timescale, in: *Paleoclimate, Tectonics and Sedimentation During Accretion, Zenith and Breakup of a Supercontinent* (G. D. Klein, ed.), Geological Society of America Special Papers, p. 288.

Ross, I. R., 1992, *Paleogeographic Information System Manual*, Earth in Motion Tech.

Ross, J. R. P., and Ross, C. A., 1997, Nealian and Lenoxian (Wolfcampian, Lower Permian) depositional sequence, fusulinid facies and biostratigraphy, Glass Mountains, West Texas, in: *Late Paleozoic Foraminifera, their Biostratigraphy, Evolution and Paleoecology and the*

*Mid-Carboniferous Boundary*, Special Publication 36 (C. A. Ross, J. R. P. Ross, and P. L. Brenckle, eds.), Cushman Foundation for Foraminiferal Research, pp. 125–128.

Royer, D. L., 2001, Stomatal density and stomatal index as indicators of paleoatmospheric $CO_2$ concentration, *Review of Palaeobotany and Palynology* **114**(1–2):1–28.

Rozman, K. S., 1999, Silurian brachiopod communities of Mongolia, in: *Paleocommunities: A Case Study from the Silurian and Lower Devonian* (A. J. Boucot and J. D. Lawson, eds.), Cambridge, Cambridge University Press, pp. 164–176.

Rubinstein, C., and Steemans, P., 2002, Miospore assemblages from the Silurian-Devonian boundary, in borehole A1–61, Ghadamis Basin, Libya, *Review of Palaeobotany and Palynology* **118**:397–421.

Rubinstein, C., Melo, J. H. G., and Steemans, P., (in press), Lochkovian (earliest Devonian) miospores from the Solimões Basin, northern Brazil, *Review of Palaeobotany and Palynology*.

Rull, V., 2000, Ecostratigraphic study of Paleocene and early Eocene palynological cyclicity in northern South America, *Palaios* **15**(1):14–23.

Russell, D. A., 1984, The gradual decline of the dinosaurs – fact or fallacy? *Nature* **307**:360–361.

Russell, D. A., 1996, The significance of the extinction of the dinosaurs, in: *The Cretaceous-Tertiary Event and Other Catastrophes in Earth History*, Special Paper 307 (G. Ryder, D. Fastovsky, and S. Gartner, eds.), Boulder, Geological Society of America, pp. 381–388.

Rutten, M. G., 1959, Paleomagnetic reconnaissance of mi-Italian volcanoes, *Geologie en Mijnbow* **21**:373–374.

Sadler, P., 1981, Sediment accumulation rates and the completeness of stratigraphic sections, *Journal of Geology* **89**:569–584.

Sadler, P. M., and Strauss, D. J., 1990, Estimation of completeness of stratigraphic sections using empirical data and theoretical models, *Journal of the Geological Society of London* **147**:471–485.

Sagemann, B., 1992, Paleoecology – a cure for sequence syndrome? *Palaios* **7**(5):485–486.

Salvador, A. (ed.), 1994, *International Stratigraphical Guide*, Trondheim, Norway, International Union of Geological Sciences, and Boulder, Geological Society of America.

Sanfilippo, A., Westberg-Smith, M. J., and Riedel, W. R., 1985, Cenozoic Radiolaria, in: *Plankton Stratigraphy* (H. M. Bolli, K. Perch-Nielsen, and J. B. Saunders, eds.), Cambridge, Cambridge University Press, pp. 631–712.

Santos-Neto, E. V., 1993, Caracterização geoquímica e paleoambiente deposicional da seqüência carbonato-pelítica superior do Membro Assistência, Formação Irati no Estado de São Paulo, Bacia do Paraná, Dissertação (Mestrado em Ciências), Rio de Janeiro, Instituto de Geociências, Universidade Federal do Rio de Janeiro, 203 pp.

Sarkis, M. F. R., 2002, Palinoestratigrafia com base em dinoflagelados do limite Cretáceo-Terciário, pedreira Poty, Bacia de Pernambuco-Paraíba, nordeste do Brasil, Tese (Doutorado), Universidade Federal do Rio de Janeiro, Instituto de Geociências, Rio de Janeiro, RJ, Brazil, 253 pp. (unpublished).

Sarkis, M. F. R., Arai, M., and Koutsoukos, E. A. M., 2002, Dinoflagelados do limite Cretáceo-Terciário (K-T), Pedreira Poty, Bacia de Pernambuco-Paraíba, nordeste do Brasil, *Boletim do VI Simpósio sobre o Cretáceo do Brasil*, UNESP-Rio Claro, pp. 271–277.

Sarna-Wojcicki, A. M., and Davis, J. O., 1991, Quaternary tephrochronology, in: *Quaternary Nonglacial Geology of the Conterminous U.S.*, Special Paper (R. B. Morrison, ed.), Geological Society of America, pp. 93–116.

Satterley, A. K., 1996, The Milankovitch theory, *Earth-Science Reviews* **40**:181–207.

Savrda, C. E., 1991, *Teredolites*, wood substrates, and sea-level dynamics, *Geology* **19**: 905–908.

Savrda, C. E., 1995, Ichnologic applications in paleoceanographic, paleoclimatic, and sea-level studies, *Palaios* **10**(6):565–577.

Savrda, C. E., Browning, J. V., Krawinkel, H., and Hesselbo, S. P., 2001a, Firmground ichnofabrics in deep-water sequence stratigraphy, Tertiary clinoform-toe deposits, New Jersey slope, *Palaios* **16**(3):294–305.

Savrda, C. E., Krawinkel, H., McCarthy, F. M. G., McHugh, C. M. G., Olson, H. C., and Mountain, G., 2001b, Ichnofabrics of a Pleistocene slope sequence, New Jersey margin: Relations to climate and sea-level dynamics, *Palaeogeography, Palaeoclimatology, Palaeoecology* **171**(1–2):41–61.

Sawyer, M. J., and Keegan, J. B., 1996, Use of palynofacies characterization in sand-dominated sequences, Brent Group, Ninian Field, UK North Sea, *Petroleum Geoscience* **2**:289–297.

Scasso, R. A., Concheyro, A., Kiessling, W., Aberhan, M., Medina, F. A. and Tagle, R., in press. A tsunami deposit at the Cretaceous–Tertiary boundary in the Neuquén Basin of Argentina, *Cretaceous Research*.

Schaefer, W., 1962, *Aktuopalaeontologie nach Studien in der Nordsee*, Frankfurt, Waldemar Kramer.

Scheihing, M. H., and Pfefferkorn, H. W., 1984, The taphonomy of land plants in the Orinoco Delta: a model for the incorporation of plant parts in clastic sediments of Late Carboniferous age of Euramerica, *Review of Palaeobotany and Palynology* **41**:205–240.

Schioler, P., Brinkhuis, H., Roncaglia, L., and Wilson, G. J., 1997, Dinoflagellate biostratigraphy and sequence stratigraphy of type Maastrichtian (Upper Cretaceous), ENCI Quarry, The Netherlands, *Marine Micropaleontology* **31**:65–95.

Schlager, W., 1992, Sedimentology and sequence stratigraphy of reefs and carbonate platforms, Continuing Education Course Notes Series. *American Association of Petroleum Geologists* **34**.

Schlanger, S. O., and Jenkyns, H. C., 1976, Cretaceous oceanic anoxic events: causes and consequences, *Geologie en Mijnbouw* **55**:179–184.

Schmitt, R. A., 1989, Cometesimal explosions in Earth's atmosphere (CEEA) and steep angle cometesimal cratering (SACC) events caused Cretaceous–Tertiary (K/T) extinction phenomena, *Abstracts of Lunar and Planetary Science Conference* **20**:962–963.

Schmitz, B., 1987, Baryum equatorial high productivity and the northward wandering of the Indian continent, *Paleoceanography* **2**:67–78.

Schmitz, B., Keller, G., and Stenvall, O., 1992, Stable isotope and foraminiferal changes across the Cretaceous–Tertiary boundary at Stevns Klint, Denmark: Arguments for long-term oceanic instability before and after bolide-impact event, *Palaeogeography, Palaeoclimatology, Palaeoecology* **96**:233–260.

Schneider, D., 1995, Paleomagnetism of some ODP Leg 138 sediments: Miocene magnetostratigraphic miscellany, in: *Proceedings of the Ocean Drilling Program*, Volume 138 (N. G. Pisias, L. A. Mayer, T. R. Janacek *et al.*, eds.), Scientific Results, College Station, TX (Ocean Drilling Program), pp. 59–72.

Schneider, D., Backman, J., Chaisson, W. P., and Raffi, I., 1997, Miocene calibration for calcareous nannofossils from the low-latitude Ocean Drilling Program sites and the Jamaican conundrum, *Geological Society of America Bulletin* **109**:1073–1079.

Schneider, R. L., Mühlmann, H., Tommasi, E., Medeiros, R. A., Daemon, R. F., and Nogueira, A. A., 1974, Revisão Estratigráfica da Bacia do Paraná, in: *Anais*... Volume 1, (SBG) Congresso Brasileiro de Geologia, 28, Porto Alegre, 1974, Porto Alegre, RS, pp. 41–65.

Scholle, P. A., and Arthur, M. A., 1980, Carbon isotope fluctuations in Cretaceous pelagic limestones: potential stratigraphic and petroleum exploration tool, *Bulletin of the American Association of Petroleum Geologists* **64**:67–87.

Schopf, J. M., Wilson, L. R., and Bentall, R., 1944, An annotated synopsis of Paleozoic fossil spores and the definition of generic groups, *Illinois State Geological Survey Report of Investigations* **91**:1–73.

Schultz, C. L., 1995, Subdivisão do Triássico do Rio Grande do Sul com base em macrofósseis: problemas e perspectivas, *Comunicações do Museu de Ciências e Tecnologia, UBEA/PUCRS, Série Ciências da Terra* **1**:25–32.

Schultz, C. L., Scherer, C. M. S., and Barberena, M. C., 1994, Uma nova proposta de zoneamento estratigráfico para o Triássico Superior sul-riograndense, in: *Anais*, Volume 1, Congresso Brasileiro de Geologia, 38, Balneário de Camboriú, pp. 107–108.

Schultz, C. L., Scherer, C. M. S., and Barberena, M. C., 2000, Biostratigraphy of southern Brazilian Middle–Upper Triassic, *Revista Brasileira de Geociências* **30**:491–494.

Schumm, S. A., 1977, *The Fluvial System*, New York, Wiley.

Schumm, S. A., and Brackenridge, G. R., 1987, River responses, in: *North America and Adjacent Oceans during the Last Deglaciation*, Volume K-3 (W. F. Ruddiman and H. E. Jr., Wright, eds.), *The Geology of North America*, pp. 221–240.

Schwartz, R. D., and James, P. D., 1984, Periodic mass extinction and the Sun's oscillation about the galactic plane, *Nature* **308**:712–713.

Schwarzacher, W., 1975, Sedimentation models and quantitative stratigraphy, in: *Developments in Sedimentology*, Volume 19, Amsterdam, Elsevier, p. 382.

Schwarzacher, W., 1987, The analysis and interpretation of stratification cycles. *Paleoceanography* **2**(1):79–95.

Schwarzacher, W., 1993, Cyclostratigraphy and the Milankovitch Theory, in: *Developments in Sedimentology*, Volume 52, Amsterdam, Elsevier, p. 225.

Scotese, C. R., Bambach, R. K., Barton, C., Voo, C. V. D., and Ziegler, A., 1979, Paleozoic base maps, *Journal of Geology* **87**(3):217–277.

Scotese, C. R., Mckerrow, W. S., Muller, R. D., Pindell, J. L., Royer, J. Y., Ross, M. I., Rowley, D. B., and Zonenshain, L., 1992, *Atlas of Phanerozoic Plate Reconstructions, International Lithosphere Project*, American Geophysical Union.

Scott, A. C., and Hemsley, A. R., 1996, Palaeozoic megaspores, in: *Palynology: Principles and Applications*, Volume 2, Chapter 18G (J. Jansonius and D. C. McGregor, eds.), Dallas, American Association of Stratigraphic Palynologists Foundation, pp. 629–639.

Scurfield, G., 1979, Wood petrification: An aspect of biomineralogy, *Australian Journal of Botany* **27**:377–390.

Scurfield, G., and Segnit, E. R., 1984, Petrifaction of wood by silica minerals, *Sedimentary Geology* **39**:149–167.

Scurfield, G., Segnit, E. R., and Anderson, C. A., 1974, Silicification of wood, in: *Proceedings, Seventh Annual Scanning Electron Microscopy Symposium*, pp. 389–395.

Seeley, H. G., 1892, Researches on the structures, organisation and classification of the fossil Reptilia, *Philosophical Transactions of the Royal Society of London* **182**:311–370.

Seilacher, A., 1959, Fossilien als Strömungsangezeiger, *Aus der Heimat, Öhringen (Württemberg)* **67**:170–177.

Seilacher, A., 1973, Biostratinomy: the sedimentology of biologically standardized particles, in: *Evolving Concepts in Sedimentology* (R. N. Ginsburg, ed.), pp. 159–177.

Sennikov, A. G., 1995, Evolution of the Permian and Triassic tetrapod communities of Eastern Europe, *Palaeogeography, Palaeoclimatology, Palaeoecology* **120**:331–351.

Sepkoski, J. J., Jr., 1990, The taxonomic structure of periodic extinction, in: *Global Catastrophes in Earth History* (V. L. Sharpton and P. D. Ward, eds.), Geological Society of America, Special Paper 247, pp. 33–44.

Sepkoski, J. J., Jr., 1994, Extinction and the fossil record, *Geotimes* March:15–17.

Servais, T., Brocke, R., Fatka, O., Le Hérissé, A., and Molyeneux, S. G., 1997, Value and meaning of the term acritarch, in: *Acritarcha in Praha*, Acta Universitatis Carolinae Geologica, 40 (O. Fatka and T. Servais, eds.), pp. 631–643.

Shackleton, N. J., 1967, Oxygen isotope analyses and paleotemperature reassessed, *Nature* **215**:15–17.

Shackleton, N. J., 1985, Oceanic carbon isotope constraints on oxygen and carbon dioxide in the Cenozoic atmosphere, in: *The Carbon Cycle and Atmospheric $CO_2$, Natural Variations Archean to Present*, Monograph 32 (E. T. Sunquist and W. S. Broecker, eds.), Washington, American Geophysical Union, pp. 412–417.

Shackleton, N. J., 1987, The carbon isotope record of the Cenozoic: history of organic carbon burial and of oxygen in the ocean and atmosphere, marine petroleum source rocks. Geological Society Special Publication 26:423–434.

Shackleton, N. J., 2000, The 100,000-year ice-age cycle identified and found to lag temperature, carbon dioxide, and orbital eccentricity, *Science* **289**:1897–1902.

Shackleton, N. J., and Kennet, J. P., 1975, Paleotemperature history of the Cenozoic and the initiation of Antarctic glaciation: oxygen and carbon isotope analysis in DSDP 277, 279 and 281, in: *Initial Reports of the Deep Sea Drilling Project* **29**:743–755.

Shackleton, N. J., Hall, M. A., and Bleil, U., 1985, Carbon-isotope stratigraphy, Site 577, *Initial Reports of the Deep Sea Drilling Project. U.S. Government Printing Office* **86**:503–511.

Shackleton, N. J., Crowhurst, S., Hagelberg, T., Pisias, N. G., and Schneider, D. A., 1995, A new Late Neogene time scale: Application to Leg 138 Sites, in: *Proceedings of the Ocean Drilling Program*, Volume 138 (N. G. Pisias, L. A. Mayer, T. R. Janacek *et al.*, eds.), Scientific Results, College Station, TX (Ocean Drilling Program), pp. 73–101.

Shackleton, N. J., Crowhurst, S. J., Weedon, G. P., and Laskar, J., 1999, Astronomical calibration of Oligocene-Miocene time, *Philosophical Transactions of the Royal Society of London* **357**:1907–1929.

Shackleton, N. J., Hall, M. A., Raffi, I., Tauxe, L., and Zachos, J., 2000, Astronomical calibration age for the Oligocene-Miocene boundary, *Geology* **28**:447–450.

Shafik, S., 1975, Nannofossil biostratigraphy of the southwest Pacific Deep Sea Drilling Project, Leg 30, in: *Initial Reports of the Deep Sea Drilling Project*, Volume 30 (J. E. Andrews, G. Parkham *et al.*, eds.) Washington, DC, Government Printing Office, pp. 549–598.

Shanley, K. W., and McCabe, P. J., 1994, Perspectives on the sequence stratigraphy of continental strata: report of a working group at the 1991 NUNA Conference on high resolution sequence stratigraphy, *American Association of Petroleum Geologists Bulletin* **74**: 544–568.

Shanley, K. W., and McCabe, P. J., eds., 1998, *Relative Role of Eustasy, Climate and Tectonism in Continental Rocks*, SEPM Special Publication **59**.

Sharpton, V. L., Dalrymple, G. B., Marin, L. E., Ryder, G., Schuraytz, B. C., and Urrutia-Fucugauchi, J., 1992, New links between the Chicxulub impact structure and the Cretaceous/Tertiary boundary, *Nature* **359**(10):819–821.

Sharpton, V. L., Burke, K., Camargo-Zanoguera, A., Hall, S. A., Lee, D. S., Martin, L. E., Suarez-Reynoso, G., Quezada-Mueton, J. M., Spudis, P. D., and Urrutia-Fucugauchi, J., 1993, Chicxulub multiring impact basin: size and other characteristics derived from gravity analysis, *Science* **261**:1564–1567.

Shaw, A. B., 1964, *Time in Stratigraphy*, New York, McGraw-Hill.

Sheehan, P. M., and Fastovsky, D. E., 1992, Major extinctions of land-dwelling vertebrates at the Cretaceous–Tertiary boundary, eastern Montana, *Geology* **20**:556–560.

Sheehan, P. M., Fastovsky, D. E., Hoffmann, R. G., Berghaus, C. B., and Gabriel, D. L., 1991, Sudden extinction of the Dinosaurs: Latest Cretaceous, Upper Great Plains, U.S.A., *Science* **254**:835–839.

Shen Yanbin, 1994, Subdivision and correlation of Cretaceous to Paleogene volcano-sedimentary sequence from Fildes Peninsula, King George Island, Antarctica, in: *Stratigraphy and Palaeontology of Fildes Peninsula, King George Island, Antarctica*, Volume 3 (Shen Yanbin, ed.), Beijing, China, pp. 1–36.

Shepard, F. P., and Moore, D. G., 1954, Sedimentary environments differentiated by coarse fraction studies, *Bulletin of the American Association of Petroleum Geologists* **38**:1792–1802.

Shipp, D. J., 1999, Well-site biostratigraphy of Danish horizontal wells, in: *Biostratigraphy in Production and Development Geology*, Special Publication No. 152 (R. W. Jones and M. D. Simmons, eds.), London, Geological Society, pp. 75–84.

Shipp, D. J., and Marshall, P. R., 1995, Biostratigraphic steering of horizontal wells, in: *GEO'94* (M. I. Al-Husseini, ed.), Selected Middle East papers from the Middle East Geoscience Conference, Bahrain, 1994, Manama, Bahrain, Gulf Petro Link, pp. 849–860.

Shishkin, M. A., Rubidge, B. S., and Hancox, P. J., 1995, Vertebrate biozonation of the Upper Beaufort Series of South Africa – a new look on correlation of the Triassic biotic events in Euroamerica and southern Gondwana, in: *Short Papers*, Sixth Symposium on Mesozoic Terrestrial Ecosystems and Biota (A. Sun and I. Wang, eds.), Beijing, China Ocean Press, pp. 39–41.

Shubin, N. H., Sues, H-D., 1991, Biogeography of Early Mesozoic continental tetrapods: patterns and implications, *Paleontology* **17**(3):214–230.

Signor, P. W., and Lipps, J. H., 1982, Sampling bias, gradual extinction patterns and catastrophes in the fossil record, in: *Geological Implications of Impacts of Large Asteroids and Comets on the Earth*, Special Publication, Geological Society of America, 190 (L. T. Silver and P. H. Schultz, eds.), pp. 291–296.

Sigurdsson, H., D'Hondt, S., Arthur, M. A., Bralower, T. J., Zachos, J. C., Van Fossen, M., and Channell, J. E. T., 1991, Glass from the Cretaceous/Tertiary boundary in Haiti, *Nature* **349**:482–487.

Sikora, P. J., Bergen, J. A., and Farmer, C. L., 1999, Chalk palaeoenvironments and depositional model, Valhall-Hod fields, southern Norwegian North Sea, in: *Biostratigraphy in Production and Development Geology*, Special Publication No. 152 (R. W. Jones and M. D. Simmons, eds.), London, Geological Society, pp. 113–138.

Simmons, M., 1998, Biostratigraphy surviving extinction, *Palaios* **13**(3):215–216.

Simões, M. G., Mello, L. H. C., Rodrigues, S. C., Leme, J. M., and Marques, A. C., 2000, Conulariid taphonomy as a tool in paleoenvironmental analysis, *Revista Brasileira de Geociências* **30**:757–762.

Simões, M. G., Ghilardi, R., Sales, A., and Holz, M., 2002, Sequence stratigraphic controls on the pattern of vertical distribution of taphonomic signatures: the Devonian of the Parana Basin, Brazil, in: Gondwana 11 International Conference, Christchurch, New Zealand, *Programme and Abstracts*.

Simpson, G. G., 1971. Clasificación, terminología y nomenclatura provinciales para el Cenozoico mamalífero, *Revista de la Asociación Geológica Argentina, Buenos Aires*. **26(3)**:281–297.

Sloan, R. E., 1987, Paleocene and latest Cretaceous mammal ages, biozones, magnetozones, rates of sedimentation, and evolution, in: *The Cretaceous–Tertiary Boundary in the San Juan and Raton Basins, New Mexico and Colorado*, Special Paper 209 (J. E. Fassett and J. K. Rigby Jr., eds.), Boulder, Geological Society of America, pp. 165–200.

Sloan, R. E., Rigby, J. K., Jr., Van Valen, L. M., and Gabriel, D., 1986, Gradual dinosaur extinction and simultaneous ungulate radiation in the Hell Creek Formation, *Science* **232**:629–633.

Sloss, L. L., 1962, Stratigraphic models in exploration, *Bulletin of the American Association of Petroleum Geologists* **46**:1050–1057.

Sloss, L. L., 1963, Sequences in the cratonic interior of North America, *Geological Society of America Bulletin* **74**(2):94–114.

Sloss, L. L., Krumbein, W. C., and Dapples, E. C., 1949, Integrated facies analysis, in: *Sedimentary Facies in Geologic History*, Memoir 39 (C. R. Longwell, ed.), Geological Society of America, pp. 91–124.

Smit, J., 1982, Extinction and evolution of planktonic foraminifera after a major impact at the Cretaceous/Tertiary boundary, in: *Geological Implications of Impacts of Large Asteroids and Comets on the Earth*, Special Paper 190 (L. T. Silver and P. H. Schultz, eds.), Geological Society of America, pp. 329–352.

Smit, J., 1990, Meteorite impact, extinctions and the Cretaceous–Tertiary boundary, *Geologie en Mijnbouw* **69**:187–204.

Smit, J., Montanari, A., Swinburne, N. H. M., Alvarez, W., Hilde-Brand, A. R., Margolis, S. V., Claeys, P., Lowriw, W., and Asaro, F., 1992, Tektite-bearing, deep-water clastic unit at the Cretaceous/Tertiary boundary in northeastern Mexico, *Geology* **20**:99–103.

Smit, J., Keller, G., Zargouni, F., Razgallah, S., Shimi, M., Ben Abdelkader, O., Ben Haj Ali, N., and Ben Salem, H., 1997, The El Kef sections and sampling procedures, *Marine Micropaleontology* **29**(2):69–72.

Smith, A., 1984, *Echinoid Palaeobiology*, London, Allen and Unwin.

Smith, A. B., and Jeffery, C. H., 1998, Selectivity of extinction among sea urchins at the end of the Cretaceous period, *Nature* **392**:69–71.

Smith, A. B., and Jeffery, C. H., 2000, Changes in the diversity, taxic composition and life-history patterns of echinoids over the past 145 million years, in: *Biotic Response to Global Change: The Last 145 Million Years* (S. J. Culver and P. F. Rawson, eds.), Cambridge, Cambridge University Press, pp. 181–194.

Smith, A. H. V., and Butterworth, M. A., 1967, Miospores in the coal seams of the Carboniferous of Great Britain, *Special Papers in Palaeontology* **1**:1–324.

Smith, R. H., 1993, Vertebrate taphonomy of Late Permian floodplain deposits in the south-western Karoo Basin of South America, *Palaios* **8**:45–67.

Smith, R. M. H., Mason, T. R., and Ward, L. F., 1993, Flash flood sediments and ichnofacies of the Late Pleistocene Homeb Silts, Kuiseb River, Namibia, *Sedimentary Geology* **85**:579–599.

Soares, M. B., 2003, The Early Late Permian Irati Formation: *mesossaurus* taphonomy and paleoecology, *Acta Geologica Hispanica* **4**:349–361.

Soliani, E., Jr., Kawashita, K., Fensterseifer, H. C., Hansen, M. A. F., and Troian, F. L., 1988, K-Ar Ages of the Winkel Point Formation (Fildes Peninsula Group) and Associated Intrusions, King George Island, South Shetland Islands, Antarctica, *Série Científica Instituto Antártico Chileno* **38**:133–139.

Sommer, F. W., 1953, Os megásporos do carvão de Santa Catarina e seu aproveitamento na correlação de camadas, *Divisão de Geologia e Mineralogia, Departamento Nacional da Produção Mineral, Boletim* **190**:1–224

Souza Cruz, C. E., Miranda, A. P., and Oller, J., 2000, Facies analysis and depositional systems of the Late Silurian–Devonian Subandean Basin, southern Bolivia and northern Argentina, in: *14th Congress on Geology, Bolivia*, La Paz, Memorias, pp. 85–90.

Speijer, R. P., and Van der Zwaan, G. J., 1994, Extinction and recovery patterns in benthic foraminiferal paleocommunities across the Cretaceous/Paleogene and Paleocene/Eocene boundaries, *Geologica Ultraiectina* **124**:19–64.

Speijer, R. P., and Van der Zwaan, G. J., 1996, Extinction and survivorship of southern Tethyan benthic foraminifera across the Cretaceous/Palaeogene boundary, in: *Biotic Recovery from Mass Extinction Events*, Special Publication 102 (M. B. Hart, ed.), London, Geological Society, pp. 343–371.

Spencer-Cervato, C., Thierstein, H. R., Lazarus, D. B., and Beckman, J.-P., 1994, How synchronous are Neogene marine plankton events? *Paleoceanography* **9**:739–763.

Speyer, S. E., and Brett, C. E., 1986, Trilobite taphonomy and Middle Devonian taphofacies, *Palaios* **1**:312–327.

Speyer, S. E., and Brett, C. E., 1988, Taphofacies models for epeiric sea environments: Middle Paleozoic examples, *Palaeogeography, Palaeoclimatology, Palaeoecology* **63**:225–262.

Speyer, S. E., and Brett, C. E., 1991, Taphofacies controls: background and episodic processess in fossil assemblage preservation, in: *Taphonomy: Releasing the Data Locked in the Fossil Record* (P. A. Allison and D. E. G. Briggs, eds.), Chapter 11, New York, Plenum Press, *Topics in Geobiology* **9**, pp. 501–554.

Spicer, R. A., 1980, The importance of depositional sorting to the biostratigraphy of plant megafossils, in: *Biostratigraphy of Fossil Plants, Successional and Paleoecological Analyses* (D. L. Dilcher and T. N. Taylor, eds.), Pennsylvania, Dowden, Hutchinson and Ross, pp. 171–183.

Spicer, R. A., 1989, Plants at the Cretaceous–Tertiary boundary, *Philosophical Transactions of the Royal Society London* **B 325**:291–305.

Spiess, V., 1990, Cenozoic magnetostratigraphy of Leg 113 Drill Sites, Maud Rise, Weddell Sea, Antarctica, in: *Proceedings of the Ocean Drilling Program*, Volume 113 (P. F. Barker, J. P. Kennett *et al.*, eds.), College Station, TX (Ocean Drilling Program), pp. 261–315.

Sprovieri, R., Bonomo, S., Caruso, A., Stefano, A., De Stefano, E., Foresi, L. M., Iaccarino, S., Lirer, F., Mazzei, R., and Salvatorini, G., 2002, Integrated calcareous plankton biostratigraphy and biochronology of the Mediterranean middle Miocene, *Rivista Italiana di Paleontologia e Stratigrafia* **108**:337–353.

Srinivasan, M. S., and Kennett, J. P., 1981, Neogene planktonic foraminiferal biostratigraphy and evolution: equatorial to subantarctic, South Pacific, *Marine Micropaleontology* **6**:499–533.

Stanley, S. M., 1984, Marine mass extinction, in: *Extinctions* (M. H. Nitecki, ed.), Chicago, University of Chicago Press, pp. 69–117.

Stanley, S. M., 1989, *Earth and Life Through Time*, New York, W. H. Freeman and Co.

Steemans, P., 1989, Palynostratigraphie de l'Eodévonien dans l'Ouest de l'Europe, *Mémoires pour servir à l'Explication des Cartes Géologiques et Minières de la Belgique, Mémoire* **27**:1–453.

Steemans, P., 1995, Silurian and lower Emsian spores in Saudi Arabia, *Review of Palaeobotany and Palynology* **89**:91–104.

Steemans, P., 1999, Paléodiversification des spores et des cryptospores de l'Ordovicien au Dévonien inférieur, *Geobios* **32**(2):341–352.

Steffen, D., and Gorin, G. E., 1993, Palynofacies of the Upper Tithonian–Berriasian deep-sea carbonates in the Vocontian Trough (SE France), *Bulletin des Centres Recherches et Exploration–Production Elf Aquitaine* **17**:235–247.

Stinnesbeck, W., Keller, G., Adatte, T., Stüben, D., Kramar, U., Berner, Z., Desremeaux, C., and Molière, E., 1999. Beloc, Haiti, revisited: multiple events across the KT boundary in the Caribbean, *Terra Nova* **11**:303–310.

Stipanicic, P. N., and Marsicano, C. A. (eds.), 2002, *Léxico Estratigráfico de la Argentina, Volumen VIII – Triásico*, Buenos Aires.

Stipanicic, P. N., and Riccardi, A. C., 2002, El Triásico en la Argentina – Introducción – 1. Escalas cronoestratigráfica y geocronológica, in: *Léxico Estratigráfico de la Argentina, Volume VIII, Triásico* (P. N. Stipanicic and C. A. Marsicano, eds.), Buenos Aires, pp. 1–4.

Stockmarr, J., 1971, Tablets with spores used in absolute pollen analysis, *Pollen et Spores* **13**:615–621.

Stoll, H. M., and Schrag, D. P., 2000, High-resolution stable isotope records from the Upper Cretaceous rocks of Italy and Spain: glacial episodes in a greenhouse planet? *Geological Society of America Bulletin* **112**:308–319.

Stott, L. D., Sinha, A., Thiry, M., Aubry, M.-P., and Berggren, W. A., 1996, Global $\delta^{13}C$ changes across the Paleocene–Eocene boundary: Criteria for terrestrial-marine correlations, in: *Correlation of the Early Paleogene in Northwest Europe*, Special Publication No. 101 (R. W. O'B. Knox, R. M. Corfield, and R. E. Dunay, eds.), London, Geological Society, pp. 381–399.

Strahler, A. N., and Strahler, A. H., 1989, *Elements of Physical Geography*, New York.

Streel, M., 1965, Techniques de préparation des roches détritiques en vue de l'analyse quantitative, *Annales de la Societe Géologique de Belgique* **88**(1–4) (1964–1965):107–117.

Streel, M., and Bless, M. J. M., 1980, Occurrence and significance of reworked palynomorphs, *Mededelingen Rijks Geologische Dienst* **32**(10):69–80.

Streel, M., and Loboziak, S., 1996, Middle and Upper Devonian miospores, in: *Palynology: Principles and Applications*, Volume 2, Chapter 18B (J. Jansonius and D. C. McGregor, eds.), Dallas, American Association of Stratigraphic Palynologists Foundation, pp. 575–587.

Streel, M., Higgs, K., Loboziak, S., Riegel ,W., and Steemans, P., 1987, Spore stratigraphy and correlation with faunas and floras in the type marine Devonian of the Ardenne-Rhenish region, *Review of Palaeobotany and Palynology* **50**:211–229.

Streel, M., Paris, F., Riegel, W., and Vangestaine, M., 1988, Acritarch, chitinozoan and spore stratigraphy from the Middle and Upper Devonian subsurface of Northeast Libya, in: *Palynostratigraphy of Northeast Libya* (A. El-Arnauti, B. Owens, and B. Thusu, eds.), Benghazi, Garyounis University Publications, pp. 111–128.

Streel, M., Fairon-Demaret, M., and Loboziak, S., 1990, Givetian-Frasnian phytogeography of Euramerica and western Gondwana based on miospore distribution, in: *Palaeozoic Palaeogeography and Biogeography*, Geological Society Memoir 12 (W. S. McKerrow and C. R. Scotese, eds.), London, Geological Society, pp. 291–296.

Streel, M., Loboziak, S., Steemans, P., and Bultynck, P., 2000a, Devonian miospore stratigraphy and correlation with the global stratotypes sections and points, in: *S.D.S. – Fossil Groups Important for Boundary Definition*, Courier Forschungsinstitut Senckenberg, 220 (P. Bultynck, ed.), pp. 9–23.

Streel, M., Caputo, M. V., Loboziak, S., and Melo, J. H. G., 2000b, Late Frasnian–Famennian climates based on palynomorph quantitative analyses and the question of Late Devonian glaciations, *Earth Sciences Review* **52**(1–3):121–173.

Streel, M., Caputo, M. V., Loboziak, S., Melo, J. H. G., and Thorez, J., 2001, Palynology and sedimentology of laminites and tillites from the latest Famennian of the Parnaíba Basin, Brazil, *Geologica Belgica* **3**(2000):87–96.

Strother, P. K., 1996, Acritarchs, in: *Palynology: Principles and Applications*, Volume 1, Chapter 5 (J. Jansonius and D. C. McGregor, eds.), Dallas, American Association of Stratigraphic Palynologists Foundation, pp. 81–106.

Suárez-Soruco, R., and Lobo-Boneta, J., 1983, La fase compresiva eohercínica en el sector oriental de la Cuenca Cordillerana de Bolivia, *Revista Técnica de Yacimientos Petroliferos Fiscales Bolivianos (YPFB)* **9**(1–4):189–202.

Suess, E., 1906, *The Face of the Earth*, Volume 2, Oxford, Clarendon Press.

Sullivan, H. J., 1965, Palynological evidence concerning the regional differentiation of Upper Mississippian floras, *Pollen et Spores* **7**:539–563.

Sullivan, H. J., 1967, Regional differences in Mississippian spore assemblages, *Review of Palaeobotany and Palynology* **1**:185–192.

Summerfield, M. A., and Hulton, N. J., 1994, Natural controls of fluvial denudation rates in major world drainage basins, *Journal of Geophysics Research* **99**(b7):13,871–13,883.

Sun Yuan-lin, and Boucot, A. J., 1999, Ontogeny of *Stringocephalus gubiensis* and origin of *Stringocephalus*, *Journal of Paleontology* **73**:860–871.

Surlyk, F., and Johansen, M. B., 1984, End-Cretaceous Brachiopod extinctions in the Chalk of Denmark, *Science* **223**:1174–1177.

Swisher, C. C., III, Grajales-Nishimura, J. M., Montanari, A., Margolis, S. V., Claeys, P., Alvarez, W., Renne, P., Cedillo-Pardo, E., Maurrasse, F. J.-M. R., Curtis, G. H., Smit, J., and McWilliams, M. O., 1992, Coeval $^{40}Ar/^{39}Ar$ ages of 65.0 million years ago from Chicxulub crater melt rock and Cretaceous–Tertiary boundary tektites, *Science* **257**:954–958.

Szaniawski, H., 1996, Scolecodonts, in: *Palynology: Principles and Applications*, Volume 1, Chapter 12 (J. Jansonius and D. C. McGregor, eds.), Dallas, American Association of Stratigraphic Palynologists Foundation, pp. 337–354.

Takaki, T., and Rodrigues, R., 1984, Isótopos estáveis do carbono e oxigênio dos calcários como indicadores paleoambientais – Bacias de Campos, Santos e Espírito Santo, in: *Anais do XXXIII Congresso Brasileiro de Geologia*, Rio de Janeiro, pp. 4750–4762.

Takaki, T., and Rodrigues, R., 1993, Isotopic evidence of the Cenomanian–Turonian anoxic event in the Sergipe-Alagoas Basin, Brazil, in: *Third Latin American Congress on Organic Geochemistry*, extended abstract, pp. 141–143.

Takhtadjan, A. L., 1991, *Evolutionary Trends in Flowering Plants*, New York, Columbia University Press.

Tappan, H., 1980, *The Paleobiology of Plant Brotists*, San Francisco, Freeman.

Tarling, D. H., 1983, *Paleomagnetism*, London, Chapman & Hall.

Taugourdeau, P., and (de) Jekhowsky, B., 1960, Répartition et description des Chitinozoaires siluro-dévoniens de la C.R.E.P.S., de la C.F.P.A. et de la S.N. REPAL au Sahara, *Revue de l'Institut Français du Petrole* **15**:1199–1260.

Tauxe, L., and Gallet, Y., 1991, A jackknife for magnetostratigraphy, *Geophysical Research Letters* **18**:1783–1786.

Tauxe, L., Gee, J., Gallet, Y., Pick, T., and Bown, T., 1994, Magnetostratigraphy of the Willwood Formation, Bighorn Basin, Wyoming: New constraints on the location of the Paleocene/Eocene boundary, *Earth and Planetary Science Letters* **125**:159–172.

Thierry, J., 1997, La Biostratigraphie, in: *Stratigraphy, Terminologie Française* (J. Rey, ed.), *Bulletin des Centres Recherches Exploration–Production Elf Aquitaine*, Memoir **19**:60–85.

Thierstein, H. R., 1981, Late Cretaceous nannoplankton and the change at the Cretaceous-Tertiary boundary, in: *The Deep Sea Drilling Project: A Decade of Progress*, Special Publications 32 (J. E. Warme, E. Winterer, and R. G. Douglas, eds.), Society of Economic Paleontologists and Mineralogists, pp. 335–394.

Thompson, P. R., 1981, Planktonic foraminifera in the western North Pacific during the past 150,000 years: comparison of modern and fossil assemblages, *Palaeogeography, Palaeoclimatology, Palaeoecology* **35**:241–279.

Thompson, P. R., 1992, Foraminiferal evidence for the sources and timing of mass-flow deposits south of Baltimore Canyon, in: *Centenary of Japanese Micropaleontology* (K. Ishizaki and T. Saito, eds.), Tokyo, Terra Scientific Publishing Co., pp. 97–128.

Thomson, M. R. A., 1992, Stratigraphy and age of the Pre-Cenozoic stratified rocks of the South Shetland islands: Review, in: *Geologia de la Antártida Occidental* (R. López- Martinez, ed.),

Simpósio T3, III Congresso Geologico de España y VIII Congresso Latinoamericano de Geologia, Madrid, pp. 75–92.

Tomlinson, P. B., 1980, *The Biology of Trees Native to Tropical Florida*, Allston, MA: Harvard Printing Office.

Traverse, A., 1988, *Paleopalynology*, London, Unwin Hyman.

Trewartha, G. T., 1969, *An Introduction to Climate*, 4th ed., New York, McGraw-Hill.

Tricart, J., and Cailleux, A., 1972, *Introduction to Climate Geomorphology*, New York, St. Martin's Press.

Trigüis, J. A., and Rodrigues, R., 2000, Chemical stratigraphy of the type Barreirinha Formation (upper Devonian), Amazon Basin-Brazil, in: *Proceedings of the 7th Latin-American Congress on Organic Geochemistry*, Brazil, Foz do Iguaçu, pp. 277–280.

Tschudy, R. H., Pillmore, C. L., Orth, C. J., Gilmore, J. S., and Knight, J. D., 1984, Disruption of the Terrestrial Plant Ecosystem at the Cretaceous–Tertiary Boundary, Western Interior, *Science* **225**:1030–1032.

Tucker, M. E., and Benton, M. J., 1982, Triassic environments, climates and reptile evolution, *Palaeogeography, Palaeoclimatology, Palaeoecology* **40**(4):361–379.

Twenhofel, W. H., 1932, *Treatise on Sedimentation*, Baltimore, MD, Williams and Wilkins.

Tye, R. S., Hewlett, J. S., Thompson, P. R., and Goodman, D. K., 1994, Integrated stratigraphic and depositional-facies analysis of parasequences in a transgressive systems tract, San Joaquin Basin, California, in: *Siliclastic Sequence Stratigraphy: Recent Developments and Applications*, Memoir 58 (P. Weimer and H. Posamentier, eds.), American Association of Petroleum Geologists, pp. 99–134.

Tyson, R. V., 1995, *Sedimentary Organic Matter. Organic Facies and Palynofacies*, London, Chapman & Hall.

Uhl, D., and Mosbrugger, V., 1999, Leaf venation density as a climate and environmental proxy: a critical review and new data, *Palaeogeography, Palaeoclimatology, Palaeoecology* **149**:15–26.

Ulmishek, G. F., and Klemme, H. D., 1992, Areal and spatial distribution and effectiveness of the world's petroleum source rocks, in: *Proceedings of the 3rd World Petroleum Congress*, Buenos Aires, pp. 121–136.

Unwin, D. M., 1987, Pterosaur extinction: nature and causes, *Memoires de la Société Géologique de France, N.S.* **150**:105–111.

Upchurch, G. R., and Wolfe, J. A., 1987, Mid-Cretaceous to Early Tertiary vegetation and climate: evidence from fossil leaves and woods, in: *The Origins of Angiosperms and their Biological Consequences* (E. M. Friis, W. G. Chaloner, and P. R. Crame, eds.), Cambridge, Cambridge University Press, pp. 75–104.

Urey, H. C., 1947, The thermodynamic properties of isotopic substances, *Journal of the Chemical Society, Chemical Communications,* **1947**:562–581.

Utting, J., 1996, Illustrations of Permian spores and pollen, in: *Palynology: Principles and Applications*, Volume 2, Chapter 18F (J. Jansonius and D. C. McGregor, eds.), Dallas, American Association of Stratigraphic Palynologists Foundation, pp. 621–627.

Utting, J., and Piasecki, S., 1995, Palynology of the Permian of northern continents, in: *The Permian of Northern Pangea, Paleogeography, Paleoclimates, Stratigraphy*, Volume 1 (P. A. Scholle, T. M. Peryt, and D. S. Ulmer-Scholle, eds.), Berlin, Springer-Verlag, pp. 236–261.

Utting, J., Keppie, J. D., and Giles, P. S., 1989, Palynology and stratigraphy of the Lower Carboniferous Horton Group, Nova Scotia, in: Contributions to Canadian Paleontology, *Geological Survey of Canada Bulletin* **396**:117–143.

Vail, P. R., 1987. Seismic stratigraphy interpretation using sequence stratigraphy, part 1: seismic stratigraphy procedure, in: *Atlas of Seismic Stratigraphy* (A. W. Bally, ed.), *American Association of Petroleum Geologists Studies in Geology* **27**(1):1–10.

Vail, P. R., Mitchum, R. M., Jr., and Tompson III, S., 1977a, Seismic stratigraphy and global changes of sea level, Part 3: relative changes of sea-level from coastal onlap, in: *Seismic Stratigraphy – Applications to Hydrocarbon Exploration*, Memoir 26 (C. E. Payton, ed.), American Association of Petroleum Geologists, pp. 63–81.

Vail, P. R., Mitchum, R. M., Jr., and Thompson, S., III, 1977b, Seismic stratigraphy and global changes of sea level, Part 4: global cycles of relative changes of sea level, in: *Seismic Stratigraphy – Applications to Hydrocarbon Exploration*, Memoir 26 (C. E. Payton, ed.), American Association of Petroleum Geologists, pp. 83–98.

Vail, P. R., Audemard, F., Bowman, S. A., Eisner, P. N., and Perze-Crus, C., 1991, The stratigraphic signatures of tectonics, eustasy and sedimentology – an overview, in: *Cycles and Events in Stratigraphy* (G. Einsele, W. Ricken, and A. Seilacher, eds.), Berlin, Heidelberg, New York, Springer-Verlag, pp. 617–659.

Vakhrameev, V. A., 1991, *Jurassic and Cretaceous Floras and Climates of the Earth*, Cambridge, Cambridge University Press.

Vallejo, C., Hochuli, P. A., Winkler, W., and von Salis, K., 2003, Palynological and sequence stratigraphic analysis of the Napo Group in the Pungarayacu 30 well, Sub-Andean Zone, Ecuador, *Cretaceous Research* **23**(for 2002):845–859.

Van Couvering, J. A., and Miller, J. A., 1971, Late Miocene marine and non-marine time scale in Europe, *Nature* **230**:559–563.

Van der Zwaan, C. J., van de Laar, J. G. M., Pagnier, H. J. M., and van Amerom, H. W. J., 1993, Palynological, ecological and climatological synthesis of the Upper Carboniferous of the well De Lutte-6 (East Netherlands), in: *Comptes Rendus XII ICC-P*, Volume 1, Buenos Aires, pp. 167–186.

Van der Zwaan, G. J., and Spaak, J., 1992, Lower to middle Triassic sequence stratigraphy and climatology of the Netherlands, a model, *Palaeogeography, Palaeoclimatology, Palaeoecology* **91**:277–290.

Van Gorsel, J. T., 1988, Biostratigraphy in Indonesia: methods, pitfalls and new directions, *Proceedings, Indonesian Petroleum Association*, 17th Annual Convention, pp. 275–300.

Vanguestaine, M., 1986, Late Devonian and Carboniferous acritarch stratigraphy and paleogeography, *Annales de la Société Géologique de Belgique* **109**:93–102.

Van Hinte, J., 1978, Geohistory analysis – application of micropaleontology in exploration geology, *American Association of Petrologists and Geologists Bulletin* **62**(2):201–222.

Van Hinte, J., 1982, Synthetic seismic sections from biostratigraphy, *American Association of Petrologists and Geologists Memoir* **34**:675–685.

Van Houton, F. B., 1964, Cyclic lacustrine sedimentation, Upper Triassic Lockatong Formation, central New Jersey and adjacent Pennsylvania, Kansas St., *Geological Survey Bulletin* **169**(2):497–531.

Van Valen, L., 1986, *Proalgae* and the significance of adaptative facies, *Evolutionary Theory* **8**:29–31.

Van Wagoner, J. C., Mitchum, R. M., Jr., Posamentier, H. W., and Vail, P. R., 1987. Seismic stratigraphy interpretation using sequence stratigraphy, part 2: key definitions of sequence stratigraphy, in: *Atlas of Seismic Stratigraphy* (A. W. Bally, ed.), *American Association of Petrologists and Geologists Stud. Geol.*, **27**(1):11–14.

Van Wagoner, J. C., Posamentier, H. W., Mitchum, R. M., Jr., Vail, P. R., Sarg, J. F., Loutit, T. S., and Hardenol, R. M., 1988, An overview of the fundamentals of sequence stratigraphy and key definitions, in: *Sea-level Changes: An Integrated Approach*, Special Publication 42 (C. K. Wilgus, B. S. Hastings, C. G. St. C. Kendall, H. W. Posamentier, C. A. Ross, and J. C. Van Wagoner, eds.), SEPM, pp. 39–46.

Van Wagoner, J. C., Mitchum, R. M. Jr., Campion, K. M., and Rahmanian, V. D., 1990, Siliciclastic sequence stratigraphy in well logs, cores, and outcrops, *American Association of Petrologists and Geologists Methods in Exploration Series* **7**.

Van Wagoner, J. C., Nummedal, D., Jones, C. R., Taylor, D. R., Jennette, D. C., and Riley, G. W., 1991, Sequence stratigraphy applications to shelf sandstone reservoirs, *American Association of Petrologists and Geologists Field Conference Guidebook*, Tulsa, Oklahoma.

Vavrdová, M., 1987, Latest Devonian miospores and acritarchs from the surface samples of the Ashkidah Formation, in: *The Geology of Libya*, Volume 4 (M. J. Salem, O. S. Hammuda, and B. A. Eliagoubi, eds.), 3rd Symposium on Geology, Libya, Tripoli, Amsterdam, Elsevier, pp. 1285–1296.

Vavrdová, M., and Isaacson, P. E., 1996, Affinities of Late Devonian acritarchs from the Madre de Díos Basin, northern Bolivia: evidence for plate tectonic interaction between Eastern Laurentia and Western Gondwana?, in: *Acritarcha in Praha*, Acta Universitatis Carolinae Geologica, 40 (O. Fatka and T. Servais, eds.), pp. 683–693.

Vavrdová, M., and Isaacson, P. E., 1999, Late Famennian phytogeographic provincialism: evidence for a limited separation of Gondwana and Laurentia, in: *Mid-Paleozoic Terranes, Stratigraphy and Biota, Abhandlungen Geologische* Bundesanstalt, 54 (R. Feist, J. A. Talent, and A. Daurer, eds.), pp. 453–463.

Vavrdová, M., and Isaacson, P. E., 2000, Palynology of selected Devonian strata, Western Gondwana, *Zentralblatt für Geologie und Paläontologie* **I**(7/8), 799–821.

Vavrdová, M., Isaacson, P.E., Díaz-Martínez, E., and Bek, J., 1991, Palinología del limite Devónico-Carbonífero entorno al Lago Titikaka, Bolivia: resultados preliminares, *Revista Técnica de Yacimientos Petroliferos Fiscales Bolivianos (YPFB)* **12**:303–313.

Vavrdová, M., Isaacson, P. E., Díaz, E., and Bek, J., 1993, Devonian-Carboniferous boundary at Lake Titikaka, Bolivia: preliminary palynological results, in: *12th International Conference on the Carboniferous Permian*, Volume 1, *Comptes Rendus*, Buenos Aires, pp. 187–199.

Vavrdová, M., Bek, J., Dufka, P., and Isaacson, P. E. 1996, Palynology of the Devonian (Lochkovian to Tournaisian) sequence, Madre de Díos Basin, northern Bolivia, *Vestník Ceského geologického ústavu* **71**(4):333–350.

Veblen, T. T., Schlegel, F. M., and Oltremari, J. V., 1983, Temperate broad-leaved evergreen forests of South America, in: *Ecosystems of the World. 10. Temperate Broad-Leaved Evergreen Forests* (J. D. Ovington, ed.), Amsterdam, Elsevier, pp. 5–31.

Veblen, T. T., Donoso, C., Kitzberger, T., and Rebertus, A. J., 1996, Ecology of the Southern Chilean and Argentinian *Nothofagus* forests, in: *The Ecology and Biogeography of Nothofagus Forests* (T. T. Veblen, R. S. Hill, and J. Read, eds.), New Haven and London, Yale University Press, pp. 293–353.

Veevers, J. J., 1991, Mid-Triassic lacuna on the Gondwanaland platform in the context of Pangean fusion and incipient fission, in: *Proceedings*, International Gondwana Symposium, 7, São Paulo, 1988, pp. 603–609.

Veevers, J. J., 1994, Pangaea: evolution of a supercontinent and its consequences for earth's paleoclimate and sedimentary environments, Geological Society of America Special Paper 288:13–23.

Veizer, J., and Jansen, S. L., 1985, Basement and sedimentary recycling –2: time dimensions to global tectonics, *Journal of Geology* **93**:625–643.

Ver Straeten, C. A., Griffing, D. H., Brett, C. E., 1994, The lower part of the Middle Devonian Marcellus "Shale," Central to Western New York State: Stratigraphy and depositional history, in: *Field Trip Guidebook*, New York State Geological Association, 66th Annual Meeting, pp. 271–325.

Villagrán, C., 1990, Glacial climates and their effect on the history of the vegetation of Chile: a synthesis based on palynological evidence from Isla de Chiloé, *Review of Palaeobotany and Palynology* **65**:17–24.

Vine, F. J., and Matthews, D. H., 1963, Magnetic anomalies over oceanic ridges, *Nature* **199**:947–949.

Virkki, C., 1946, Spores from the Lower Gondwanas of India and Australia, *Proceedings of the National Academy of Sciences, India* **15**:93–176.

Vogel, K., 1966, Eine funktionsmorphologische Studie an der Brachiopodengattung *Pygope* (Malm bis Unterkreide), *Neues Jahrbuch für Geologie und Paläontologie Abhandlungen* **125**:423–442.

Vogel, K., 1984, Lebensweise und Umwelt fossilere Tiere, in: *Biologische Arbeitsbücher*, Volume 39, Heidelberg, Quelle und Meyer.

Volkheimer, W., Melendi, D. L., and Salas, A., 1986, Devonian chitinozoans from northwestern Argentina, *Neues Jahrbuch für Geologie und Paläontologie Abhandlungen* **173**:229–251.

Vonhof, H. B., and Smit, J., 1997, High-resolution late Maastrichtian-early Danian oceanic $^{87}Sr/^{86}Sr$ record: Implications for Cretaceous-Tertiary boundary events, *Geology* **25**(4):347–350.

Voorhies, M. R., 1969, Taphonomy and population dynamics of an Early Pliocene vertebrate fauna, Krox County, Nebraska, *Contributions to Geologia.* Special Paper 1:69.

Wakeham, S. G., Farrington, J. W., Gagosian, R. B., Lee, C., de Barr, H., Nigrelli, G. E., Tripp, B. W., Smith, S. O., and Frew, N. M., 1980, Organic matter fluxes from sediment traps in the equatorial Atlantic Ocean, *Nature* **286**:798–800.

Walker, R. G., 1979, Facies and facies models. 1. General introduction, in: *Facies Models*, Geoscience Canada Reprint Series 1 (R. G. Walker, ed.), pp. 1–7.

Walker, R. G., 1992, Facies, facies models and modern stratigraphic concepts, in: *Facies Models: Response to Sea-levels Change* (R. G. Walker and N. P. James, eds.), St John's, Newfoundland, Geological Association of Canada, pp. 1–14.

Walker, R. G., and James, N. P. (eds.), 1992, Facies Models: Response to Sea-levels Change, St. John's, Newfoundland, Geological Association of Canada.

Walliser, O. H. (ed.), 1996, Global events and event stratigraphy in the Phanerozoic: results of international interdisciplinary cooperation in the IGCP Project 216 "Global Biological Events in Earth History". Berlin and Heidelberg, Springer-Verlag.

Walsh, S. L., 1998, Fossil datum and paleobiological event terms, paleontostratigraphy, chronostratigraphy, and the definition of land mammal "age" boundaries, *Journal of Vertebrate Paleontology*.

Walter, H., and Box, 1976, Global classification of natural terrestrial ecosystems, *Vegetatio* **32**:75–81.

Wang Yu, Boucot, A.J., Rong Jia-yu, and Yang Xue-chang, 1987, Community paleoecology as a geologic tool: The Chinese Ashgillian–Eifelian (Latest Ordovician through Middle Devonian) as an example, Geological Society of America, Special Paper 211.

Wanless, H. R., and Shepherd, F. P., 1936, Sea level and climate change related to late Paleozoic cycles, *Geological Society of America Bulletin* **47**:1177–1206.

Ward, L. W., and Blackwelder, B. W., 1975, *Chesapecten*, a new genus of Pectinidae (Mollusca: Bivalvia) from the Miocene and Pliocene of eastern North America, U.S., Geological Survey Professional Paper 861.

Ward, P., Wiedeman, J., and Mount, J. F., 1986, Maastrichtian molluscan biostratigraphy and extinction patterns in a Cretaceous/Tertiary boundary section exposed at Zumaya, Spain, *Geology* **14**:899–903.

Wardle, J., 1967, Biological flora of New Zealand. 2. *Nothofagus menziesii* (Hook.*f.*) Oerst. (Fagaceae) Silver beech, *New Zealand Journal of Botany* **5**:276–302.

Warrington, G., 1996, Permian spores and pollen, in*: Palynology: Principles and Applications*, Volume 2, Chapter 18E (J. Jansonius and D. C. McGregor, eds.), Dallas, American Association of Stratigraphic Palynologists Foundation, pp. 607–619.

Wasmund, E., 1926, Biocoenose und thanatocoenose, *Archiv für Hydrobiologie* **17**:1–116. *Der Geologe* 1927:1069–1076.

Webb, L. J., 1959, A physiognomic classification of Australian Rain Forests, *Journal of Ecology* **47**:551–570.

Weedon, G. P., 1993, The recognition and stratigraphic implications of orbital-forcing of climate and sedimentary cycles, *Sedimentology Review* **1**:31–50.

Wegener, A., 1915, *Die Entstehund der Kontinente und Ozeane*, 1er. ed., Brauschweig, Friedrich Vieweg & Sohn.

Wegener, A., 1924, *Die Entstehung der Kontinente und Ozeane*, Braunschweig.

Weigelt, J., 1927a, *Rezente Wirbeltierleichen und ihre paläobiologische Bedeutung*, Leipzig, Verlag M. Weg.

Weigelt, J., 1927b, *Über Biostratonomie. Eine Betrachtung zu Dollos siebzigstem Geburtstag*.

Weller, J. M., 1956, Argument for diastrophic control of late Paleozoic cyclothems, *Bulletin of the American Association of Petroleum Geologists* **40**:17–50.

Westerlund, , S. F. G., Anderson, L. G., Hall, P. O. J., Iverfeldt, A., Rutgers Van der Loeff, M. M., and Sundby, B. 1986, Benthic fluxes of cadmium, copper, nickel, zinc and lead in the coastal environment, *Geochimica et Cosmochimica Acta* **50**:1289–1296.

Westermann, G. E. G., 2000 Biochore classification and nomenclature in paleobiogeography: An attempt at order, *Palaeogeography, Palaeoclimatology, Palaeoecology* **158**:1–13.

Whatley, R., 1990, Ostracoda and global events, in: *Ostracoda and Global Events* (R. Whatley and C. Maybury, eds.), London, Chapman & Hall, pp. 3–24.

Wheeler, H. E., 1958, Time-stratigraphy, *Bulletin of the American Association of Petroleum Geologists* **42**:1047–1063.

Wheeler, H. E., 1959a, Note 24 – unconformity-bounded units in stratigraphy, *Bulletin of the American Association of Petroleum Geologists* **43**:1975–1977.

Wheeler, H. E., 1959b, Stratigraphic units in time and space, *American Journal of Science* **257**:692–706.

Whitaker, M. F., 1984, The usage of palynostratigraphy and palynofacies in definition of Troll Field geology, in: *Sixth Offshore Northern Seas Conference and Exhibition, Stavanger 1984*, Paper G6.

Whitaker, M. F., Giles, M. R., and Cannon, S. J. C., 1992, Palynological review of the Brent Group, UK sector, North Sea, in: *Geology of the Brent Group*, Special Publication 61 (A. C. Morton, R. S. Haszeldine, M. R. Giles, and S. Brown, eds.), London, Geological Society, pp. 169–202.

Whittaker, A., Cope, J. C. W., Cowie, J. W., Gibbons, W., Hailwood, E. A., House, M. R., Jenkins, D. G., Rawson, P. F., Rushton, A. W. A., Smith, D. B., Thomas, A. T., and Wimbledon, W. A., 1991, A guide to stratigraphical nomenclature, *Journal of the Geological Society of London* **148**:813–824.

Whittaker, J. E., Jones, R. W., and Banner, F. T., 1998, *Key Mesozoic Benthic Foraminifera of the Middle East*, London, Natural History Museum.

Wicander, R. 1974, Upper Devonian–Lower Mississippian acritarchs and prasinophycean algae from Ohio, U.S.A., *Palaeontographica Abteilung B* **148**:9–43.

Wicander, R., and Wood, G. D., 1981, Systematics and biostratigraphy of the organic-walled microphytoplankton from the Middle Devonian (Givetian) Silica Formation, Ohio, U.S.A., *American Association of Stratigraphic* Palynologists, *Contributions Series* **8**:1–137.

Wicander, R., and Wood, G. D., 1997, The use of microphytoplankton and chitinozoans for interpreting transgressive/regressive cycles in the Rapid Member of the Cedar Valley Formation (Middle Devonian), Iowa, *Review of Palaeobotany and Palynology* **98**:125–152.

Widmark, J. G. V., 1997, Deep-sea benthic foraminifera from Cretaceous-Paleogene boundary strata in the South Atlantic – taxonomy and paleoecology, *Fossils & Strata* **43**:1–94.

Widmark, J. G. V., and Malmgren, B., 1992, Benthic foraminiferal changes across the Cretaceous-Tertiary boundary in the deep sea; DSDP Sites 525, 527, and 465, *Journal of Foraminiferal Research* **22**:81–113.

Wignall, P. B., 2001, Large igneous provinces and mass extinctions, *Earth-Science Reviews* **53**:1–33.

Wignall, P. B., and Benton, M. J., 1999, Lazarus taxa and fossil abundance at times of biotic crisis, *Journal of the Geological Society, London* **156**:453–456.

Wilf, P., 1997, When are leaves good thermometers? A new case for Leaf Margin Analysis, *Paleobiology* **23**:373–390.

Wilf, P., 2000, Late Paleocene–early Eocene climate changes in southwestern Wyoming: paleobotanical analysis, *Geological Society of America Bulletin* **112**:292–307.

Wilf, P., Wing, S. L., Greenwood, D. R., and Greenwood, C. L., 1998, Using fossil leaves as paleoprecipitation indicators: An Eocene example, *Geology* **26**:203–206.

Wilgus, C. K., Hastings, B. S., Kendall, C. G. St. C., Posamentier, H. W., Ross, C. A. and Van Wagoner, J. C., eds., 1988, Sea-level changes: an integrated approach, Society of Economic Paleontologists and Mineralogists Special Publication 42.

Williams, G., 1992. Palynology as a palaeoenvironmental indicator in the Brent Group, northern North Sea, in: *Geology of the Brent Group*, Special Publication 61 (A. C. Morton, R. S. Haszeldine, M. R. Giles, and S. Brown, eds.), London, Geological Society, pp. 203–212.

Williams, G. D., and Stelck, C. R., 1975, Speculation on the Cretaceous paleogeography of North America, in: *The Cretcaeous System of the Western Interior of North America*, Special Papers 13 (W. G. E. Caldwell, ed.), Geological Association of Canada, pp. 1–20.

Williams, G. L., and Bujak, J. P., 1985, Mesozoic and Cenozoic dinoflagellates, in: *Plankton Stratigraphy* (H. M. Bolli, J. B. Saunders, and K. Perch-Nielsen, eds.), Cambridge, Cambridge University Press, pp. 847–964.

Williams, J. L., 1891, On cycles of sedimentation, *American Geologist* **8**:315–324.

Williamson, M. A.,1987, Quantitative biozonation of the Late Jurassic and Early Cretaceous of the East Newfoundland Basin, *Micropaleontology* **33**:37–65.

Willis, B. J., and Behrensmeyer, A. K., 1994, Architecture of Miocene overbank deposits in northern Pakistan, *Journal of Sedimentary Research* **B64**(1):60–67.

Wilson, D. S., 1993, Confirmation of the astronomical calibration of the magnetic polarity time scale from sea-floor spreading rates, *Nature* **364**:788–790.

Wilson, K. M., Pollard, D., Hay, W. W., Thompson, S. L., and Wold, C. N., 1994, General circulation model simulation of Triassic climates: preliminary results, Geological Society of America Special Paper 288:91–116.

Wilson, L., 1973, Variations in mean annual sediment yield as a function of mean annual precipitation, *American Journal of Sciences* **278**:335–349.

Wilson, M. V. H., 1988, Taphonomic processes: information loss and information gain, *Geoscience Canada* **15**:131–148.

Wing, S. L., and Bown, T. M., 1985, Fine scale reconstruction of the late Paleocene–early Eocene paleogeography in the Bighorn basin of northern Wyoming, in*: Paleography of West-central United States: Denver, Colorado, Rocky Mountains Sector* (R. M. Flores and S. S. Kaplan, eds.), SEPM, pp. 93–106.

Wing, S. L. and DiMichele, W. A., 1995, Conflict between local and global changes in plant diversity through geological time, *Palaios* **10**:551–564.

Wing, S. L., Alroy, J., and Hickey, L. J., 1995, Plant and mammal diversity in the Paleocene to Early Eocene of the Bighorn Basin, *Palaeogeography, Palaeoclimatology, Palaeocology* **115**:117–155.

Wise, S. W., Jr., Tjalsma, R. C., and Dailey, D. H., 1991, Late Paleocene carbonate deposition on the Southwest Indian Ridge, in: *Geological Evolution of Antarctic* (M. R. A. Thomson, J. A. Crame, and J. W. Thomson, eds.), Cambridge, Cambridge University Press, pp. 633–638.

Wolbach, W. S., Lewis, R. S., and Anders, E., 1985, Cretaceous extinctions: evidence for wildfires and search for meteoritic material, *Science* **230**(4722):167–170.

Wolbach, W. S., Gilmour, I., and Anders, E., 1990, Major wildfires at the Cretaceous–Teriary boundary, in: *Global Catastrophes in Earth History* (V. L. Sharpton and P. D. Ward, eds.), Geological Society of America, Special Paper 247:391–400.

Wolfe, J. A., 1990, Palaeobotanical evidence for a marked temperature increase following the Cretaceous/Tertiary boundary, *Nature* **343**:153–156.

Wolfe, J. A., 1991, Palaeobotanical evidence for a June 'impact winter' at the Cretaceous/Tertiary boundary, *Nature* **352**:420–423.

Wolfe, J. A., 1992, Plants at the K/T boundary, *Nature* **356**, 296.

Wolfe, J. A., 1993, A method of obtaining climatic parameters from leaf assemblages, *U.S. Geological Survey Bulletin* **2040**:1–71.

Wolfe, J. A., 1995, Paleoclimatic inferences from Tertiary leaf assemblages, *Annual Review of Earth and Planetary Sciences* **23**:119–142.

Wolfe, J. A., and Upchurch, Jr., G. R., 1986, Vegetation, climatic and floral changes at the Cretaceous–Tertiary boundary, *Nature* **324**:148–151.

Wolfe, J. A., and Upchurch, G. R., 1987, North American non marine climates and vegetation during the Late Cretaceous, *Palaeogeography, Palaeoclimatology, Palaeoecology* **61**:33–77.

Wood, G. D., 1984, A stratigraphic, paleoecologic and paleobiogeographic review of the acritarchs *Umbellasphaeridium deflandrei* and *Umbellasphaeridium saharicum*, in: *9ème Congr. Int. Strat. Géol. Carb.*, Volume 2, Washington and Champaign-Urbana, Compte Rendu, pp. 191–211.

Wood, G. D., 1994, *Togachitina*, a new bilayered chitinozoan genus from the Devonian of Sierras Subandinas region, Bolivia, *Palynology* **18**:195–204.

Wood, G. D., 1997, The acritarch *Proteolobus wallii* gen. et sp. nov. from the Devonian Iquiri Formation of Bolivia: a possible coenobial alga, *Micropaleontology* **43**:325–331.

Wood, G. D., Gabriel, A. M., and Lawson, J. C., 1996, Palynological techniques – processing and microscopy, in*: Palynology: Principles and Applications*, Volume 1, Chapter 3 (J. Jansonius and D. C. McGregor, eds.), Dallas, American Association of Stratigraphic Palynologists Foundation, pp. 29–50.

Worbes, M., 1989, Growth rings, increment and age of trees in inundation forests, savannas and a moutain forest in the neotropics, *IAWA Bulletin* n.s. **10**(2):109–122.

Wright, J. D., Miller, K. G., and Fairbanks, R. G., 1992, Early and middle Miocene stable isotopes: Implications for deep water circulation and climate, *Paleoceanography* **7**:357–389.

Wright, V. P., and Marriott, S. B., 1993,The sequence stratigraphy of fluvial depositional systems: the role of floodplain sediment storage, *Sedimentary Geology* **86**:203–210.

Yao, Z.-Q., 1983, Ecology and taphonomy of Gigantopterids, *Bulletin of the Nanjing Institute of Geology and Palaeontology*, Chinese Academy of Sciences **6**:63–83 (in Chinese with English abstract).

Yao, Z.-Q., Liu, L.-J., and Zhang, S., 1994, Permian wood from western Henan, China: implications for palaeoclimatological interpretations, *Review of Palaeobotany and Palynology* **80**:277–290.

Yolkin, E. A., Kim, A. I., Weddige, K., Talent, J. A., and House, M. R., 1997, Definition of the Pragian/Emsian Stage boundary, *Episodes* **20**(4):235–240.

Young, S. W., Basu, A., Mack, G., Darnell, N., and Suttner, L. J., 1975, Use of size-composition trends, in: *Holocene Soil and Fluvial Sand for Paleoclimatic Interpretation*, Proceedings 1, 9th International Sedimentologic Congress, pp. 201–206.

Zachariasse J. W., 1999, Present status of the astronomical (polarity) time-scale for the Mediterranean Late Neogene, *Philosophical Transactions of the Royal Society London* **A357**:1931–1947.

Zachos, J. C., Aubry, M. P., Berggren, W. A., Ehrendorfer, T., Heider, F., Lohmann, K. C., 1992, Chemobiostratigraphy of the Cretaceous/Paleocene boundary at Site 750, southern Kerguelen Plateau Wise, *Proceedings of the Ocean Drilling Program, Scientific Results* **120**:961–977.

Zastawniak, E., 1990, Late Cretaceous Leaf Flora of King George Island, in: *Proceedings of the Symposium Paleofloristic and Paleoclimatic Changes in the Cretaceous and Tertiary* (E. Knoblock and Z. Kvacek, eds.), Prague, Geological Survey Publisher, pp. 81–85.

Zhivkovich, A. E., and Chekhovich, P. A., 1985, Paleozoiskie formatsii i tektonika Ufimskogo Anfeatra. Izdatelskvo "Nauka", Moskva, Akad. Nauk SSSR, Institut Litosferi.

Ziegler, A. M., Scotese, C. R., and Barret, S. F., 1983, Mesozoic and cenozoic paelogeographic maps, in: *Tidal Friction and the Earth's Rotation II* (P. Brosche and J. Sundermann, eds.), Berlin, Springer-Verlag, pp. 240–252.

Ziegler, A. M., Huelver, M. L., and Rowley, D. B., 1996, Permian Paleogeography (small maps illustrate continental configuration, with 50 X 50 reference grids, in: *Late Glacial and Postglacial Environmental Changes – Quaternary, Carboniferous–Permian and Proterozoic* (I. P. Martini, ed.), New York, Oxford University Press.

Zikui, Z., Jie, Y., Huamai, L., Zhenhua, Z., and Zheng, Y., 1991, Extinction of the dinosaurs across the Cretaceous–Tertiary boundary in Nanxiong Basin, Guandong Province, *PalAsiatica* **29**:13–29.

Zinsmeister, W. J., and Feldmann, R. M., 1996, Late Cretaceous faunal changes in the high southern latitudes: A harbinger of global biotic catastrophe? in: *Cretaceous–Tertiary Mass Extinctions: Biotic and Environmental Changes* (N. MacLeod and G. Keller, eds.), New York, W. W. Norton & Company, pp. 303–325.

Zinsmeister, W. J., Feldmann, R. M., Woodburne, M. O., and Elliot, D. H., 1989, Latest Cretaceous/earliest Tertiary transition on Seymour Island, Antarctica, *Journal of Paleontology* **63**(6):731–738.

Zweers, G. A., and Vanden Berge, J. C., 1997, Trophic transitions in birds before and beyond the K–T boundary, *Journal of Morphology* **232**(3):346.

Zweers, G. A., Vanden Berge, J. C., and Berkhoudt, H., 1997, Evolutionary patterns of avian trophic diversification, *Zoology* **100**:25–57.

# Appendix: Color Version of Figures

*Figure 7.3*

*Figure 7.6*

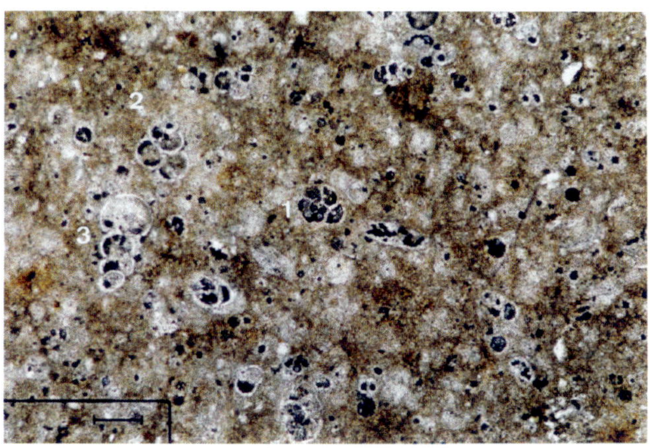


*Figure 7.7*

A

*Figure 9.1A*

B

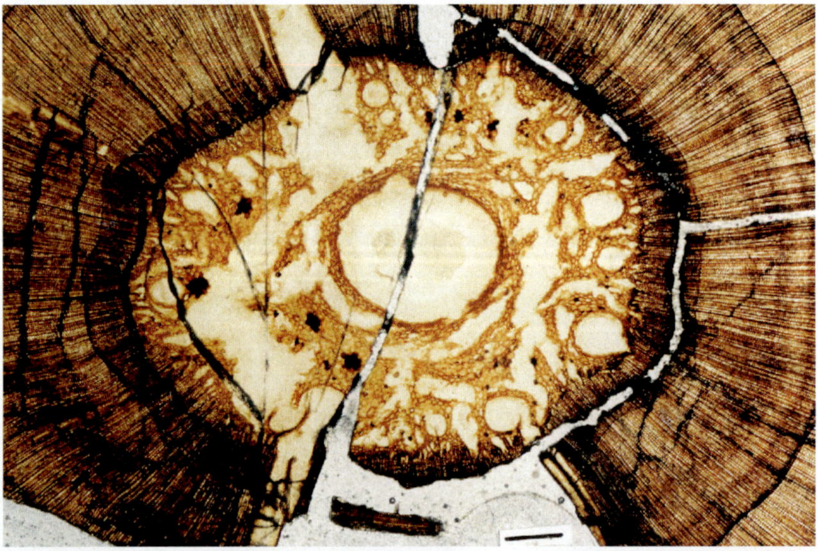

*Figure 9.1B*

*Figure 9.2*

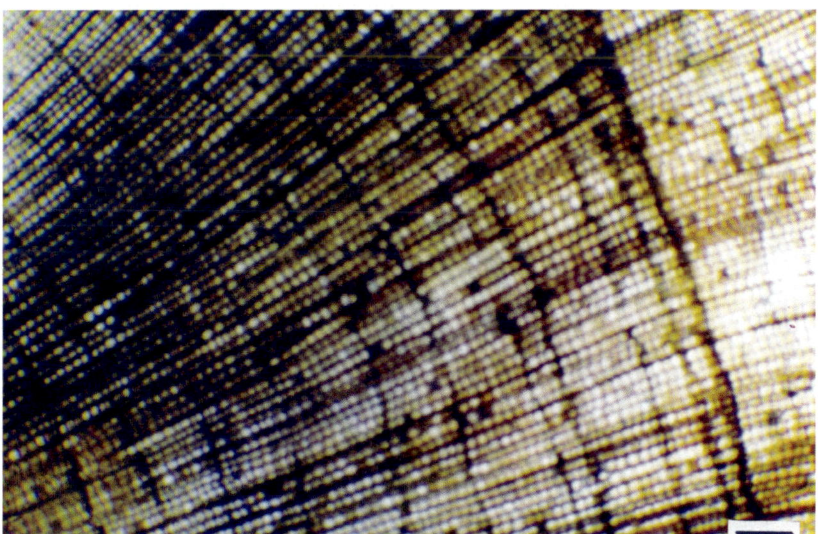

*Figure 9.3*

*Figure 9.4*

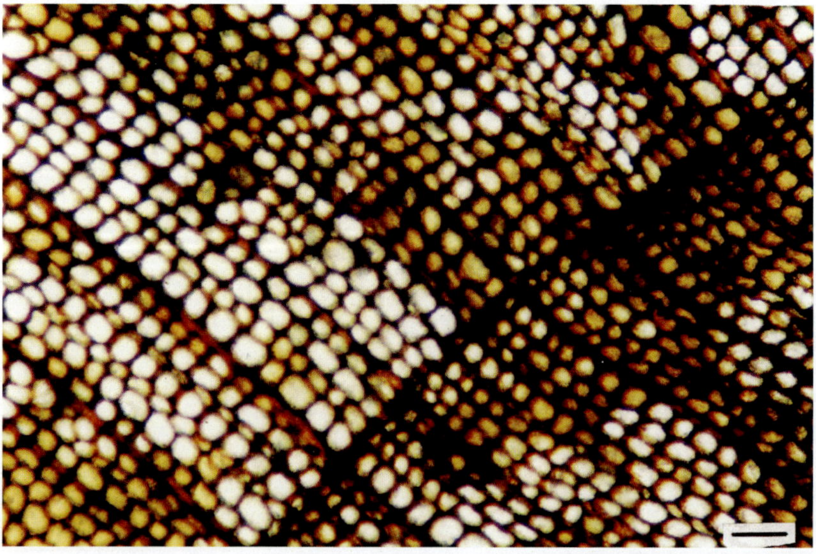

*Figure 9.5*

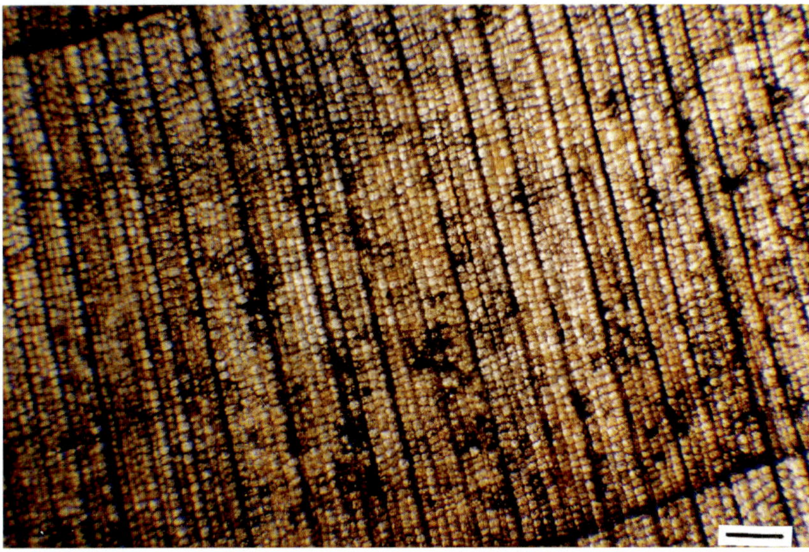

*Figure 9.6*

*Figure 9.7*

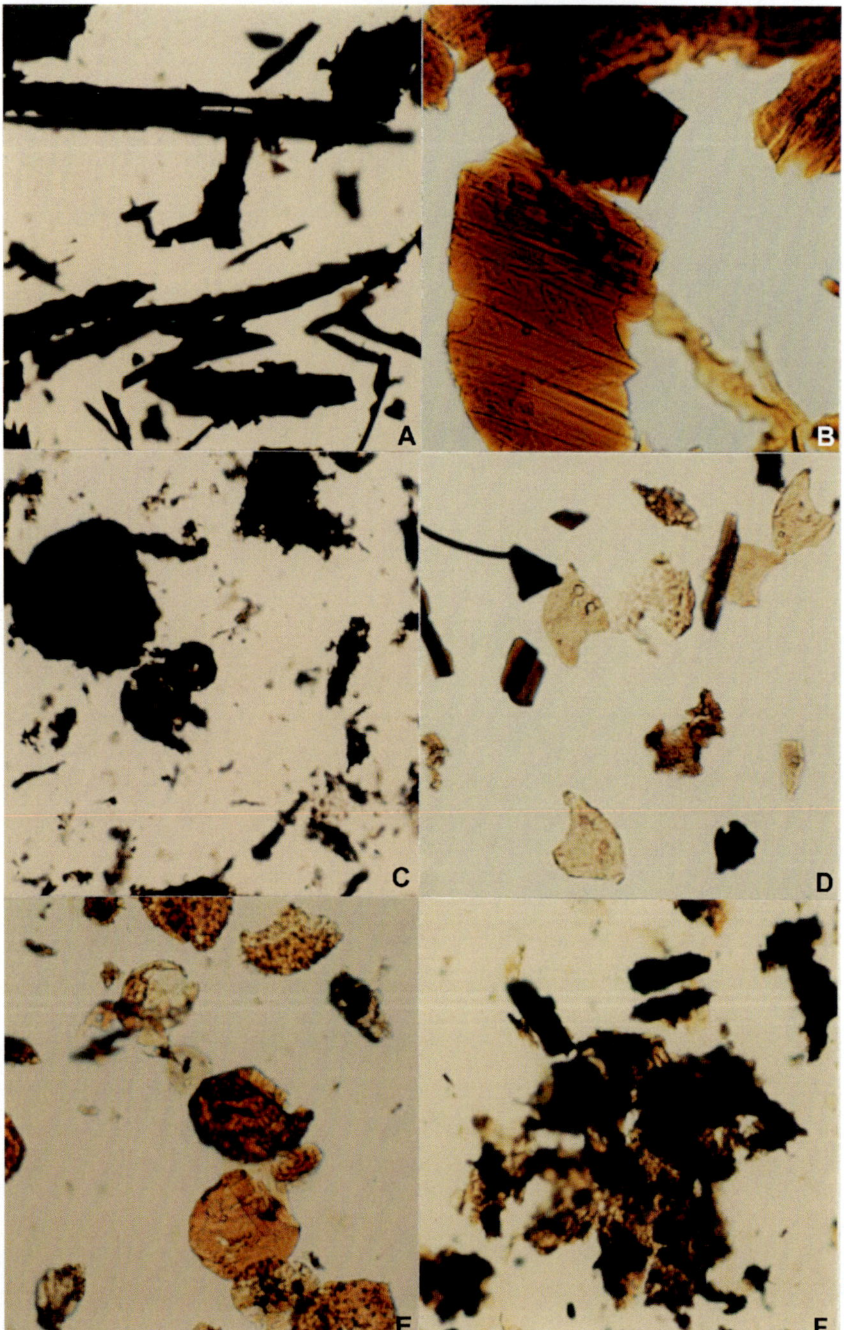


*Figure 10.2*

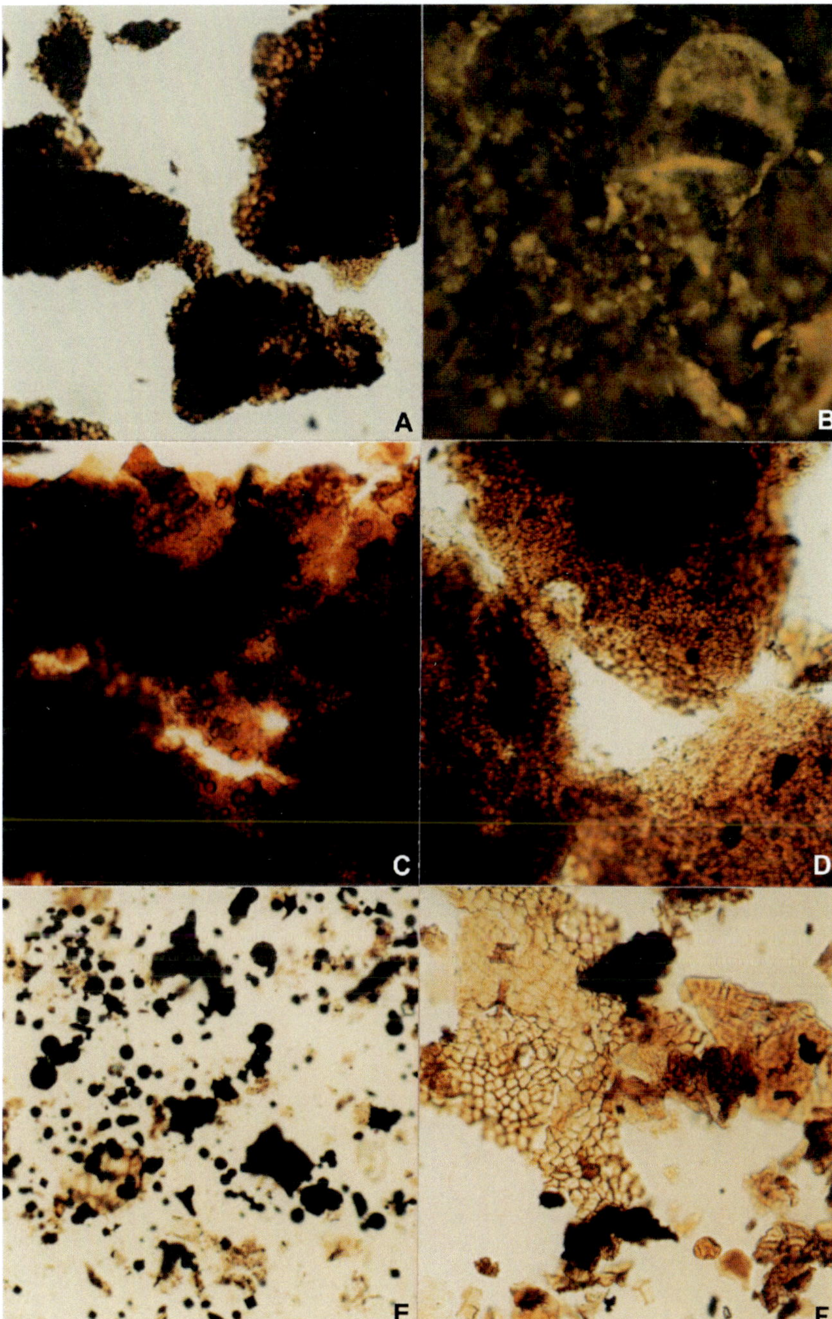


*Figure 10.3*

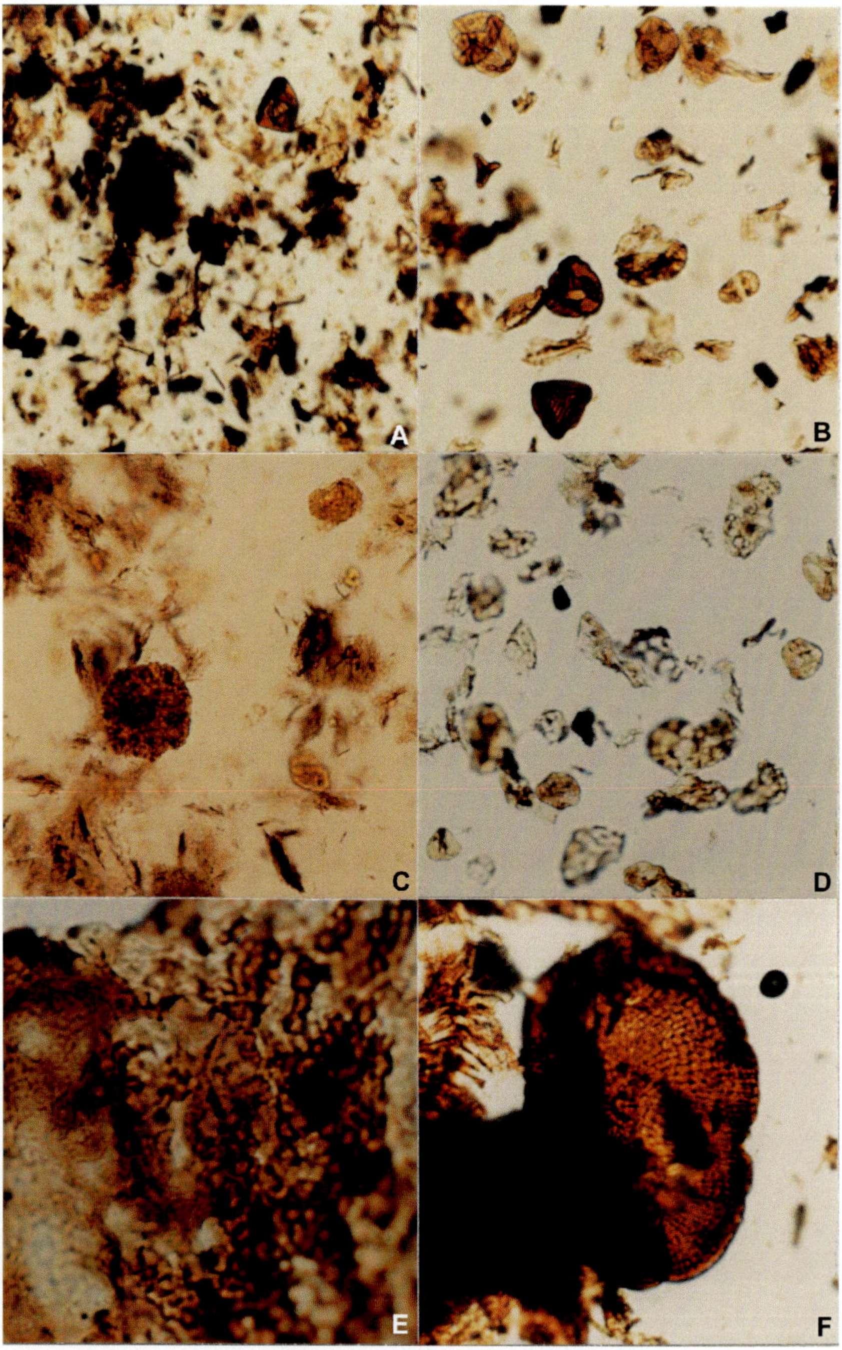


*Figure 10.4*

## Stratigraphic Reference: Piston Core IG41-17

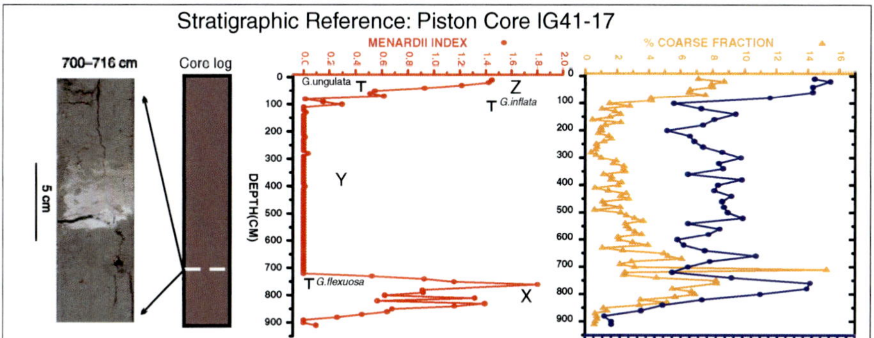


White volcanic ash generally occurs as a layer within hemipelagic clays or within burrows. The ash is generally coincident with the first downhole occurrence of *Globorotalia flexuosa* and the top of the X zone.

## Downslope Transport: Timing and Sediment Sourcing

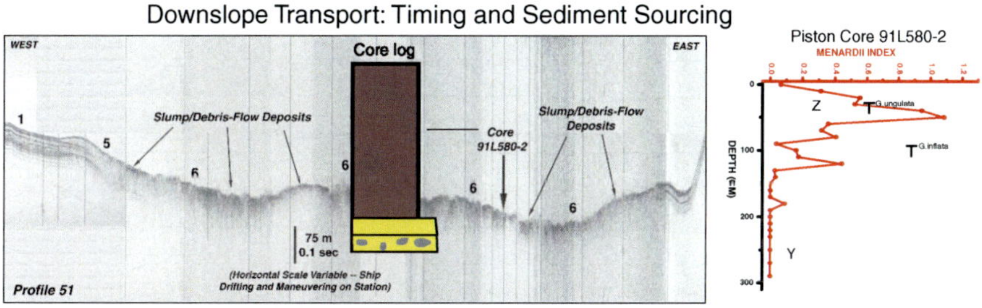


3.5 kHz record across the middle portion of the western lobe of the East Breaks Slide Complex. The mass-transport deposits of the slide complex return hyperbolic echoes (Type 6 facies), which transition into prolonged echoes (Type 5 facies) at the edges of the complex. Well-stratified sediments (Type 1 facies) are abruptly truncated at the edges of the slide complex. Location of and log of piston core 91L580-2, containing basal sandy sediments, is shown on the profile. At right, stratigraphy for 91L580-2 demonstrates that sandy deposition occurs during glacial periods (Zone Y) and sediments are likely sourced from sediments of glacial intervals because they do not contain *Globorotalia menardii* complex forms. Below left, piston core IG38-15 data shows an example of cross-overs of the coarse fraction and carbonate signals that occur within sandy intervals deposited during these interglacials. Below right, piston core 91L575-2B demonstrates that some of the sandy sediments deposited during the Y zone are sourced from previous interglacial sediments because the sandy sediments contain *Globorotalia flexuosa,* a foraminifera characteristic of X interglacials and older. The thickness of units and the nature of the samples indicate that the *G. flexuosa* forms are reworked, rather than stratigraphically in place.

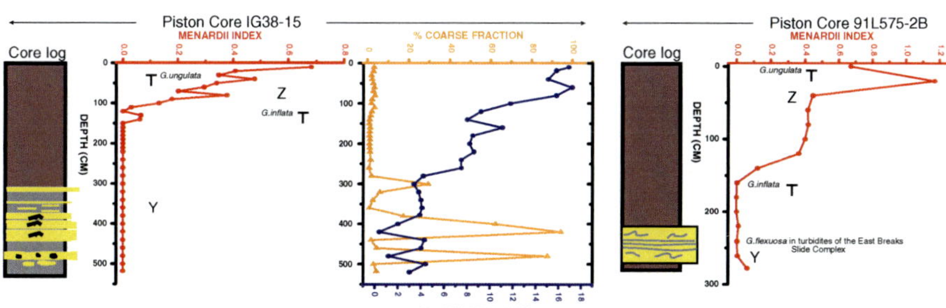


*Figure 11.6*

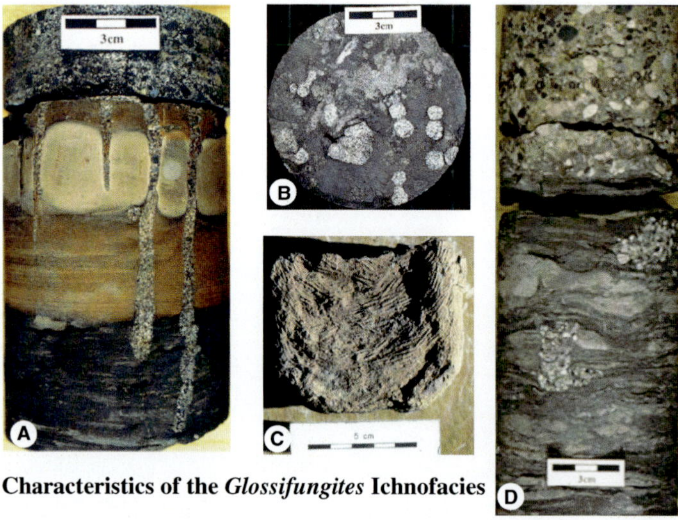


*Figure 13.6*

*Figure 13.7*

Glossifungites-enhanced Permeability
TRIASSIC SAG RIVER FORMATION, ALASKA
1. Erosion
2. Firmground Exhumation and Colonization
3. Infilling of Open Burrows
4. Enhanced Permiability

*Figure 13.12*

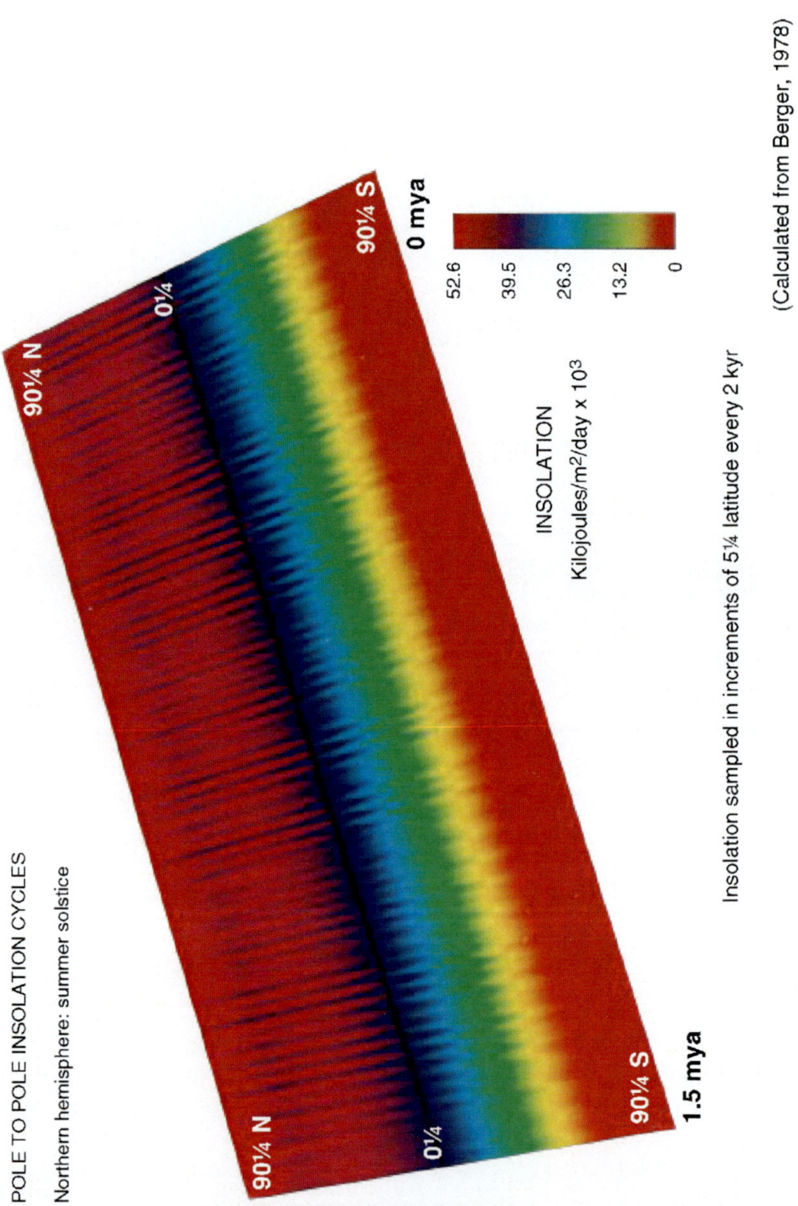
POLE TO POLE INSOLATION CYCLES
Northern hemisphere: summer solstice
90¼ N
0¼
90¼ S
0 mya
1.5 mya
INSOLATION
Kilojoules/m2/day x 103
52.6
39.5
26.3
13.2
0
Insolation sampled in increments of 5¼ latitude every 2 kyr
(Calculated from Berger, 1978)

*Figure 14.18*

(a) JULY: PRESENT

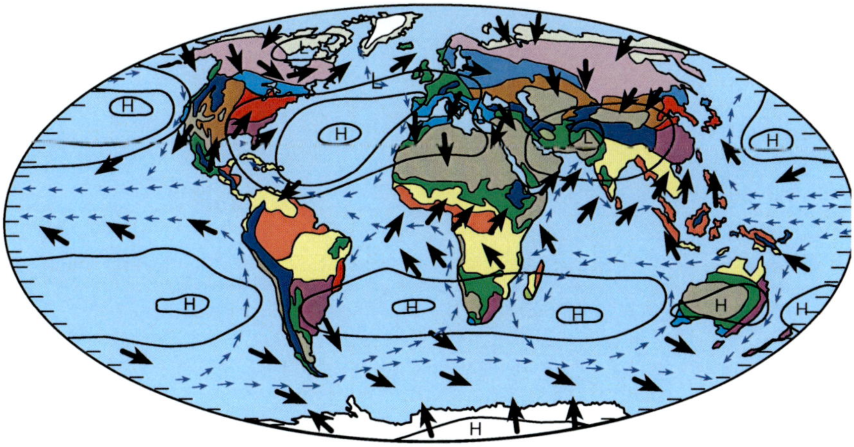

(b) JULY: CLIMATIC MINIMUM

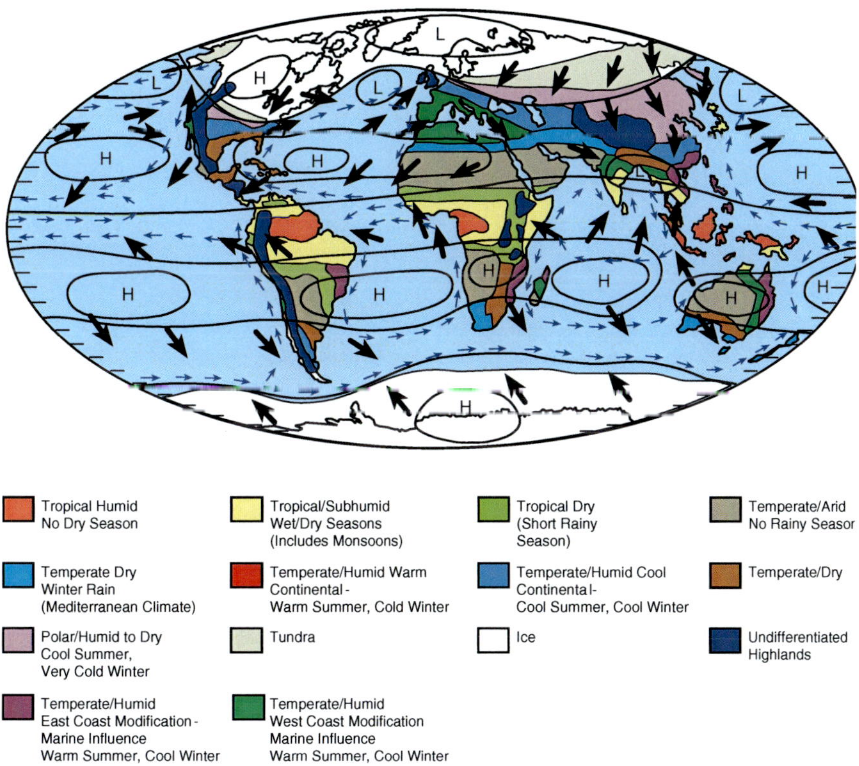


**(Perlmutter and Matthews, 1989)**

*Figure 14.20 (a, b)*

***Zones of Similar Climate Ranges***

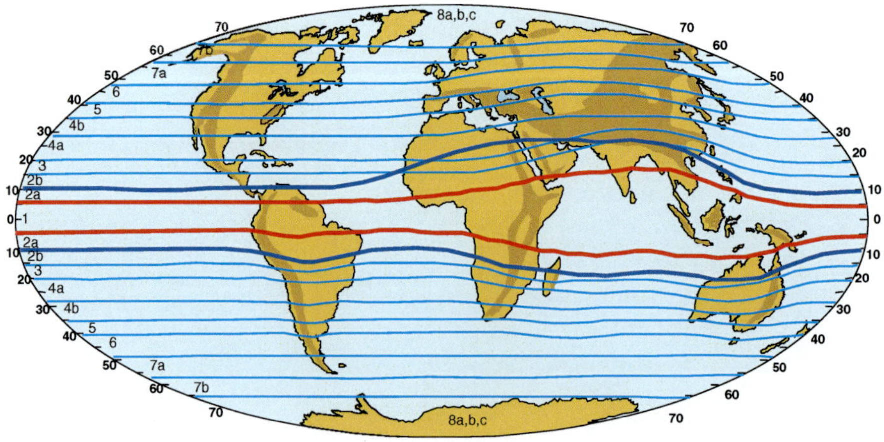


**0–3 MYBP**

**(Perlmutter *et al.*, 1998)**

*Figure 14.21a*

Early Permian
277 mya

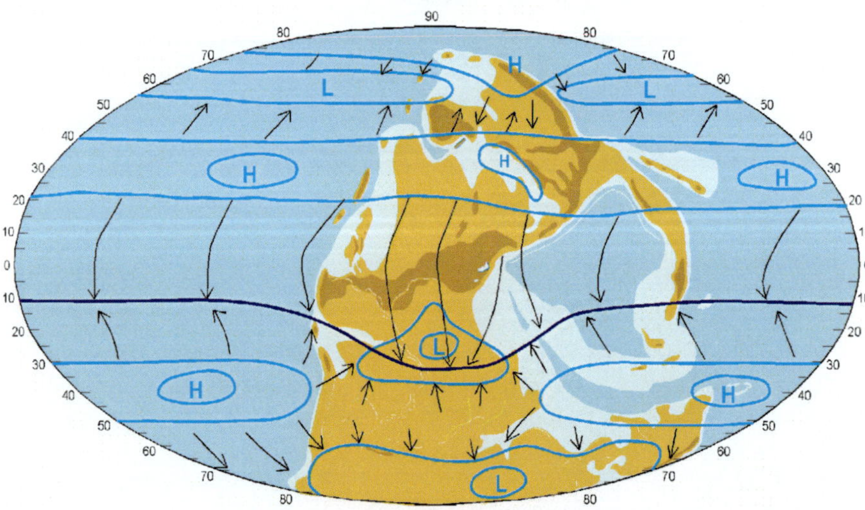


Climatic Maximum Southern Hemisphere Summer

**(Perlmutter and Plotnick, 2002)**

High Sea Level

*Figure 14.22a*

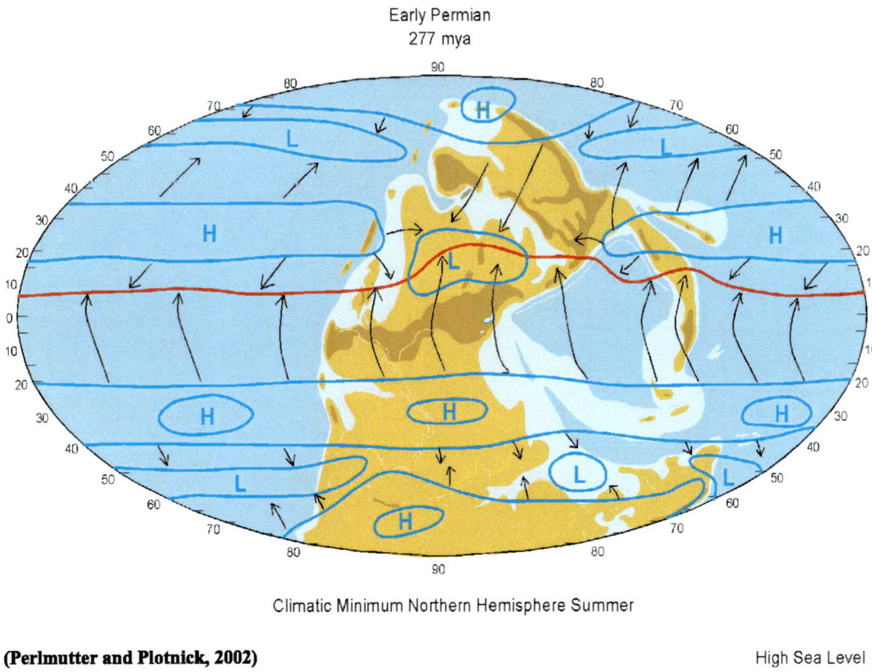


*Figure 14.22b*

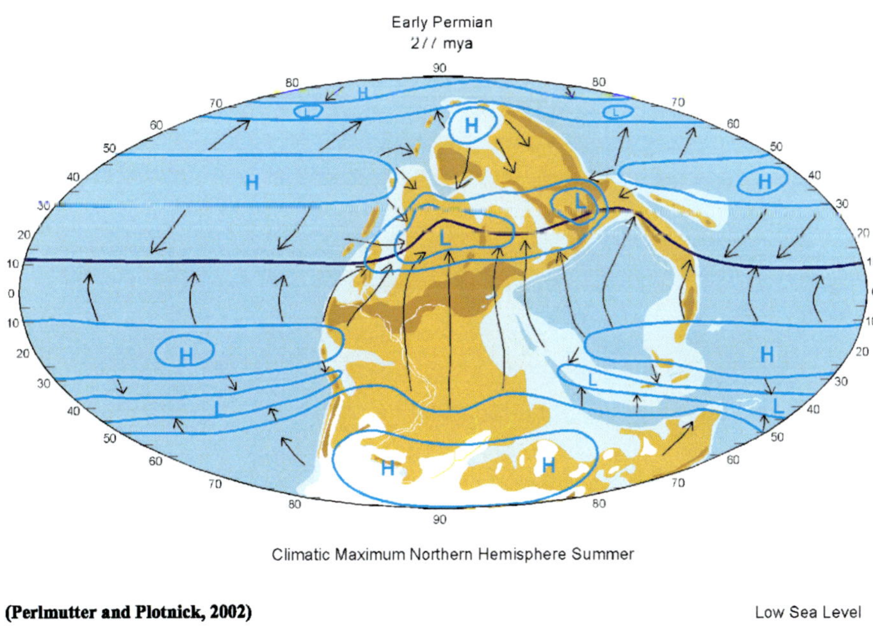


*Figure 14.22c*

Early Permian
277 mya

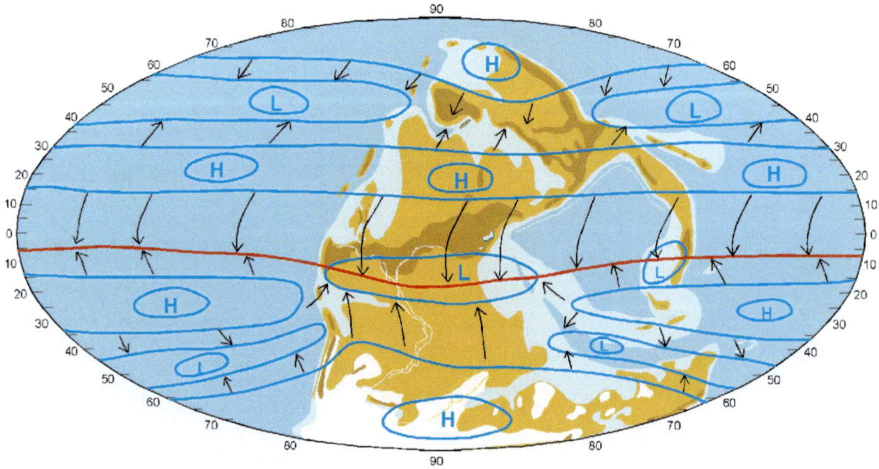


Climatic Minimum Southern Hemisphere Summer

**(Perlmutter and Plotnick, 2002)**

Low Sea Level

*Figure 14.22d*

# Index